AutoCAD®
for Architecture

• •

Release 14

AutoCAD®
for Architecture

• •

Release 14

Alan Jefferis

Michael Jones

 An International Thomson Publishing Company

Albany • Bonn • Boston • Cincinnati • Detroit • London • Madrid • Melbourne
Mexico City • New York • Pacific Grove • Paris • San Francisco • Singapore • Tokyo
Toronto • Washington

NOTICE TO THE READER

Trademarks

AutoCAD, AME, 3D Studio, AutoCAD® Designer, and AutoCAD® logo are registered trademarks of Autodesk, Inc. Autovision is a trademark of Autodesk Inc. Windows is a trademark of Microsoft Corporation. All other product names are acknowledged as trademarks of their respective owners.

Delmar Staff

Publisher: Alan E. Elken
Aquisitions Editor: Sandy Clark
Production Coordinator: Jennifer L. Gaines
Art and Design Coordinator: Mary Beth Vought
Editorial Assistant: Christopher C. Leonard

Cover Credits: Taj Mahal 3D and wireframe images courtesy of VR Technologies (P) Ltd., 3d@vrrt.com, http://www.vrrt.com Wireframe image photoenhanced by Scott Keidong's Image Enterprises. Bottom left and top left ©1998 Digital Stock.

The ITP logo is a trademark under license
Printed in the United States of America

Copyright © 1998
Delmar Publishers
Autodesk Press imprint
an International Thomson Publishing Company
For more information, contact:

Autodesk Press
3 Columbia Circle, Box 15-015
Albany, New York USA 12212-5015

International Thomson Publishing Europe
Berkshire House 168-173
High Holborn
London WC1V7AA
United Kingdom

Thomas Nelson Australia
102 Dodds Street
South Melbourne, 3205
Victoria, Australia

Nelson Canada
120 Birchmount Road
Scarborough, Ontario
Canada M1K 5G4

International Thomson Publishing Southern Africa
Building 18, Constantia Park
240 Old Pretoria Road
P.O. Box 2459
Halfway House, 1685 South Africa

International Thomson Editores
Campos Eliseos 385, Piso 7
Colonia Polanco
11560 Mexico D.F. Mexico

International Thomson Publishing GmbH
Königswinterer Strasse 418
53227 Bonn Germany

International Thomson Publishing France
1 rue St. Georges
75009 Paris France

International Thomson Publishing-Japan
Hirakawacho Kyowa Building, 3F
2-2-1 Hirakawa-cho Chiyoda-ku
Tokyo 102 Japan

International Thomson Publishing Asia
221 Henderson Road
#05-10 Henderson Building
Singapore 0315

2 3 4 5 6 7 8 9 10 XXX 03 02 01 00 99 98

Library of Congress Cataloging-in-Publication Data

Jefferis, Alan.
    AutoCAD for architecture : R14 / Alan Jefferis, Michael Jones.
        p.   cm.
    ISBN 0-7668-0335-X
    1. Architectural drawing--Data processing--Programmed instruction.
  2. Computer-aided design--Programmed instruction. 3. AutoCAD
  (Computer file)   I. Jones, Michael, 1964-   . II. Title.
  NA2728.J45   1998
  720'.28'402855369--dc21
                                                    98-4706
                                                       CIP

v

CONTENTS

PREFACE

AutoCAD for Architecture is a practical comprehensive text and workbook that is easy to use and understand. The contents are designed to teach the skills needed to master the 2D drawing commands used on most construction-related drawings. The content may be used as presented to provide an introduction to the use of AutoCAD Release 14. This text can also be used with older releases of AutoCAD. All concepts are presented using architectural examples so that each command can be correlated to a specific skill that will be required of architects, engineers, and drafters. Commands and drawing skills are presented in a sequence that is easy to understand and follow with each chapter building on the skills presented in the previous chapter. Other specific features of this text include:

- explanation of how a skill will relate to office practice;
- professional drawings by architects, engineers, and residential designers;
- chapter exercises which review chapter contents as well as reinforce previous material;
- each drawing chapter contains drawing problems that will reinforce the chapter concepts.

Individual chapters will provide knowledge to master the following drawing skills:

- drawing of lines and geometric shapes
- varied linetype and width
- editing of basic geometric shapes
- combining skills to produce drawings such as floor and foundation plans, elevations, details, and section
- editing of an entire drawing to produce other similar drawings
- dimensioning methods and techniques
- placement of text within a drawing
- creation of varied text and dimensioning styles
- creating symbols libraries to increase drawing speed and efficiency
- provide a user-friendly environment to master skills that will lead to a productive career using tools that have long since replaced the T-square, parallel bar, or drafting machine.

In addition, the authors have written an Instructor's Guide that contains supplemental drawing exercises, which are also available on disk. For more information, instructors can contact Delmar Publishers.

INTRODUCTION TO DRAWING COMMANDS

AutoCAD is a very powerful program that contains hundreds of commands to meet a variety of drawing needs. In this text, only the command options that are typically used within the construction field will be explored. Several methods are available to execute each drawing command. Data entry can be made by using toolbars, a keyboard, a mouse, or pull-down menu. The selection method will depend on individual preference and the hardware that is available. Each command will be presented initially through the use of the keyboard, with available options used to supplement the basic procedure. By presenting the keyboard as the primary option, the entire command sequence can be presented, displaying the many options that are available within a specific command.

Commands, options, or values that must be typed at the keyboard are listed with **BOLDFACE** capital letters. Once the command is entered, pressing the enter or return key (⏎) is required and will be represented by the symbol ⏎ throughout the text. Command sequences will be presented in the following manner:

Command: **LINE** ⏎
From point: **4,8** ⏎
To point: **8,8** ⏎
To point: ⏎

When input is required to complete a command, and the selected point can be any point, it will be presented using italics in the following manner:

Command: **LINE**
From point: (*Pick a point.*)
To point: (*Pick another point.*)
To point: ⏎

Some commands may allow any key to be selected to allow a specific sequence to continue. *Select any key* refers to striking any key on the keyboard to continue. There is no key labeled "select any key" on the keyboard, but many novice users spend a lot of time looking for it.

PREREQUISITES

Although this text can serve as an excellent reference for experienced professionals, no knowledge of architecture or construction is required to learn the command structure of AutoCAD. Some of the advanced projects do require a knowledge of basic residential architectural standards.

ONLINE COMPANION

The Online Companion™ is your link to AutoCAD on the Internet. We have compiled supporting resources with links to a variety of sites. Not only can you find out about training and education, industry, and the online community, we also point to valuable archives compiled for AutoCAD users from various Web sites. In addition, there are pages specifically for users of **AutoCAD for Architecture**. These include an owner's page with updates, a swap bank where you can share your drawings with other AutoCAD students, and a page where you can send us your comments. You can find the Online Companion at:

http://www.autodeskpress.com/onlinecompanion.html

When you reach the Online Companion page, click on the title **AutoCAD for Architecture**.

ABOUT THE AUTHORS

Mike Jones is the CAD Systems Manager and a drafting instructor at Clackamas Community College, which is an Autodesk Premier Training Center in Oregon City, Oregon. Mike has introduced new users and experienced professionals to the commands and options of AutoCAD for over 14 years. His training center experience is enhanced by 17 years of professional drafting and CAD system management experience in private industry.

Alan Jefferis has been a drafting instructor at Clackamas Community College for 18 years. His professional experience includes eight years of drafting experience with structural engineers, and 16 years as a professional building designer. He is currently the owner of Residential Designs, A.I.B.D., and is co-author of *Architectural Drafting and Design, Print Reading for Architectural and Construction Technology, Commercial Drafting and Detailing, AutoCAD for Architecture Release 12, and AutoCAD for Architecture Release 13*. All are available through Delmar Publishers.

ACKNOWLEDGMENTS

The quality of this text has been greatly enhanced by the donations of many professional offices and individuals within private and municipal agencies. My special thanks to:

Kenneth D. Smith
Kenneth D. Smith Architects
El Cajon, CA

Tom Sponsler, Director
Department of Environmental Services
City of Gresham, OR

Jessie O. Kempter
International Business Machines Corporation

F.D. Lee
Lee Engineering, Inc.
Oregon City, OR

G. Williamson Archer, A.I.A.
Archer & Archer P.A.
Meridian, MI

Ron Lee, A.I.A.
Architects Barrentine, Bates & Lee, A.I.A.
Lake Oswego, OR

Thomas J. Kuhns, A.I.A.
Michael & Kuhns Architects, P.C.
Portland, OR

Ned Peck,
Peck, Smiley, Ettlin Architects

Havlin G. Kemp, P.E.
Van Domelen / Looijenga / McGarrigle / Knauf Consulting Engineers
Portland, OR

David Rogencamp
kpff Consulting Engineers
Portland, OR

Sylvia Sullivan
College of the Canyons

Joseph Yamello
Greater Lowell Regional Vocational
Technical School

Robert Kamenski
Lenape A.V.T.S.

Walter Eric Lawrence
Gwinnett Technical Institute

Jason Kuhns
Clackamas Community College

All of these people have made substantial contributions to the success of this text.
We're also especially thankful for the work and efforts of Sandy Clark of Delmar
Publishers; and Suzanne Pescatore, Michael Pescatore, and Ralph Pescatore of TKM
Productions. No one has been more helpful and supportive than Janice Ann. Thanks
for the help and support.

TO THE STUDENT

AutoCAD for Architecture is designed for students in architectural or engineering programs. The contents are presented in an order that has been tested within the classroom. Information is based on common drafting practices from the offices of architects, engineers, and designers. To make the best use of the information within the text, students should follow these few helpful hints.

Read the Text.   I know it sounds so simple, but if you'll read a chapter prior to sitting at a computer, you'll find your drawing time will be much more productive. Read through the chapter and make note of things that you're not quite sure of. With the reading complete, now try to mentally work through the commands that you've been reading about.

Study the Examples.   One of the many features of this text are the many illustrations and examples of how to complete each command sequence. Carefully study and compare the command sequence with the illustration of how to complete the command.

Practice at a Workstation.   Once you've read the chapter and studied the illustrations, complete the command sequences at a computer workstation. Practice each command several times and don't be afraid to explore all of the options within a command. Because each chapter is based on the previous chapters, don't hesitate to reread past material. Above all else, don't be afraid to experiment. It's very common for new computer users to be filled with a fear of breaking something. If you save your drawings often, there is very little that can go wrong by experimenting. Relax and have fun!

SECTION 1

SYSTEM COMPONENTS

CAD System Components

Exploring Windows

. . . .

CHAPTER

1

CAD SYSTEM COMPONENTS

. . . .

Whether their drawings were painted on walls, methodically scribed into clay, or etched on papyrus, people have been planning structures and equipment to better their lives. Most recently, those charged with converting a sketch into a usable drawing have progressed from T squares, to parallel bars, to drafting arms, to tract machines, to computers.

In the early '80s, computers became available to large firms that could afford their hefty price tags. Mechanical and Electrical engineering firms became computerized, but many architectural firms spurned computers because their drawings just did not reflect the artistic trends in the field of architecture. By the mid '80s, prices had dropped and programs had been developed that could produce drawings with artistic flair. By the late '80s, many architectural firms had discovered that computers might have a place in their office.

Because of the tremendous drop in price, increased speed of the machines, and a wide variety of program aids, computers are now found in most architectural and engineering firms. This probably has a lot to do with why you're reading this book. Computer-aided drafting, or CAD, has changed the marketplace greatly. If you have experience in drawings, you might find yourself forced to learn this new system. Relax! It's not really that bad. CAD drafting, and AutoCAD in particular, tend to become addictive. People who have used manual drawing methods for 20 years and swore someone would have to pry the pencils from their fingers, now need the plug pulled to get them to go home. CAD may seem frustrating for the first few weeks, but I think you'll love it.

BENEFITS

If you're new to the field, you may have an easier time learning the basics of computer drawing, since you will not have to "unlearn" some old habits. In either case,

CAD has many benefits that will make its use invaluable. Among the benefits that make AutoCAD so popular are the accuracy, speed, consistency, and neatness that can be achieved with each drawing.

Accuracy

As you will see in the next few chapters, the accuracy of drawings can be enhanced greatly by using AutoCAD. This is because you will be able to control the exact placement of objects in the drawing, enlarge the size of an object for better viewing, or use a series of drawing aids that accurately place an object exactly were it was intended to go. Figure 1–1 shows an example of an architectural drawing created using AutoCAD.

Drawing Speed

Speed in the office is a great asset as long as quality can be maintained. CAD drafting offers both. Although you may be able to manually draw some objects more quickly, AutoCAD allows you to make changes much more quickly than manual drafting. AutoCAD also excels in repetitive functions. For example, several different command options are available so that 10 similar bolts can be drawn almost as quickly as one.

Consistency

Multiple objects not only can be drawn quickly, they can be drawn consistently. This consistency is a major consideration in the office setting, not just within your own drawings, but within the drawings of the entire staff. Each employee can letter exactly the same as fellow staff members. Other important drawing qualities, such as line weight, linetype, and lettering angle, can be controlled easily by AutoCAD.

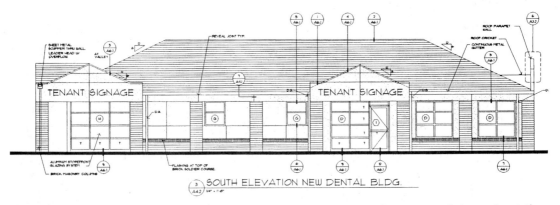

FIGURE 1–1 AutoCAD and programs created by third-party vendors are used throughout the construction field to create drawings quickly and economically. (Courtesy Scott R. Beck Architect.)

FIGURE 1–2 Once the basic drawing skills are mastered, add-on programs such as Big D Rendering Software by Graphics Software Inc., can be used to produce realistic drawings. (Courtesy Graphics Software Inc.)

The Challenge

Of all the features of AutoCAD, perhaps its best one is that it is challenging. As you develop your manual skills, you can learn to use a drafting machine in an hour. You can develop the manual skills to get quality linework and lettering by the time you've completed your first few drawings. You will be able to draw simple objects using AutoCAD with little effort. By the end of just a few more chapters, you will be doing complex drawings. Because AutoCAD is always changing and very complex, it may take you several months before you feel totally at ease with it. You can become very proficient with the core of the AutoCAD program in a short time and then choose to expand your knowledge as needed.

THE EQUIPMENT

Not only does CAD drafting offer new benefits to the architect, engineer, and designer, it also brings new tools to complete the drawing. These tools are the hardware and software. Software is the program or instructions that makes the computer perform specified tasks such as word processing or computer-aided drafting. Software will be discussed in Chapter 2. Hardware is the equipment that you will be using to run the software. One is useless without the other. Before you can begin to draw or design, you need a basic understanding of the main components of hardware found at a computer workstation. These include the computer, a monitor, input devices such as a keyboard and a mouse, and output devices such as a printer or plotter.

The Computer

Computers typically are referred to from largest to smallest as mainframe computers, minicomputers, and microcomputers.

Mainframe Computers.   Mainframe computers are used by large firms, which may have several hundred employee terminals connected to the same computer. A mainframe can run several large, sophisticated programs at the same time. Your local building and planning department is probably served by a mainframe computer. The receptionist, zoning department, plans examiner, document control department, and fee collectors all can run different programs on individual terminals that are connected to the mainframe computer.

Minicomputers.   A minicomputer allows several people to use the same program at one time. An office with several drafters and engineers may provide each with a terminal served by one computer.

Microcomputers.   A microcomputer or desktop computer is typically found in most educational and small office settings. It is also known as a PC (personal computer) since it is designed to be used by one person at a time. Microcomputers can be interconnected or networked for better efficiency of class or office personnel. An example of a microcomputer can be seen in Figure 1–3. Its key components include a systems board, central processing unit, memory, and disks.

Laptop Computer.   A laptop compter is a small, portable computer that draws its power from a battery. Because of its small size, it can be easily carried, allowing productive use of time when the user must be away from an office. Laptop computers are excellent for use at a construction site because of their mobility. Many professionals believe the use of laptop computers might allow the transfer of information electronically from the architect or engineer to the contractor at the job site and reduce the amount of paper now required to convey ideas.

Systems Board

The systems board in Figure 1–4 contains the electronic chips and boards that perform the basic functions of the computer. It is referred to as the motherboard because other components are all connected to it. In addition to the CPU and memory chips, the motherboard contains terminals or bus slots for control cards. A control card similar to Figure 1–5 is used to control the monitor, printers, plotters, and other external devices.

FIGURE 1–3 A typical work station includes a central processing unit, monitor, keyboard and mouse. Most systems now come with CD-Rom and sound systems. (Courtesy International Business Machines Corporation.)

Central Processing Unit

The central processing unit (CPU) is a chip located on the motherboard where the software program is processed. Once information is processed, the chip relays messages to other parts of the computer. The CPU chip is given a number that determines the class or type of computer. In the 1980's microcomputers were divided into the classes XT and AT. Early models of computers were dubbed XTs to show compatibility with the IBM-XT. Typically, XT computers contained an 8088 CPU chip by Intel. As technology advanced, PCs were classed as ATs, to represent "advanced technology" and to show compatibility with IBM-AT machines. AT machines were referred to by their bit processing ability, using the numbers 80286, 80386, 80486. Newer machines are based on the Intel Pentium® chip and are referred to by name rather than by number. Intel's newest machines are the Pentium®, Pentium Pro®and Pentium II®.

A bit is a binary number of 0 or 1 and is the smallest unit of data that the computer can recognize. XT machines typically had 8-bit processors. AT machines typically

FIGURE 1–4 The systems board contains all of the cards and chips that perform the basic functions of the computer. (Courtesy International Business Machines Corporation)

FIGURE 1–5 A control card is installed onto the systems board to control each piece of equipment attached to the computer. (Courtesy Hercules Computer Tech)

have 8-, 16-, or 32-bit processors, which allow them to process information much faster. Many users of advanced applications of AutoCAD work on machines with 64-bit processors. The system's power typically is measured by the speed of its processor in megahertz (MHz). A megahertz is a unit of frequency equal to 1 million cycles per second. AutoCAD Release 14 requires a minimum of a 486 or higher machine with 32 MB of RAM and between 43 and 141 MB of hard-disk space. A typical installation requires 73 MB. A minimum of 64 MB of disk swap space is also required for AutoCAD. Speeds as slow as 20 MHz can be found on older machines, but machines typically range from 66 to 100 MHz for 486 machines. Common Pentium processing speeds included 60, 90, 133, 166, and 200 MHz for the Pentium and Pentium Pro chips, and 233, 266, and 300 MHz for the Pentium II chips. Autodesk recommends using a Pentium 90 as the minimum speed processor to run AutoCAD Release 14.

Math Coprocessor

The math coprocessor is a chip that is added to the mother board to aid the CPU in processing information. Figure 1–6 shows a math chip. Older computers required the math coprocessor be added to the mother board to enhance program speed. Math

FIGURE 1–6 Although small in size, chips control each of the basic functions of the central processing unit. (Courtesy International Business Machines Corporation)

chips were referred to by a number that related to the machine number. A 386 computer had a 387 math chip. As the 486 line of machines were marketed, the math chip was built into the 486DX but had to be added to 486SX machines. The math coprocessor is built into the mother board of Pentium based computers.

Memory

Each computer must have memory to load, run, and save the programs it uses. Memory chips are located on the motherboard. Read only memory (ROM) and random access memory (RAM) are the two major types of memory typically specified when referring to the power of a computer.

ROM is the memory built into electronic chips on the motherboard that enable basic commands to be stored. When the computer is shut down, information in ROM is maintained by battery. Your disk operating system (Windows 95 or NT) is stored in ROM. Each time your computer is started, or booted as it is typically called, ROM is checked for specific commands in DOS that tell the computer what functions to perform with the needed internal components.

RAM is memory that is used as a program is being run. It provides temporary storage, so that any information that has not been saved is lost each time the computer is turned off. RAM should be thought of as the computer's work space. If you were a manual drafter, it would be your desktop. The bigger the desktop, the more "stuff" you can look at. The more complicated the task that needs to be done, the more RAM space will be required. Each program you purchase will specify how much RAM is required to run it. RAM is measured in kilobytes (kilo = thousand), such as 640 K (kilobytes). Many program memory reuirements are expressed in megabytes (MB), which represent 1 million bytes each. AutoCAD requires a minimum of 32 MB of RAM.

The terms extended memory and cache are often used when referring to RAM memory. Base RAM memory is 640 K. This barrier can be overcome by using extended memory, which is in the 1MB to 64MB or more range. Extended memory enables you to run programs requiring you to use large amounts of data, which could not be run efficiently otherwise in conventional memory.

Cache memory is used by your computer to achieve the highest possible data transfer rate between the main memory and the CPU. Cache memory holds just those items in memory that are for imminent use. This allows the computer to function 5 to 10 times faster than when it works in main memory. Cache memory is a very high-speed, small storage area in the CPU. Main memory is a much larger area, but information stored there takes longer to access because of the area's size. When a byte of information is requested, your CPU will check cache memory first. If the information is found there, the next request can be processed. If it's not found in cache memory, the CPU will examine main memory. The 486 line of computers typically have 256K cache memory with newer machines having 512K cache memory.

STORAGE DEVICES

Before AutoCAD can be used effectively, you'll need an understanding of how information is stored. Chapter 5 will deal with the process of saving drawings. This chapter will deal with where drawing files are stored and proper handling procedures to ensure that your drawings are usable once the computer is shut down. For students, your two common choices for storage include the hard drive or floppy disk drives. For professional CAD operators, you'll also have the option of storing drawings on tape backup and ZIP drives. Neither of these methods will be discussed in this chapter.

Most personal computers are purchased with hard disk and floppy disk drive units already installed. Although the size and number of drives will vary, their basic functions remain the same.

Hard Drives

Depending on your computr, you may have one or more hard drives and one or more floppy drives. The hard drives similar to Figure 1–7 are mounted inside the computer. Occasionally they may be referred to as fixed drives since they are fixed permanently inside the machine. There is no physical opening in the computer case to allow access to the hard drive. This drive is accessed by electronic signals processed by the other computer components. AutoCAD is installed on your hard disk. Hard disks typically range in storage capacity from 100 MB to several gigabytes. Computers with 300 MB to 2.4 GB (gigabytes) are typically sold at computer outlets as standard equipment. Because the disk contains so much space, this space can be divided into partitions in much the same way that a file cabinet is divided into drawers. This

FIGURE 1–7 The hard disk is located inside the CPU. Many computers provide disk drives for reading 3½" and 5¼" disks, as well as CD ROMs. (Courtesy International Business Machines Corporation)

allows the user to put valuable files in one area of the disk, and work in another area of the disk, without the chance of mistakenly entering an error into the base program. Placing one partition in the hard disk essentially gives you two hard disks. Since the size of the partition is smaller than the whole disk, information can be accessed much faster. It's hard to imagine that a fraction of a second saved while working on a project could make a difference, but when you consider the hundreds of thousands of commands that will be entered to draw a large commercial project, these milliseconds will mount up. If you have a hard drive measured in gigabytes, partitioning can greatly reduce the time spent searching for information.

DISKETTES

Diskettes are referred to as floppy disks, and they are read in a floppy drive. Most PCs are equipped with one or two floppy disk drives. Figure 1–8 shows a typical arrangement of drives. Notice that floppy drives have an access slot for inserting and

FIGURE 1–8 Drives for reading 3½" and 5¼" drives are typically provided with each computer. (Courtesy International Business Machines Corporation)

retrieving floppy disks, a light to indicate that the drive is in use, and a knob for securing the floppy disk in the drive. Typically, two types of diskettes are used with PCs.

Floppy and Rigid Disks

In the late 1980s floppy disks were the most common type of diskettes. These are 5¼", vinyl-covered flexible diskettes. With the increase in size of the hard drives and the use of CD-ROM, many new computers are no longer installing 5¼" drives as standard equipment.

Because of their ability to store more information with less chance of damage, 1.44 MB rigid diskettes have become the popular choice for storage of information. Rigid diskettes are 3½" across and are made with a rigid plastic protective shell. A metal shield provides protection against dust getting into the diskette surface. Components can be seen in Figure 1–9. Storage capacity ranges from 720 KB for the standard capacity to 1.44 MB for the high-density disk. Each type of disk may need to be formatted before it can be used, although most disks are preformatted. Drives that read CDs, tapes, or ZIP drives are also popular methods of storing information.

Alternative Drives.   In addition to the two possible floppy drives, most computers now come equipped with internal drives which read CDs. Drives for writing information to CDs are also available. External drives that can be moved between individual work stations are also available for reading CDs. Many professional offices currently use tape drives for storing drawings for archival purposes. Floppy and hard drives are often used for the day-to-day storage of drawing files because of their

FIGURE 1–9   Storage capacity of 3½" disks ranges from 720 KB for standard capacity to 1.44 MB for high-density disks. (Courtesy International Business Machines Corporation)

access speed. Once the job is completed, it can be transferred to tape storage where it is less subject to damage from poor handling. Tape drives offer the ability to economically store large quantities of drawings at a relatively low price. A ZIP drive or portable drive can be attached to a computer. ZIP disks are similar to 3½" disks, but store 100 MB of information. ZIP drives can be easily transported between work stations, and can carry whole programs as well as many drawing files.

DISKETTE HANDLING PROCEDURES

Diskettes typically serve to back up projects that have been saved on the hard disk. Because of limited memory in many school PCs, you may not be allowed to save on the hard disk. This will make your diskettes even more valuable. Because diskettes can be damaged easily, extreme care must be taken with them. The biggest destroyers of information on a diskette are dust, smoke, magnets, and folding.

Dust and smoke interfere with the head in the disk drive as it attempts to interpret the diskette. Magnetic fields from such common items as a stereo speaker or telephone can scramble the information on a disk into an unusable form. In addition to destroying information, a fold may hinder retrieval of a diskette from the drive unit. Disk labels that are not affixed well to the disk also can cause the disk to jam in the drive.

Guidelines for Safe Diskette Handling

- ALWAYS keep the disk labels in the outer corners of the disk.
- ALWAYS label diskettes clearly, indicating the contents.
- ALWAYS store diskettes in their protective envelope or carrying case.
- ALWAYS store diskettes in a protective container in a cool, dry location.
- ALWAYS store diskettes between 50 and 125° Fahrenheit (10–52° centigrade).
- NEVER touch the inner diskette surface.
- NEVER place diskettes on top of a monitor.
- NEVER clean or expose the inner diskette.

Protecting Disks

Diskettes can be protected so that no one can accidentally enter information or change the original contents. There is a notch in the top right-hand side of the disk when looking at it from the front. Turn the disk over and slide the black tab to its upper position. This will allow you to see through the notch and make the diskette write protected.

DRIVE IDENTIFICATION

The various drives within your computer are designated by a letter followed by a colon. Typically, C is used to designate the hard drive. Large hard drives can be subdivided into smaller areas using partitions. If partitions have been placed in the hard drive, a letter is assigned to each partition on the hard drive. The letter A is used to designate the floppy disk drive. If you have a second floppy disk drive, it would be accessed by the letter B. Depending on how your machine is configured the 3½" drive may be either A: or B: with the unused letter assigned to the 5¼" drive. Additional letters can also be assigned to represent a CD-ROM or a ZIP drive. A work station equipped with a CD-ROM, two floppy disks, a ZIP drive, and a hard disk with two partitions would include the following drive letters:

A: 3½" drive

B: 5¼" drive (optional)

C: hard drive

D: & E: partions within the hard drive

F: CD-ROM

G: ZIP drive

MONITORS

The monitor is a display device that allows users to see the results of their input into the software program (see Figure 1–10). Older monitors were monochrome, but 486 and Pentium machines require color monitors. Depending on the manufacturer, a color monitor with AutoCAD can display between 256 to 16 million colors on a white or black background. Monitors are specified by screen size and display ability. Most PCs come with 15", 17", 19", or 21" monitors measured diagonally across the screen. Screen size can be 35" diagonally, or larger for display to large groups.

The clarity or resolution of the monitor is influenced by the numbers of pixels. A pixel is a specific point on the screen that can be lighted or remain unlit. Objects are displayed on the screen as pixels are lit. A typical monitor will have 640 × 480 pixels. As the number of pixels is increased, the resolution increases as well. Monitors are refered to as *EGA (enhanced graphics array), VGA (video graphics array), or SVGA (super video graphics array)*. SVGA is the norm for Pentium machines. VGA monitors display up to 256 colors simultaneously with 640 × 480 pixels. High-resolution monitors display 1024 × 768 or more pixels and display over 16 million colors. Your monitor will come with a graphics card that goes inside the computer on the systems board that translates the computer output to the display on the monitor. A graphics accelerator board can produce resolutions up to 1600 × 1200.

INPUT DEVICES

The two most common input devices used with AutoCAD are the keyboard and the mouse.

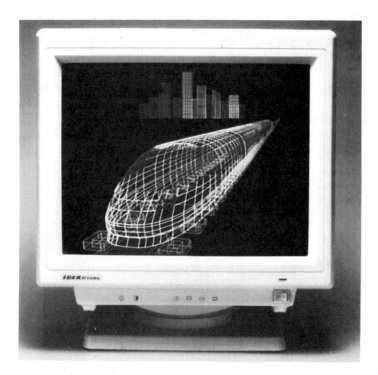

FIGURE 1–10 The monitor is a display device that allows the user to see the result of input with the software. (Courtesy Idek/Iiyama North America, Inc.)

Keyboard

The keyboard of your computer is similar to a typewriter's keyboard. Some variation will be found between manufacturers, but basic keyboard configuration can be seen in Figure 1–11. You will notice that across the top of the standard keys is an additional row of keys called function keys. These keys can be used as shortcuts within specific programs rather than having to type out an entire command. On the right side of the keyboard you will see directional keys and a 10-key layout for math functions. An alternative to a standard keyboard can be seen in Figure 1–12. Keyboards with the keys placed to meet the natural position of the hands and wrist can be a great aid at over coming fatigue caused from long hours of typing. Even with all of the icons and toolbars provided by Windows and AutoCAD for most of the commands that you will use in AutoCAD, the keyboard is the fastest and easiest method of executing the command. This is true even if you're a two-finger typist. Throughout this text, commands will be discussed by providing examples of keyboard entry. For example, once you are in AutoCAD the only typing required to start the LINE command would be to type **L** [enter] or **LINE** [enter] at the command prompt. L represents the

FIGURE 1–11   Many of the commands to run AutoCAD can be entered by keyboard. In addition to the standard keyboard, numeric keys and special function keys are provided.

FIGURE 1-12   The use of an ergonomic keyboard such as the Natural Keyboard by Microsoft can greatly ease the stress placed on the wrist and fingers. (Courtesy Microsoft Corporation.)

LINE command, and ⏎ represents pressing the enter key. The command sequence for keyboard entry for the LINE command will be displayed as:

Command: **L** ⏎
LINE From point:

Mouse

A mouse, shown in Figure 1–13, is the pointing device used with windows applications. The two most common types are a roller and optical mouse. In addition to presenting commands by keyboard input, executing commands by use of a mouse will also be discussed as an entry method. On a two-button mouse, the left button is used for the select button and the right button to terminate a command. Chapter 2 will introduce additional functions of each button. The mouse can be used to select commands from menus that are displayed beside, above or inside the drawing area, or to select starting and ending points while in a command. When a command is to be selected from a menu, the text will refer to *picking* a selection from a menu. This refers to the process of going to a menu, placing the cursor on the desired option and pressing the left button.

OUTPUT DEVICES

The two types of output devices are printers and plotters.

Printers

The most common types of printers are dot matrix, laser, and inkjet. Each provides a quick method of getting a paper sample (hardcopy) of a drawing or a list of commands. Printers are typically used in an office for check prints.

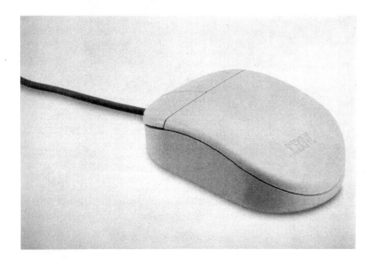

FIGURE 1–13 A mouse provides an inexpensive method for inputting drawing commands. (Courtesy International Business Machines Corporation)

Dot matrix printers produce drawings comprising a series of dots. The number of dots per inch is reflected in the quality of the machine. Dot matrix printers have a quick output rate and are relatively inexpensive to operate. Prints can be produced in sizes up to 18" × 24".

Laser printers can produce a very sharp graphic image allowing the user to set the density of lines to be produced. Laser printers allow for a quick, quiet print to be made, although the size is often limited. The typical size of hardcopy from a laser is 8½" × 11" with a resolution of 300 to 600 dots per inch (dpi). A laser printer can be seen in Figure 1–14. Inkjet printers offer the highest quality of prints at prices that are often below the cost of laser printers. These printers make use of color cartridges and provide resolutions of 300 to 600 dpi.

Inkjet printers that produce color prints at inexpensive prices are also available. (See Figure 1–15.) Many inkjet printers are able to produce drawings using resolutions that match photo-quality color images. Portable laptop printers are also available for use with a laptop computers. Notebook printers similar to the printer in Figure 1–16 allow prints to be made, by use of infrared signals, with no printer cables connected to the computer .

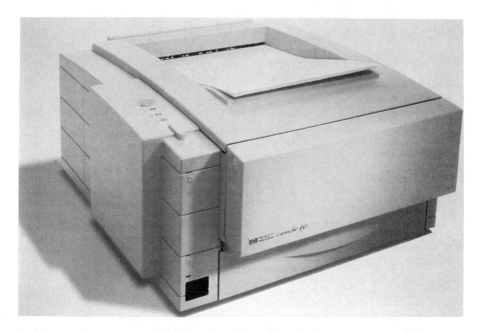

FIGURE 1–14 Lasers printers such as the HP LaserJet 6P produce a fast, economical method of reproducing check prints. (Courtesty Hewlett Packard Company.)

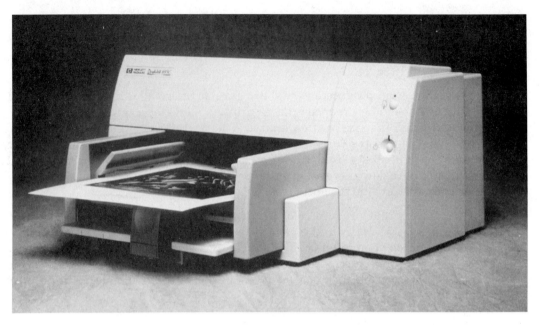

FIGURE 1–15 Inkjet printers such as the HP DeskJet 690 produce photo-quality color images. (Courtesy Hewlett Packard Company.)

FIGURE 1–16 Notebook printers allow quick prints to be made at a job site using wireless technology with printers such as the HP DeskJet 340. (Courtesty Hewlett Packard Company.)

Plotters

Plotters are typically used to get a final print of a completed drawing. The most common types of pen plotters are flatbed plotters and drum plotters. Both types of plotters use a variety of pen tips to draw on vellum, Mylar®, or fine matte paper. Each type of plotter allows for the use of color. As CAD drawings are completed, they typically are reproduced onto vellum and then multiple copies are blueprinted.

A flatbed plotter can be seen in Figure 1–17. Material ranging in size from 8½" × 11" through 18" × 24" can be used. The plot is produced by the pen moving over the drawing material in horizontal and vertical directions. Many offices will use small flatbed plotters for check prints only because the plotter that handles larger-sized paper takes up more space. A drum plotter can be seen in Figure 1–18. These function by moving the pen and the paper. The pen moves across a track as the paper is moved back and forth. Drum plotters can handle almost any size of paper. As technology advances and prices descend, many professional offices now use electrophotographic plotters, which offer quick, quiet plots of up to 400 dpi. Ink-jet plotters have also become a popular choice of professional CAD users. Laser quality production and speed are now available in many plotters similar to the plotter shown in Figure 1–19. These plotters use LED technology and dry toner cartridges to produce plots up to "E" size media— including paper, 4-mil matte film, or roll-feed media in lengths up to 20 feet.

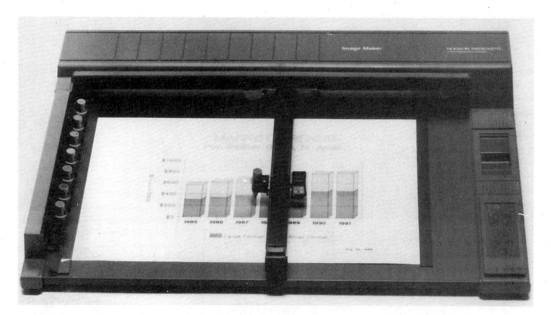

FIGURE 1–17 A flatbed plotter produces drawings by moving a pen across paper. (Courtesy Houston Instrument®, A Summagraphics Company™)

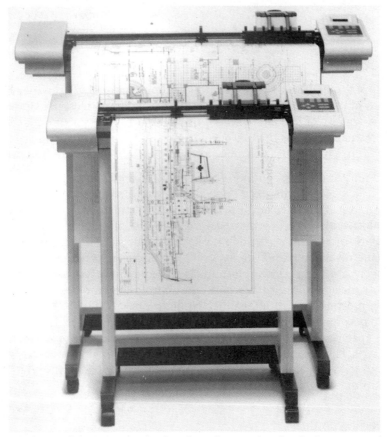

FIGURE 1-18 A drum plotter produces drawings by moving both the pen and the paper. (Courtesy Houston Instrument®, A Summagraphics Company™)

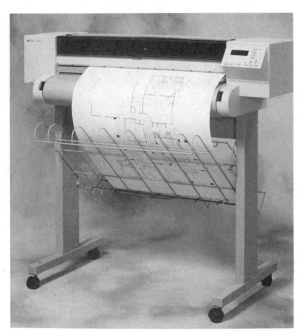

FIGURE 1-19 Laser quality plots can be made in seconds using 400 dpi resolution on plotters similar to the CalComp Solus 4. (Courtesy of CalComp Corporation.)

These plotters use color cartridges to produce prints with a resolution of up to 600 dpi with a speed that is as much as ten times faster than a pen plotter. Plotters are also available that feature direct image technology to produce drawings up to "E" size material with up to 800 dpi resolution. These direct image plotters are approximately twice as fast as pen plotters and are often used on a network or to meet other high volume plotting demands.

Although not common in schools or small offices because of their expense, electrostatic plotters are a third type of plotter. These plotters are much faster and quieter than pen plotters and offer much higher resolution in full-color printing.

CHAPTER 1 QUIZ

1. List two types of memory and explain their uses.

2. List five components that comprise a computer.

3. List four advantages of CAD over manual drawing.

4. What units are used to measure memory?

5. What are the components of a well-equipped workstation?

6. What are three different categories of computers and how could they be used?

_____

_____

_____

7. What are two tools that can be used to input information into the computer?

_____

_____

_____

8. You've just purchased a new monitor and display card. What does the display card typically connect to?

_____

_____

_____

9. Using three local suppliers, research the various speeds available for Pentium computers, as well as their price range.

_____

_____

_____

10. Using three local suppliers, research hard disk size and price range.

_____

_____

_____

11. What is added to the systems board to enhance speed of graphics programs?

_____

_____

_____

12. List two types of disk drives.

_____

_____

_____

13. What two categories are used to describe monitors?

14. What type of plotter holds the paper while the pen moves?

15. How is information entered into the hard disk if no access door is provided?

16. List the storage capacity of 3½" diskettes.

17. Describe the process to write-protect a 3½" disk.

18. List three places disks should not be placed.

19. What letter is typically used to describe the hard disk?

20. Why would it be useful to partition the hard disk?

21. What is the major difference between Windows 95 and Windows NT?

22. Select the My Computer icon from the startup display. Select the Printer icon and determine which printers or plotters are configured to the current workstation.

23. Select Program Explorer from the Programs menu. Select the AcadR14 folder and determine the amount of folders and the total size of the program loaded on your workstation.

24. Compare the cost and features of printers made by at least three different manufacturers that can be used to obtain high quality colored prints.

25. Verify with a minimum of three local retailers the cost of setting up a work station using a 200 MHz Pentium or faster machine with MMX technology and equipped with Windows 95, a CD-ROM multi-media kit, 512K cache memory, and 128 MB Extended memory. Get estimates to include the following equipment:

    16 and 32 MB of extended memory.

    An internal ZIP drive.

    3, 15, and 17 inch monitors.

    33.6–Kbps internal fax modem.

    1.6, 2, 3, 4 gigabyte hard disk drive.

    1.44 MB floppy drive

    5¼" 1.2 MB floppy drive

••••••••••••••

CHAPTER

2

EXPLORING WINDOWS

•••••••••••••••••••••••••

Before you can explore AutoCAD, an introduction to the platform used to run each program will be explored. This chapter will introduce you to Microsoft Windows 95. Methods of creating, storing, and accessing files will then be discussed. Remember that this is a book about using AutoCAD, and not the operating systems. Only a brief introduction will be given to Windows 95 and Windows NT. With a few exceptions, once in AutoCAD, the program functions the same no matter what platform is used by your computer to organize information. Other resources by Delmar Publishers® can be consulted for a detailed explanation of how each of these operating systems function.

AN INTRODUCTION TO COMPUTER PLATFORMS

In Chapter 1, the components of a work station were reviewed. Neat stuff, but these gadgets can do nothing without software. Software consists of programs that are loaded into the computer to tell it what to do. The focus of this book is the software program of AutoCAD Release 14. The platform coordinates the computer and the software so that each can function. Two platforms or operating systems are available to run AutoCAD Release 14. The operating systems that coordinate the software programs in your computer include Microsoft Windows 95 or Windows NT. It's important to remember that these platforms are just tools used to operate AutoCAD.

Windows 95

Windows 3.1 worked well to introduce many users to the world of computing, but MS DOS and its commands were still lurking in the background waiting to intimidate the user. Microsoft Windows 95 solved the need to know DOS, because it is an independent icon-driven disk operating system that does not rely on typed commands. If you purchase a PC (personal computer), it will come loaded with Windows 95 as the controlling platform. Unlike Windows 3.1 that works with DOS, Windows 95 is the operating system. Windows 95 is compatible with DOS, and many DOS programs can be run from inside Windows 95, but DOS is not required to operate in Windows 95. Unlike older versions of AutoCAD that could run on any of the operating systems, Release 14 requires Windows 95 or Windows NT as its operating platform. When a computer is booted up using Windows 95, a screen display similar to Figure 2–1 will be displayed. Methods of accessing programs will be discussed later in this chapter. Chapter 3 will discuss the Windows Explorer of Windows 95 and Windows NT.

Windows NT

Although many home users of PCs have been very happy with Windows 95, many business users, including those who use the advanced functions of AutoCAD, have been disappointed in the compatibility of AutoCAD and Windows 95. Windows 95 has been known for crashes. A *crash* is an unplanned termination of a program.

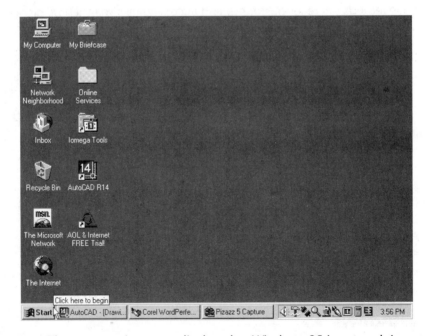

FIGURE 2–1 The program icons are displayed as Windows 95 is entered. Icons are used throughout Windows 95 and AutoCAD to represents specific programs, or components within a program.

When Windows 95 crashes it often disrupts all open files. Windows NT, short for network, solves many of the compatibility problems between AutoCAD and Windows, which often cause computer crashes. AutoCAD is a 32-bit program. For most of us, this prompts a rousing "So". To computer programmers, this is of great significance. NT is a true 32-bit program which can communicate effectively with AutoCAD with minimal disruptions. Combined with true compatibility, and a better method of linking multiple work stations to a network, NT is becoming the popular favorite of many schools and CAD-operated business. Both Windows 95 and Windows NT allow two programs to be opened and displayed on the monitor simultaneously as seen in Figure 2–2. Rather than having to minimize and maximize windows, Windows 95 and Windows NT allow two windows to be created. This can be especially helpful when trying to coordinate two different drawings. Windows NT looks just like Windows 95 when it is entered.

A WINDOWS SURVIVAL GUIDE

Before the AutoCAD program is explored, the basis of Windows must be understood. No matter which version of Windows you use to run AutoCAD, they each have common features that must be understood to function in Windows. If you have experi-

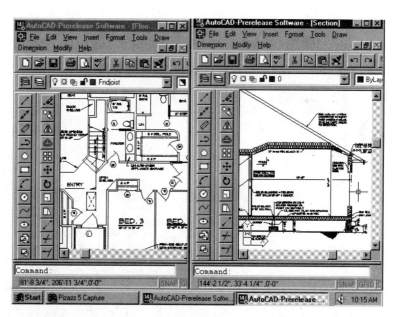

FIGURE 2–2 Windows 95 and Windows NT allow two windows with different AutoCAD programs to be viewed at the same time, providing a quick comparison between drawings.

ence working with a Windows-based program, skip to the next section which discusses the storage devices. If you're a new computer user it's important that you understand the basics of Windows including icons, clicking, double-clicking, and dragging before you tackle AutoCAD. There are also several common tools provided in each window. These tools include the title bar, the task bar, the menu bar, dialog boxes, control buttons, the scroll bar, and alternatives to displaying multiple windows. Each of the window options will be demonstrated using the Windows Explorer. The Windows Explorer is used to edit, sort, and store programs, folders, and files. The Windows Explorer is opened by selecting the Start button in the lower portion of the Windows screen. Your screen should resemble Figure 2–1 as you enter Windows 95. Select the Start button by moving the mouse until the arrow points to the word Start. Pressing the Start button once with the left mouse button will produce the menu shown in Figure 2–3. Move the arrow until the Programs option is highlighted by a blue bar. Moving the arrow to the right of the Programs listing will display the options shown in Figure 2–4. Move the arrow down the list until the Windows Explorer option is highlighted. With the option highlighted, pressing the left mouse button once will open the Windows Explorer to a display similar to the one shown in Figure 2–5.

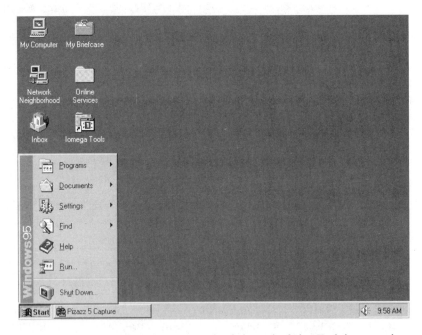

FIGURE 2–3 Selecting the Start button at the left end of the Task bar produces the Windows 95 menu.

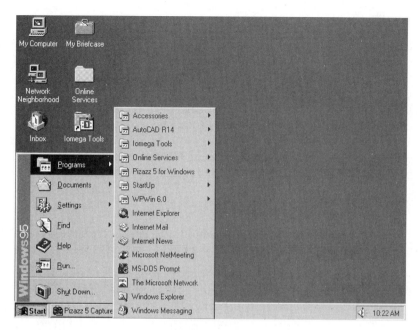

FIGURE 2–4   Moving the cursor to the Programs listing will display a new menu.

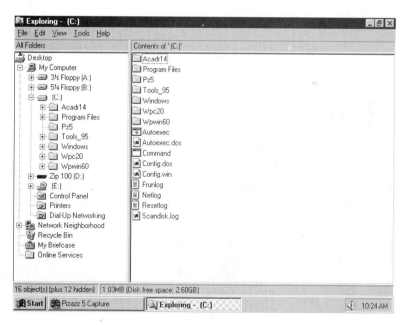

FIGURE 2–5   Highlighting the Windows Explorer listing and making a pick with the left mouse button will display a listing of the drives, folders, and files of your computer.

Icons

Pictures called *icons* are displayed to identify what program is to be activated. Windows 95 and Windows NT use an abundance of icons. The icons allow greater productivity and are used in place of commands that had to be memorized by DOS users. Figure 2–1 shows the icons displayed when Windows 95 is opened. Icons are also used within AutoCAD to identify specific portions of the programs to be executed. Figure 2–6 shows the icons from the standard tool bar of AutoCAD Release 14. Chapter 3 will discuss the use of each of the icons. This text will provide an introduction to the Windows Explorer and AutoCAD R14 icons and their contents. Consult the Help menu of Windows or a related text to explore other options of Windows.

Clicking and Double-Clicking

Pressing a mouse button as you would a touch-tone phone button is referred to as *clicking* or *picking*. A single click of a mouse button is referred to as a *pick*.

Menu items are typically selected with a single click. If a menu is displayed with six options, picking one of the options with a single click will highlight that menu option. Figure 2–7 shows the Windows Explorer with the ACADR14 option highlighted by a single click. Another option such as Run, Copy or Move could now be applied to this menu option.

Pressing a button twice in rapid succession is referred to as *double-clicking*. If a double-click is entered too slowly, it will be interpreted as two single clicks. A double-click is used to select a program, folder, or file for an immediate use. If the ACADR14 option had been double-clicked, a display of the contents will be displayed similar to Figure 2–8. Double-clicking the ACADR14 icon of the original Windows display shown in Figure 2–1 will remove that display and start the program. If you find yourself unsure about when to click or double-click, start with a single click. If the desired results are not achieved with a single click, use a double click. If you accidentally double-click when you should have clicked, press the Ctrl key and the Z key at the same time to undo the damage.

Most commands and options are executed using the left mouse button to make selections in Windows and AutoCAD. The right mouse button can produce several responses depending on what you're doing. Clicking a blank portion of the screen will typically reveal a hidden menu. Selecting a blank portion of the Windows Explorer with the right button will produce the menu shown in Figure 2–9. Clicking

FIGURE 2–6 Icons of the Standard tool bar from AutoCAD R14 can be used to perform specific functions.

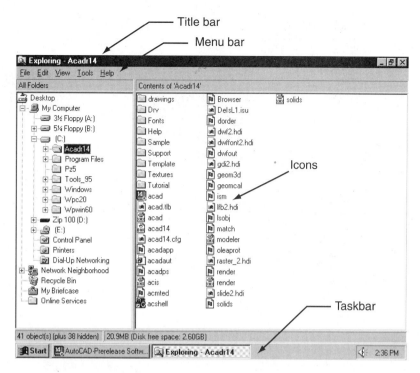

FIGURE 2–7 Selecting a folder with a click of the left mouse button selects the folder for use with another command.

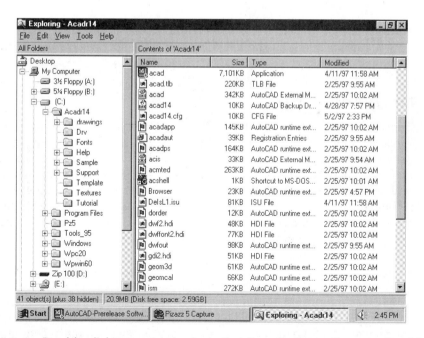

FIGURE 2–8 Double clicking an icon selects that item for immediate use. Double-clicking the AutoCAD icon in the left column will display the contents of the AutoCAD folder. The information that is displayed will depend on the settings of the View menu.

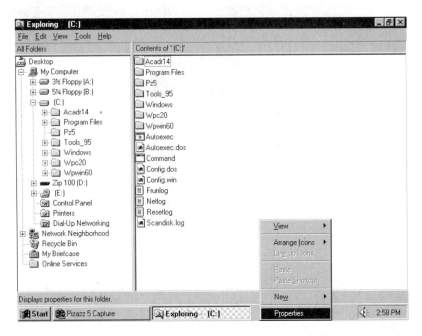

FIGURE 2–9 Selecting a blank portion of the task bar with a click of the right mouse button will produce a pop-up menu. Clicking the left button will remove the menu.

with the left button will remove the menu. Other uses of the right mouse button produce a similar result as the left button. The selections will be introduced with the command they affect.

Drag and Drop

These terms relate to the use of the mouse to select and move windows, folders, icons, or other portions of a program from one section of the screen to another. To drag a menu item, push and hold the select button. To move the Exploring window, move the cursor until it touches the blue bar across the top of the window. Next press and hold the select (the left) button on the mouse. While holding the select button, if the mouse is moved, so is the menu. When you're satisfied with the new location, release the button. Drag and drop can be used to move menus, icons, windows, and many of the components in AutoCAD.

Cursors

The term *cursor* refers to the pointing device that is placed on the screen to indicate the current mouse position. The cursor displayed on your monitor moves relative to the movement of the mouse. The cursor also changes shape depending on where it is located. Initially the cursor is an arrow as you enter Windows, but may be displayed as an hourglass, pointing finger, flashing vertical line, a double arrow, or a four-sided arrow. Each cursor can be seen in Figure 2–10.

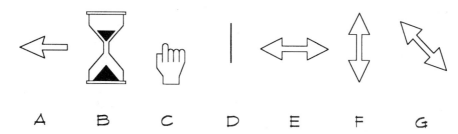

A   B   C   D   E   F   G

FIGURE 2–10   The cursors of a windows platform include (a) the arrow—used to make selections; (b) the hourglass— indicating a command is being processed; (c) the pointing finger—for making a selection; (d) the flashing bar—for entering text; (e) the horizontal double arrow—for altering a window size in the left/right direction; (f) the vertical double arrow—for altering a window in the top/bottom direction; and (g) the diagonal double arrow—for altering a window in two directions in one movement.

The Hourglass.   The cursor changes to an hourglass each time Windows is processing information. While the hourglass is present, Windows is unable to process additional information. For example, if you save a file, the hourglass will be displayed, indicating no new information can be processed. Once the command is executed, the hourglass reverts to the previous cursor, and new commands can be processed. The hourglass may also be displayed beside the arrow cursor. The arrow means you can still keep working but the hourglass is telling you that Windows is preforming a task while you're doing another. Your current program may run slightly slower, but full function will be restored when Windows is done.

Pointing Finger.   The pointing finger is the politically correct symbol Windows uses to indicate that there is additional information available for a specified command or option. Clicking the mouse will display additional information related to the designated topic.

Flashing Bar.   The arrow cursor changes to a flashing bar cursor when the cursor is moved into a portion of the screen where text input is required. Any text that is entered at the keyboard will be entered just to the left of the cursor bar.

Double Arrow.   As the arrow is placed on the border of a window, it will change to a double arrow. When placed on a side border of a window, the cursor will be a horizontal double arrow. When placed on the top or bottom border, the cursor will be a vertical arrow. If placed in a corner it will be a diagonal double arrow. If the select button of the mouse is picked and held, the size of the window can be enlarged or decreased depending on the movement of the mouse. Using the arrow to move a side border will alter the horizontal size. Picking the bottom border will alter the vertical size. If the pointer is placed in the corner of a window and the select button is held, the size of the window can be altered in both directions in one movement. The size of a window can be altered throughout a drawing session.

Title Bar

The title bar is located across the top of every window. It displays the icon and name of the current program as well as the name of the current file. The title bar also serves as a handle to drag the window across the screen. If multiple windows are open the title bar of the active window is highlighted with a blue bar while inactive windows are displayed in another color. Figure 2–6 shows the title bar for AutoCAD. Notice that the current drawing name is [Drawing] because the drawing file has not been saved yet. Storage locations will be discussed later in this chapter, and saving methods will be introduced in Chapter 5. At the far right end of each title bar are three buttons that can be used to control the size of the current window. These buttons are the minimize, the maximize, the restore and the close file button. Each can be seen in Figure 2–11. The minimize and close buttons will always be displayed, but only the maximize or the restore button is displayed depending on the current window setting. You'll notice that sometimes there are two rows of buttons. Although they function in a similar manner, the buttons at the end of the title bar control the display of the current window, and the buttons at the end of the menu bar control the display of the current file. The second row will be further explained in the next chapter.

The Minimize Button.   The minimize button is the horizontal bar icon on the left side of the three buttons (See Figure 2–11a) . Selecting the minimize button with the mouse pick button will remove the current program from the display screen and place the program icon in the Task bar. The task bar will be discussed later in this chapter. It's important to remember that if you select the minimize button and remove the file from the screen, the file still exists, it's just been placed out of your view. It hasn't been saved, or destroyed, just moved for convenience so you can do something else. To restore the window, use the mouse and a single click to select the program icon in the task bar.

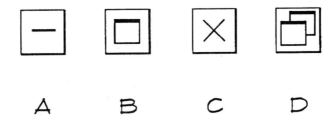

FIGURE 2–11 Four buttons for controlling windows are located on the right end of the menu bar. Selecting the minimize button (a) will remove a window from the screen and place the title in the task bar. Selecting the maximize button (b) will increase the window to it's largest possible size. Selecting the close button (c) will close a window and remove the program icon from the task bar. The restore button (d) is only displayed when a window is maximized. Selecting the restore button will return the window to its previous size.

The Maximize Button.   The maximize button is the file icon in the center of the three buttons. (See Figure 2–11b.) Selecting a window corner or border of a window with the cursor allows the size of the window display to be altered. When the size of a window has been reduced, picking the maximize button will restore the current window to its largest size. Other windows that may have been open are still open, they've just been covered by the current window. Notice that as soon as the window is maximized, the maximize icon is changed to the restore icon.

The Restore Button.   The restore button shown in Figure 2–11d is displayed only when a window is maximized. Selecting the restore button allows the window to be reduced to its previous size.

The Close Button.   Selecting the button with the X on the far right side of the title bar (Figure 2–11c) will close the current program. When the close button is selected while in a program such as AutoCAD, a box will be displayed with options to save or not to save the current file, or cancel the close command. When the close button is used in programs such as Solitaire, the program is closed without any other prompts.

The Task Bar

The task bar is located across the bottom of the screen display. The Start button is located on the left end and the current time is located on the right end of the task bar. The taskbar keeps track of all open folders by displaying the icon and name of each program. Figure 2–12 shows a task bar and indicates that AutoCAD, WordPerfect, Pizazz, Explorer, and Solitaire are open. As more windows are opened, the taskbar automatically reduces the size of each display. When the minimize button is selected, the current window is placed in the taskbar. By clicking an icon from the taskbar, the current window will be replaced by the selected program. Clicking the icon on the taskbar of the current program with the right mouse button will produce a menu. These menus will be discussed later in this chapter. Clicking the icon with the left mouse button removes the menu. Picking the Start button displays the Start menu and allows another program to be opened. Clicking the speaker on the right end allows the volume of your sound systems to be adjusted. Double-clicking the speaker icon displays a menu that controls the sound card's mixer. Placing the cursor over the time display produces a display of the day and date.

FIGURE 2–12 The task bar shows programs that are currently opened and the time. Five programs are presently opened, with Windows Explorer the current program

The Menu Bar

The menu bar is located below the title bar and is used in each window to store menus specific to that program. The menu bar of the Explorer can be seen in Figure 2–5 and contains the listings of File, Edit, View, Tools, and Help. Each word of the menu bar represents a separate menu with options related to the key word. Moving the cursor to one of these words and clicking will display the corresponding pull-down menu. Figure 2–13 shows the View pull-down menu for the Windows Explorer. Notice that the Line up Icons listing appears different from other listings. This option is inactive and is not currently available for selection. Other menus such as the Arrange Icons have a triangle on the right side that indicates selecting this option will produce another menu. The options do not have to be selected with the mouse pick button. Moving the mouse to the Arrange Icons listing will produce the menu shown in Figure 2–14. A check on the left side of a menu listing indicates that the option is currently on. For instance, in the View pull-down menu, the Status bar is currently set to ON and the Toolbar option is OFF. Pick the Toolbar option with the right mouse button and the toolbar shown in Figure 2–15 will be displayed.

Control Keys and Hot Keys.   Close the View menu by placing the cursor on View. Once the menu is highlighted, a click of the left button will close the menu. Move the cursor to the Edit menu and open the menu. Although many of the options are not active, you'll notice that several of the menu options have shortcut keys listed next to the command. The Undo command has a shortcut of Ctrl+Z. The Ctrl+Z symbol

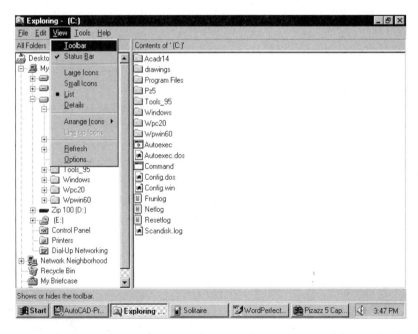

FIGURE 2–13   The menu bar for the Windows Explorer. A menu bar can be found in each program window just below the program name.

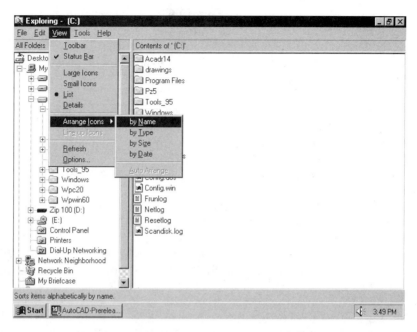

FIGURE 2-14 Moving the cursor to a listing followed by a triangle will produce another menu. Since the cursor was moved to Arrange Icons, a submenu is displayed.

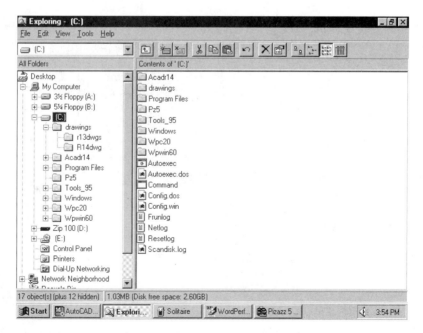

FIGURE 2–15 Picking the toolbar option of the View pull-down menu displays a tool-bar option below the menu bar.

represents pressing the Ctrl key and the Z key at the same time. The shortcut can be entered through the keyboard rather than opening the Edit menu and then picking the Undo option. In each menu you'll also notice that a letter of each menu option is underlined. Entering that letter by keyboard will produce the same effect as selecting the option with the mouse.

DIALOG BOXES.   Display boxes are used throughout windows to allow interaction with the current program. These boxes each contain methods of making selections as well as other menus and displays. Common components that will be encountered within display boxes include text, list, drop-down, and check boxes. Chapter 3 will discuss how each component relates specifically to AutoCAD.

Text Boxes.   A text box is a place that requires text to be provided. An example of a text box can be seen in the Find: All Files dialog box shown in Figure 2–16. The box is displayed by selecting Files or Folders... from Find of the Tools menu. As the dialog box is displayed, the arrow cursor is changed to the flashing bar cursor in the text box. The name of a folder or file to be found can now be entered into the text box by keyboard.

List Boxes.   A list box contains options that are to be selected by highlighting one of the options. Open the Start menu and pick Control Panel from the Settings menu. Double-click the Display icon and the Display properties dialog box will be displayed. The box contains two list boxes that can be see in Figure 2–17. A menu

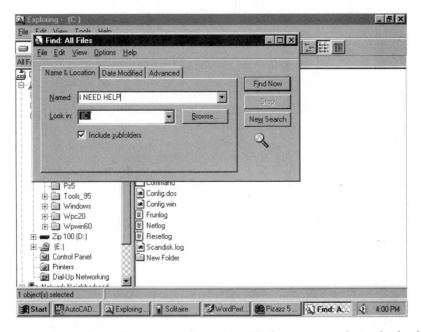

FIGURE 2–16   A text box allows you to interact with the program from the keyboard. Move the cursor to the text box and pick the box with a click to activate the box.

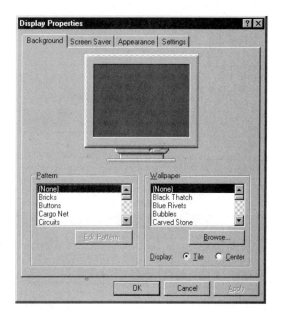

FIGURE 2–17 A list box is a window that contains a list of options. The Pattern and Wallpaper list boxes can each be used to select options for the desktop display.

option that is followed by a triangle that points downward will produce a drop-down list box. Selecting Map Network Drive...from the Tools menu of the Explorer will produce a display shown in Figure 2–18a. Picking the arrow at the right end of the Drive: box will produce the drop-down list box shown in Figure 2–18b.

Check Boxes. A fourth method of providing information in a dialog box is by picking a check box. A check box is placed next to a menu option. If you place the cursor in the box and pick the box, a check or an X will be placed in the box indicating that the selected option is active. An example of check boxes can be seen in Figure 2–19. This window can be accessed by selecting Options...from the View menu of the Explorer.

Control Buttons

In addition to the endless bars, borders and boxes that are used to control a window are the control buttons. The three most common are the OK, Cancel, and Help buttons found in most windows. Selecting the OK button will accept all options or settings that have been made in that window. Cancel will terminate the window and ignore any settings that might have been made. The Help button will provide access to a menu that can explain why or how a specific command should function.

Scroll Bars

The scroll bar is a vertical bar located on the right side of a window and a horizontal bar located below a window that has more material than can be displayed on the screen at one time. (See Figure 2–15.) The vertical bar consists of an up arrow at the

FIGURE 2–18A A box with a triangle that points downward indicates that more options are available. Pick the arrow to display the options.

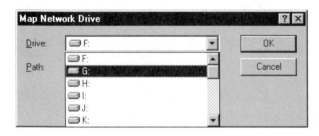

FIGURE 2–18B Picking the arrow in Figure 2-18a will produce the additional options.

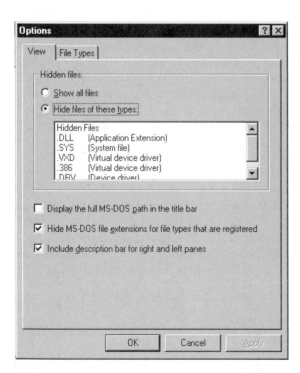

FIGURE 2–19 Many options contain check boxes to select settings. The three boxes below the display are check boxes. Two are currently active and one is inactive. The cicular buttons above the display are radio buttons. A radio button can also be used to activate or disable a setting.

top of the bar, a down arrow at the bottom of the bar, and a slide box that changes it's position relative to the location of the cursor to the start or end of the display. The horizontal bar has directional arrows for moving the slide box to the left or right. Each scroll bar provides a quick method of moving through a long list or a wide display. Keeping track of the location of the scroll box will provide an estimate of how near you are to the beginning or end of a list. By moving the cursor to the box and picking and holding the select button, the box can be moved up, down or across the scroll bar. As the box moves up the bar, the display shown on the screen scrolls up. A horizontal scroll bar is also placed on the bottom of the window allowing scrolling from left to right. Scroll bars are used on many of the displays in AutoCAD.

Displaying Windows

Working in a Windows program gives you the ability to open one or several windows at one time. A *window* is a display that allows you to see some aspect of a program. You can have one window active that covers the entire monitor, or open several different displays. Using the same methods that you used to open the Windows Explorer, open AutoCAD R14 from the Program menu. Open Solitaire and Calculator from Accessories of the Program menu. Each program is now listed in the task bar along the bottom of the screen. Picking the Calculator icon from the menu bar will display the calculator. Once you're done with Calculator, it can be returned to the task bar or closed and another program can be viewed. Unlike older versions of Windows, Windows 95 and Windows NT allow two or more programs to be displayed on the screen at the same time. Pick the AutoCAD icon from the task bar. With AutoCAD displayed on the screen, select and hold an edge of the border to decrease the size of the window. Another window with AutoCAD showing a different drawing can also be opened and displayed at the same time as seen in Figure 2–2 or a completely different program can be opened as shown in Figure 2–20. By selecting an open area of the task bar with a single click of the right mouse button, the menu shown in Figure 2–21 will be displayed. This menu can be used to alter the method used to display windows. A cascading or tiled display is often used when two or more windows are to be displayed.

Cascade. Selecting Cascade from the Windows menu displays the program windows similar to those shown in Figure 2–22. Cascading windows are stacked over each other in descending order. Programs are displayed in the order that they are selected. The Accessories window is displayed over the other three because it was selected last. To display one of the lower windows, clicking the menu bar of the desired window will move it to the top of the stack. By clicking and holding the menu bar, the window can be dragged anywhere on the screen.

Tile. Selecting Tile from the Windows menu allows program windows to be arranged as shown in Figure 2–23. Tiled windows are arranged in a side-by-side order, allowing each window to be viewed at the same time. Depending on the number of windows displayed, all program icons may not be displayed.

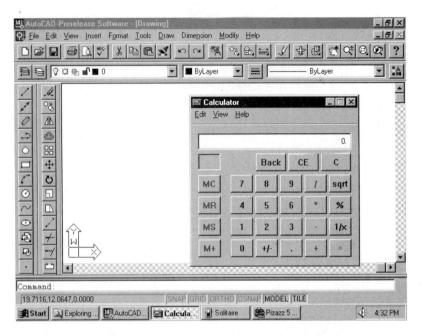

FIGURE 2–20 The Calculator window is open and displayed at the same time as AutoCAD R14.

FIGURE 2–21 Selecting the menu from the task bar allows options for the display of multiple windows to be altered.

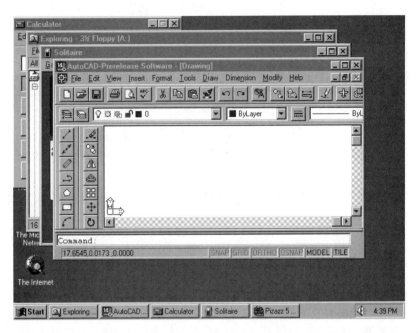

FIGURE 2-22   The five open windows can each be seen by selecting the Cascade option. A program can be moved to the top of the list by picking the title bar.

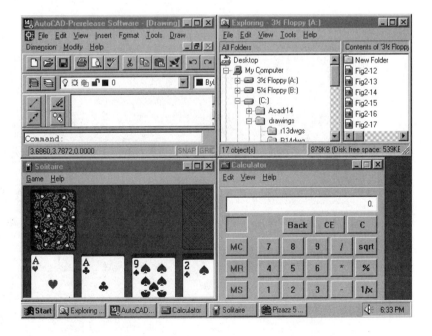

FIGURE 2-23   Multiple windows can also be displayed as individual tiles. Picking the maximize button will display one of the windows over the other three. Picking the restore button will place it back in a tile display.

STARTING AUTOCAD

Throughout this chapter you've been introduced to the basics of Windows. The balance of this chapter will introduce you to opening AutoCAD Release 14. Although some windows have been opened throughout this chapter, opening AutoCAD will be reviewed one last time to make sure you feel totally at ease with accessing AutoCAD. AutoCAD can be opened from the Program menu, the Windows Explorer or the Run menu. Each method is accessed by picking the start button. The easiest method is to double click the AutoCAD R14 icon from the original menu shown in Figure 2–2.

Using the Program Manager

Picking the Start button will produce the menu shown in Figure 2–3. Move the cursor to the Programs icon. As the cursor is moved to the right end of the listing, the menu shown in Figure 2–4 is displayed. Move the cursor to the listing for AutoCAD R14 and the menu shown in Figure 2–24 will be displayed. Each of these menus can be displayed by moving the mouse, with no clicking of the select button required. With the AutoCAD R14 icon highlighted, click once with the left mouse button. This will produce the AutoCAD display graphics momentarily, followed by the AutoCAD drawing screen shown in Figure 2–25. If you're sitting at a work station, and just entered AutoCAD, take a few minutes to explore the icons displayed on the screen. Each screen component will be discussed in Chapter 3, but it rarely hurts to peek.

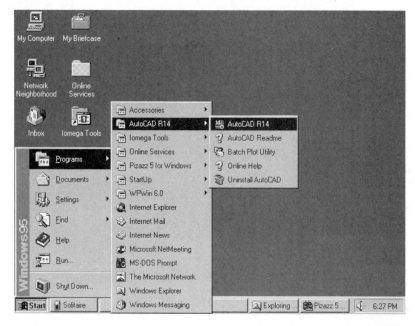

FIGURE 2-24 Double-clicking the AutoCAD R14 icon will open and display AutoCAD.

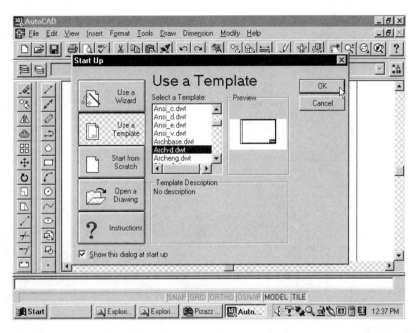

FIGURE 2-25 The start-up screen for AutoCAD R14. Each component will be introduced in Chapter 3.

As long as you keep liquids away from your computer, you'll be pretty safe. After you've spent a little time exploring, invest some time in Chapter 3, so you can get a better idea of just how powerful AutoCAD R14 is.

Using The Run Menu

Selecting a program from the Program menu is the easy way. The Run menu is for DOS users, who can't quite admit that icons are more than a passing fad. Picking the Start button produces the menu shown in Figure 2–3. Picking the Run icon from the menu produces the menu shown in Figure 2–26. Move the cursor to the text box and enter **C:ACADR14** at the keyboard. This entry assumes AutoCAD is stored on the C drive. If you've stored AutoCAD in another drive, enter the appropriate drive letter. If you're unsure of the location, use the Explorer to find where AutoCAD is stored. When you're satisfied with the correct location, pick the OK button and a display similar to Figure 2–27 is displayed. Double-clicking the Acad 14 icon opens the program and produces the display seen in Figure 2–25. This method is much more troublesome than double-clicking the AutoCAD R14 icon, but the designers of Windows believe in providing lots of options.

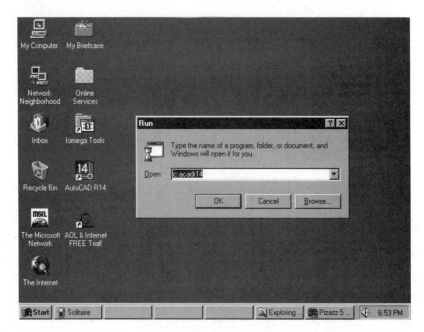

FIGURE 2–26 An alternative method of opening AutoCAD can be found by opening the Run menu. Type the location of the ACADR14 folder in the edit box.

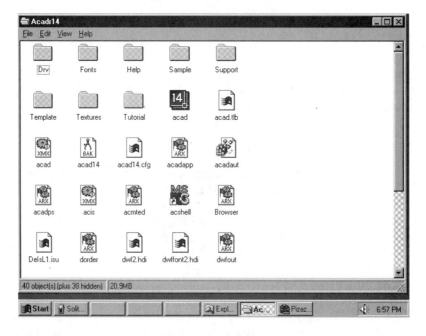

FIGURE 2–27 Entering **C:ACADR14** and picking the OK button will produce this menu. Picking the ACAD icon will produce the screen display shown in Figure 2–25.

Using the Explorer to Enter AutoCAD

Press the Start button and enter the Programs menu. Rather than selecting the AutoCAD R14 icon, move the cursor to highlight the Windows Explorer icon. Selecting the Explorer icon with a single click produces a display similar to Figure 2–28. Selecting the Acadr14 folder with a single click produces a display similar to Figure 2–29. Double-clicking the Acadr14 icon opens AutoCAD and produces the display shown in Figure 2–25. Sure it's not as fast as selecting AutoCAD icon, but it's an option.

Starting AutoCAD in Windows NT

Starting AutoCAD in Windows NT is very similar to starting the program in Windows 95. When a computer is booted up using Windows NT, the Microsoft icon is displayed. This icon is quickly replaced with a login display. Because the system is designed to be run on a network, your access name and password must be entered before you are allowed to proceed. Once you have logged in, selecting the start button in the lower left corner produces the menu shown in Figure 2–30. Selecting Programs will produce a list of programs loaded onto your machine similar to Figure 2–31. If AutoCAD R14 is selected, the AutoCAD menus group shown in Figure 2–32 is displayed. Picking the AutoCAD R14 icon with a single click will load the program and produce the screen display shown in Figure 2–33.

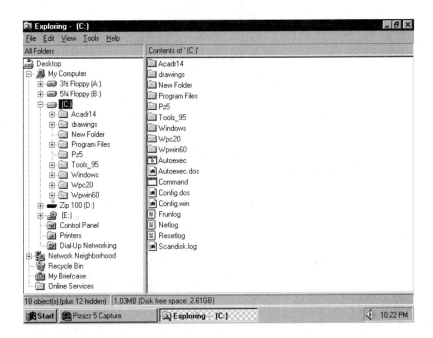

FIGURE 2–28 Selecting the Explorer icon with a click from the Start menu will display the contents of the hard drive.

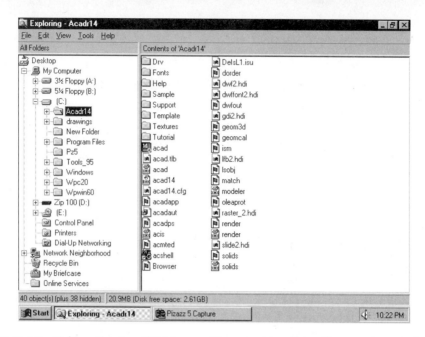

FIGURE 2–29 Picking the AcadR14 icon with a double-click will open AutoCAD.

FIGURE 2–30 Selecting the Start button at the bottom of the initial screen display of Windows NT produces the Windows NT Workstation menu. (Courtesy Floyd Miller.)

FIGURE 2–31 Selecting Programs from the menu shown in Figure 2–30 produces a listing of programs loaded in Windows NT. (Courtesy Floyd Miller.)

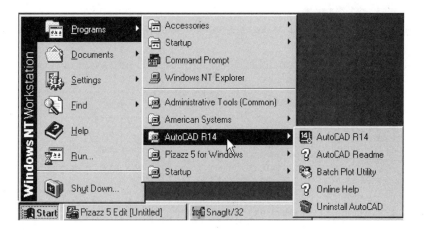

FIGURE 2–32 Selecting AutoCAD R14 from the menu shown in Figure 2–31 produces the AutoCAD program menu. (Courtesy Floyd Miller.)

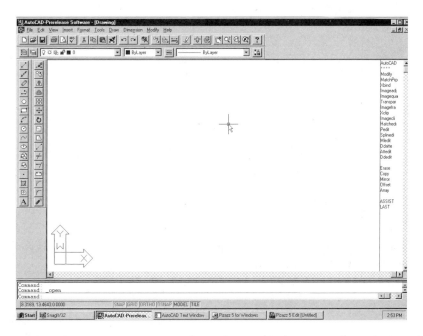

FIGURE 2–33 Selecting AutoCAD R14 from the menu shown in Figure 2–32 produces the R14 drawing screen displayed in Microsoft Windows NT. Each portion of the screen will be discussed in Chapter 3. (Courtesy Floyd Miller.)

CHAPTER 2 QUIZ

1. What is the major difference between Windows 95 and Windows NT?

   _____

   _____

2. Explain the difference of a click and a double-click.

   _____

   _____

3. List six cursors you'll encounter in Windows and AutoCAD.

   _____

   _____

   _____

   _____

   _____

4. What steps would be required to boot your computer and get into AutoCAD if you are using Release 14 with no batch files?

   _____

   _____

   _____

   _____

5. List five advantages of using Windows 95 or Windows NT instead of older disk operating systems.

   _____

   _____

   _____

6. List and describe the function of each AutoCAD icon that is displayed when AutoCAD R14 is selected from the Programs menu.

   _____

   _____

   _____

7. Where is the Acad R14 folder located on your current work station?

_____

_____

_____

8. Is a click or double click required to start AutoCAD using the Programs menu?

_____

_____

_____

9. Describe when single and double clicks are typically used.

_____

_____

_____

10. Describe key functions of the left and right mouse buttons.

_____

_____

_____

_____

11. List five menu options of the Explorer menu bar.

_____

_____

_____

12. Describe two different methods to reduce the size of a window.

_____

_____

_____

13. An hourglass is displayed beside the arrow cursor. Will you be able to enter a command?

_____

_____

_____

14. List and describe the four bars found in a typical window.

_____

_____

_____

15. List and explain the menu found in the Task bar.

_____

_____

_____

16. AutoCAD and Solitaire are both open. AutoCAD is the current program. Describe the easiest method to take a break and play a game of solitaire.

_____

_____

_____

17. List and describe at least four options found in the Windows Help menu.

_____

_____

18. List and describe each of the three types of dialog boxes found in Windows.

_____

_____

_____

19. Describe the advantage of using scroll bars versus the directional keys on the keyboard.

_____

_____

_____

20. Describe the process for ending a work session in Windows.

_____

_____

_____

SECTION

2D DRAWING CONSTRUCTION

Exploring AutoCAD

Creating Drawing Aids

Opening, Retrieving, and Saving Drawings

Coordinate Entry Methods

Drawing and Controlling Lines

Drawing Geometric Shapes

Drawing Tools

Drawing Display Options

· · · · · · · · · · · · ·
CHAPTER

3

EXPLORING AUTOCAD

· ·

T he drawing display shown in Figure 2–25 is displayed on the monitor as Release 14 is entered. To use the program effectively, an understanding of each of the methods of interacting with the software must be understood. This would include a knowledge of the drawing area displays, toolbars, types of menus available, common components of a dialog box, function keys, and control keys. Don't let all of the different menus, boxes and buttons confuse you. Each will become quite easy to use if you are willing to explore the program. Push a few keys, explore some options. As long as you keep liquid away from the keyboard, there is very little you can do to hurt the hardware, software or yourself.

The drawing area display seen in Figure 3–1a is displayed as AutoCAD is brought to the screen. Picking the OK button will produce the display shown in Figure 3-1b. Chapter 4 will introduce other options for opening AutoCAD drawings. AutoCAD will provide either a black or light gray drawing screen, depending on the configuration of your monitor, which contains three important areas. These areas include: tools and displays; toolbars; and menus.

DRAWING AREA TOOLS AND DISPLAYS

Each drawing screen provides the user with many tools and displays to input information into the program, and to determine the status of the current drawing file. These tools include the title bar, User Coordinate icon, scroll bar, cursor, and the status bar.

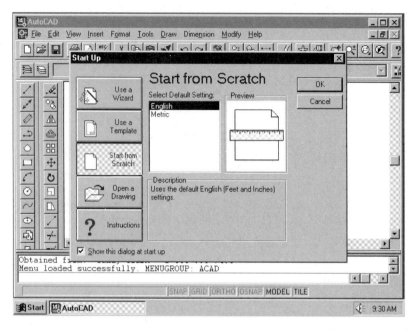

FIGURE 3–1a The Start Up menu of R14 is provided as AutoCAD is entered to help establish drawing parameters.

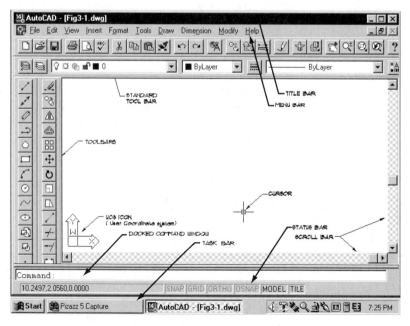

FIGURE 3–1b Key components of the AutoCAD drawing screen.

Title Bar

The Title bar lists the name of the current drawing file that is displayed in the drawing area. See Chapter 2 for a review of the buttons found in the title bar. As AutoCAD is entered, the current filename is listed as (UNNAMED). Once a drawing file is saved, the listing of UNNAMED is changed to display the current filename.

User Coordinate Icon

The Coordinate System icon describes the coordinate system that is currently being used and provides a method of locating point locations. In its standard setting, as you draw in AutoCAD, you're working in model space. *Model space* can best be thought of as drawing at full scale. If a structure is to be drawn that is 300 feet long, instead of working at a scale so the structure will fit on the screen, the area displayed on the screen is altered so that the entire building can be seen. When a drawing is to be plotted, it is often plotted using paper space. *Paper space* is space created so that one or several drawings at one or more scales can be plotted on one sheet of paper. Model and paper space will be discussed in detail in Chapter 23. Figure 3–2 shows the model and paper space icons.

Scroll Bar

A scroll bar is located on the right side and across the bottom of the AutoCAD drawing display area. Each can be used to adjust the portion of the drawing area that is displayed. Moving the pick box of the side bar to the bottom of the scroll bar will display the bottom portion of the drawing on the monitor. Moving the bottom pick box to either side will scroll the drawing from side to side.

Cursor

Figure 3–1 shows the graphics cursor used by AutoCAD. As the mouse is moved, the location of the cross-hairs is moved throughout the drawing display area. The cursor provides a method of entering the location for drawing entities. When commands such as LINE command are started, the cursor is the primary method of selecting the From point: and To point: for indicating the location of a line. New to AutoCAD

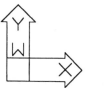

FIGURE 3–2 The user coordinate system (UCS) icons for model and paper space.

MODEL SPACE   PAPER SPACE

14R is the size of the cursor. The size of the cursor can be altered by typing **CURSORSIZE** [enter] at the command prompt. This will produce the prompt:

Command:**CURSORSIZE** [enter]
New Value for Cursorsize <5>:

The five represents the percentage of the screen the cursor will occupy. Entering **10** [enter] at the prompt will double the size of the cross-hairs. Entering **2.5** [enter] at the prompt will decrease the size of the prompt to half of the original size. Entering **100** [enter] at the command prompt will cause the lines forming the cross-hairs to cross the entire screen. The longer cross-hairs can be very helpful in aligning multiple drawings. The cursor changes from the cross-hairs to a small square called a *pick box* for commands that require an object to be selected. For commands that require text to be placed in the drawing, the cursor is displayed as a flashing bar. Several areas of the display contain groups of icons. Each of these groups of icons are referred to as *toolbars*. As the cross-hairs are moved into a toolbar, the cursor is changed to an arrow. Moving the arrow to the desired icon and clicking the select button will activate the selected icon.

Status Bar

The Status bar is located below the command prompt. Key elements of the status bar can be seen in Figure 3–1b; these include the drawing aids of SNAP, GRID, ORTHO and OSNAP that have been activated, the current time, the current viewport mode, the current type of space, and a coordinate display to describe the location of the cross-hairs. The cross-hairs show the entry point in the drawing editor and are selected by moving the mouse in the drawing area. Drawing aids will be discussed in Chapter 5; coordinate control will be introduced in Chapter 6; Viewports will be discussed in Chapter 10; and model space/paper space will be discussed in Chapter 23.

COMMAND SELECTION METHODS

AutoCAD contains more than 250 commands related to drawing or editing a drawing. Each can be entered by use of the keyboard. A list of AutoCAD commands can be found in Appendix B. Six options are available as an alternative to typing, which include: toolbars, pull-down menu, screen menu, dialog boxes, icon menus, and button menus. Each method has it's advantages. Throughout the balance of the text, commands will be presented in the order that tends to be the fastest method of entry. For most commands the fastest entry method is by keyboard, followed by selecting an icon from a toolbar. Remember, the key word is fastest, not best. As you're learning each command, take the time to explore each option. You may decide the keyboard is the slowest method of command entry. As you begin to work with AutoCAD menus, you will see common terms used repeatedly. You will need to understand these so you can progress through AutoCAD. Common terms include:

Command—An instruction to be carried out by the computer, such as LINE or ARC. Commands will be written in capital letters throughout this text.

Default—A value that will remain constant until a new value is entered. Default values are shown on the command line in brackets: < 3 >.

Option—A portion of a command that requires a selection.

Pick—To use a pointing device to make a selection.

Enter— ⟨enter⟩ is used to denote pressing the appropriate key (keynames vary, depending on the keyboard manufacturer) after inputting a request at the command line.

Select—Making a choice from a menu of commands or options.

KeyBoard Entry

The command window displays the prompt "Command:" and allows you to enter your request into the program using the keyboard. After typing the desired command, press the ⟨enter⟩ key to activate the command. AutoCAD also will ask you on this line for information to complete a command. Users new to AutoCAD often fail to read the command prompt and miss AutoCAD's attempt to communicate with them. To enter a command by keyboard, type the name of the command. To draw a line, type the word **LINE** or the command alias of **L** ⟨enter⟩. A complete list of aliases is listed in Appendix A. The text will automatically be entered at the command prompt. Once the command is entered, pressing the ⟨enter⟩ key will activate the specified command. Notice that once the ⟨enter⟩ key is pressed, the command line display will be altered. The command sequence will be:

Command:**LINE** ⟨enter⟩
From point:

Use the cursor to select a starting point for a line. Once a point is selected, the prompt will request a To Point. Use the mouse to indicate the location of the line end point and select the point with the left mouse button. The prompt will continue to request additional points after each new line start point has been entered until you choose to stop drawing lines. You will notice that LINE is a continuous command. As you draw one line, the command will continue and allow you to add on to the existing line. To terminate the line, press the ⟨enter⟩ key or the right mouse key. To start another line in a different location, move the mouse and select the right mouse button or ⟨enter⟩ key. The pattern will allow you to make an unlimited number of line segments without having to select LINE from the menu for each line. To stop and go on to your next command, press the ⟨enter⟩ key. After providing the required information, the command will be performed and the command prompt will reappear.

As the program is entered, the command window will display only a single command line. Additional lines of command text can be viewed by picking the top edge of the window and dragging it to a new location. As the size of the window is increased, a scroll bar is added to the window to allow viewing of previous commands. The size of the command bar can also be altered by selecting Preferences... from the Tools pull-down menu and altering the value for the docked command line. Up to 100 lines of text may be displayed. The scroll bar on the right edge of the display can be used to show additional lines of text. Figure 3–3 shows an example of an enlarged command window. In addition to displaying information about the current command sequence, AutoCAD can display a list of commands used throughout the drawing. Pressing the F2 button will produce a display of the AutoCAD Text window that contains the drawing history. (See Figure 3–4.) The window can be used to display all of the commands used to create the current drawing file. The display can be removed by pressing the F2 button again. Like other windows, the drawing history display can be moved to a new location or resized.

TOOLBARS

A toolbar contains icons that represent commands. In its default display, the standard toolbar is displayed above the drawing area just below the menu bar, and the Modify and Draw menus are displayed on the left side of the drawing screen. Because of their ease of use, and availability, for most AutoCAD users the toolbars will become the primary method of selecting commands. AutoCAD Release 14 offers

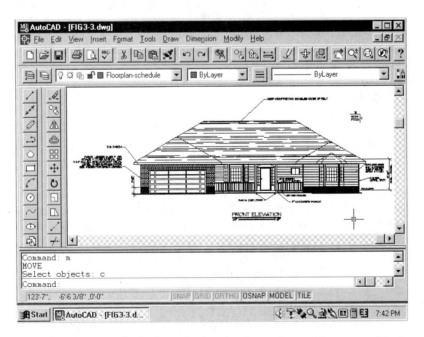

FIGURE 3–3 The command window display can be enlarged from one line to several by picking the top border of the window and selecting a new location. (Courtesy of Tereasa Jefferis)

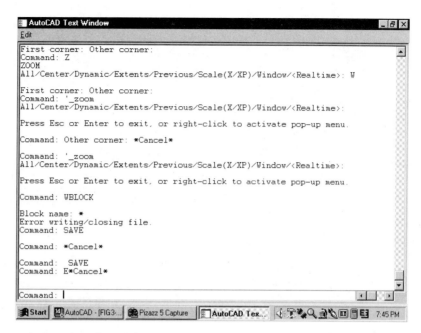

FIGURE 3–4 Pressing the F2 function button of the keyboard will display the drawing history.

17 different toolbars that can be displayed to meet various drawing needs. Additional toolbars can be displayed by selecting Toolbars from the View pull-down menu. This will produce the menu shown in Figure 3–5. Selecting Dimensioning will remove the menu display and produce the Dimension toolbar shown in Figure 3–6. Each of the toolbars listed in Figure 3–5 will be discussed as specific commands are introduced throughout the text. As you get started using AutoCAD it is more important that you feel comfortable using the three toolbars that are currently displayed. To remove the Toolbars dialog box, click the close button in the upper right hand corner. The Dimension toolbar can also be removed from the screen by clicking the close button in the toolbox title bar.

As the cursor is moved from the drawing screen to a tool bar, it changes from a crosshair to the arrow cursor. The cursor can now be used to pick the desired icon. Move the cursor to the icon in the left corner beside the drawing window. As the cursor is placed above the icon, the title of the icon will be displayed below the cursor, similar to the display in Figure 3–7. The title is referred to as a Tooltip. Tooltips provide a written display of the results that will occur if the icon is picked. In this case, the cursor is over the Line icon. Picking the Line icon with a single-click will start the LINE command and display the following prompt at the command line:

Command:_line From point:

FIGURE 3–5 Additional toolbars can be displayed by selecting Toolbars from the View pull-down menu.

FIGURE 3–6 The Dimension toolbar is displayed by making the selection started in Figure 3–5.

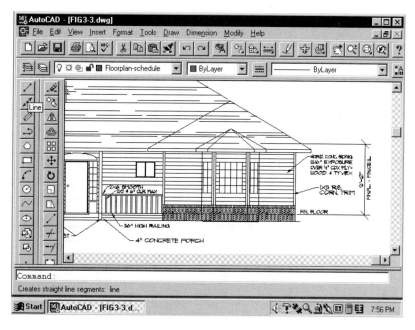

FIGURE 3–7 Resting the cursor over an icon provides a written description of the icon's function called a Tooltip. Notice below the command line a brief description is also displayed.

The program is now waiting for you to select a starting point for a line. Once started, the command will function just as it did when started by keyboard. End the command by pressing the [enter] key.

Several of the icons have a small triangle in the lower left corner. Clicking and holding an icon with a small black triangle in the lower right corner will produce a flyout menu. A *flyout menu* is a submenu of the selected menu. Figure 3–8 shows an example of a flyout menu.

Altering Toolbars

Toolbars are the main method of entering commands because they are extremely flexible. In addition to being able to display multiple toolbars on the screen, each toolbar can be moved, docked or floating, resized, and the contents altered.

Moving Toolbars. A toolbar can be moved to any convenient area of the drawing screen. The toolbar is moved by moving the cursor to the title bar of the menu and picking and holding the title bar. While pressing the select button and moving the mouse, the location of the toolbar is altered.

Floating Toolbars. A toolbar is referred to as a *floating toolbar* once it is moved from the border of the drawing display. Once a toolbar is floating, it can be resized, docked, or the contents altered. When the Dimension toolbar was activated (see Fig-

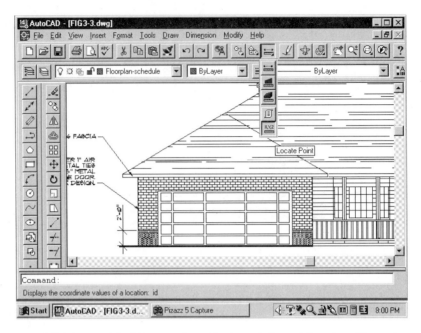

FIGURE 3–8 Clicking and holding an icon that contains a triangle will produce a Flyout menu.

ure 3–6), it was placed in the middle of the drawing screen, making it a floating tool bar. Picking the title bar and holding the select button allows the toolbar to be moved to any desired location.

Resizing Toolbars. If you're really bored, the size of a toolbar can be altered. The shape of a toolbar is altered by placing the cursor anywhere on the border of the tool-bar and dragging it in the direction you want the bar to be resized. Figure 3–9 shows examples of the resized Draw toolbar.

Docked Toolbars. A toolbar is docked by picking and holding the title bar and moving the toolbar to a position above, below, or on either side of the drawing screen. Figure 3–10 shows the Draw toolbar floating. Figure 3–8 shows the toolbar docked on the left side of the drawing area. An alternative to docking a toolbar is to totally remove it. A toolbar can be removed from the screen by picking the close button in

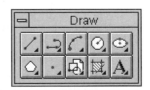

FIGURE 3–9 The shape of floating toolbars can be adjusted to please the user.

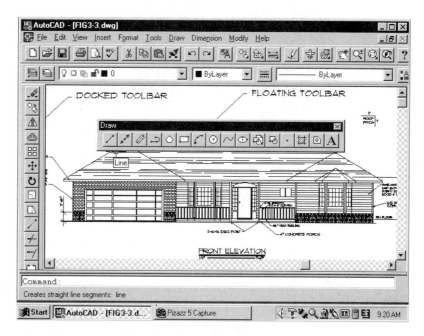

FIGURE 3–10 Floating toolbars can be docked above, below or on either side of the drawing area. The toolbar is docked by picking and holding the Title bar, and then dragging the toolbar to the desired edge.

the upper right corner. To restore the window, select the name of the desired toolbar from Toolbars in the View pull-down menu.

Refloating Docked Toolbars. A docked toolbar can be refloated by placing the cursor anywhere on the border of the toolbar and dragging and dropping the box in its desired position on the drawing screen.

Standard Toolbar

The standard toolbar contains twenty-four icons for controlling AutoCAD. Toolbar icons from left to right include:

> **New.** This icon can be used to start a new drawing. (See Chapter 4.)

> **Open.** The Open option allows an existing drawing to be opened. (See Chapter 4.)

> **Save.** Selecting this option allows a drawing file to be saved to the desired source. (See Chapter 4.)

> **Print.** This option allows drawing files to be exported to a printer or plotter. (See Chapter 24.)

> **Print Preview.** This provides a display of how a print will appear when complete.

Spelling. This icon will start a spell check program. (See Chapter 18.)

Cut to Clipboard. This option will remove a specified portion of the current drawing file and store it in the Windows clipboard.

Copy to Clipboard. This option copies a selected portion of a drawing file and stores it on the Windows clipboard.

Paste from Clipboard. Using this option takes a portion of a drawing file that is attached to the Windows clipboard and adds it to the current drawing file.

Match Properties. This option will copy the properties from one object to one or more objects.

Undo. The Undo command reverses the effect of the last command.

Redo. This option can be used to restore something that was just undone.

Launch Browser. Selecting this icon provides a method to launch an Internet web browser from inside of AutoCAD. (See Chapter 26.)

Tracking Flyout. Selecting the Tracking icon provides a flyout menu with object selection methods. These options provide alternatives for selecting specific points of an object and will be discussed in Chapter 9.

USC Flyout. The USC icon provides a flyout menu that presents options for managing the user coordinate system. The user coordinate system is a moveable coordinate system that serves as a home point of reference for creating and viewing drawing. The user coordinate system will be discussed in Chapter 4.

Distance Flyout. The distance icon presents a flyout menu of options for obtaining information about drawing entities. Options include area, mass properties, list, and locate point. Each option will be introduced in Chapter 17.

Redraw All. This option can be used to refresh the display of all the viewports. Options will be discussed in Chapter 10.

Aerial View. This option provides additional icons for selecting commands and options that control how the current drawing is viewed on the monitor. (See Chapter 10.)

Named View Flyout. This option provides alternatives for changing the drawing view that is displayed. (See Chapter 10.)

Pan Realtime. Selecting this option provides options for scrolling across the drawing area. (See Chapter 10.)

Zoom Realtime. Allows for increasing or decreasing the window display by moving the mouse. This option allows for altering the size of the display area presented on the monitor. Zoom will be discussed in Chapter 10.

Zoom Window Flyout. Selecting this option displays each of the options for the ZOOM command. (See Chapter 10.)

Zoom Previous. Selecting this option will return the screen display to the previous status prior to using a pan or zoom. The command will be discussed in Chapter 10.

Help. Selecting this icon will display the on-line help menu for AutoCAD, discussed later in this Chapter.

Pull-Down Menu

As the cursor is moved into the menu bar, a key method of interacting with AutoCAD is presented. The menu bar displays the names of ten menus used to control AutoCAD Release 14. As with other window menus, selecting a specific name will produce a display of that pull-down menu.

Picking a pull-down menu displays a complete listing of related commands. Moving the cursor to the Draw menu and picking Draw displays the menu shown in Figure 3–11. If you accidently pick a wrong menu, move to the desired menu. Once a menu has been opened, other menus will automatically open and close as the mouse is moved. A complete listing of pull-down menus can be found in Appendix C.

Once in the Draw menu you'll notice that two types of options are listed. Selecting one of the first eight options removes the menu and returns the cross-hairs to the

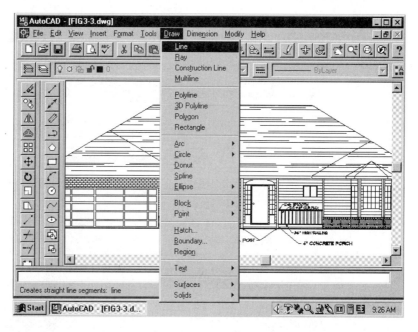

FIGURE 3–11 Moving the cursor to the Menu bar and picking Draw will display the Draw pull-down menu.

screen so that the selected command can be executed. Picking Line from the menu will remove the menu from the screen and start the LINE command. Notice the command line has now changed, and "_line From point:" is displayed. AutoCAD now waits for you to pick a point with the mouse.

Exit the LINE command by pressing the ⏎ key and return to the Draw pull-down menu. Many of the options on the lower portion of the menu are followed by a triangle. The triangle means that by selecting this command, another menu will be displayed. Picking Arc of the Draw menu will display the menu shown in Figure 3–12. Picking 3 Points or any of the other options will remove the menu, return the crosshair cursor, and allow the selected command option to be executed.

Although not found in the Draw menu, there is a third type of display found in the pull-down menus. Move to the left and pick the Tools menu. Notice that several of the options are followed by three periods. Selecting an option such as Drawing Aids... with a single click will produce the dialog box shown in Figure 3–13. Using dialog boxes will be introduced later in this chapter and explained throughout the balance of the text as specific commands are introduced. For now, to exit the Drawing Aids dialog box, move the arrow to the Cancel button and press the pick button to return to the drawing screen.

Screen Menu

In its default setting, no screen menu is displayed in AutoCAD Release 14. Because of the availability of toolbars, the screen menu is considered obsolete by some users.

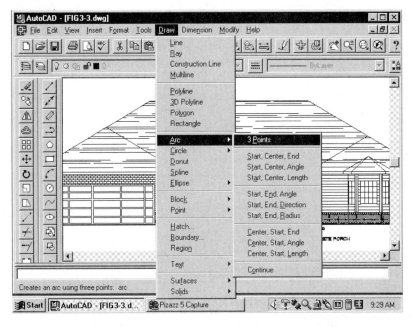

FIGURE 3–12 The submenu box produced by picking the Arc option.

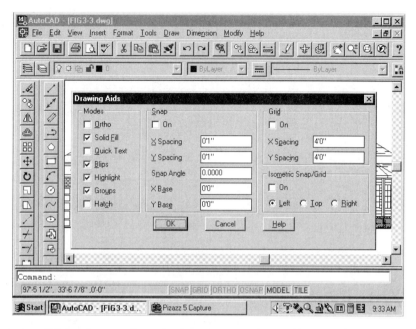

FIGURE 3–13 Dialog boxes allow the functions performed by certain commands to be adjusted.

For users of previous versions of AutoCAD who may be accustomed to selecting commands and options from a screen menu, Release 14 allows the screen to be activated by selecting Preferences... from of the Tools pull-down menu. This will display the Preferences dialog box shown in Figure 3–14a. Picking the Display tab will display the menu shown in Figure 3–14b. Moving the cursor to the top box and pressing the select button will place a check in the Display AutoCAD screen menu in drawing window selection. Selecting the OK button with a single-click will remove the dialog box and display the screen menu shown in Figure 3–15. This menu is known as the *Root menu*. Although it is not AutoDESK's preferred method of activating commands, you may find it very useful.

Choices are made by moving your mouse to the right side of the drawing area until the desired command is highlighted. Press the pick button while the command in the menu is highlighted to execute the desired command.

Select the DRAW 1 option from the screen menu. The menu will change to reflect the display seen in Figure 3–16. The Draw 1 options begin with Line and are grouped by function. Selecting a listing such as Arc will produce another menu showing a list of options, which can be seen in Figure 3–17. To return to the Root menu, pick the AutoCAD listing at the top of the menu. Return to the Root menu and select the Draw 1 option. Now select the LINE option. The menu display will change to display

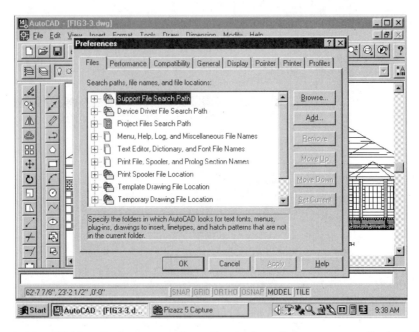

FIGURE 3–14a Picking Preferences… from the tools pull-down menu displays the Preferences dialog box. This dialog box can be used to adjust the software to meet Individual needs.

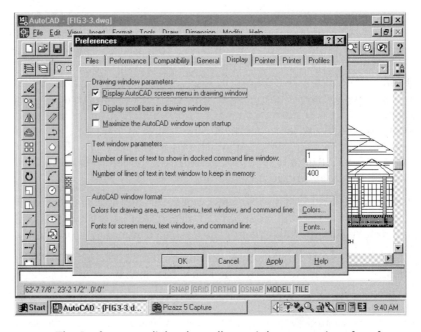

FIGURE 3–14b The Preferences dialog box allows eight categories of preferences to be preset and saved to meet the need of the individual using the workstation.

FIGURE 3–15 Picking the *Display AutoCAD screen menu in Drawing window* box and the OK button will display the screen menu on the right edge of the screen.

FIGURE 3–16 Commands from the screen menu are specified by moving a pointing device to the desired command, so that the command name is highlighted. Use the pick button to activate the command. Selecting Draw 1 from the Screen Menu will produce a menu showing Draw options.

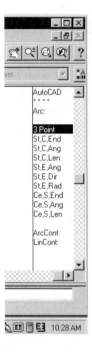

FIGURE 3–17 Selecting Arc: from the Draw 1 menu will display a sub-menu with new options.

the menu seen in Figure 3–18. Not only will the menu change, but with this selection, the command line is now displaying

        Command: _line From point:

as AutoCAD waits for your mighty command. When selected from the screen menu, the LINE command will function exactly as it did when entered by keyboard, toolbar, or pull-down menu. Notice at the top of the screen menu is the listing:

        AutoCAD
        * * * *

If you pick AUTOCAD, you'll be returned to the root menu seen in Figure 3–15 no matter what menu or submenu you're in. If you select the **** option at the top of the menu, the menu will change to display object snap selection modes that are used for accurately selecting drawing objects. They can also be found in the Object Snap toolbar. (See Toolbars… in the View pull-down menu.) Each method will be further discussed in Chapter 9.

New users of AutoCAD tend to confuse commands and menus. Selecting from the screen menu will lead to another menu. As you know from valuable years of reading fast-food restaurant menus, menus are listings of available options. Commands are orders given to AutoCAD to execute a specific drawing function. Figure 3–19 shows a listing of the root menu, the Draw1 menu, and the related Arc: command.

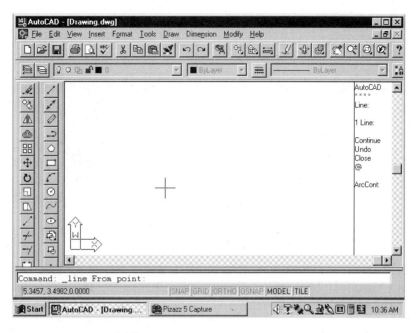

FIGURE 3–18 Selecting a command such as LINE from a menu will produce another listing of options as well as a prompt on the Command line. AutoCAD is now waiting for your input.

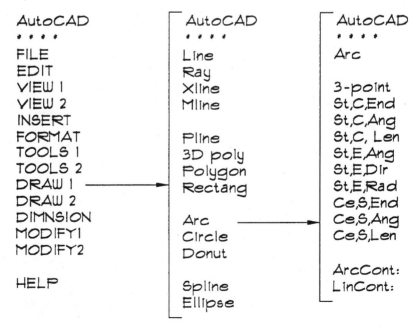

FIGURE 3–19 Selecting the Draw 1 option from the screen menu will display a submenu. Selecting Arc: from the submenu will produce a listing of the options for the ARC command.

DIALOG BOXES

Although not a menu of commands, dialog boxes provide a convenient way to adjust certain command parameters. You encountered your first dialog box when AutoCAD opened and the Start-up dialog filled the screen. A dialog box provides options for controlling commands, options, or portions of the drawing program. Figure 3–20 is a display of the dialog box for controlling LAYERS. This dialog box can be accessed by selecting the Layer icon in the Object Properties toolbar or by selecting the Layer... option from the Format pull-down menu. Dialog boxes can also be accessed by typing in a command that begins with the letters DD, which represent dynamic dialog. Typing **DDLMODES** [enter] (dynamic dialog layer modes) at the command prompt will display the box shown in Figure 3–20. Layers will be discussed in Chapters 7 and 15. Don't worry about what a layer is or "DDwhat?" for now. Each will be explained as the skill is needed. For now, what's important is understanding some of the common features found in dialog boxes.

As a dialog box is selected, the cross-hairs of the drawing area change to an arrow. The arrow can be used to select an option by placing the arrow in the box or button by the desired option and pressing the pick button. More specific information regarding the actual use of each dialog box will be discussed as the controlling command is introduced. Common components of a dialog box are seen in Figure 3–21 and include

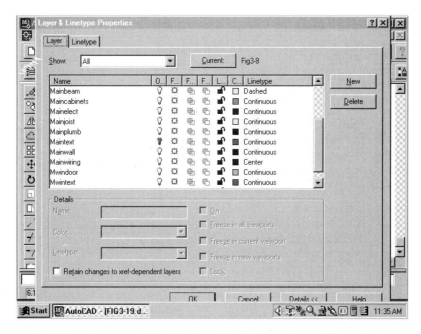

FIGURE 3–20 Dialog boxes can be displayed by typing the command name preceded by the letters DD. DDLMODES will produce the dialog box for Layer and Linetype control.

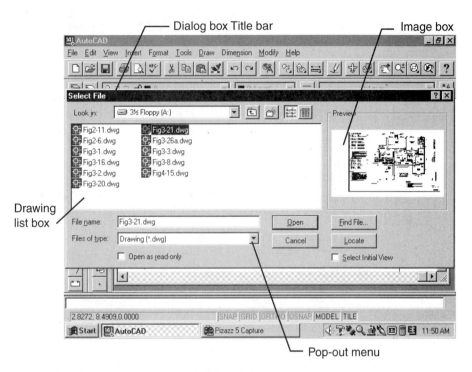

FIGURE 3–21 Common components of a dialog box.

OK/Cancel buttons, scroll bar, buttons, radio buttons, check boxes, edit boxes, image tiles, and alert.

Scroll Bar

Some dialog boxes may contain long lists of options. Dialog boxes that contain many options typically have a scroll bar, which will allow for moving rapidly between options. By selecting either arrow and pressing the pick button, the list will scroll one line in the direction of the arrow picked. Moving through a list of 40 layer options is a very slow process using the arrow box one pick at a time. Picking the slide box and holding the pick button down, as you move the box toward either arrow will scroll rapidly through the listings shown on the screen. Moving the slide box to the bottom of the scale will place the last entries of the listing on the screen. Moving the slide box to the middle of the scale will place the middle entries of the listing on the screen.

OK/Cancel

As you enter a dialog box, you will notice that the cursor no longer responds to the drawing area. To return to the drawing and remove the dialog box, select either the

OK or Cancel options. Pressing the [enter] key is equal to selecting OK. Pressing the **ESC** key has the same effect as picking cancel.

Buttons

Several buttons often are included in a dialog box in addition to OK and Cancel. Examples can be seen in Figure 3–22. Picking one of these buttons with the pick button will cause the selected option to be performed. Buttons with a wide border are default options. A button with an option followed by ... will display another dialog box that must be addressed before proceeding. A button with an option followed by < indicates that an action needs to be performed in the graphics area before proceeding. This might include picking an object or point in the drawing field. When an action button is selected, the dialog box is removed temporarily to allow for selecting an object from the drawing.

Radio Buttons

Radio buttons are buttons that list mutually exclusive options. Examples can be seen in Figure 3–22. Display, Extents, Limits, and Window are each radio buttons found in the Print/Plot Configuration dialog box. Each will present a different area to be plotted, but only one of the four options can be used per plot. These options may be selected by moving the cursor to the desired button and pressing the pick button.

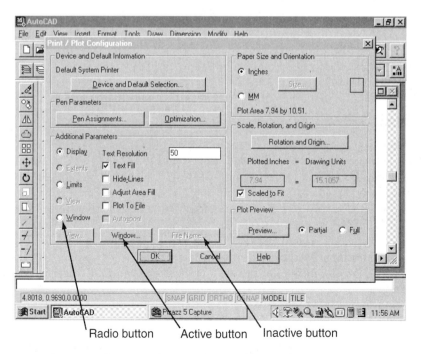

FIGURE 3–22 Common buttons that can be used to control options in a dialog box.

Check Boxes

Check boxes serve as toggle switches and can be seen in Figure 3–23. In the Drawing Aids... dialog box, each of the Modes is represented by a check box. Each mode is independent of the other listings. Selecting the Ortho mode to be on only allows vertical or horizontal line to be drawn. With Ortho in the Off mode, lines can be drawn at any angle. Each of these options, as well as many others, will be presented throughout the text. For now, remember that if a box has an "x" or check in it, that option is enabled. If the box is empty, the option is off.

Edit Boxes

An edit box is a single line text entry box that allows for information to be entered by keyboard. Figure 3–21 shows an example of an edit box. When the edit box is selected, it contains a vertical line (a cursor) that flashes slowly. The flashing cursor means AutoCAD is waiting to be told the name of the file. Typed information can be entered into the box, in this case the "File:" box. If errors are made in typing, the cursor can be moved by using the left and right arrow keys. The text in the box is activated by pressing [enter] twice.

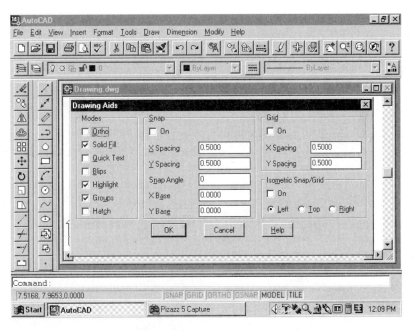

FIGURE 3–23 Check boxes in a dialog box serve as toggle switches. In the current state, Ortho is OFF. Picking the Ortho box would place a check in the box and activate Ortho.

Edit Keys

Several keys can be used to aid input into a dialog box. These include text keys, arrow keys, and control keys. These functions include:

TEXT—Symbols from the keyboard are inserted at the cursor. As a letter is inserted, the cursor and existing text move to the right. When the cursor reaches the right side limit, the cursor remains and text is moved to the left.

LEFT ARROW—This will move the cursor to the left without changing the existing text.

RIGHT ARROW—This will move the cursor to the right without changing the existing text.

DEL—This will delete the character at the cursor and move all remaining text to the left. The cursor will remain stationary.

BACKSPACE—This will delete the character to the left of the cursor while the cursor and all remaining text is moved to the left.

Image Tiles

An image tile is a portion of a dialog box that shows an image of the selected item. Figure 3–24a shows the image tiles used to show how text will be placed when dimensions are placed. This box can be found by picking the Format... button of the

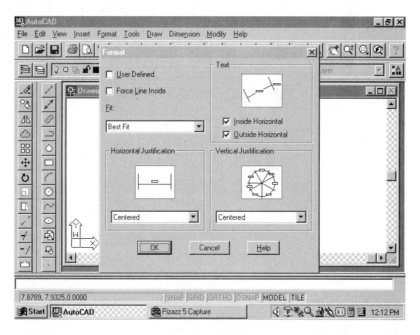

FIGURE 3–24a An image tile is a portion of a dialog box that displays an image of the selected item.

Dimension Styles… option found in the Format pull-down menu. Touching the highlighted bar of one of the three options displays a pop-up menu. Each time an item is selected from the pop-up menu, the image displayed in the tile will change to reflect the effects of the choice. Dimension styles will be further discussed in Chapters 19 and 20.

An image button is similar to an image tile, but it displays options for selections rather than the effects of selections. Selecting one of the images shown in Figure 3–24b from Layout… of the Tiled Viewport submenu of the View pull-down menu will highlight the corresponding text in the list box and define the results of the selection. Viewport layout will be discussed further in Chapter 10.

Alerts

An alert is a message that is displayed in the lower left corner of a dialog box, or in a separate box that is superimposed over the dialog box. Alerts typically warn the user that they have made a mistake by displaying a message such as "Invalid filename." Another common type of alert is a warning that a file is about to be changed, for example,

The specified file already exists.
Do you want to replace it?
Yes   No

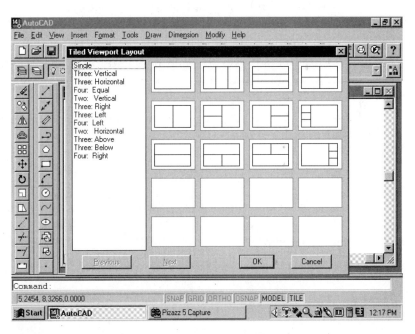

FIGURE 3–24b An image button shows options for selections.

FUNCTION KEYS

Twelve function keys are located across the top of the keyboard or to the left of the keyboard. The actual layout depends on the keyboard manufacturer. Seven function keys are used by AutoCAD to provide access to commands. Each is used as a toggle from ON to OFF and will be explained further in Chapter 5. Other keys can be programmed to perform specific commands. Keys used by AutoCAD are:

F1— Picking. This key clears half of the screen and displays the HELP menu. Using the Help options of AutoCAD will be discussed later in this chapter.

F2—Flip Screen. Pressing F2 in the middle of a drawing will remove the drawing from the screen and display a written list of the commands that were used to generate the drawing. See Figure 3–25b.

F3— Object Snap. Pressing the F3 key will display the dialog box shown in Figure 3–26. These options can be used to select specific portions of a drawing entity and will be discussed in Chapter 9.

F5—Isoplane crosshair mode. This is used when drawing isometric drawings. This button toggles between the top, left, right, and bottom planes of an isometric square. These planes will be discussed further in Chapter 25.

F6—Coordinate display. The coordinate display can be seen on the status line in Figure 3–1. Key F6 will switch the coordinated display on or off.

F7—GRID display. In Chapter 5 you will learn how to create a grid to aid in drawing layout as seen in Figure 3–27. Once you have selected the size of the grid to be displayed, the grid can be displayed or removed by key F7.

F8—ORTHO. You have explored the Ortho option earlier in this chapter. When it is on, only vertical and horizontal lines can be drawn. When off, lines at any angle may be drawn. These options will be discussed further in Chapter 5. By switching ortho off, angular features can be drawn.

F9—SNAP mode. This key activates the ability to move the cross-hairs within the drawing area at specific intervals. Snap will be explained in Chapter 5.

F10—This key toggles the status bar between ON/OFF.

The key thing to remember about the function keys is that they can be used at any time throughout the life of a drawing. Although the Ortho in the ON setting will make a floor plan much easier to draw, it can be quickly toggled OFF to allow for lines representing an angled wall to be drawn, and then reset to the ON position to finish other perpendicular walls.

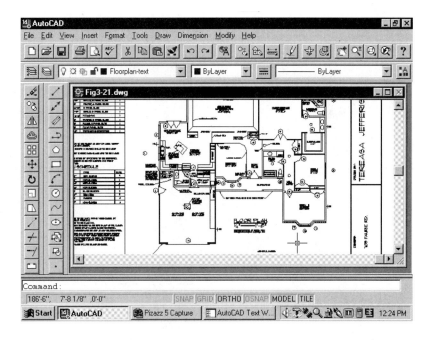

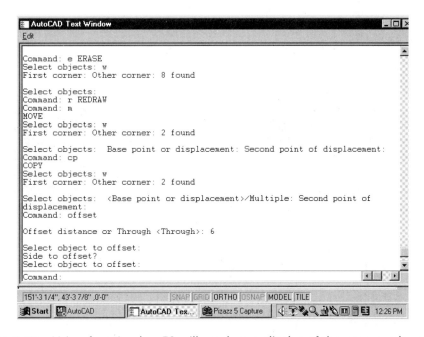

FIGURE 3–25 Using function key F2 will produce a display of the commands used to produce a drawing. Pressing F2 once will produce the graphic display. Pressing F2 again will restore the previous drawing screen.

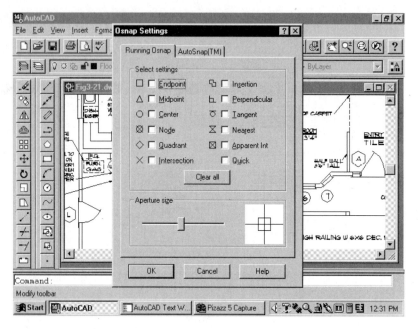

FIGURE 3–26 Pressing the F3 key will display the Osnap Settings dialog box. The options can be used to accurately select portions of an object for editing.

CONTROL KEYS

Like many other software programs, AutoCAD uses control keys to perform common functions. These functions are achieved when the control key is pressed in conjunction with a specified letter key. Many of these functions duplicate the functions performed by the function keys. The control keys typically are located in the lower corners of the keyboard and may be labeled CTRL. The control key functions are:

       CTRL + B—Snap mode toggle ON/OFF.

       CTRL + C—Copies selection to the clipboard.

       CTRL + D—Coordinate display of status line toggle ON/OFF.

       CTRL + E—Cross-hair in isoplane position toggle left/top/right.

       CTRL + G—Grid toggle ON/OFF.

       CTRL + H—Same as using the backspace arrow.

       CTRL + L—Ortho mode toggle ON/OFF.

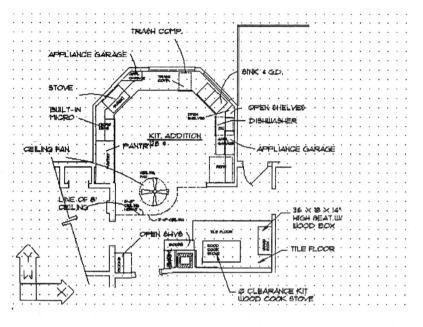

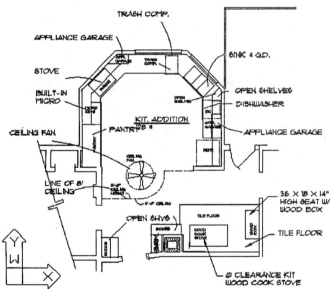

FIGURE 3–27 Using function key F7 will toggle the grid display ON/OFF.

CTRL + O—Used to open a new file.

CTRL + T—Tablet mode toggle ON/OFF.

CTRL + V—Paste clip.

CTRL + X—Cut clip.

CTRL + Z—Undo.

HELP

One of the best features of AutoCAD is that it offers free advice with no ridicule. You're going to be exposed to hundreds of commands, options, and menus and sometimes it's easy to feel lost or overwhelmed. If you can avoid panic in the first 30 seconds you'll probably do fine. Your chances of success are even better if you remember that there has been a help button in every menu you've looked at.

The fastest method to get help is to press the F1 function key. The HELP command can be entered by typing **HELP** [enter] at the command prompt. Each method will produce the dialog box seen in Figure 3–28. Picking HELP from the standard toolbar will also activate the Help menu. The HELP command can also be executed in the middle of another command sequence because it is a transparent command. A *transparent command* is a command that can be completed in the middle of another com-

FIGURE 3–28 Typing **HELP** [enter] at the command line, picking the Help icon from the standard toolbar or pressing the F1 key will produce the Help menu.

mand sequence. A transparent command is started by typing an apostrophe before the command. To get help in the middle of the LINE command would require the following entry:

> Command: **LINE** [enter]
> LINE
> From point: (*Select desired starting point.*)
> To point: '**HELP** [enter]

The F1 key can also be used in the middle of another command. When the HELP command is accessed in the middle of a command, the HELP command automatically provides command-related help. Asking for help in the middle of the LINE command will produce the Help display shown in Figure 3–29 that is specific to the command.

The Help menu offers help in the areas of Contents, Index, and Find. The Contents folder is opened as the Help menu is opened. If you use the Index folder and then exit Help, it will be the open folder the next time the Help is accessed.Each menu has other listings that will produce information related to that topic. In each of the menus that will be viewed, you'll find a Back and History button, as well as commands that are surrounded by double brackets. Selecting Back will take you back through other help menus that you've already viewed. Selecting History will list

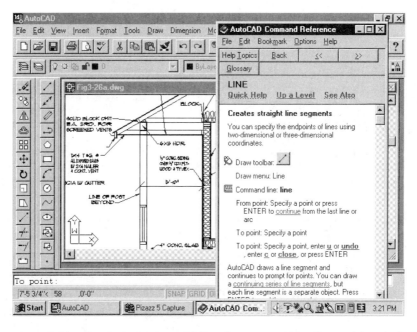

FIGURE 3–29 Asking for help in the middle of a command will provide a help menu specific to the command.

sequentially the help topics that have already been viewed. Selecting either pair of the double brackets will sequentially display a list of topics that can be selected for a help search. Selecting the << button will display previous portions of the menu. Selecting the >> button will show the next menu listing. Most menus also provide a method of reaching a glossary of AutoCAD terms.

Contents

The Contents folder of the Help menu can be used to obtain information about the Help program, or individual portions of the AutoCAD program. The Contents folder contains a How to Guide and a Users Guide to provide instruction on using each aspect of the program. The Command Reference portion of the folder provides information on each command used by AutoCAD. To receive information about the LINE command, enter the Help menu by picking the F1 button or by typing **HELP** [enter] at the command line. Selecting Command Reference with a double-click will produce the menu shown in Figure 3–30 with seven new options. Double-click the Commands listing and the display will be changed to the display shown in Figure 3–31. The menu provides a list of commands presented in alphabetical order and two methods of moving thru the menu. At the top of the menu is a partial keyboard. Selecting a letter from the keyboard will scroll the menu display to that portion of the listing. On the right side of the display is a scroll bar that can be used to move through the list. Use the keys to move to an area of the listing, and then use the scroll bar to refine the portion of the listing to be viewed.

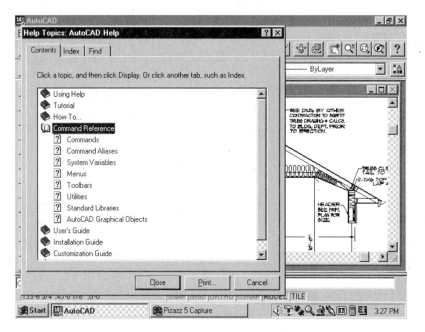

FIGURE 3–30 Selecting Command Reference from the Contents folder will provide seven options for seeking help about AutoCAD commands.

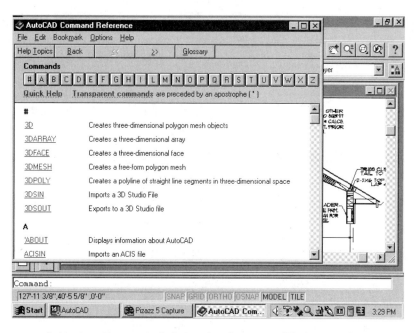

FIGURE 3–31 Selecting Commands from the Command Reference listing provides access to an alphabetical listing of AutoCAD's commands.

To find information on the LINE command, pick the L key from the menu display. This will produce the display shown in Figure 3–32, and provide a brief description of the LINE command. A detailed description of the command can be found by moving the cursor to the command name and selecting the name with a single click. Figure 3–33 shows a portion of the Line command description. Pick the F1 key or the close button to return to the drawing screen.

Selecting the System Variables option of the Contents folder will provide a listing similar to the Command listing. System variables control how a specific command will function and will be discussed as each command is introduced. Selecting the Menus option will produce the display shown in Figure 3–34. These options can be used to introduce each of the menus listed in the Menu bar. Selecting the Draw menu listing will display the listing shown in Figure 3–35.

Selecting the Toolbars option will provide a list of all of AutoCAD's toolbars. Selecting one of the names will produce a listing of the options contained in the toolbar. By selecting the Draw listing, the menu shown in Figure 3–36 will be displayed to provide a brief description of each command. The Utilities option provides an introduction to three applications to help set-up AutoCAD to convert files and data for use in other programs.

Selecting the Standard Library option provides a menu of each of the libraries in AutoCAD. A *library* is a set of predetermined symbols designed to save the user

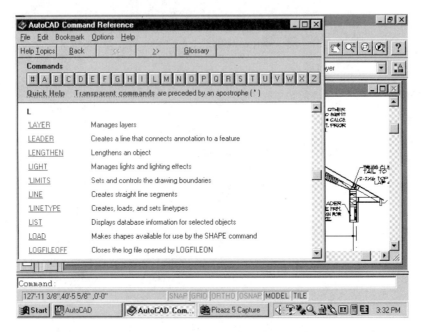

FIGURE 3–32 Picking the L button will scroll the menu to the start of that portion of the commands list.

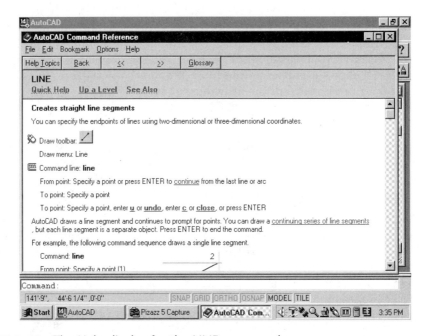

FIGURE 3–33 The Help display for the LINE command.

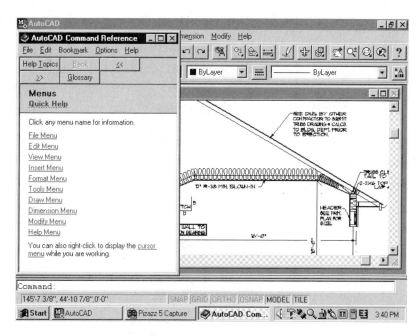

FIGURE 3-34   Selecting Menus from the Command Reference listing provides access to listing of each of the pull-down menus.

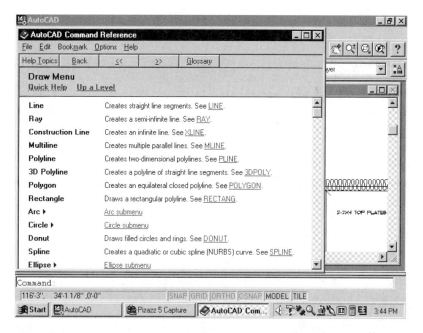

FIGURE 3-35   Selecting a menu name will provide a list of each command and a brief description of that command.

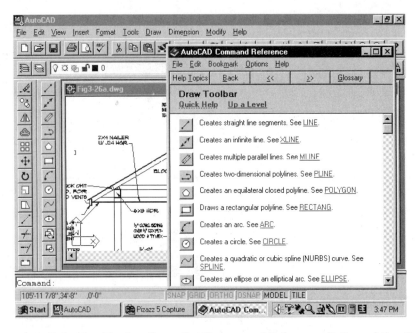

FIGURE 3–36 Selecting Toolbar from the Command Reference listing of the Contents folder of Help provides a list of the toolbars. Selecting a specific toolbar name will display each of the icons found in that toolbar and a brief description of that command.

drawing time. If you're going to draw a floor plan, you'll need to draw toilets, sinks, doors, windows, as well as many other common symbols. Each of these symbols are examples of symbols that can be placed in a library to reduce drawing time. Many useful symbols have been created by Autodesk and can be viewed from this menu. These symbols include common line types (see Chapter 7), hatch patterns (see Chapter 16), text fonts (see Chapter 18), and geometric shapeses (see Section 3).

The final listing of the command references is AutoCAD Graphical Objects. Each shape that is created in a drawing such as a circle, arc or square is a distinct AutoCAD object type. This menu is a listing of each shape and a reference to obtaining help or a listing about related commands or variables related to that shape.

Index

Selecting the Index tab will produce the menu shown in Figure 3–37. This menu is like the index found at the end of a book. Look up a key term and the book index will list several pages where information can be found. With the Help index, you don't even have to know the word you're looking for. By entering the first few letters of a command or option, the help menu will display a list of commands and sources of information about that command. If time is no object, the scroll bar can be used to

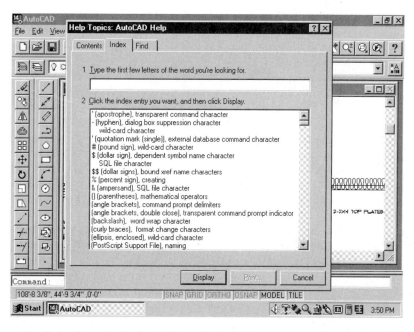

FIGURE 3–37 Selecting the Index folder of the help menu produces a listing of commands similar to an index in a book.

scan through the list until your subject is displayed. Once found, double-clicking the option will produce the display shown in Figure 3–38. This display box can be used to select information about the command. Entering the name of the command in edit box 1 will produce the help window shown in Figure 3–39. This window can be reduced in size and moved beside the drawing so that it serves as a tutorial for the specified command.

Find

The Find option of Help is similar to an index, but it's also faster than a speeding bullet. Figure 3–40a shows how the Find box will appear as the folder is opened. There will be a slight delay as the dictionary is loaded into memory. Once active, to obtain information about the LINE command, type the letter **L** into the edit. The upper windows will change to display the L portion of the index and the lower window will change to display related information about the selected word. Figure 3–40b shows the changes in the display as LINE is entered into the edit box. By selecting the word LINE in box 2 (the matching word box), the search is restricted to the LINE command, as opposed to other line options. As the search field is restricted in box two, box three is updated. Selecting Draw Menu from the topic box (box 3), the display shown in Figure 3–41 is displayed. This is a result of asking for information about the LINE command as it relates to the Draw pull-down menu.

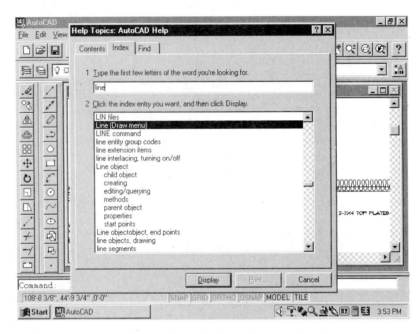

FIGURE 3–38 As commands are entered in the edit box, the display options related to line (the subject) are listed.

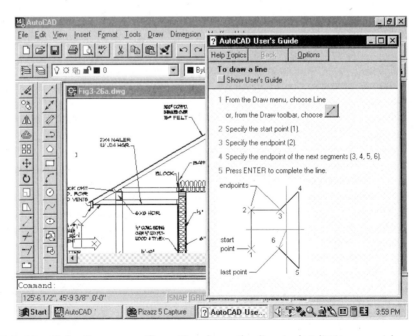

FIGURE 3–39 Selecting a specific option from the line index listing provides a help box.

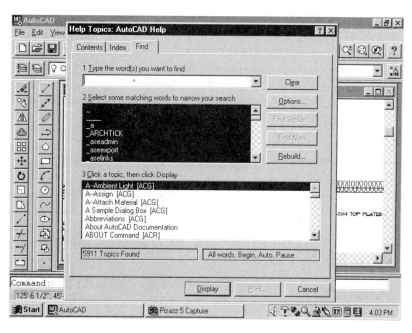

FIGURE 3-40a   Using the Find option of the Help command alters the display listing as each letter is entered into the edit box.

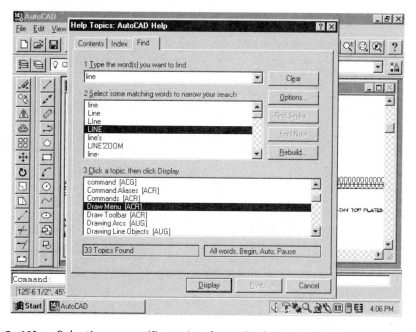

FIGURE 3-40b   Selecting a specific option from the line index listing provides a list of related subjects.

WHAT IT ALL MEANS

By now, you might be feeling overwhelmed by all of the bells and whistles of AutoCAD. In this chapter you've been introduced to several methods of entering a command, and shown a multitude of buttons, boxes, and controls for altering the commands and options. Each has been introduced using the LINE command. By the time you explore each of these controls in relations to the hundreds of commands, options, and icons, your brain may be on overload. The good news is you don't have to memorize every command, option, and control method. You only have to remember your favorites, and the ones you use on a regular basis. Use the Help menu (and this book) to help keep track of options that you don't use on a regular basis. The other key point of this chapter is that there will usually be several ways to do one thing. Generally, one method is as good as another, but don't limit yourself to the first method you find. The key to becoming a successful AutoCAD operator is to explore. The following quiz will require you to do just that. Many of the questions will deal with information not introduced in this chapter. This is your first real opportunity to explore AutoCAD.

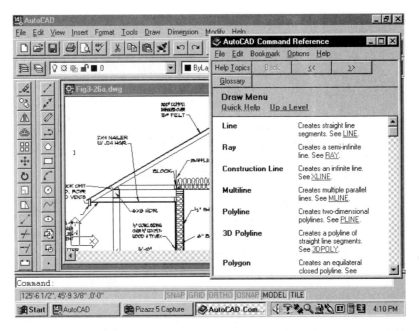

FIGURE 3–41 Picking the desired subject displays the related menu. By picking the Line Draw Menu listings, a display of the contents of the Draw menu are provided.

CHAPTER 3 QUIZ

1. List three different AutoCAD menus used for selecting commands and tell how they are accessed.

   _____

   _____

   _____

2. List the commands found in the Draw toolbar. What addditional commands are listed in the Draw pull-down menu.?

   _____

   _____

   _____

   _____

   _____

3. You've entered a dialog box with 40 options. Describe two methods that can be used to move through the list to the end.

   _____

   _____

4. What menu and option would allow you to find information about a drawing file while working in AutoCAD?

   _____

   _____

5. List the four screen areas displayed after entering AutoCAD.

   _____

   _____

   _____

   _____

6. Explain the following terms.

   Default _____

   Pick _____

   Command _____

7. List in order the options for the LINE command when picked from the screen menu. Explain how the display changes when the LINE command is picked from the pull-down menu.

_____

_____

8. You have just finished drawing a line and would like to draw an arc. Explain how to get the arc options using the pull-down menu.

_____

_____

_____

_____

9. What is the last arc option in the pull-down menu listed?

_____

10. List the options in the File pull-down menu.

_____

_____

_____

_____

11. You've entered the Insert pull-down menu, but you wanted the View menu. How can you get the View display box to replace the Setting box?

_____

_____

_____

12. List seven displays that may be seen on the status line.

_____

_____

13. Is ELLIPSE of the draw menu a command or an option? How can you be sure?

_____

_____

_____

14. An icon showing an airplane is in the standard toolbar. What is the icon, and what does it do?

    _____

    _____

15. What are two methods of altering command options?

    _____

    _____

16. What key will execute the following request:

    Display grid _____

    Ortho _____

    Snap mode _____

17. Someone has asked you what the last three commands were that you used on your drawing. How could you get a record if you don't remember?

    _____

    _____

18. List five items that can be found in a dialog box.

    _____

    _____

19. Search through each menu and find the location of

    Erase _____

    Tablet _____

    Osnap _____

20. What letter key would be needed when using the control key to complete the following options?

    Grid toggle _____

    Save _____

    Backspace _____

    Coordinate display _____

    Move viewports _____

21. List three methods for entering the Help menu.

    _____

    _____

22. Explain the procedure to float the Draw toolbar.

_____

_____

_____

23. Describe the process to display a hidden toolbar.

_____

_____

_____

24. You've drawn 100 lines and would now like to quit. How can you terminate the command?

_____

_____

_____

25. You've discovered after terminating the command, that you actually wanted to draw one more line. What is the quickest way to reenter the LINE command?

_____

_____

_____

26. How many different types of arcs are listed in the Draw pull-down menu?

_____

_____

_____

27. How many sides will be drawn if the default value is selected for a polygon?

_____

_____

_____

28. What is the first and last listing in the Line type library?

_____

_____

_____

29. What is the effect of a 0 and a 1 setting on Tooltips?

_____

_____

_____

30. What, according to AutoCAD, is a Zombie Timeouts?

_____

_____

_____

4

CREATING DRAWING AIDS

By now you've become familiar with the hardware, and entered the software. It's time to start building your drawing files. In Chapter 3 you opened AutoCAD and were exposed to the Start Up window. To quickly get you drawing, you picked the OK button and made the window go away. This window offers four alternatives to help set up a drawing file. Selecting the Instructions button will provide a brief explanation of each of the three buttons in this window. For new users of AutoCAD, the Use a Wizard, and Use a Templet options offer help to organize the drawing area. The default option of Use a Wizard offers helpful prompts to quide new users in drawing setup. The Wizard contains the Quick Setup and Advanced Setup options for determining how the new drawing environment will be established. Selecting the Use a Templet button will allow predetermined drawing parameters to be used to control the drawing environment.

As you gain an understanding of the options that are available to you, you can create your own templets. Once you've created and stored your own templet, each time you start a drawing, the drawing parameters will be preset to meet your standards. Chapter 5 will help you create your own templet. For experienced users, the Start from Scratch option can be used to enter common drawing controls. The Start from scratch option is the equivalent of grabbing a blank sheet of paper with no preset drawing controls. It is best suited for users who like to do things the long way, but it does not allow AutoCAD to help make their drawing tasks easier. The Open a Drawing option is used to open an existing drawing file and will be introduced in Chapter 5. No matter which option you chose to open a drawing, several tools must be understood before we explore the methods to set these options.

DRAWING TOOLS

These tools include UNITS, LIMITS, SNAP, GRIDS, and ORTHO. Two additional skills—STATUS and HELP—will provide information about your present drawing setup as well as an aid if you get stuck.

Units

If you have previous manual drawing experience, you've worked with a scale similar to that shown in Figure 4–1. Many architectural and structural projects are drawn using 1/4" = 1'-0". Figure 4–2 shows a floor plan drawn in architectural units and scaled down to fit the paper limitations.

One of the unique features of AutoCAD is that you will be drawing at full scale. The drawing will not be scaled until you print or plot your drawing. Figure 4–3 shows portions of Figure 4–2 printed at 1/4" = 1'-0". If you want to draw a structure that is 300 feet long, you will draw it full size, and then choose a scale as you plot, using AutoCAD's paper space command. These will be covered in Chapters 23 and 24. Units will allow you to choose which measuring device you will use to establish size. As your computer is currently set, you are measuring lines in four-place decimal units. This can be confirmed by looking at the coordinate display in the lower left corner of the status bar. To describe the current location of the mouse, two four-place decimal numbers are used. The drawing units may be set using the Quick Wizard, the Advanced Wizard and with the methods used in Start from Scratch as the drawing is started. Each of the unit controls may also be changed at any time throughout the life of the drawing to ease in setup using the methods described in Start From Scratch.

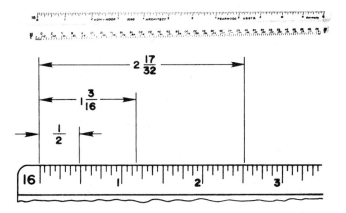

FIGURE 4–1 Architectural scales have been used for years to reduce or enlarge construction component. (Photo courtesy of Koh-I-Noor Rapidograph, Inc.)

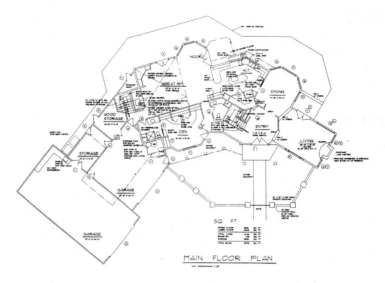

MAIN FLOOR PLAN

FIGURE 4–2 Architectural projects such as floor plans are typically drawn manually at a scale of ¼"=1'-0". AutoCAD will allow you to draw at full scale by altering the size of the display shown on the screen. The "scale" can then be set as the drawing is plotted.

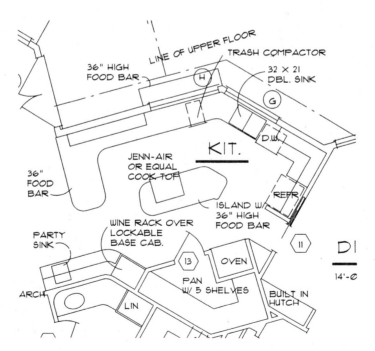

FIGURE 4–3 The floor plan shown in Figure 4–2 can be enlarged by using the ZOOM command to view a specific area of a drawing or by changing the scale factor during plotting.

Unit Selection. Unless you are using a templet, setting the UNITS should be the first thing you do once you have entered the drawing editor. Notice that the coordinate display on the task bar shows the default value is for decimal inches. This would be fine for a drawing such as a beam connector shown in Figure 4–4. Most architectural projects are better suited to the Architectural option. This will measure objects in feet and inches. The Engineering option would be the best option for working on drawings such as site plans or street layouts, where larger tracts of land are involved, as seen in Figure 4–5. Before proceeding, select the type of units that are appropriate for your drawing. For now, select Architectural units.

Unit Accuracy. Once the drawing units have been selected, you'll also have to consider the degree of accuracy of your drawing. For Architectural units the precision is now set at 0'-0 1/16". This will work great for drawing a small detail, but will be of little help for drawing a 300' long structure. Setting the unit accuracy in whole inches will allow you to draw lines that are an even number of inches in length. The drawing accuracy can be changed throughout the drawing, but should be kept as large as possible to facilitate drawing speed. The unit accuracy may be set using the Advanced Wizard or by the methods described in Start from Scratch.

Area. Selecting the Area defines the area that future settings will control. Later in this chapter you'll be introduced to Snap and Grid. The area that is specified in the drawing setup will determine the location where these features will be displayed. The Area setting also adjusts the default settings for features such as text height and linetype scaling. The drawing area may be set using the Quick Wizard or the

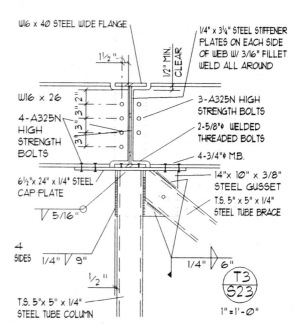

FIGURE 4–4 Although some drawings can be drawn using Decimal or Engineering Units, most are created using Architectural Units.

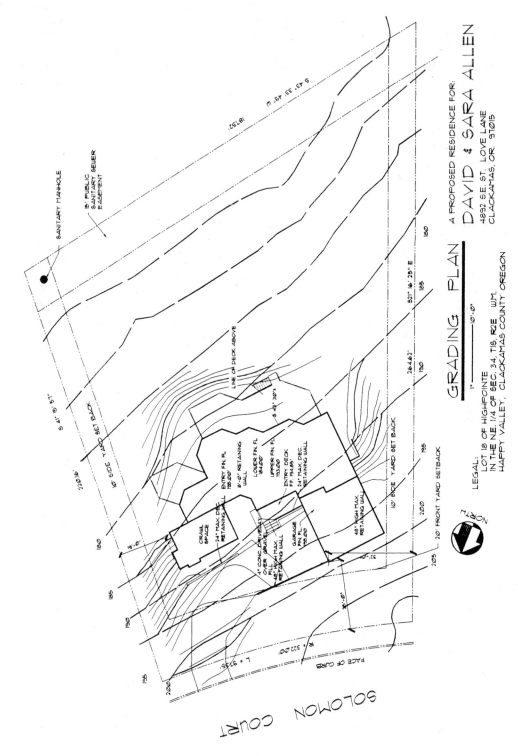

FIGURE 4-5 Large parcels of land are often created using Engineering Units. (Courtesy Residential Designs)

Advanced Wizard. The area can also be adjusted throughout the life of the drawing with the methods used in Start from Scratch to adjust the drawing limits.

Angle Measurement.   If your drawing will include lines drawn at an angle other than vertical or horizontal, you'll need to decide how the angle increment will be displayed. Common angle units seen on construction drawings include angles measured in degrees/ minutes/ seconds and angles measured in degrees located from compass points.

Angle Measurement

The Decimal Degrees option is best suited for plan views, such as shown in Figure 4–6, and will display angles in degrees and decimal parts of a degree. The Deg/Min/ Sec option is best suited for land measurements, such as a site plan as shown in Figure 4–5. This will display angles in degrees, minutes, and seconds. The Surveyor option presents surveyor's units, which also could be used to plot land, street, or sewer layouts, with angles displayed as bearings displayed as <N/S> <angle> <E/ W>. The angle is based on north or south and will always be less than 90°. An example of common bearings is listed in Figure 4–7. The method of angle measurement may be set using the Advanced Wizard or be adjusted throughout the life of the drawing with the methods used in Start from Scratch.

Angle Direction

Once the accuracy is selected, you can decide the starting point for angular measurement. Picking the Direction... button at the bottom of the Units Control dialog box will display the Direction Control dialog box shown in Figure 4–8 shows the visual representation of the angle menu. Standard AutoCAD layout of angles places zero degrees at 3 o'clock and then moves counterclockwise.

You may specify that angle measurement start in any location. This will be particularly helpful if you want to work on a site plan where north does not lie at the top of the paper. To place north in another location, enter the new starting angle at the prompt. The starting angle that is being specified is the location of east. Once the starting place is determined, you are given an option to decide the direction of angle measurement. The default direction is to measure angles counter-clockwise. The angle direction may be set using the Advanced Wizard or be adjusted throughout the life of the drawing with the methods used in Start from Scratch.

LIMITS

Setting UNITS is equal to deciding which scale to use in manual drafting. Setting the LIMITS would be equal to deciding what size drawing paper to use. As the drawing screen is configured now it is equal to a 12″ × 9″ piece of paper. Great for drawing a bolt, but very difficult for drawing an office structure. Remember, AutoCAD is drawing in real size, not at scale as in manual drafting.

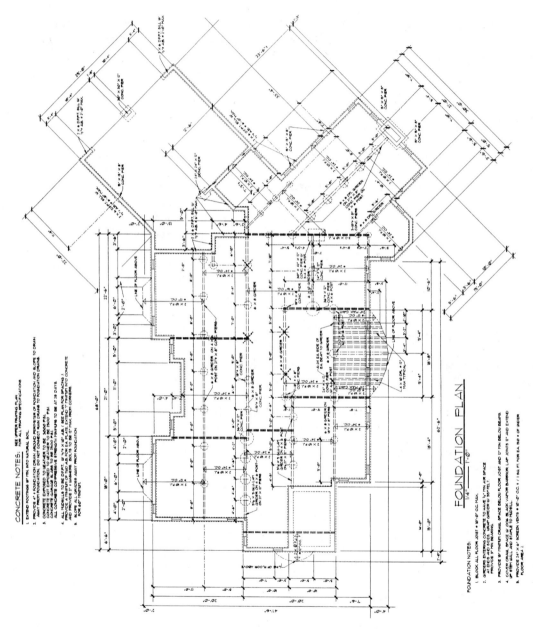

FIGURE 4-6 Angles on a drawing such as this foundation plan can be measured conveniently using option 1, Decimal Degrees.

Ø° = E
45° = N 45d Ø'Ø"E
90° = N
135° = N 45d Ø'Ø"W
180° = W
225° = S 45d Ø'Ø"W
270° = S 45d Ø'Ø"E
315° = N 45d Ø'Ø"E

SURVEYOR'S UNITS

FIGURE 4–7 Bearings are useful on site-related drawings showing all angles based on either north or south.

Drawing limits are expressed by a pair of X,Y coordinates that represent the lower left corner, and another pair that represent the upper right corner of the drawing area. The drawing limits may be set using the Advanced Wizard or be adjusted throughout the life of the drawing with the methods used in Start from Scratch.

GRIDS

Another helpful tool for drawing layout is GRID. This will provide the equivalent of the nonrepro blue grid that is preprinted on drafting vellum. The GRID command will produce a visible grid of dots at any desired spacing. The grid size can be adjusted throughout the life of the drawing, depending on the size of the object being designed. The grid also can be displayed or removed to aid visual clarity, but will not

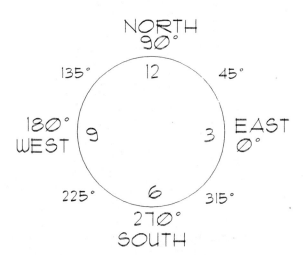

FIGURE 4–8 A visual representation of the angle menu comparing compass and angular measurement.

be produced when the drawing is printed or plotted if displayed on the screen. Figure 4–9 shows an example of a floor plan with the grid display. Options for grid settings are similar to snap settings and include ON, OFF, Snap, and Aspect. Each will be discussed as the drawing is set up using the methods described in Start from Scratch.

SNAP

When working on a modular component it is helpful to be able to control the accuracy of lines entered with the mouse. The SNAP command sets up an imaginary rectangular grid that controls placement of the drawing cursor in precise intervals.

The distance between cursor movements is defined as snap resolution. Snap intervals can be set at any interval and may be changed throughout the life of the drawing. A change in the snap grid affects only the points entered after the change in resolution. Points entered prior to the change in resolution may no longer line up with the new snap location. The size of the grid setting and the size of the object to be drawn will effect the snap value. If the grid is set at 2', a snap value of 6" or 12" will be helpful in producing accuracy. If a detail of a beam to post is drawn, a snap of 1/2" might be used to accurately move through the drawing. If a 200-foot-long structure is to be drawn, a 4" or 6" snap would be more helpful than the 1/2" accuracy used for a detail. The snap value can be adjusted throughout the life of the drawing using the methods described in Start from Scratch. The current status is displayed in the Task bar. Snap settings include toggling between ON/OFF, Aspect, Rotate, and Style.

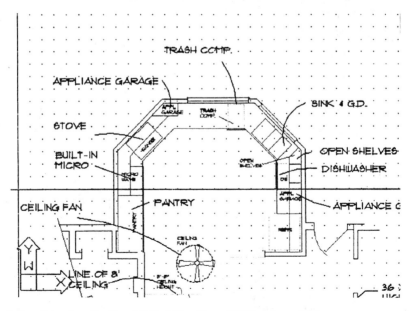

FIGURE 4–9   Drawing limits of 15', 12' with a GRID value of 6".

ORTHO

Most of the structures that you will draw are composed of perpendicular lines. Enter **LINE** at the command prompt and draw a few lines. You'll notice that as a line is drawn from the "From" point to the "To" point, a line is extended to the cross-hairs and moves as the cross-hairs move. This line is referred to as a rubberband line. The ORTHO command will allow you to place the end of the rubberband line in an exact horizontal or vertical position based on the current snap or grid pattern. If the snap and grid are rotated, ORTHO will respond by producing perpendicular lines based on the new grid. The residence shown in Figure 4–6 has a grid set to 45° for the right edge of the structure. As ORTHO was used on the right half of the structure, lines were automatically set to 45°, 135°, 225°, and 315°.

DRAWING SETUP USING A WIZARD

Four methods are available to set up a drawing. Three are listed in the Start Up window. The fourth method is to close the Start Up window, and begin drawing. This would be the equivalent to taking a piece of velum with no other tools and starting to draw. This system worked well in kindergarten but is not suitable for the construction industry. AutoCAD R14 offers the two new methods of Use a Wizard and Use a Templet, as well as the traditional Start from Scratch approach. Each can be seen in Figure 4–10a. Picking the OK button will display the screen shown in Figure 4–10b.

The default method of starting a drawing in AutoCAD is to use a wizard to help set the drawing parameters. Two options are provided based on the amount of help desired. Choosing Quick Startup will assist you in setting the drawing units and the drawing area. The Advanced setup will provide seven startup options.

Quick Setup

The default option of Use a Wizard is the Quick Startup option. Using this procedure will produce a drawing area in model space. This option works well for drawing, but will require some adjustment for plotting. Picking the OK button will produce the display shown in Figure 4–11 and allow the selection of the drawing units. Use the mouse to place the cursor in the desired setting and press the select button.

This will place a black dot in the button indicating the selection is activated. See Figure 4–11b. Pick the Next >> button at the window to continue the drawing setup. This will produce the display shown in Figure 4–12 showing a default value of 1' × 9". The drawing area should be based on what you expect to draw. For now accept the default value and pick the Done button. This will display the drawing screen shown in Figure 4–13. Notice that the grid is confined within the drawing area of 12" × 9" even thought the screen displays a larger area. To explore other setup options, move the cursor to File of the pull-down menu and pick New. This will produce the display shown in Figure 4–14. AutoCAD is giving you the reminder

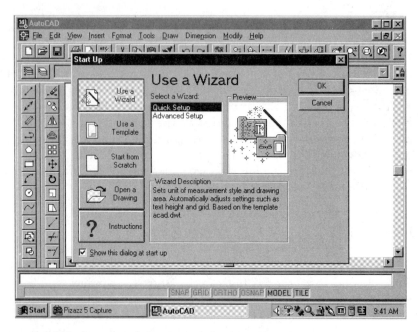

FIGURE 4–10a The Start Up dialog box is displayed as AutoCAD is entered to provide help in setting drawing parameter.

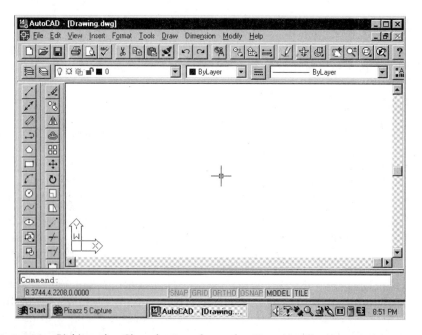

FIGURE 4–10b Picking the Close button from the Start Up display provides access to the drawing area with limited parameters. Drawing aids can be added at anytime throughout the life of the drawing.

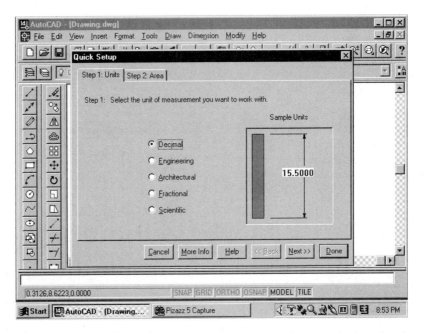

FIGURE 4–11a Selecting the Use a Wizard button provides two choices for drawing setup. Choosing Quick Setup will produce this display to aid in the selection of the drawing units.

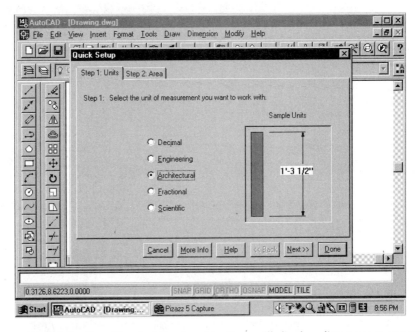

FIGURE 4–11b Selecting the Architectural button will display distances measured in units of feet and inches.

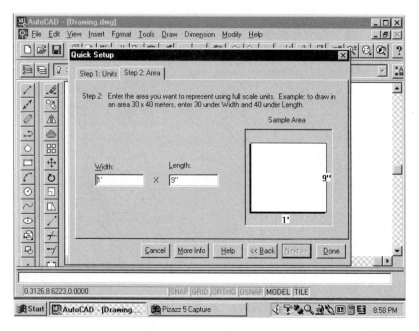

FIGURE 4–12 Setting the area defines how much space will be displayed on the screen. Set the area slightly larger than what you think you will need to complete the drawing.

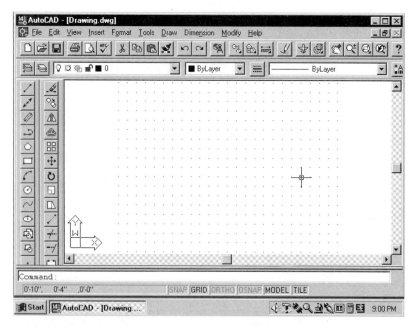

FIGURE 4–13 Accepting the default value of 1' x 9" and picking the Done button will display the drawing screen with a grid filling the default area.

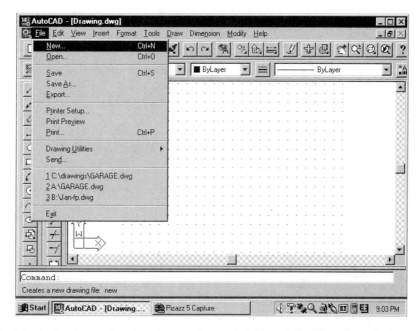

FIGURE 4–14   Explore other Startup options, pick New from the File pull-down menu.

that if you continue, your drawing will vaporize. For now, pick the No option and continue so the Advanced setup option can be explored.

Advanced Setup

With this button active, the AutoCAD wizard will help you set seven controls related to the units of measurement, angle of measurement, direction of angle, area, title block, and layout. These controls will open a new drawing and provide a title block that is viewed in either paper or model space (See Chapter 3) depending on the decisions made in step 7 of the setup.

Units.   The Units control folder is displayed as the Advanced Setup is opened. The default value is decimal. Choose the value that best meets the need of the drawing to be started. Figure 4–15 shows the value set to record Architectural units and displays a sample of how measurements will be displayed.

Unit Accuracy.   Notice that at the bottom of the Units column is an edit box for setting the precision of the units. For Architectural units the precision is now set at 0' -0 1/16". To change the setting, pick the arrow on the right side of the box. This will display the accuracy menu shown in Figure 4–16. Picking the down arrow will scroll through the menu allowing a greater degree of accuracy to be selected. Picking the up arrow sets the accuracy in larger units. For now, highlight the 0'-0" setting, and then select the option with the mouse. This will return you to the units box

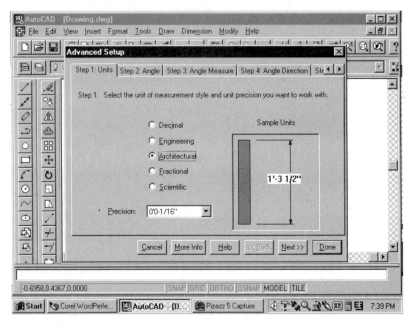

FIGURE 4–15 The first step in using the Advanced Wizard is to set the drawing units. Notice this method allows the precision of the units to be measured.

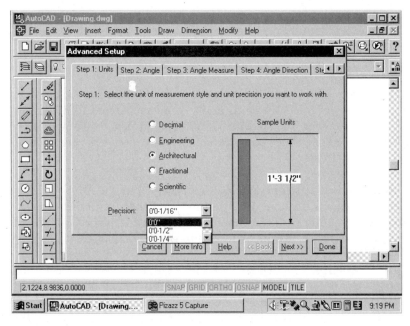

FIGURE 4–16 Picking the Precision button displays this menu of units. Accepting the default value will display units with an accuracy of 1/16 inch. Selecting 0'0" will measure units in whole inches.

and allow you to continue the drawing setup. Notice as the unit precision was changed from 1/16" to 0" the display of units is also changed. The units will now display 1'-4". If you're satisfied with the units, pick the Next >> button to continue the drawing setup. If you change your mind, use the << Back button to move back through the menu. The units can also be changed at any point during the life of the drawing using the methods that are presented in Start from Scratch that will be presented later in this chapter. Selecting the Done button will set the Units, exit the wizard, and place the drawing area on the screen. The Done button may be used throughout the seven steps advanced setup.

Angles.   Once the Units are selected, the folder for Step 2 will be displayed allowing the type of measurement for angles to be selected. Figure 4–17 shows the Angle display. The benefits of each type of measurement has been presented earlier in this chapter. For now accept the default option of Decimal Degrees. The precision for measuring angles can be set using the same method used to set the unit precision. Picking the down arrow will provide options for zero to eight decimal places, with the default of 0. Normally two place accuracy is sufficient for architectural drawings, while civil drawings might require a greater accuracy. For now, accept the default by picking the Next >> button to proceed to step 3.

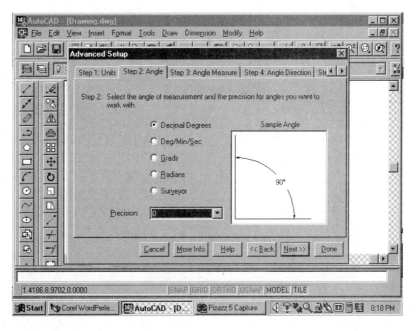

FIGURE 4–17   The second step of the Advanced Setup menu allows the units for measuring angles to be selected.

Angle Measurement.   Step 3 of the Advanced Wizard allows the direction of angle measurement to be selected. The default of east and the options are shown in Figure 4–18. To change the starting point of angles to north, pick the North radio button with the mouse select button. If you select north as the new starting direction, 0° will now be placed at north, 90° will be west, 180° is south and east will be at 270°. An angle other than one of the four compass points can be selected by entering the desired starting angle in the edit box and then picking the Other button. Entering a starting angle of 45° and picking Other will place 0 half way between north and east. When the desired starting point has been selected, selecting Next >> will allow Step 4 to be completed.

Angle Direction.   Figure 4–19 shows the folder for selecting the directions that angles will be measured, beginning at the starting point that was selected in step 3. The AutoCAD default for angle measurement is counterclockwise. Picking the Clockwise button will change the direction of angle measurement. For now, accept the default. Remember, the defaults are there because they fit the needs of most users. The direction, like each of the other variables, can be altered at any time throughout the life of the drawing. Once the direction is selected, pick the Next >> button to proceed to Step 5.

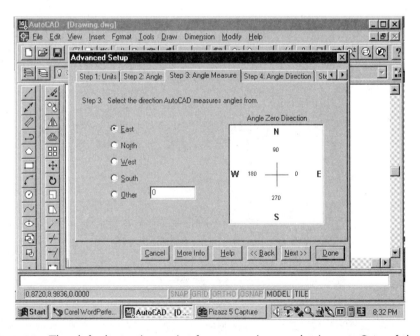

FIGURE 4–18 The default starting point for measuring angles is east. One of the other three quadrants can be selected, or any a point in between can be picked by selecting Other and providing an angle.

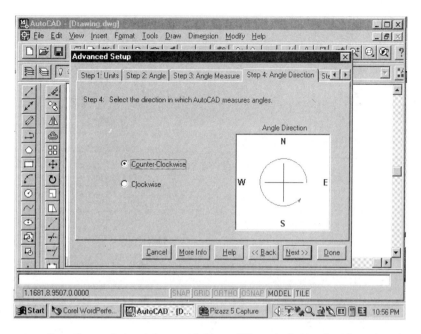

FIGURE 4–19 Step four of the Advanced Setup Wizard allows the direction for angle measurement to be entered.

Area. Step 5 of the Advanced Wizard setup allows the drawing area to be specified using the folder shown in Figure 4–20. In the default setting, the drawing will be 1' × 9". Enter the values for the width and length of the area of the intended drawing to be completed in the appropriate edit box. The values entered will determine the space between the borders of the drawing area that is being created.

Selecting Drawing Limits

Your drawing limits will be determined by the size of the object to be drawn and the number of notes and dimensions to be placed. Drawing limits may be set arbitrarily and adjusted throughout the life of the drawing. For instance, you may start a plan view of a lot with the limits of 300', 250'. Halfway through the project you may decide you need to draw the adjoining lot and may need to change the limits to 750', 475'.

Limits are also set by the size of paper the project will be printed on. Common paper sizes include A: 8½ × 11, B: 12 × 18, C: 18 × 24, D: 24 × 36, E: 36 × 48. Remember that these sizes are in inches. If you want to draw a structure that will be printed on "C" size paper at a scale of ⅛" = 1'-0", there are eight feet per inch of paper. Multiply 24 × 8 and 18 × 8 to determine the limits. This would produce limits of 192' × 144'. This will be explained in more detail in Chapter 5 as title blocks are added to a drawing. See Chapters 5 and 24 for a listing of common scale, paper size limits, and suggestions for setting up drawing templates. Once the area values have been entered, pick the Next >> button to proceed.

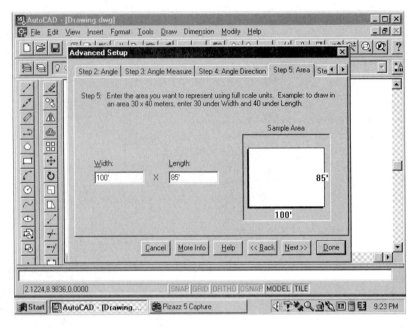

FIGURE 4–20 Entering drawing units of 100' by 85' will allow for the layout of a floor plan to be started.

Title Block. Step 6 of the setup wizard allows for choosing the use of a title block for the drawing. Figure 4–21a shows the window for the selection of a title block. You can decide to have no title block, one of several from the AutoCAD file, or a title block of your own creation. In the next chapter you'll begin work on your own title block. If you choose no title block and select the Done button you'll find a border displayed in the drawing area as shown in Figure 4–21b. This border will not be displayed when a print is made. It merely defines the limits of model and paper space. Displaying a title block on a print will be discussed in the next portion of this chapter.

For now, open a new drawing with no border. You'll notice that as the cursor is placed inside the border it's in model space, and the cursor is displayed as a cross hair. When the cursor is moved outside the border, the cursor is now in paper space and is displayed as an arrow. Remember when you're in model space you're working at full scale. For now, open the Title Block Description edit box, move through the list, and select Arch/Eng (in). This will produce the display shown in Figure 4–21c. Figure 4–21d shows the results of the selection. When the desired selection has been entered, pick the Next >> button to move to the final stage of the setup.

Layout. The final option in the Advanced Setup is to choose the layout of the drawing space. Your choices include model or paper space, with the display folder shown in Figure 4–22. Paper space allows the drawing area to be divided into areas called viewports. The viewports can then be used to display drawings that are to be plotted at different scales. Using model space will create one viewport, requiring all

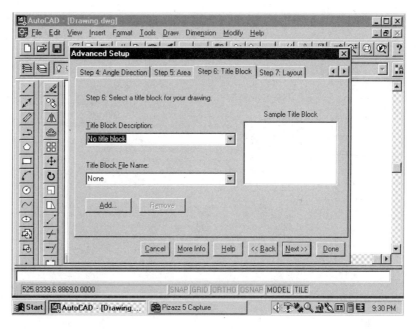

FIGURE 4–21a A title block can be added to a drawing, or with the default option no title block will be displayed in the drawing area.

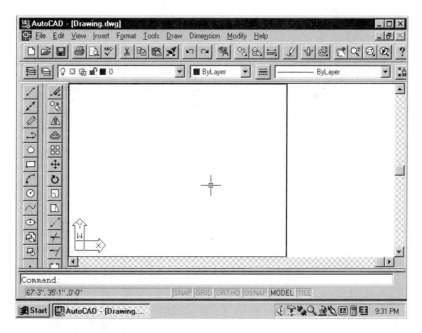

FIGURE 4–21b If the default is accepted, when the drawing is displayed an outline is placed around the specified drawing area.

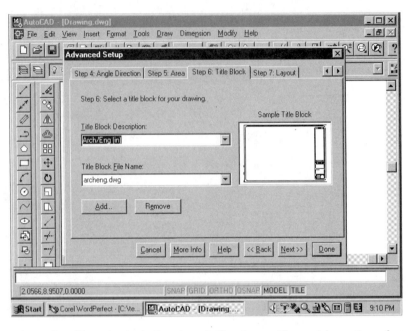

FIGURE 4–21c Scrolling through the description box will provide options for various title blocks. Selecting the Arch/Eng (in) option will place a title block and border around the drawing area.

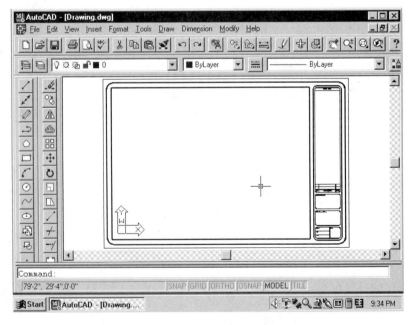

FIGURE 4–21d A title block suitable for basic architectural or engineering projects. Chapter 5 will introduce options for creating and saving your own title block.

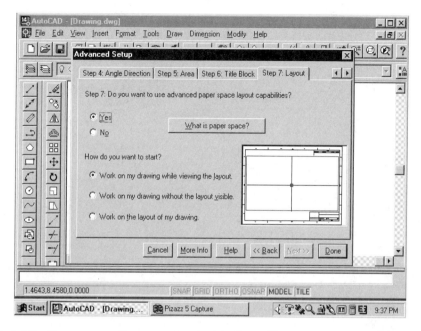

FIGURE 4–22 Step seven allows you to decide if you'll use paper or model space. Chapters 23 and 24 will introduce the use of paper space. Most drawings are created in model space and plotted in paper space.

details on the page to be plotted at the same scale. If you accept the default values for paper space and the viewing of the layout, the display will be similar to Figure 4–21d, depending on the border selected. If you choose no border, in Step 6, and no visible border, in Step 7, the display will be similar to Figure 4–10b. The area within the title block will be model space, and the outer area will be paper space. The Advanced Setup menu also allows setting the drawing up in model space. For now, take your pick and experiment.

DRAWING SETUP USING A TEMPLET

Rather than use the Setup wizard, drawing templets with predetermined drawing values can be used to start the drawing. The templet contains values for drawing units, limits, and other values that can be altered throughout the life of a drawing. A templet supplied by AutoCAD can be customized to meet specific requirements and saved for future use. A templet can be used by picking the Use a Templet button from the Startup window. This will produce the display shown in Figure 4–23. Accepting the default setting of ACAD.DWT and picking the OK button will produce a display as shown in Figure 4–21. The drawing area will be approximately 12" × 9" (size varies depending on monitor size) with units measured in 4 place decimal inches. The limits can be adjusted as needed. Scroll through the list of templets and select Archeng.dwt and the OK button. This will produce a drawing display as

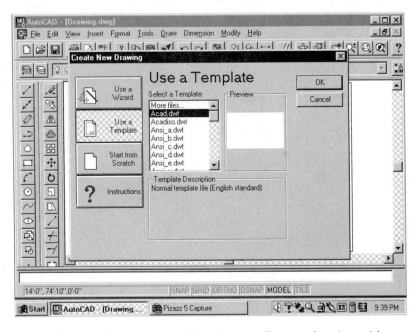

FIGURE 4–23 Selecting the Use a Template button allows a drawing with preset drawing parameters to be used. Selecting More files... allows user-created templets to be accessed.

shown in Figure 4–21d. In addition to using the templets supplied with AutoCAD, you may develop you own templet to meet specific needs of a project. Since most plan views in residential architecture are drawn at a scale of 1/4"=1'-0", a templet can be established for use with all drawings that will be printed at the scale. In addition to each of the drawing controls that were introduced earlier in this chapter, the templet could also include common text fonts, linetypes, layers, and other related features. Each of these components will be introduced throughout the text. As each is introduced, features that you find useful can be added to your templet. Methods for creating and saving templets will be introduced in Chapter 5. For now, if you choose to use a templet to enter the drawing area, accept the default and OK.

STARTING FROM SCRATCH

This option allows a drawing to be started with a minimum of predetermined values. Once the decision between inches or metric is made, a drawing screen similar to Figure 4–24 is displayed. Units are displayed in 4-place-decimal units, with drawing limits of 12" × 9" (varies based on monitor size). Rather than being prompted by a wizard or restricted by a templet, Start from Scratch allows drawing controls to be set as needed from the command prompt or a dialog box.

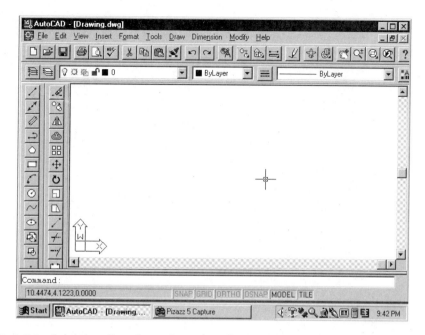

FIGURE 4–24 Selecting Start from Scratch will provide a blank screen with no preset values.

Setting Units

The units of a drawing may be altered at any time during the life of the drawing using a dialog box or the command prompt.

Setting Units Using a Dialog Box. Units can be altered by selecting Units... from the Format pull-down menu or by typing **UN** [enter] or **DDUNITS** [enter] at the command prompt. Each method will produce the Unit Control dialog box shown in Figure 4–25. The desired units are selected by highlighting the corresponding radio button. Selecting the arrow on the right side of the Precision edit box will display the menu shown in Figure 4–26. The desired accuracy can be selected by highlighting the fraction. Once selected, the accuracy will be displayed in the edit box and the menu will be closed. The box can also be used to select the method for measuring angles and for setting the angle precision. Angle precision accuracy is set using the same methods used to set the unit precision. Picking the Direction... button will display the menu shown in Figure 4–27. This menu allows for the angle starting point and the angle direction to be set, by picking the appropriate radio button or highlighting the desired accuracy.

Setting Units Using the Command Prompt. The units of a drawing can easily be changed by typing **UNITS** [enter] at the command prompt. This will produce the prompt shown in Figure 4–28a and 4–28b.

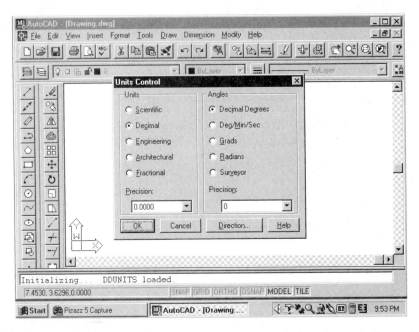

FIGURE 4–25 Drawing units can be altered at any time using the Unit Control dialog box. The dialog box is displayed by selecting Units... from the Format pull-down menu or by typing **UN** [enter] or **DDUNITS** [enter] at the command prompt.

FIGURE 4–26 The Unit Control dialog box allows the accuracy of measurements to be adjusted to meet the need of a drawing.

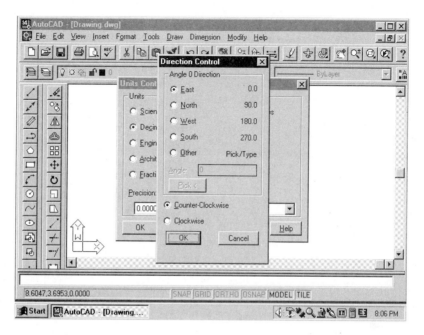

FIGURE 4–27 Picking the Direction... button allows the orientation for angle direction to be altered.

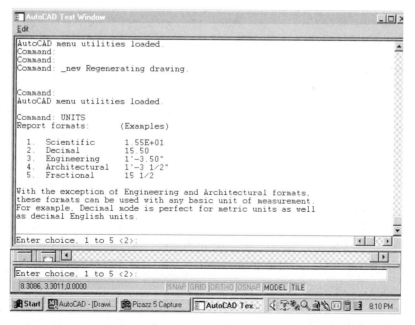

FIGURE 4–28a Each of the drawing units can be altered by entering **UNITS** at the command prompt.

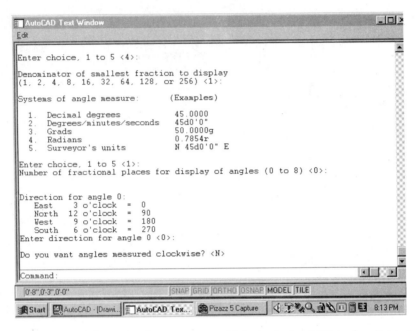

```
AutoCAD Text Window                                        _|□| x
Edit

Enter choice, 1 to 5 <4>:

Denominator of smallest fraction to display
(1, 2, 4, 8, 16, 32, 64, 128, or 256) <1>:

Systems of angle measure:      (Examples)

  1.  Decimal degrees          45.0000
  2.  Degrees/minutes/seconds  45d0'0"
  3.  Grads                    50.0000g
  4.  Radians                  0.7854r
  5.  Surveyor's units         N 45d0'0" E

Enter choice, 1 to 5 <1>:
Number of fractional places for display of angles (0 to 8) <0>:

Direction for angle 0:
  East    3 o'clock  =  0
  North  12 o'clock  =  90
  West    9 o'clock  =  180
  South   6 o'clock  =  270
Enter direction for angle 0 <0>:

Do you want angles measured clockwise? <N>

Command:

 0'-8",0'-3",0'-0"            SNAP GRID ORTHO OSNAP MODEL TILE

 Start    AutoCAD - [Drawi...   AutoCAD Tex...   Pizazz 5 Capture    8:13 PM
```

FIGURE 4–28b Once the unit of measurement is selected, additional unit controls can be set from the prompt.

Setting Limits

Changing the limits alters the size of the drawing displayed on the screen. Before you change the limits draw a square inside the drawing area similar to Figure 4–29a. This will help you to visualize the change that occurs when limits are set.

To change the size of your "paper," type **LIMITS** [enter] at the command line. This will produce the display:

> Reset Model space limits:
> ON/OFF/<Lower left corner> <0'-0",0'-0">:

The acceptable responses are ON, OFF, and [enter].

ON will keep the current values (1'-0",0'-9") and activate limit checking. Limit checking will not allow objects to be drawn that are outside the limits. Think of limit checking as a border on your drawing paper to keep your project on the paper.

OFF will disable limit checking but retain the limit values for future use of limit check.

Pressing [enter] retains the current location of the lower left drawing limit of (0.0000,0.0000). This will be the most typical response. Pressing the return key will in turn produce the prompt:

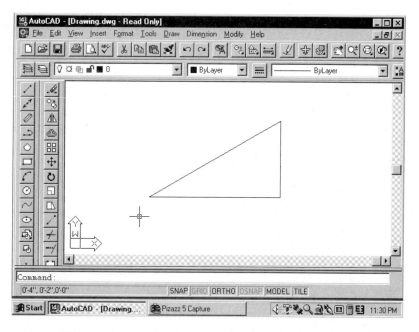

FIGURE 4–29a With the drawing limits set at the default units of 1'-0", 9" an object may appear quite large.

Upper right corner (1'-0", 0'-9"):

The default value is 1'-0" × 9", with the first number representing the horizontal dimension and the second number representing the vertical dimension. This is your opportunity to change the size of the drawing area to fit your current project. Enter 15',12' [enter], and the command prompt will return with no apparent change. Type **ZOOM** [enter] at the keyboard, then **A** [enter]. Now the effect of your new limits can be seen in Figure 4–29b. Remember, the box is still the same size but it appears smaller because the paper size has increased. The limits can also be adjusted by selecting Drawing Limits from the Format pull-down menu.

Setting Grids

Grid can be set by dialog box or from the command line. Once the values have been set, the grid can be toggled ON/OFF by double-clicking the Grid button in the status bar. The grid can also be toggled ON/OFF in the middle of another command by using the F7 key or by pressing the Ctrl + G keys.

Assigning Grid Values Using a Dialog Box. Grid values can be set using the Drawing Aids dialog box. The box can be accessed by picking Drawing Aids... from

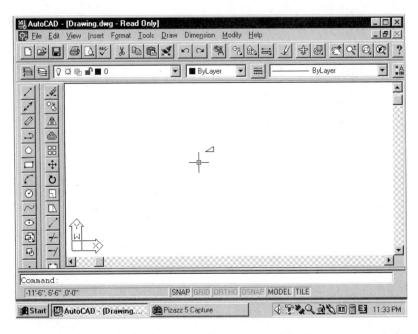

FIGURE 4-29b By changing the location of the upper right-hand corner, and pressing the [enter] key, the screen will allow viewing of a larger area. In this example, the limits are set to 15′, 12′.

the Tools pull-down menu. This will produce the dialog box shown in Figure 4–30. The dialog box can also be displayed by typing **RM**[enter] or **DDRMODES** [enter] at the command prompt. Values for the X and Y spacings can be entered in the appropriate edit box. Once selected, the grid can be toggled ON/OFF by picking the On box. The dialog box also offers methods of altering the angle used to display the grid. Picking the On button in the Isometric Snap/Grid box places the grid on a 30° angle from horizontal. Other angles can be used to place the grid by entering the desired angle in the Snap Angle edit box in the Snap portion of the dialog box. Figure 4–31 shows an example of a floor plan that would require an inclined grid. Isometric grids will be discussed further in Chapter 25.

Setting Grids at the Command Line. The grid values can be established, altered, and displayed from the Command prompt. The controls for GRID can be accessed at the command prompt by typing **GRID** [enter]. The command line will now read:

Command: **GRID** [enter]
Grid spacing(X) or ON/OFF/Snap/Aspect <0′-0″>:

Enter **12** [enter]. This will return the command prompt and display the grid field with dots at 1′-0″ spacing on the drawing screen.

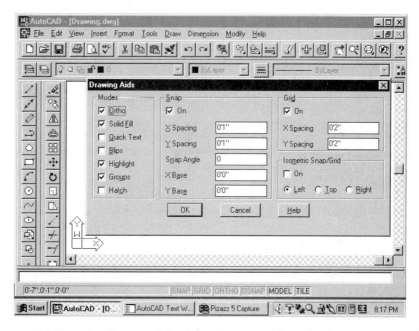

FIGURE 4–30 Values for Snap and Grid can be altered by selecting Drawing Aids... from the Tools pull-down menu or by typing **RM** [enter] or **DDRMODES** [enter] at the command prompt.

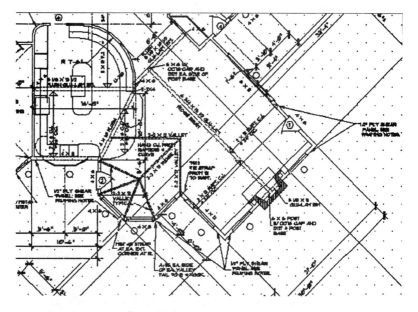

FIGURE 4–31 SNAP Rotate allows the invisible snap grid to be rotated to any angle to ease layout of an irregular shape.

Spacing.   Grid is typically set as a multiple of the snap setting. For instance, when drawing a foundation plan, the grid might be set at 12" and the snap at 2". This will provide great accuracy and speed. If the grid value is set at zero, the grid spacing automatically adjusts to the snap resolution. To specify the grid spacing as a multiple of the snap resolution, enter **X** [enter] after the grid value. For instance, if the snap value is 2", entering 6x at the command prompt would produce a grid of 12".

OFF.   The grid may be removed by typing **GRID** [enter] at the command prompt. As the menu is displayed, type **OFF** [enter].

Snap.   Selecting the Snap option sets the grid spacing to the current snap resolution. If the snap resolution is changed, the grid spacing will be changed as well. Snap may be set by typing **S** [enter] at the command prompt.

Grid spacing(X) or ON/OFF/SNAP/ASPECT <1'-0">: **S** [enter]

Aspect.   Aspect allows the grid to be set with a different value for the horizontal (X) and vertical (Y) directions. To do this, type **GRID** [enter]. The standard grid prompt is repeated:

Grid spacing(X) or ON/OFF/SNAP/ASPECT <1'-0">: **A** [enter]

Type Aspect or "A" and the prompt will display the additional listings:

Horizontal spacing (X) <1'-0">: [enter]
Vertical spacing (X) <1'-0">: **24** [enter]
Command:

An **X** may follow either of these variables to make it equal to the Snap value. Figure 4–32 shows an example of a floor plan with different X and Y values.

Setting Snap

Snap can be set by dialog box or from the command line using similar methods to adjust the grid. Once the values have been set, snap can be toggled ON/OFF by double-clicking the Snap button in the status bar. Snap can also be toggled ON/OFF in the middle of another command by using the F9 key or by pressing the Ctrl+B keys.

Assigning Snap Values Using a Dialog Box.   Snap values can be set using the Drawing Aids dialog box. The box can be accessed by picking Drawing Aids... from the Tools pull-down menu or by typing **RM** [enter] or **DDRMODES** [enter] at the command prompt. See Figure 4–30. Values for the X and Y spacings can be entered in the appropriate edit box. Once selected, Snap can be toggled ON/OFF by picking the On box. Entering an angle in the Snap Angle edit box controls the display angle for Snap and Grid. Entering a value in the X or Y Base edit boxes allows the base point

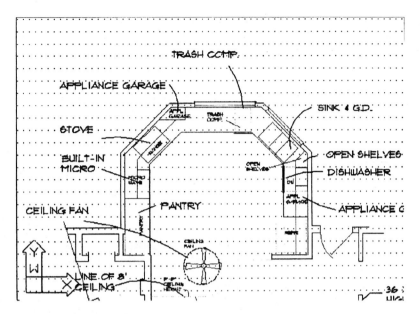

FIGURE 4–32 SNAP aspect allows the spacing between the invisible grid of horizontal and vertical units to be altered.

for the snap pattern to be established. The dialog box also offers methods of altering the angle that the grid is displayed. Picking the On button in the Isometric Snap/ Grid box places the grid on a 30° angle from horizontal. Other angles can be used to place the grid by entering the desired angle in the Snap Angle edit box in the Snap portion of the dialog box. Isometric grids will be further discussed in Chapter 25.

Setting Snap at the Command Line. The snap values can be established, altered and displayed from the command prompt. The controls for SNAP can be accessed at the Command prompt by entering **SN** [enter] or **SNAP** [enter]. This will produce the prompt:

Snap spacing or ON/OFF/Aspect/Rotate/Style <0'1">:

Entering a numeric value will set the spacing of the grid for entry points. Type the number **6** [enter]. The command line returns and allows for drawing. Type **L** [enter] and draw a line. You'll notice that the cross-hairs jump across the screen in 6″ intervals rather than moving smoothly as they had before. Figure 4–15 shows the grid arrangement that would result from selecting a 6″ snap grid.

OFF. To terminate the use of snap, type **SN** [enter]. Once the menu is presented, type **OFF** [enter]. The command line will return and allow for free movement of the cross-hairs.

ON. Snap can be activated at any time by typing **SN** [enter]. Once the SNAP menu is presented, type **ON** [enter]. Notice that the last value used is presented as the default. If you would like to activate the snap with a different spacing, just enter the desired numeric value and [enter].

Aspect. Some projects are set up when a different value for the horizontal (X) and vertical (Y) directions is used. To do this, type **SN** [enter]. Type **A** [enter] and the prompt will display the additional listings:

Horizontal spacing <0'-6">: [enter]
Vertical spacing <0'-6">: **12** [enter]
Command:

Rotate. Because not all projects are rectangular, such as the floor plan in Figure 4–31, it may be helpful to rotate the snap grid to match the drawing. Type **SN** [enter] and enter **R** [enter] at the prompt. This will produce the additional display:

Snap spacing or ON/OFF/Aspect/Rotate/Style <6>: **R** [enter].
Base point <0-0", 0'-0">:

AutoCAD is asking for the point of the drawing on which you would like the snap grid to be rotated. To select a location move the cross-hairs to the desired location in the drawing and press the pick button. This will now provide the prompt:

Base point <0-0", 0'-0">: Rotation angle <0>:

A number between –90 and 90 may be entered. Using a negative angle will rotate the snap grid counterclockwise. Using a positive angle will rotate the snap grid clockwise. Enter a value and [enter] the snap grid will be rotated around the base point just selected.

Style. The format of the snap grid is adjusted by using the style option. By entering **S** [enter] at the snap menu, the prompt will display:

Snap spacing or ON/OFF/Aspect/Rotate/Style <6>: **S** [enter].
Standard/Isometric <S>:

For orthographic drawings the style will not need to be set. For isometric drawings enter **I** [enter].

ORTHO

Most of the structures that you will draw are composed of perpendicular lines. Enter the line command at the prompt and draw a few lines. You'll notice that as a line is drawn from the "From" point to the "To" point, a line is extended to the cross-hairs and moves as the cross-hairs move. This line is referred to as a rubberband line. The

ORTHO command will allow you to place the end of the rubberband line in an exact horizontal or vertical position based on the current snap or grid pattern.

Ortho may be activated from the status bar, dialog box, by control keys, or from the command prompt. Double-clicking the Ortho button will toggle Ortho ON/OFF. Ortho can also be set by picking the Ortho button in the Drawing Aids dialog box. (See Figure 4–29). The dialog box can be accessed by picking Drawing Aids... from the Tools pull-down menu. The F8 button and Ctrl+L can also be used to toggle Ortho ON/OFF. Ortho is started at the Command prompt by typing **ORTHO** (enter).

This will display:

ON/OFF <off>:

Entering **ON** (enter) will activate ORTHO. Now any line drawn will be perfectly horizontal or vertical. You will be unable to draw inclined lines as the cursor is now set. In Chapter 6 you will learn how to draw inclined lines while ORTHO is activated by entering coordinates. To terminate ORTHO, type **ORTHO** (enter) to produce the basic display and type **OFF** (enter).

If the snap and grid are rotated, ORTHO will respond by producing perpendicular lines based on the new grid. The residence shown in Figure 4–6 has a grid set to 45° for the right edge of the structure. As ORTHO was used on the right half of the structure, lines were automatically set to 45°, 135°, 225°, and 315°.

STATUS

As you set up future drawing templets, you may lose track of the current drawing parameters. The STATUS command will provide an updating of the units, limits, snap, and grid setting, ortho status, and the settings of several other features that you have yet to explore. To get a listing of current drawing values, type **STATUS** (enter) on the command line. This will produce a screen display similar to the one in Figure 4–33. Notice that all displays are shown in the format specified when limits were established. The status text window will also be displayed by picking Status from Inquiry of the tools pull-down menu. Pressing F2 will restore the drawing screen.

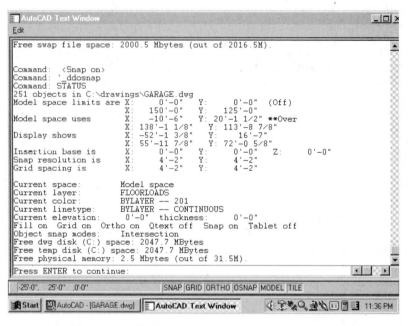

FIGURE 4–33 Typing **STATUS** [enter] at the command prompt will provide a listing of current drawing values.

CHAPTER 4 QUIZ

1. What command will set an invisible grid to aid layout? _____

2. List the five options of the units. _____

_____

_____

_____

_____

3. What is the default unit for angle measurement? _____

4. What is the default for angle direction? _____

5. Your boss would like a drawing of a street and sewer layout based on the surveyor's notes. Describe the process for setting up the proper method of two-place angle measurement by keyboard.

_____

_____

_____

6. List three methods to toggle between GRID ON/OFF.

_____

_____

_____

7. List three methods to toggle between SNAP ON/OFF.

_____

_____

_____

8. Access the Users Guide. Research and briefly describe Using Commands Transparently.

_____

_____

9. List the first and last listings in the HELP menu and explain how to find the last listing.

_____

_____

_____

10. What is the main difference between GRID and SNAP?

_____

_____

11. You've just entered AutoCAD and don't want to establish any drawing parameters. How can you get to the drawing screen?

_____

_____

_____

12. When would Start from Scratch be a useful method of entering a drawing?

_____

_____

_____

13. What does Area control?

_____

_____

_____

14. A site plan must be drawn with angles measured in a format such as N42° 30'-30" E. How can this be done if you forgot to adjust the angle measurement as you set up the drawing?

_____

_____

_____

15. A friend wants to draw a house, but has no idea how big it will be. What would you recommend setting for drawing limits as the drawing is started?

_____

_____

_____

16. How do GRID and SNAP relate to each other?

_____

_____

_____

17. What is the default method of entering AutoCAD?

_____

_____

18. How can you use the Advanced Wizard to start a drawing and not have a border or an outline?

_____

_____

_____

19. What command is used to access the Drawing Aids dialog box from the command prompt?

_____

_____

_____

20. What is the effect of entering **DDUNITS** [enter] at the Command prompt?

_____

_____

········· • • • •

CHAPTER

5

OPENING, RETRIEVING, AND SAVING DRAWINGS

• •

In Chapter 3 you discovered a little about the drawing environment of AutoCAD, and in Chapter 4 you started to create great works of art. You now need to understand how to open, save, exit, and store drawings. In Chapter 4 you were introduced to methods of entering the AutoCAD drawing area when a new drawing file is to be created using the wizard or a templet. This chapter will introduce you to methods for creating a templet drawing, opening existing drawings, saving drawings, and managing drawing files.

OPENING A NEW DRAWING

In Chapter 4 you explored the start up options using the wizards or a templet. The drawing area was also accessed by closing the Start Up window without setting any drawing parameters. A new drawing can also be opened from inside of AutoCAD by typing **NEW** ⏎ at the command prompt or by picking New... from the File pull-down menu. Figure 5–1 shows the contents of the AutoCAD File pull-down menu. A new drawing file can also be opened from inside of AutoCAD by selecting the New icon at the far left end of the standard toolbar. Each method will remove the existing drawing file and begin the process of opening a new file. If the current drawing file has been altered in any way since it was opened, attempting to open a new drawing will produce the alert message shown in Figure 5–2. To proceed, one of the three buttons must be selected. Picking the No button will remove the alert box and produce the Start Up dialog box so that the parameters for the new drawing can be set. Picking the Cancel option will remove the alert box and return you to the current

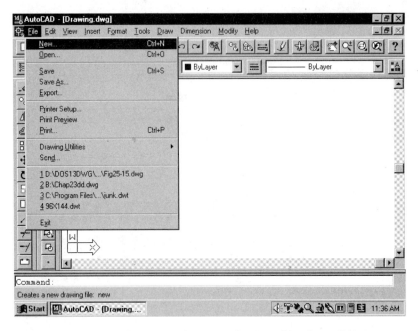

FIGURE 5–1 A drawing can be opened or saved using the File pull-down menu.

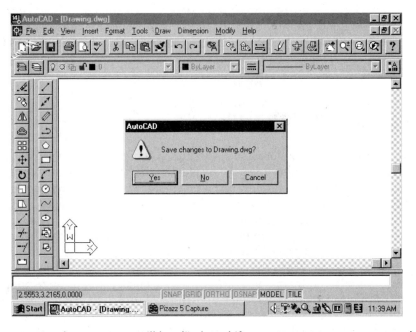

FIGURE 5–2 An alert message will be displayed if you attempt to open a new drawing without saving the existing drawing.

drawing. Selecting the Close button in the Alert window will have the same effect as picking the Cancel button. Selecting the default value of Yes will produce the Save Drawing As dialog box shown in Figure 5–3. This box can be used to select a location and a name for the current drawing before the new drawing is started. Methods of saving a drawing file will be discussed later in this chapter.

OPENING AN EXISTING DRAWING

Much of your time in an office will be spent adding to or editing existing drawings. An existing drawing can be brought to the screen by typing **OPEN** [enter] at the command prompt, by picking the Open icon from the standard toolbar, of by selecting Open... from the File pull-down menu. Each method will produce the dialog box shown in Figure 5–4a. This box can be used to list and select a drawing file from any of your storage locations. By default, files will be selected from the AutoCAD R14 folder. Pick the arrow beside the Look in: edit box to begin the selection process for finding the desired drawing file to be opened. As the arrow is selected, a flyout menu will be displayed reflecting the contents of the Windows Explorer. (See Figure 5–4b.) Clicking the desired drive to be searched will display a list of drawing folders contained in the selected drive. Double-clicking a folder will produce a list of files contained in that folder. Clicking the desired file name will place the file name in the File name: edit box. If the selected drawing was created in Release 13 or 14, the

FIGURE 5–3 If the Yes option of the alert box is selected, the Save Drawing As dialog box will be displayed to allow the existing drawing to be saved. Methods of saving a drawing will be presented later in this chapter.

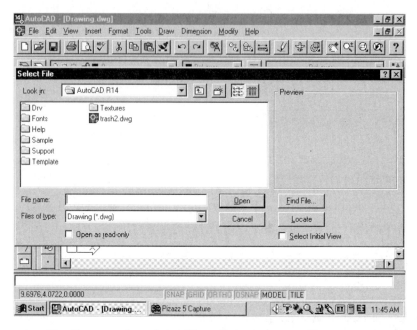

FIGURE 5–4a The Open option of the File pull-down menu allows an existing drawing to be accessed using the Select File dialog box. Files can be accessed from any of the folders listed in the Windows Explorer.

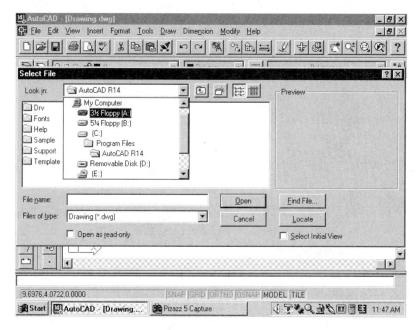

FIGURE 5–4b Slecting the arrow beside the Look in: edit box will display a listing of the Windows Explorer and allow you to determine what drive is to be searched for a drawing file. Notice that the A:drive will be searched for the desired file.

drawing will display in the Preview box. See Figure 5–4c. If you're satisfied with the results of the selection, pressing the Open button will remove the dialog box and open the selected drawing on the drawing area similar to Figure 5–4d. If the selected file is not what you expected, press the Find File... button to use the Browse/Search dialog box shown in Figure 5–5 to locate the desired drawing file. You can also use the Up One Level icon to work backward through the Explorer menu to select a different file.

SAVING A DRAWING

Once you have created your drawing, it will need to be saved. If you are in an educational setting this will mean saving your drawings on floppy disks or on a network. Check with your instructor to see if saving on the hard disk is allowed. If so, set up a folder so that all of your drawings can be accessed easily. If not, pick the letter of the appropriate drive so that files will be saved to the floppy disk. Generally you should avoid saving a file to a floppy drive except when the drawing session will be ended. Typically, a file should be saved to the hard drive throughout the drawing session to save time and avoid a possible problem caused by a full disk.

If you are learning AutoCAD in an office setting you may be saving on the hard drive, tape drives, ZIP drives, as well as on floppies. Because working on the hard drive is so much faster than accessing floppies, you will be saving to the hard drive

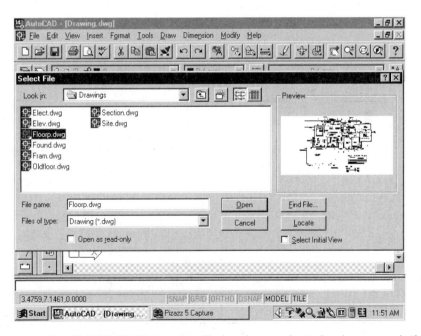

FIGURE 5–4c The FLOOR.DWG drawing file has been selected to be opened. If the drawing to be opened was created using AutoCAD R13 or 14, the drawing will be displayed in the preview box.

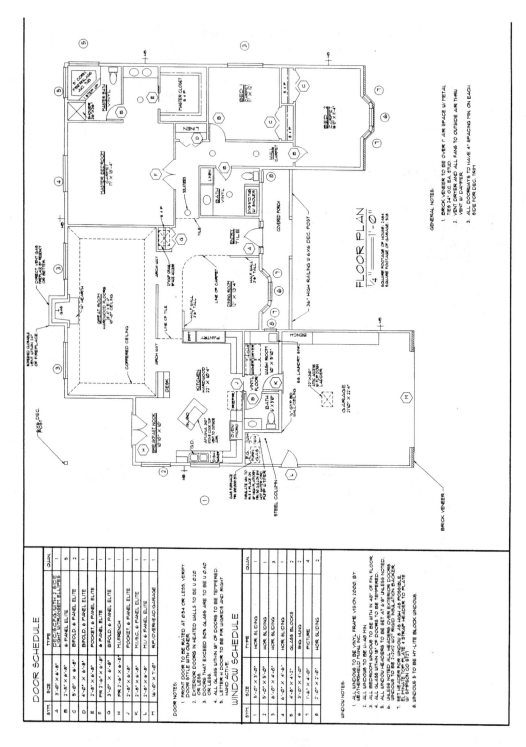

FIGURE 5–4d Pressing the Open button in the Select File dialog box will open the specified drawing. (Courtesy Tereasa Jefferis.)

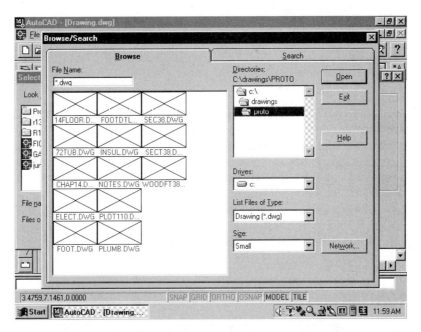

FIGURE 5–5 Pressing the Find File... button in the Select File dialog box will display the Browse/Search display and allow files to be located. Drawings created in versions older than Release 13 will only be displayed as a box. The current display shows the drawings names for each of the files in the PROTO folder. The PROTO folder is contained in the DRAWINGS folder of the hard drive (C:\).

during the drawing session, and saving on the floppy and the hard drive at the end of the drawing session. Many professionals save on the hard drive and make two floppy backups, which are each stored in different places. The open drawing file should be saved to the hard drive approximately every 15 minutes. When the drawing session is done, the file should again be saved to update the existing file on the hard disk. The file should also be saved on another source such as a floppy disk. This is especially true for students who must transport their drawing between home and school.

No matter where you save your drawing, the most important thing to remember is to save often. Saving every 15 to 20 minutes will help you avoid timely redraws if the power supply to your computer is interrupted. As you start saving your drawings on disk, you will have four options: SAVE, SAVEAS, QSAVE, and QUIT. SAVE and SAVEAS can be accessed from the File pull-down menu. Each command can also be entered by keyboard at the command prompt.

The SAVE Command

Using the SAVE command will store your drawing in a selected location but still leave it on the screen. This will allow you to continue working on the drawing. If your computer should crash, only work done after the last save would be lost. To save a drawing, type **SAVE** [enter] at the command prompt, pick the Save icon in the standard toolbar, or pick Save from the File pull-down menu. Each will produce a similar result and produce the Save Drawing As dialog box shown if Figure 5-6a. By picking the arrow on the right side of the Save in: edit box, the contents of the Windows Explorer will be displayed similar to Figure 5–4b. When the destination drive has been selected, the contents of the specified drive will be listed. Enter the desired file name in the File name: edit box and press the Save button. Guidelines for file names will be given later in the chapter. For now, limit the file name to eight letters. The .DWG drawing extension is added to the drawing file and it is saved to the designated drive. This will be the desired method of saving for most AutoCAD users. In Figure 5–6b three other alternatives for saving a file are presented. These alternatives allow drawings created in Release 14 to be saved so that the drawing file will be compatible with different versions of AutoCAD. These options can be especially useful for students whose home machines are not as current as those at school or for architectural and engineering firms with different versions of the software. To save

FIGURE 5–6a   The Save Drawing As dialog box is displayed by selecting the Save icon in the standard toolbar, picking Save or Save As from the File pull-down menu or by typing **SAVE** [enter] or **SAVEAS** [enter] at the command prompt. The dialog box is only displayed the initial time the drawing is saved. The second time the drawing is saved, the SAVE will occur automatically using the name that was assigned during the initial save. Current drawing file names are displayed. The name of the file to be saved must be typed in the File name: edit box.

FIGURE 5–6b The Save as type: edit box allows drawings to be saved and be compatable with older versions of AutoCAD.

a drawing created in R14 to be used on a machine running R13, pick the AutoCAD R13/LT95 Drawing (*.dwg) option.

Open a drawing file, draw a few lines and then save the file with a name of JUNK. Picking the floppy disk icon is the fastest method of saving a drawing. As you assign a name, you do not have to type the .DWG at the end of the drawing file since it will be added automatically by AutoCAD. Once the drawing is saved, add a few more lines to the drawing and save the drawing again. If you save the drawing using the Save icon or the File pull-down menu, the Save Drawing As dialog box will not be displayed. The current file of JUNK.DWG is automatically updated. If you enter the command by keyboard, the Save Drawing As dialog box will be displayed allowing a new drawing name to be assigned to the drawing.

The SAVEAS Command

The SAVEAS command is similar to the SAVE command except that it allows the name of the file to be altered. SAVEAS can only be accessed from the File pull-down menu or from the command prompt. Enter **SAVEAS** [enter] at the command prompt. This will produce the Save Drawing As dialog box, seen in Figure 5–6a. The default name will be the name you selected the last time the drawing was saved. If you like that name, pick the Save button. If you want to change the name of the drawing file, enter the new name in the File: edit box.

Be sure that the correct drive destination is current. Note that the destination is listed in the Drives: edit box. To change the current drive, use the arrow beside the Drives: edit box to display the desired drive. Pressing the Save button will save the drawing to the specified drive. When a drawing is saved, it will be placed in the current drive. AutoCAD will assign the extension .DWG or .BAK following the filename to help identify contents of your directory.

QSAVE

Typing **QSAVE** [enter] at the command prompt will save the current drawing using its given name without displaying the dialog box. If QSAVE is used on an unnamed drawing file, the Save Drawing As dialog box of Figure 5–6a will be displayed. With QSAVE, SAVE, and SAVEAS, your drawing will be cleaned up as it is saved. AutoCAD will remove from the drawings data base internal records that have been marked for deletion.

Automatic Saving

By default, the existing drawing will be saved every 120 minutes SAVETIME can be used to alter the time between saves. The time between SAVES can be altered by typing **SAVETIME** [enter] at the command prompt. This will produce the prompt

> Command: **SAVETIME** [enter]
> New value for SAVETIME <120>:

allowing for a value between 1 and 120 minutes to be entered. The valued entered will start the timer when a change is made to the current drawing file. The commands SAVE, SAVEAS, and QSAVE will reset the timer. With a value of 20, your drawing file will be automatically saved every 20 minutes with a name of AUTO.SV$. If this file needs to be accessed, the Windows Explorer can be used to rename the file from a .SV$ file to a .DWG file. SAVETIME can be disabled by entering a value of **0** but this could prove to be a huge mistake if your computer ever crashes.

CREATING AND SAVING A DRAWING TEMPLET

In Chapter 4, the templets contained in AutoCAD 14 were introduced. Just as plastic templets were used by manual drafters to make drawing easier, templets can be used to make each new drawing that you start. The templet can be created using the Start Up dialog box, saved, and then used each time a drawing requiring those parameters is needed. Several of the parameters were introduced in Chapter 4. Other useful values—such as linetype, lettering size and styles, paper size, and dimensioning styles and format—will be introduced in succeeding chapters. Necessary values can be added to the drawing templet for future use. These templets drawings are often referred to as prototype drawings by many professionals. Once created and saved, they can be opened to form the base for a new drawing file and provide a great time savings. Creation methods and suggestions for the contents of a templet drawing will be introduced later in this chapter.

A templet drawing can be created using methods used with any another drawing. Open a drawing and establish the drawing units, limits, snap, and grid values. Special consideration will need to be given to the drawing limits as you establish a drawing templet. In Chapter 4, as the limits were established common paper sizes were introduced. The size of paper that the completed project will be printed or plotted on must be considered as the drawing templet is established. Although the drawing limits that are shown on the screen can be altered at anytime throughout the life of the drawing, establishing limits that match standard paper sizes will aid in the use of the templet. To establish a templet for use with drawings completed at 1/4" = 1'-0", drawing values will need to be multiplied by a factor of 48 to determine the size to be used in the templet. The scale factor for 1/4" = 1'-0" or .25 = 12 is determined by dividing 12 by .25. If the drawing is to be placed on "D" size material, limits would need to be 1152" × 1728" or 96' × 144' . The limits can also be determined by multiplying the desired paper size by the number of feet in one inch of paper space. At a scale of 1/4"=1'-0", an inch of paper will contain 4' of drawing, resulting in the same 96' × 144' limits. Other common limits for "D" material are shown in Figure 5–7.

Once each of the values for the templet have been established, the drawing can be saved and reused whenever suitable values are used. Save the temple using SAVEAS with the Drawing Templet File (*.dwt) option selected . (See Figure 5–6b.)

DRAWING SCALE FACTORS

ARCHITECTURAL SCALE FACTORS				ENGINEERING SCALE FACTORS			
SCALE	LIMITS		TEXT SCALE	SCALE	LIMITS		TEXT SCALE
	18 × 24	24 × 36			18 × 24	24 × 36	
1" = 1'-0"	18 × 24	24 × 36	12	1"= 1'-0"	18 × 24	24 × 36	12
3/4"= 1'-0"	24 × 32	32 × 48	16	1"= 10'-0"	180 × 240	240 × 360	120
1/2" = 1'-0"	36 × 48	48 × 72	24	1"= 100'-0"	1800 × 2400	2400 × 3600	1200
3/8"=1'-0"	48 × 64	64 × 96	32	1"= 20'-0"	360 × 480	480 × 720	240
1/4" =1'-0"	72 × 96	96 × 144	48	1"= 200'-0"	3600 × 4800	4800 × 7200	2400
3/16" =1'-0"	96 × 128	128 × 192	64	1"= 30'-0"	540 × 864	864 × 1080	360
1/8" = 1'-0"	144 × 192	192 × 288	96	1"= 40'-0"	720 × 960	960 × 1440	480
3/32" = 1'-0"	192 × 256	256 × 384	128	1"= 50'-0"	900 × 1200	1200 × 1800	600
1/16" = 1'-0"	288 × 384	384 × 576	192	1"= 60'-0"	1080 × 1440	1440 × 2160	720

NOTE: ALL LIMIT SIZES ARE GIVEN IN FEET.

FIGURE 5–7 The scale that a drawing will be plotted at must be considered as a templet is constructed. The drawing limits will be affected by the scale to be used and the paper size.

Enter the name of the drawing just as you would with any other drawing. A name that describes the scale or paper size such as 14DSIZE or 18CSIZE will be helpful for sorting templets. Once the name has been entered, pick the SAVE button. Provide a brief description of the templet, such as 1/4" templet. This description is displayed whenever you select this templet from the Create New Drawing dialog box. Enter the measurement system as English or Metric for the templet. The final step to save the templet is to pick the OK button. The templet can now be reused by selecting it as a new drawing is opened. Chapter 23 will introduce an easier method of establishing a drawing templet.

Leaving AutoCAD

Once a drawing is saved, you may want to leave the drawing and start another project. The QUIT command will exit the drawing and close AutoCAD. If you have unsaved changes that have been made to the drawing and you type QUIT, you will see the alert box shown in Figure 5–2.

If you pick the Yes button, AutoCAD will save the changes to the default file. Selecting No will exit the drawing, close AutoCAD, and return you to other opened programs. Any changes made since the last SAVE will be destroyed. CANCEL will terminate the QUIT command and return you to the drawing.

Filenames

As you save drawings, you may misname a drawing. If you haven't hit the enter key, just use the backspace key to remove the unwanted name. If the name is entered, the file can be renamed using the Windows Explorer. Occasionally you will assign a drawing file a name that already has been used. In this case you will see a display similar to the one in Figure 5–8. If you proceed, you will destroy the existing file. The name will remain in the folder and your present drawing will be saved using the old filename, but the existing drawing will vanish. This is a fine way to update an old drawing, but be sure you can live without the old drawing.

Back-Up Files

Earlier you discovered that when AutoCAD saves a file it adds an extension of .DWG or .BAK. The .DWG stands for a drawing file. AutoCAD also adds .BAK as an extension to drawing files. The .BAK is the extension for a back-up file. These are created automatically by AutoCAD each time you end a drawing. When a drawing is terminated using SAVE, the existing .DWG file becomes the .BAK file and a new .DWG file is created. The .BAK file may not contain all of your recent changes, but it should be only 15 minutes out of date if you've been saving at proper intervals. What a movitator to keep saving your drawings! Back-up files are also great if your disk has been damaged. The .BAK file can be renamed using the Windows Explorer and then reloaded. If the .BAK file cannot be converted to a .DWG file, or if it was scrambled, this will serve as one last reminder to save your files on more than one floppy disk and store them in separate places.

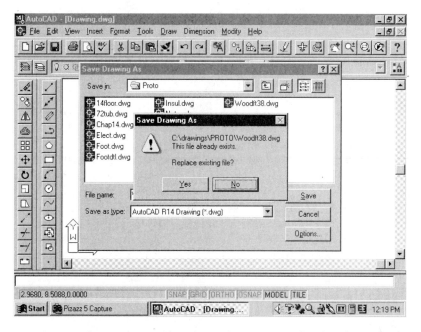

FIGURE 5-8 Attempting to save a drawing using a name that has been assigned to another file will result in this alert message. If the yes option is accepted, the drawing will be saved using the selected name, and the existed drawing will be altered.

MANAGING FILES AND FOLDERS

No office could function if the drafters took their finished drawings and threw them into a room. If a drawing was needed, some poor drafter would have to wade in and hunt through thousands of sheets of paper to find the specific project. Architecture and engineering offices typically use well-organized filing systems to keep track of each project for quick, easy retrieval.

Files

No matter what method is used to save a drawing, information that is to be kept on your disk must be stored in some orderly manner to allow quick, easy retrieval. Disks are ordered in files, folders, and paths. Your drawings will be stored as files and will need names to distinguish them from other files. Filenames for Windows 95 or Windows NT can contain up to 255 characters, although drawings names for files created in AutoCAD can still only be eight characters long. Characters recognized by AutoCAD include $ (dollars), - (hyphen), and _ (underscore). Other characters found on a keyboard, such as %, @, /, \, <, *, may not be used in a filename because they are used within the program to activate special routines.

As you store your files, three-letter extensions will be added to titles by the Windows Explorer or AutoCAD to aid retrieval. For example, a drawing saved by AutoCAD might be filed as FLOOR.DWG. FLOOR is the drawing name, and .DWG is the extension.

Folders

Files should be organized in folders to provide efficient storage. Directories are the equivalent of dividers in a file cabinet drawer. By default your drawings will be stored in the AutoCAD R14 folder. Information about files is kept automatically as files are stored, including files stored, file size, time of creation, and the time of last revision. Folders are usually created by software programs as they are loaded onto the hard drive. Folders can also be created as needed to organize information based on client names, class requirements, or any other similar criteria. A folder can be created by using the Windows Explorer. Folder names should follow the same guidelines used to name drawing files. Names should be used that will quickly define the contents such as DRAWINGS, MAPS, REPORTS, or CLIENTS.

Folders can be placed inside other folders to better organize information. Subfolders can be created and named using the same command and guidelines used to create a folder.

Figure 5–9 shows an example of how folders are set up within folders created by the Windows Explorer in a hierarchical method. This use of paths in saving material will help in retrieval.

Office Procedure

Many offices keep each client in separate folders or disks and label the disk by the client name. Drawings are then saved by contents such as floor, found, elev, sects, specs, or site. Some offices assign a combination of numbers and letters to name each project. Numbers are usually assigned to represent the year the project is started, as well as a job number with letters representing the type of drawing. For example, 9853fl would represent the floor plan for the fifty-third project started in 1998. Another common practice is to name the drawing file by the page number it will occupy in the drawing set based on AIA (American Institute of Architects) page numbering guidelines. When this system is followed, a drawing file named A4.01 would represent a detailed floor plan. Common page numbers and file names include:

A0.01   Index, symbols, abbreviations, notes, and location maps

A1.01   Demolition, site plans and temporary work

A2.01   Plans, and schedules such as room material, door, or windows and keyed drawings

A3.01   Sections and exterior elevations

A4.01   Detailed, large-scale floor plans

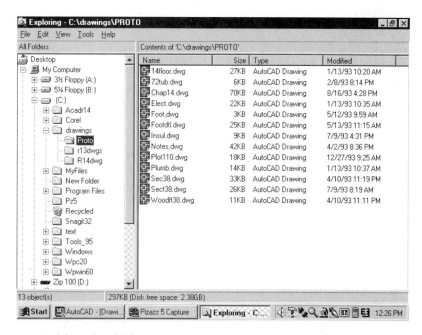

FIGURE 5–9 Folders should be created to contain similar drawing files to ease retrieval. Notice in the C: drive all drawings are stored in the DRAWING folder. The DRAWING folder is further broken down into folders to ease retrieval.

A5.01 Interior elevations

A6.01 Reflected ceiling plans

A7.01 Vertical circulation drawings such as stair, elevator, and escalator drawings

A8.01 Exterior details

A9.01 Interior details

If plans for a five-level building were to be drawn, each floor could be saved as a separate drawing file by using drawing names such as:

A2.01 —level one

A2.02 —level two

A2.03 —level three

A2.04 —level four

A2.05 —level five

A few offices will save an entire house plan in one file. The drawings are separated by layers, which will be discussed in Chapter 15. The drawback to saving large amounts of information in one file is that more time is needed to load and process the information. Chapter 23 will introduce External Reference as another method of saving drawing files.

STORAGE PROBLEMS

One of the great thrills for a new CAD operator is seeing the message, "DISK FULL." It's not a serious problem, but it is discouraging. As a drawing is started, AutoCAD examines the disk to find a suitable storage space. Because AutoCAD creates space for several files, it will need to perform expected drawing commands and there may not be enough space to create the intended drawing. In the middle of a command you may find that a DISK FULL error is displayed. Typically your drawing will be terminated, but all work up to that point will be saved. You will typically only encounter this problem if you save to a floppy drive in the middle of a drawing session.

Occasionally a full disk will cause FATAL ERROR to flash across the screen just before your machine destroys two hours worth of drawing. Sure, you know to save every 15 minutes, but sometimes you've got to learn the hard way. The unfortunate aspect of a fatal error is that it may disrupt or destroy other files on your floppy. You'll find that you typically only have FATAL ERROR when you're 99 percent done with a project.

Correcting File Errors

No matter how carefully you follow the guidelines in Chapter 1 for handling a disk, they will occasionally become corrupted. The technical terms is sick. AutoCAD provides the AUDIT and RECOVERY commands in Utilities of the File pull-down menu to attempt to make the corrupted drawing files useable.

Examining a Drawing File with Audit. This command can be used to examine the current drawing file for errors and to correct any errors that exist. These errors are not spelling mistakes or drawing errors, but errors that occasionally are made by the software. This command will detect errors, generate extensive descriptions of the problems, and recommend actions to correct the errors. Audit can be used by selecting Audit of Drawing Utilities File in the pull-down menu, or by entering **AUDIT** [enter] at the command prompt. The command sequence is:

> Command: **AUDIT** [enter]
> Fix any errors detected? <N>

Responding with a **N** [enter] will produce a report of the file and list the errors, but will not repair the errors. The final listing of the report resembles:

> Total errors found 3   fixed 0

Responding **Y** 〔enter〕 at the prompt will produce a report of the file and list the errors. The listing resembles:

3	Blocks audited
Pass 1	12 entities audited
Pass 2	4 entities audited
Total errors found 2	fixed 2

With the AUDITCTL system variable set to 1, an Audit report file will be created using the filename with .adt as the extension. This file will be placed in the same directory containing the original drawing file and will list all of the functions required to fix file errors.

Recovering a Drawing. If errors are discovered that cannot be repaired by AUDIT, the RECOVER command can be used to try and salvage the file. RECOVER can be accessed by selecting the Recover... option from Drawing Utilities of the File pull-down menu. AutoCAD will automatically attempt to repair damaged drawings as they are opened. If the drawing can't be opened, a message will be displayed that the drawing must be recovered. Once the file to be recovered has been selected, AutoCAD will begin a series of diagnostic procedures. When the process is complete, a listing of problems and the corrective measures taken by the program will be listed. Generally the recovery process will open a damaged drawing. If the RECOVERY command fails, it's time to get out the back-up file. This is one last opportunity to remind you that no matter how fast and good computers have become, they occasionally fail. They generally only fail at times of great stress such as the end of a school term or the due date of a big project. Back-up files can greatly reduce the stress of a file that can't be recovered.

FORMATTING A FLOPPY DISK

If you purchased an unformatted disk, you will need to format it before information can be stored on it . Most disks sold are now preformatted. Occasionally information on a disk will become unreadable due to poor disk care. Although unusable files can be deleted using the Delete option of the File pull-down menu in the Windows Explorer, another common method of removing information on a disk is to format the disk. Formatting a disk places the disk operating instructions, or DOS, on the disk. The format process divides the disk space into tracks and sectors and then checks for preexisting damage to the disk.

A disk can be formatted by selecting My Computer from the Start menu. This will produce the menu shown in Figure 5–10. Select the icon that represents the drive containing the disk to be formatted. Typically this will be the 3½" Floppy (A) drive icon. With the icon highlighted, click the File pull-down menu. Picking Format... displays the Format Dialog box shown in Figure 5–11. Before the simple command is used, it is extremely important to understand what it does. FORMATTING REMOVES ANYTHING THAT IS STORED ON THE DISK. This is fine if you take

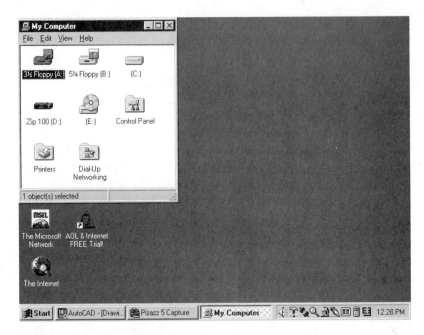

FIGURE 5–10 A disk can be formatted to remove unwanted material by selecting the icon of the drive containing the disk. With the icon highlighted, open the File pull-down menu.

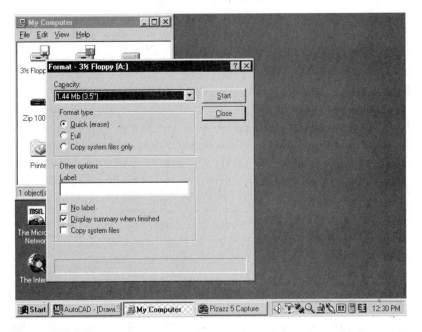

FIGURE 5–11 The Format dialog box allows type and size of the disk, as well as the type of format to be preformed to be selected.

a new disk out of the box and format it. Formatting takes an empty disk and makes it usable. If a disk with several weeks of drawings is formatted, you'll have an empty disk! Before you format a floppy disk, determine the contents using the Windows Explorer. If the disk has not been formatted, you will see the following display:

No files found

If material is on the disk, the contents will be listed similar to Figure 5–9. Now you can determine whether you want to save this information or remove the contents by formatting.

There are three important keys to formatting a disk. Always verify the drive letter that is displayed. In this case the format will affect the disk in the A drive. DO NOT FORMAT THE HARD DRIVE UNLESS YOU'RE SURE YOUR INSTRUCTOR OR BOSS WANTS TO LOAD A NEW OPERATING SYSTEM. Once the drive is selected, verify the capacity of the disk being formatted. The default option is for 1.44 MB. Picking the option arrow displays the option for 720K. In the Options box, you are given the option of labeling the disk. You can provide a name for the disk if you know what you will be storing on the disk. If you're unsure of what the disk will contain, skip this option. Disk names can be up to 256 characters, and names can include spaces. You also have the option of selecting a quick or full format. During the formatting process, the disk is examined for bad sectors. Selecting the quick format option eliminates the scanning function. Generally the extra few seconds required to scan the disk will prove worthwhile. Once all options have been selected, pick the OK button to begin the formatting process. Picking the Cancel button will allow you to rethink the need to format the disk. When the formatting process is complete, a display will be given showing the total disk space, the available disk space and a prompt for formatting more disks. Picking the No button ends the process, and allows you to continue with another task. Selecting the Yes button allows you to format other disks.

CHAPTER 5 EXERCISES

E-5-1. Format a blank floppy disk.

E-5-2. Start a new drawing. Draw a square, a rectangle, and a triangle. Save the drawing on your floppy using the name EXER5–2.

E-5-3. Start a new drawing called EXER5–3 that has the same drawings contained on EXER5–2. Add another circle and a rectangle. Save as 5–3EXER.

E-5-4. Edit exercise 5–3EXER by adding any figure that you would like. Save as EXER 5–4.

E-5-5. Get a directory of your floppy disk; list the following:
   a. The contents.
   b. The time your first exercise was finished.

     c. The date of your last exercise.

     d. Number of bytes used.

     e. Number of bytes free

E-5-6. Open the Archeng.dwt templet.

    1. What is the current snap resolution? X _____ Y _____

    2. What is the current grid setting? X _____ Y _____

    3. What layer will new lines be added to? _____

    You will be using this exercise to set up a drawing for a plan view of a room that will be drawn at 1/4" = 1'-0" and plotted on "D" size paper. Two-place decimal angles are desired.

    4. Set the units at the proper value. What option number will be required? _____ What was the default? _____

    5. The smallest fraction to be worked with will be ¼". What denominator should be used? _____

    6. Give the current system for measuring angles. _____

    7. Will the current setting allow for two-place decimals? _____

    8. In which direction will angles be drawn? _____

    9. What are the current drawing limits? _____

    10.What limits will be needed to achieve the listed parameters of this exercise? _____

    11.What is the current snap value? _____

    12.Set the snap value to 6".

    13.What is the current grid default value? _____

    14.Set the grid to 12".

    15.List two other methods of listing the 12" value.

    SAVE THIS EXERCISE ON YOUR FLOPPY DISK AS 1–4.DWT.

E-5-7. Start a new drawing.

    You will be using this exercise as a base for a drawing of a small subdivision. The drawing will be drawn at a scale of 1" = 30' and plotted on "D" sized paper. North will be at the top of the paper (12 o'clock). Set all applicable defaults to two-place decimals.

    1. What unit setting should be used for this drawing? _____

    2. Set the limits at the proper value. What option number will be required? _____

    3. Many of the property lines are described with angles measured in degrees, minutes, and seconds. Set the angle measurement system in the appropriate system. What default number was picked? _____

CHAPTER 5 QUIZ

1. What prompt will be displayed if you try to read an empty floppy disk?

2. What happens to existing information on a disk when it is formatted?

3. What are the capacity options for saving to a floppy disk?

4. You're working on the fifth level of a multilevel project. Suggest three different names that could be used to save the file.

5. What command will store the drawing on a disk but leave the drawing on the screen?

6. How many characters may be used for a volume label?

7. What command will store the drawing on a disk and terminate the drawing?

8. List three different methods to access the SAVE command.

9. You have a drawing on your floppy disk labeled SECTION. You draw a roof detail and start to save it as SECTION. What options will AutoCAD give you and how will they affect your drawing?

_____

_____

_____

10. List six forms of information listed when the dialog box is used to save a drawing.

_____

_____

11. What two extensions are assigned to a drawing file?

_____

12. What does QUIT do differently from SAVE?

_____

_____

13. What is a fatal error?

_____

_____

14. Give two reasons for backing up a drawing in multiple locations.

_____

_____

15. List two reasons why you might receive an alert message while saving a drawing.

_____

_____

16. You've opened an existing drawing named FLOOR.DWG, made some changes, and now want to save the drawing while keeping FLOOR.DWG intact. How can this be done?

_____

_____

17. You've saved a file named LOST.DWG some where on your hard drive, but you're not sure where. How can you find the file?

_____

_____

18. Explain the difference between SAVE and SAVEAS.

_____

_____

19. What is the default value for SAVETIME and is this value adequate?

_____

_____

20. List 10 factors that can be added to a templet drawing.

_____

_____

6

COORDINATE ENTRY METHODS

· ·

As you enter AutoCAD, you are using a system of coordinates called the world coordinate system. For two-dimensional drawing, points on a plane can be entered using a Cartesian coordinate system. Points for objects will be entered by keyboard or a pointing device to provide coordinates. These coordinates will be based on AutoCAD's user coordinate system and can be entered using absolute, relative, and polar coordinate entry methods.

USER COORDINATE SYSTEM

By now you should recognize the drawing screen, toolbars, pull-down menu, status line, and command prompt. As you start to complete drawings in this chapter, it will be important to understand the user coordinate system (UCS) of AutoCAD. The UCS is based on a Cartesian coordinate system dividing space into four quadrants. Figure 6–1 shows the division of space based on the Cartesian coordinates. Points in these quadrants are located in turn by using their points in a horizontal (X) and vertical (Y) direction along the plane. Points are always specified by listing the X coordinate followed by the Y coordinate.

Figure 6–2 shows the relationship of the drawing screen to the Cartesian coordinates. As you begin to draw, AutoCAD will ask you to specify a "From point." This may be entered by moving the cursor to any point on the screen and pressing the pick button. As you move to the "To point," the coordinate display at the left end of the status line will reflect the cross-hair movement. To activate coordinate display, double-click the coordinate display button, or use the F6 key, or CTLR +D.

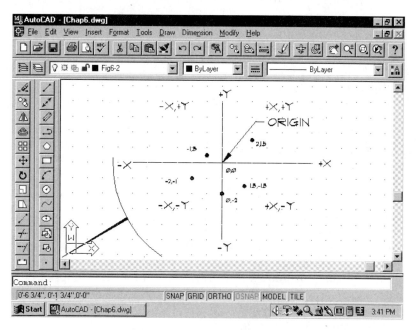

FIGURE 6–1 The division of space using the Cartesian coordinate system. The intersection of the X and Y coordinates is assumed to be 0,0 which is the lower left corner of the screen. The grid has been moved to the center of the screen to represent negative X and Y locations.

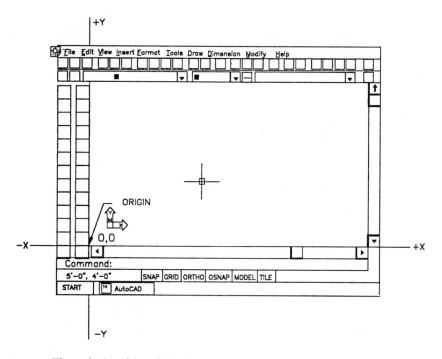

FIGURE 6–2 The relationship of the drawing screen to Cartesian coordinates with the Limits set at the lower left = 0,0, and the upper right values set as two positive values.

Because the fields of engineering and architecture tend to deal with very specific locations, a means of controlling the "To point" is often desirable. This can be done by using absolute, relative, or polar coordinates for point-entry methods.

ABSOLUTE COORDINATES

The absolute coordinate system locates all points from an origin assumed to be 0,0. The 0,0 origin point is assumed to be the same 0,0 point that the limits, snap, and grid are based on, although it can be changed. The axis for the X and Y coordinates intersects at 0,0. Each point on the screen is located by a numeric coordinate based on the original axis. Objects are entered by entering the coordinates of X and Y separated by a comma. In the example in Figure 6–3 a line is drawn between 2,2 and 2,4. Notice the coordinates that were typed in at the command line or compare the line with the grid pattern. The next "To point" is 4,4 based on the cursor location and the current coordinate display. To complete the drawing in Figure 6–4 would require the following command sequence.

Command: **L** (enter)
From point: **4,1.5** (enter)
To point: **6,1.5** (enter)
To point: **6, 3.5** (enter)

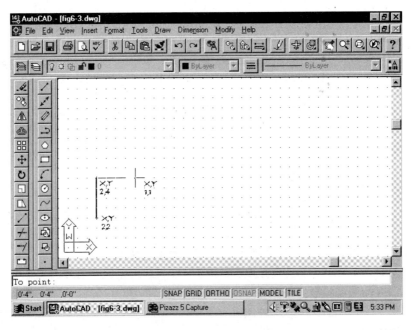

FIGURE 6–3 With absolute coordinates, and the 0,0 location as the lower left corner of the drawing screen, the line drawn starts at 2" over (+X) and 2" up (+Y) and extends to a location 2" over (+X) and 4" up (+Y).

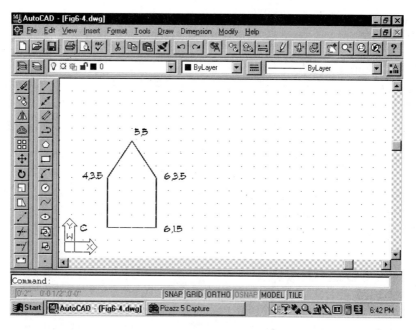

FIGURE 6–4 A simple shape and the absolute coordinates required to describe it.

To point: **5,5** [enter]
To point: **4,3.5** [enter]
To point: **C** [enter]
Command:

Notice that **C** [enter] was entered as the last command rather than numeric coordinates. **C** [enter] will end the LINE command sequence at the point where the line sequence was begun.

Because absolute coordinates are based on an origin of 0,0, they are not often used for most architectural and engineering projects. Many CAD users find relative coordinates more suited to a construction project.

RELATIVE COORDINATES

Relative coordinates locate "to" points based on the last point instead of the 0,0 origin. Relative coordinates are entered using a similar method as absolute coordinates. Absolute coordinates were entered X,Y. Relative coordinates are entered @ X,Y. In addition to preceding the coordinate value with the @ symbol, relative coordinates require you to be mindful of positive and negative values.

Figure 6–5 shows an object drawn using relative coordinates. Positive coordinates will move the cross-hairs to the right or above the origin point. Negative coordinates will move the cross-hairs down or to the left of the last entry point. The object in Figure 6–5 was drawn using the following command sequence:

Command: **L** enter
From point: **4,1.5** enter
To point: **@2,0** enter
To point: **@0,2** enter
To point: **@–1,1.5** enter
To point: **@–1,–1.5** enter
To point: **@–0,–2** enter

The first point was entered using absolute coordinates, with the following coordinate points entered by relative coordinates. The final coordinates could have been omitted and replaced with **C** enter.

POLAR COORDINATES

Polar coordinates are similar to relative coordinates. Polar coordinates use the last entry point as the origin and combine a distance and an angle. This is useful on

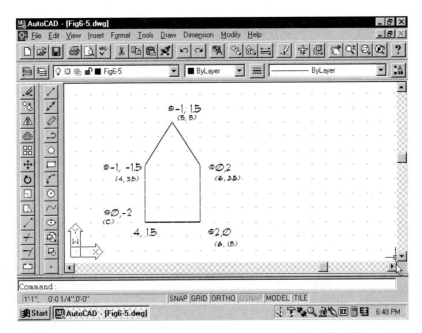

FIGURE 6–5 Relative coordinates describe the location of a point based on the last point. Relative coordinates must be preceded by the @ symbol. The value @2,0 would extend a line 2" to the right with no vertical rise. Relative coordinates are a common entry choice for many projects.

drawings such as sections for drawing a roof or site plans with angled property lines. Polar entries are entered as:

@3<27.5

This would produce a line 3" long at a 27½° angle, as shown in Figure 6–6. To draw the opposite side of this "roof truss" would require @3<332.5. Remember, polar coordinates are based on the current origin of the cross-hairs. In Figure 6–7 the location of line two at 332½° is determined by subtracting 27½° from 360°.

To complete the triangle started in Figure 6–7, type **C** ⌷enter⌷. C ⌷enter⌷ may be used to close a parallelogram begun using absolute or relative coordinates and will return the cross-hairs to the original starting position. Figure 6–8 shows an object drawn using polar coordinates. The required command sequence would be:

Command: **L** ⌷enter⌷
From point: **4,1.5** ⌷enter⌷
To point: **@2,<0** ⌷enter⌷
To point: **@2<90** ⌷enter⌷
To point: **@1.8<124** ⌷enter⌷
To point: **@1.8<236** ⌷enter⌷
To point: **C** ⌷enter⌷

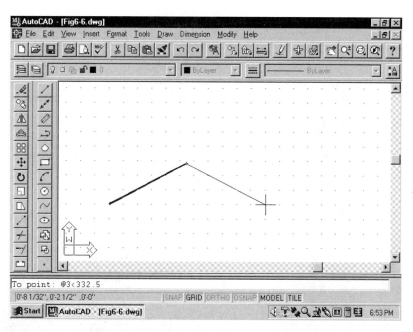

FIGURE 6–6 Polar coordinates describe a point relative to the last position using a length and an angle. The @ symbol precedes the length, and the < symbol precedes the angle specification in degrees. Polar coordinates are useful in drawing angled sides of a parcel of land.

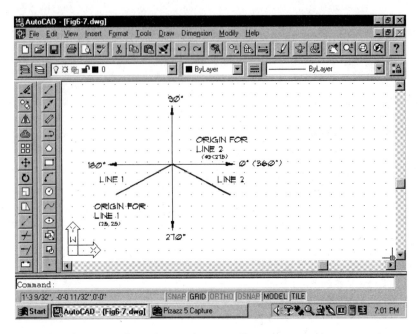

FIGURE 6–7 Specifying angles using polar coordinates.

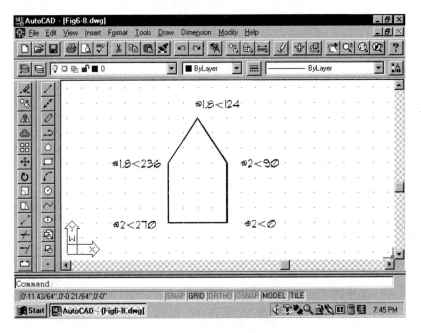

FIGURE 6–8 The use of polar coordinates to establish a simple shape.

CHAPTER 6 EXERCISES

Use the disk that was formatted in Chapter 5 to save the following exercises. Set grid, snap, units, and units of angle measurement to best suit the given project. Set the limits at the default setting.

E-6-1. Draw the following outline using the coordinates that are provided.

Point of beginning: 1,1

point b   @4.2426<45
point c   @3.1623<342
point d   @2.5<0
point e   @4.0311<120
point f   @4.2720<159
point g   @3.9051<230
point a   @4.000<270

Save your drawing on your floppy disk as E-6-1.

E-6-2. Draw the following outline using the coordinates that are provided.

Point of beginning: 2,1

point b   @2.5<N80dE
point c   @3.5<N38d23'15"E
point d   @2.5<S40d20'40"E
point e   @1,1
point f   @1,–1
point g   @4<N
point h   @–6,1
point i   @3.5<S39d7'W
point a   back to the true point of beginning

Save your drawing on your floppy disk as E-6-2.

E-6-3. Draw the following outline using the coordinates that are provided. Save the drawing as E-6-3.

Point of beginning: 9,9

point b   @–5,–2.5
point c   @0,–1.5
point d   @4<S39d48'W
point e   @–1.5,–2
point f   4.5,1.5
point g   @1.5<90
point h   @2<S65dE
point i   @2.5<N30d15'15"W
point j   @5.5<S60d20'40"E
point k   @1,1

point l   @–2.5,1
point m   @–.5,–.5
point n   @3<N61d25'10"W
point o   @3.8125<E
point a   @4.5<N to the true point of beginning

CHAPTER 6 QUIZ

Use the drawing below to provide the answers for Problem 1 by providing absolute coordinates. Axis = 1", Grid = .5", Snap = .25"

1. a. 2.5,2   e. _____   i. _____

   b. _____   f. _____   j. _____

   c. _____   g. _____   k. _____

   d. _____   h. _____   l. _____

                                                a. _____

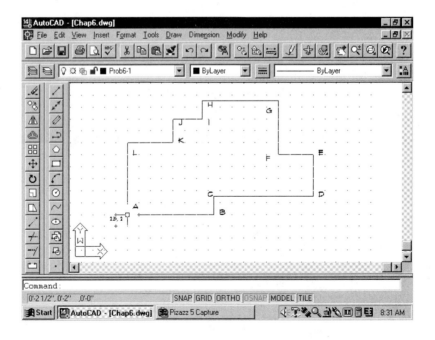

Use the drawing from Problem 1 to provide the answers for Problem 2 by providing relative coordinates.

2. a. 2.5,2   e. _____   i. _____

   b. _____   f. _____   j. _____

   c. _____   g. _____   k. _____

   d. _____   h. _____   l. _____

                                                               a. _____

Use the drawing from Problem 1 to provide the answers for Problem 3 by providing polar coordinates.

3. a. 2.5,2   e. _____   i. _____

   b. _____   f. _____   j. _____

   c. _____   g. _____   k. _____

   d. _____   h. _____   l. _____

                                                                a. _____

Use the drawing below to provide the answers for Problem 4 by providing relative coordinates.

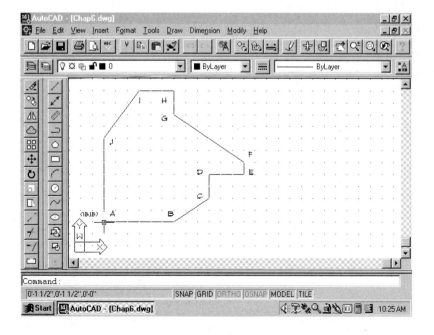

4. a. 1.5,1.5   e. _____   i. _____

   b. _____   f. _____   j. _____

   c. _____   g. _____   a. _____

   d. _____   h. _____

Use the drawing from Problem 4 to provide the answers for Problem 5 by providing absolute coordinates.

5. a. 1.5,1.5   e. _____   i. _____

   b. _____   f. _____   j. _____

   c. _____   g. _____   a. _____

   d. _____   h. _____

Use the drawing from Problem 4 to provide the answers for Problem 6 by providing polar coordinates. Use relative coordinates where insufficient information is available to determine polar entries.

6. a. 1.5,1.5   e. _____   i. _____

   b. _____   f. _____   j. _____

   c. _____   g. _____   a. _____

   d. _____   h. _____

7. What entry method uses the X,Y format to enter coordinates? _____

8. What entry method uses the @X<Y format to enter coordinates? _____

9. What entry method uses the @X,Y format to enter coordinates? _____

10. Several entry coordinates are listed below. Circle the coordinates that will not work.

   a. 3,4   f. −3,<27   k. 27'6

   b. −3,−4   g. @1'−0",<27.5   l. 25',13'6

   c. 3,4   h. @ 27'6<n3615'e   m. 14'<W30N

   d. @3,<45   i. @ 36'<27.5   n. 20'4,S27d36,15E

   e. @−3,−5   j. N27d36'27"e,15'   o. @3'6,7'4

11. Can an angled line be drawn with snap and grid both on? _____

12. Describe the line that is formed between 2,0 and 4,0.

_____

_____

13. Describe the line that is formed between 2,0 and 4,2.

_____

_____

14. List two methods to activate the coordinate display.

_____

_____

15. What settings must be changed to use absolute, relative, and polar coordinates on the same drawing?

_____

_____

_____

CHAPTER

7

DRAWING AND CONTROLLING LINES

Now that you're able to set up drawing units and limits, basic drawing components can be explored. The main components of the drawings you will be working with will consist of lines. This chapter will introduce you to methods for making effective use of the LINE command as well as introducing you to using varied linetypes. As varied linetypes are added to a drawing, a method of managing these linetypes must also be considered. Most professional users of AutoCAD sort linetypes by the use of layers. The LAYER and LINETYPE commands will be introduced in this chapter, with a detailed discussion presented in Chapter 15. In addition to exploring the methods of drawing lines, it will be useful to explore a few of the basic methods available to edit a drawing.

AN INTRODUCTION TO EDITING DRAWINGS

This chapter will introduce two methods of removing drawing errors from a drawing file. These methods include the ERASE and OOPS command. The REDRAW command will also be introduced as a method of cleaning the drawing display

ERASE

Since Chapter 3 you have been experimenting with the LINE command. You might have even made an error or two. The ERASE command will allow you to remove unwanted drawing objects. To remove objects type **E** enter or **ERASE** enter. At the command prompt or pick the Erase icon from the Modify toolbar. This will display the prompt:

Command: **E** [enter]
Select objects:

As the prompt is displayed, the cross-hairs turn into a pick box as seen in Figure 7–1. The dimension line for the depth of the footing has been selected for removal. The pick box can be used to select lines to be removed. Place the box on the line to be removed and pick it with the pick button. The selected line will now appear highlighted as in Figure 7–2. You may pick an unlimited number of lines to be erased. The command line will continue to show "Select objects:" until the [enter] key is picked. Pressing [enter] will remove the selected lines and restore the command prompt as seen in Figure 7–3.

ERASE allows for the last drawing entity to be erased without having to pick the entity. This can be done by typing **E** [enter] **L** [enter]. The command line will appear as:

Command: **E** [enter]
Select object: **L** [enter]

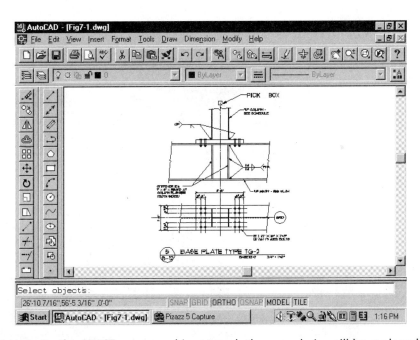

FIGURE 7–1 As the ERASE command is entered, the cross-hairs will be replaced by a pick box. Placing the pick box on the desired item to be erased and picking it, will select that object for removal. Notice that the line in the center of the column has been selected for removal. (Courtesy Van Domelen / Looijenga / McGarrigle / Knauf Consulting Engineers)

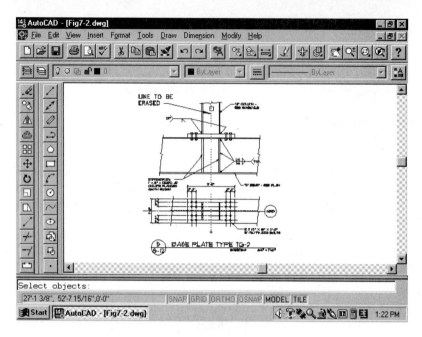

FIGURE 7–2 Once an item has been selected for removal from the drawing, it will be displayed as short dashed lines. (Courtesy Van Domelen / Looijenga / McGarrigle / Knauf Consulting Engineers)

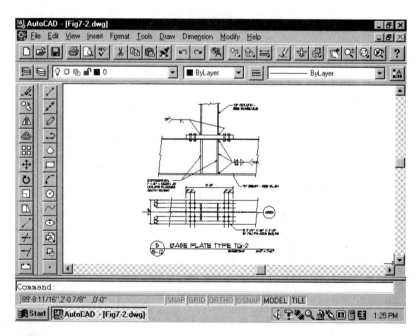

FIGURE 7–3 Pressing the [enter] key or the right mouse button will remove the selected object from the drawing screen. (Courtesy Van Domelen / Looijenga / McGarrigle / Knauf Consulting Engineers)

Select objects: [enter]
Command:

As the command line is restored, the last entity drawn is eliminated. If you would like to remove the last sequence of entities type:

Command: **E** [enter]
ERASE
Select objects: **L** [enter]
Select objects: (*Pick object.*)
1 found
Select objects (*Pick object or* [enter] *to end sequence.*) [enter]
Command:

Picking lines to be erased can be continued indefinitely until the entire drawing is removed; however, this process is not the most efficient way to remove multiple objects. Chapter 11 will provide in-depth coverage of ERASE and other editing options.

OOPS

Occasionally an object will be erased by mistake. The OOPS command will restore the object removed by a previous ERASE command. Once another command is started, OOPS cannot be used to retrieve the erased entity. Chapter 11 will provide other options for removing elements that cannot be removed by OOPS.

Cleaning the Screen

One of the neat features of AutoCAD is its many variables. One of these variables that can be used to mark space is a blip. A *blip* is a marker placed to identify the ends of lines or a selection point. If you pick a line to be erased, a temporary blip in the shape of a plus sign (+) will be placed where the line to be erased is picked. The default setting for AutoCAD is for blips to be OFF. The placement of blips is controlled by typing **BLIPMODE** [enter] at the command prompt. If the variable is set to ON, each time a drawing is edited, a blip is placed on the screen. When the erase command is used, blips are left on the screen marking the end points of the removed line. As these blips accumulate on the screen, they may become a distraction. The blips may be removed by toggling the GRID to on/off/on. The screen may also be cleaned by entering **REDRAW** [enter] or entering **REGEN** [enter], which will be covered in Chapter 10. For now, type **R** [enter] to clear the screen. The **R** represents the command REDRAW.

LINES

The basic components of every drawing are lines, which are used to represent topography, grades, pipelines, easements, walls, roads, property lines, and an endless list of other construction materials. Figure 7–4 shows the lines required to represent a

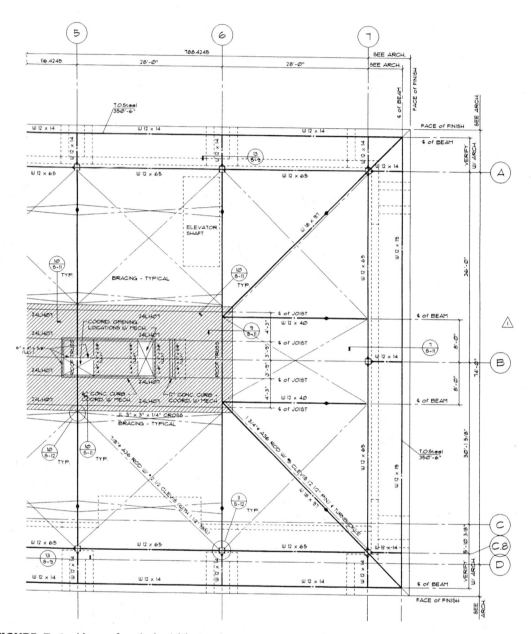

FIGURE 7–4 Lines of varied widths and patterns are used on most construction drawings to represent varied materials. Methods of changing line type and width will be discussed in Chapter 7, 13, 14 and 15. (Courtesy Van Domelen / Looijenga / McGarrigle / Knauf Consulting Engineers)

portion of a framing plan for a multilevel steel structure. Although many of these lines are represented with different line patterns or thicknesses, each can be drawn using the LINE command. Lines are used to represent different types of information within a drawing. Common types of lines used on construction drawings can be seen in Figure 7–5. Establishing varied linetypes will be introduced later in this chapter, and will be discussed in detail in Chapters 13, 14, and 15.

Lines are drawn by typing **L** [enter] or **LINE** [enter] at the command prompt and then entering the desired coordinates or endpoints. The sequence for using the LINE command using coordinates would be:

Command: **L** [enter]
LINE From point: **3,3** [enter]
To point: **@3,0** [enter]
To point: [enter]

This will produce a 3"-long horizontal line. In addition to the horizontal line, a rubberband cursor is displayed connecting the last line drawn to the cross-hairs. This

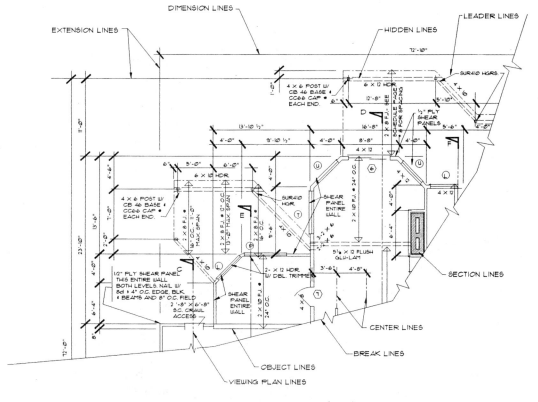

FIGURE 7–5 Common linetypes used on a construction drawing.

can be seen in Figure 7–6. The rubberband line will help you to visualize the placement of the next line to be drawn and to experiment with its placement. The rubberband line will remain as the cross-hairs are moved until the next "To point" is selected or the line command is terminated. To terminate the line command press the right button on the mouse or [enter]. In addition to providing specific endpoints for lines, remember that lines can also be entered using absolute, relative, or polar coordinates. As you start to experiment with drawing lines, draw objects combining all entry methods.

Alternate Line Entries

Other than typing **L** [enter] at the command line, the next fastest way to start the LINE command is to pick the LINE icon with a single click from the Draw toolbox. Lines can also be drawn by selecting the LINE command from the Draw pull-down menu. Each was introduced in Chapter 3. By selecting Line, the menu will disappear and the command prompt will show:

    Command: _line From point:

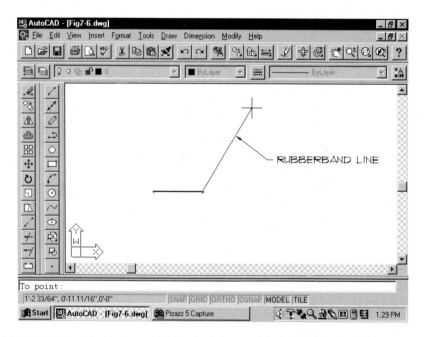

FIGURE 7–6 Notice that when a "To point" is provided, a rubberband line extends from the "To point" to the present cross-hair location. This line indicates the line that will be drawn if the current cross-hair location is used for the next "To point."

Notice that four other options for lines are presented in the Draw menu—the Construction Line, Ray, Multilines, and Polylines commands. Each will be discussed in detail in Chapter 13 and 14. A fifth method of drawing lines is SKETCH. The SKETCH command allows free hand drawing methods to be reproduced and will be introduced in Chapter 14. Creating lines with the Sketch command will be introduced in Chapter 14.

Multiple Lines

No matter what method is used to enter the Line command, the process works the same. You'll be prompted for a From point and a To point. The command will continue to prompt for additional line segments until the command is terminated by picking the right mouse button. Each line that is drawn is an individual drawing entity, even though the lines are connected.

Although not listed in a menu, AutoCAD allows for an unlimited number of unconnected line segments to be drawn. This is done by typing **MULTIPLE** (enter) at the command prompt. The multiple followed by a space and the name of a command can be used to repeat most commands. The sequence for line would be:

Command: **MULTIPLE** (enter)
Multiple command: **L** (enter)
LINE From point:

Notice that you'll be allowed to draw an unlimited number of connected segments. Once the LINE command is terminated, the command prompt returns to:

to point: (enter)
LINE From point:

allowing a new line segment to be drawn. To terminate the sequence for drawing line segments press the ESC key. This command could be useful if you're creating a drawing consisting of mostly straight line segments such as the plan view seen in Figure 7–7. An alternative to MULTIPLE is to use the LINE command to draw the desired segment and then pressing (enter) to repeat the last command. The sequence to continue the line command is:

Command: **L** (enter)
LINE From point: (*Select a point.*)
To point: (*Select a point.*)
To point: (*Select a point.*) (enter)
Command: (enter) (*Enter restarts the last command.*)
LINE From point: (*Select a point.*)
To point: (*Select a point.*)
To point: (*Select a point.*) (enter)
Command:

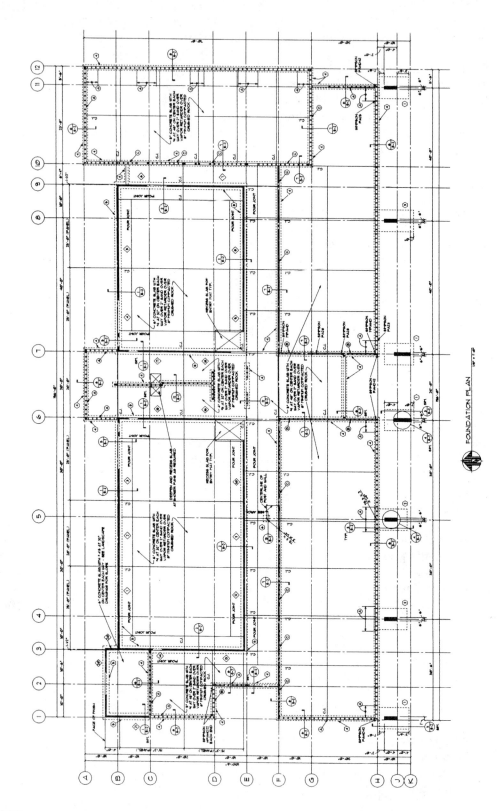

FOUNDATION PLAN

FIGURE 7-7 Because many construction drawings consist of line segments, the MULTIPLE LINE command can frequently be used. This command allows segments to be drawn and repeats the command so that other attached segments can be drawn. (Courtesy Tom Kuhns, Michael & Kuhns Architects, P.C.)

Line Continuation

Once the LINE command has been terminated, it can be resumed from the end of the last line segment that was drawn. This can be done even if other commands were used. The horizontal line segment drawn in Figure 7–8 was drawn prior to the circle. The circle was drawn by typing **C** enter at the command prompt and picking a center and radius point. This command will be further explored in the next chapter. By entering **L** enter and then entering enter for the "From point:" the new line will begin at the end of the last line segment. Picking @ from the screen menu will have the same effect as the continue option. The @ option returns the cursor to the end point of the last line and uses that point as the starting point for a new line segment.

Line Closure

Although it was mentioned in the last chapter, it is important to remember that if a sequence of lines will form a closed polygon, the last line can be drawn without providing endpoints or coordinates for the termination point. Rather than struggling to hit the exact end point of the first line, typing **C** enter at the "To point:" will close the polygon precisely. This can be seen in Figure 7–9 and in the following command sequence:

Command: **L** enter
LINE From point: **3,3** enter

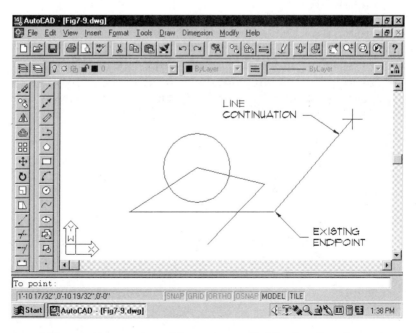

FIGURE 7–8   A line can be continued from the end of the last line drawn by pressing enter for the "From point."

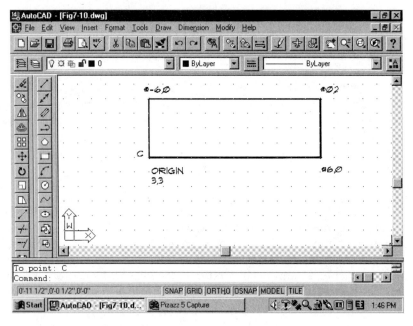

FIGURE 7–9 Entering **C** [enter] for the last "To point" will close the polygon.

> To point: **@6,0** [enter]
> To point: **@0,2** [enter]
> To point: **@–6,0** [enter]
> To point: **C** [enter]

Typically the polygons you will draw for projects will be much more complex. It's not uncommon to have to interrupt the drawing of the shape to execute another command, such as ERASE. Once the polygon has been interrupted, **C** [enter] cannot be used. This can be seen in Figure 7–10 and in the following command sequence:

> Command: **L** [enter]
> LINE From point: **4,2** [enter]
> To point: **@2,0** [enter]
> To point: **@0,2** [enter]
> To point: **@2,2** [enter]
> To point: **@–2,2** [enter]
> To point: **@–2,–2** [enter] (*Wrong coordinate entered and a line drawn.*)
> To point: [enter]
> Command: **E** [enter]
> ERASE
> Select objects [enter]
> Command: **L** [enter]
> From point: [enter] (*This uses the endpoint of the last line drawn as the new endpoint.*)

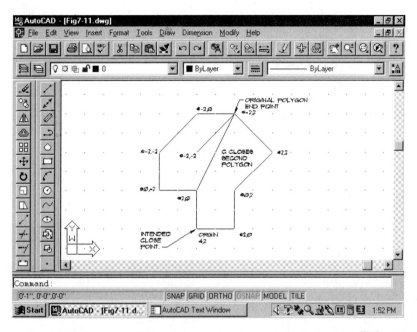

FIGURE 7–10 If the drawing of the polygon has been interrupted, **C** [enter] will termi-nate the line at the original end point rather than at the close of the total polygon.

To point: **@–2,0** [enter]
To point: **@–2,–2** [enter]
To point: **@0,–2** [enter]
To point: **@2,0** [enter]
To point: **C** [enter]

Although the intended close is at the origin point of 4,2, the polygon will close at the start of the second sequence of lines that were drawn. Chapter 9 will offer other drawing tools to pick exact points of a drawing entity as alternatives to **C** [enter].

In Figure 7–10 an error was made and the line command was exited to correct the mistake. A better way to correct the mistake would have been to use the OOPS command when the "To point: **@–2,–2** [enter]" was selected. This would have returned the origin point to the end of the last segment drawn (–2,–2) and the polygon could have been continued, as shown in Figure 7–11.

Line Undo

An alternative to ERASE or OOPS is UNDO. UNDO will remove the last line drawn or command that was executed, and allow you to continue from the previous line. In the example shown in Figure 7–10 an error was made at "To point: @–2,–2." This

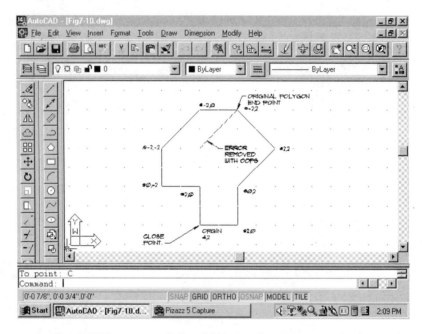

FIGURE 7–11 The OOPS command should be used to remove mistakes rather than existing the LINE command and using the ERASE command.

could have been corrected easily with UNDO, using the following commands sequence:

To point: **@–2,–2** enter
To point: **U** enter
To point: **@–2,0** enter

AN INTRODUCTION TO LINETYPE.

Earlier in this chapter you were introduced to the varied types of lines that are required to produce construction drawing. AutoCAD contains many of the linetypes found in the construction industry in the Standard linetype library. These linetypes are stored in the ACAD.LIN file. These lines can be accessed by picking the Linetype icon from the Object Properties toolbar, by typing **LINETYPE** enter at the command prompt, or by selecting Linetype... from the Format pull-down menu. Each method will produce the Layer and Linetype dialog box shown in Figure 7–12. The Linetype library is displayed by picking the Load... button. Figure 7–13a shows the linetype display. Figure 7–13b shows a complete listing of the lines found in the linetype library.

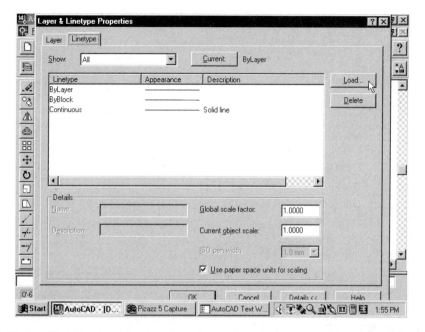

FIGURE 7–12 The Linetype Properties tab can be used to change and load linetypes. Press the Load... button to display the linetype library.

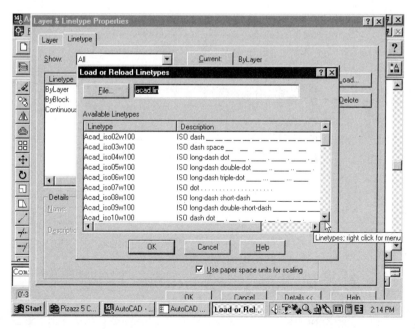

FIGURE 7–13a The Linetype library displays each AutoCAD linetype. Use the scroll bar to see the balance of the menu.

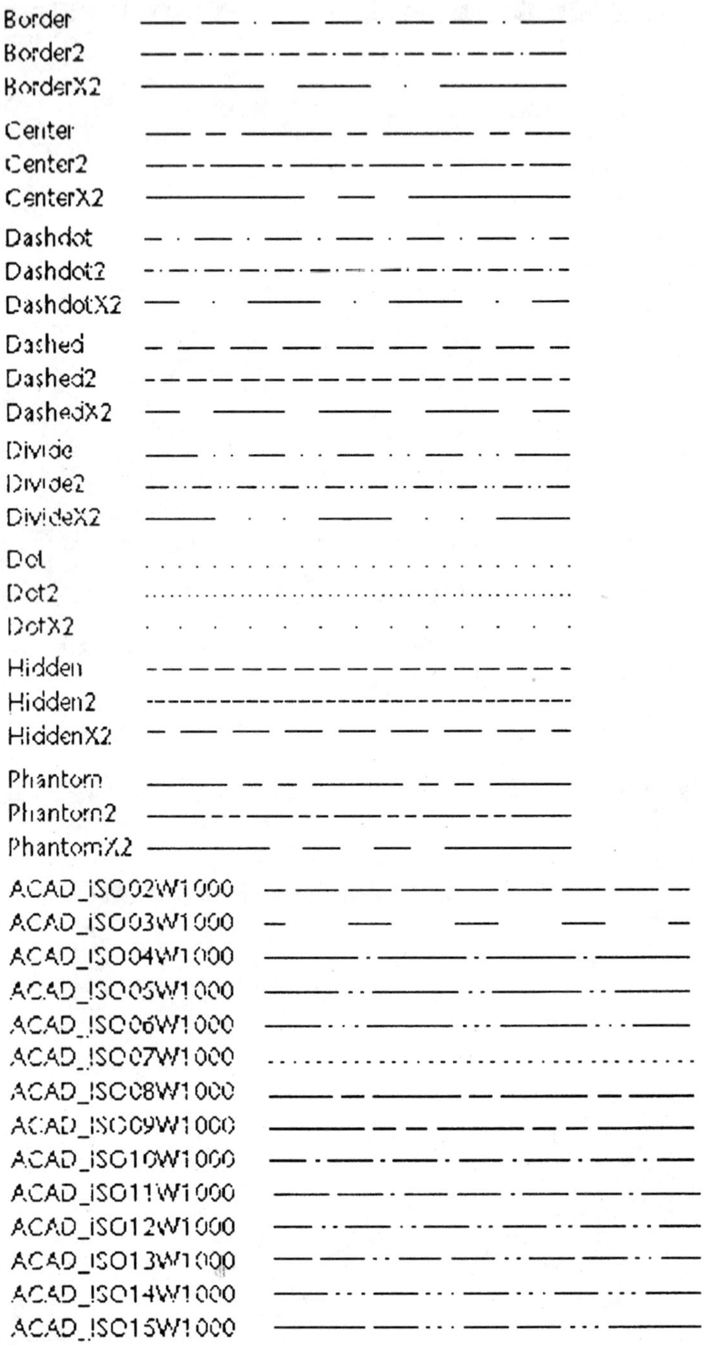
Border
Border2
BorderX2
Center
Center2
CenterX2
Dashdot
Dashdot2
DashdotX2
Dashed
Dashed2
DashedX2
Divide
Divide2
DivideX2
Dot
Dot2
DotX2
Hidden
Hidden2
HiddenX2
Phantom
Phantom2
PhantomX2
ACAD_ISO02W1000
ACAD_ISO03W1000
ACAD_ISO04W1000
ACAD_ISO05W1000
ACAD_ISO06W1000
ACAD_ISO07W1000
ACAD_ISO08W1000
ACAD_ISO09W1000
ACAD_ISO10W1000
ACAD_ISO11W1000
ACAD_ISO12W1000
ACAD_ISO13W1000
ACAD_ISO14W1000
ACAD_ISO15W1000

FIGURE 7–13b The complete list of linetypes can be viewed by picking the Load button, or by selecting Standard Linetypes from the Standard Libraries of Command Reference of the Help menu.

Linetype Characteristics

The LINETYPE command will allow you to vary the linetype pattern for each entity you draw. Most offices assign a specific linetype to a specific layer, but several different colors can also be assigned to a layer. As you start to assign linetypes to drawing entities, you need to remember two key elements regarding linetypes.

Affects. The only entities that will be affected by a change of linetype are Line, Circle, Arcs, and 2D Polylines. Any other drawing entity will be displayed as a continuous line. That will be good news as you start to add dimensions or text to a drawing and you're working on a layer drawn with center lines or another line pattern as the current linetype.

Loading. A second concept to keep in mind as you work with linetypes is that you can't use a LINETYPE unless it has been loaded into your drawing. The linetypes of Figure 7–13b exist in a library file called ACAD.LIN, which is contained in the \ACADR14\SUPPORT folder. If you are entering linetypes by keyboard, the linetypes that you are going to use in a specific drawing must be loaded into the drawing file. They will be loaded automatically when LINETYPE is entered as a function of the dialog box for LAYER, which will be covered later in this chapter.

Setting Linetypes

Once the LINETYPE prompt has been displayed, the pattern for the existing layer—in this case the 0 layer—can be altered. Once altered, all entities governed by LINETYPE will reflect the new line pattern. Objects drawn prior to the selection will not be affected. Objects drawn prior to the selection can be transformed to a different linetype using the CHANGE command, which will be discussed in Chapter 15. The fastest method for loading and changing a linetype is by selecting the Linetype icon from the Object Properties toolbar and then selecting the Load... button from the Layer and Linetype properties dialog box. To load the dashed linetype displayed in the menu, pick either the name, description or the line. Either selection will highlight the name. To load more than one linetype, press and hold the Ctrl button while picking the names of the desired linetypes with the select button of the mouse. When you're through picking the linetypes to be loaded into a drawing, press the OK button. The menu will be removed, and the selected linetypes will be displayed in the Linetype Edit window of the Layer and Linetype Properties dialog box similar to Figure 7–14.

Remember that all you've done so far is to load linetypes into the drawing. To use the linetype you must select one of the loaded linetypes as the current linetype. As a default, you've been drawing with a continuous black line. Pick the desired linetype to be used, pick the Current button, and then the OK button. This process will make the selected linetype the current linetype to be used, close the dialog box, and return you to the drawing screen. Depending on the linetype selected, an alert box may be displayed about the ISO pen width. For now, pick the OK button. Altering

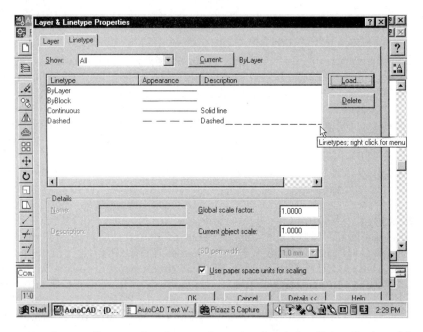

FIGURE 7–14 Once a linetype has been selected to load, it will be displayed in the active linetypes. Select the desired linetype and the Current button to use the linetype in the drawing.

the pen width will be introduced in Chapter 15 as linetypes are discussed in detail. Notice the name and pattern of the selected line will now be displayed in the Object properties toolbar as seen in Figure 7–15. Enter the LINE command and draw several lines. The new linetype should now be reflected on the drawing screen. If the line still appears as a continuous line, check the Object Properties toolbar to verify that you did pick the current button. If the name and linetype show the desired linetype, reenter the Layer and Linetype Properties dialog box and alter the global scale factor. If you're working on a templet set for 1/4"=1'-0" plotting, change the global scale factor to 48 and press the OK button. As you return to the drawing screen, lines should now reflect the selected pattern as shown in Figure 7–16.

AN INTRODUCTION TO COLOR

Up to this point, all of your drawings have been done with black lines on a white background or white lines on a black background. The choice of screen background can be altered by selecting the Display tab from Preferences... of the Tools pull-down menu. Picking the Colors... button will produce the AutoCAD Window Colors dialog box shown in Figure 7–17. The box allows for the color of each of the drawing window displays to be altered. To change the color of the drawing area, select the desired color tile, press the OK button to close the Color box and then press the OK button

FIGURE 7–15 The current linetype is displayed in the Object Properties toolbar.

of the Preferences box. The dialog box will be closed and the screen will display the selected background color.

AutoCAD also allows the color of each drawing entity to be altered. For most CAD users though, a whole new world is about to be revealed. You should be able to start drawing with 256 colors. Most users will have millions of shades to chose from. The number of colors available will depend on the hardware you are working with and how the software is configured. Both will be addressed in later chapters.

Color Numbers and Names

Each drawing object or layer may be assigned a color. The seven basic colors used by AutoCAD are listed in Figure 7–18. Most monitors can display 256 colors with the selection based on the color displayed in the color palette.

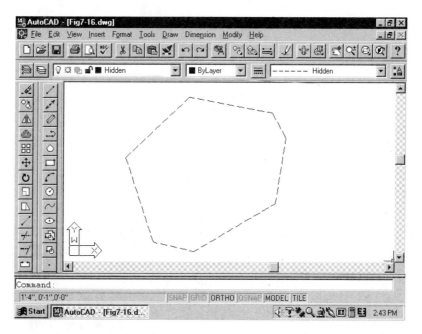

FIGURE 7–16 Lines drawn with the current linetype.

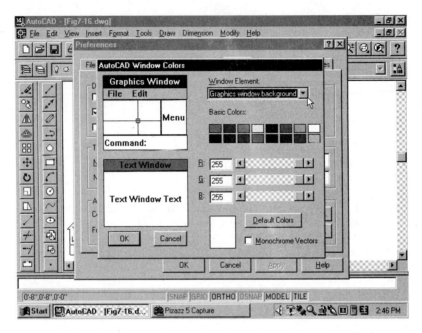

FIGURE 7–17 Picking the Colors... button will produce the Colors dialog box and allow the various aspects of the display to be altered.

COLOR NUMBER	COLOR NAME
1	RED
2	YELLOW
3	GREEN
4	CYAN
5	BLUE
6	MAGENTA
7	WHITE

FIGURE 7–18 The major colors of AutoCAD.

Setting Entity Color

The COLOR command can be used to assign colors for individual objects or entire layers of entities. Setting colors for layers is the default and will be explained as you proceed through layer options later in this chapter. Setting colors can be done by selecting the Color Control icon from the Object Properties toolbar, by selecting Color... from the Format pull-down menu, or by typing **COLOR** at the command prompt. The easiest method to assign color is to pick the Color icon from the middle of the Object Properties tool bar. You may pick the color tile, the color name or the down arrow. Each will display the menu shown in Figure 7–19. Picking one of the listed colors, will close the menu and return the drawing display. Now if a line is drawn, it will be drawn using the selected color. Using the current settings, the line would be a red line drawn with a hidden linetype. Pick the Color icon again to reopen the color menu. Notice the last menu option is Other. Picking this option will display a menu similar to Figure 7–20. (It's really impressive in color, but Figure 7–20 will have to do.) The Select Color dialog box allows shades of the standard colors to be selected. Notice the current color is listed as red. Picking any of the color tiles will alter the current color display tile and list the color by number. Picking the OK button will close the display and make that the current color. Using the LINE command now will create lines using the selected color.

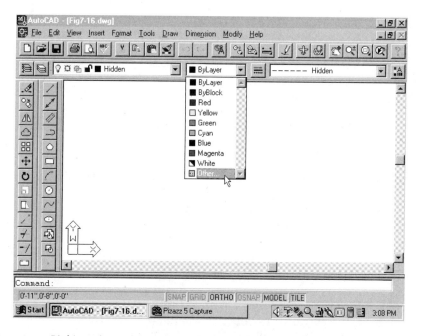

FIGURE 7–19 Picking the color tile from the Object Properties toolbar will display the color menu and allow the color of the current layer to be changed.

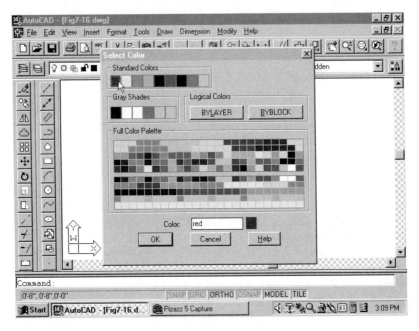

FIGURE 7–20 Selecting Other... from the color menu displays the color pallet and allows a wide variety of color tints to be selected for drawing objects.

The color of lines can also be altered by entering **COLOR** [enter] at the command prompt.

The command sequence is:

    Command: **COLOR** [enter]
    New object color <BYLAYER>:

Either a number or a standard color name may be entered at the prompt:

    New entity color <BYLAYER>: **RED** [enter]

or

    New entity color <BYLAYER>: **1** [enter]

will produce the same results, with all entities drawn after this change drawn with red lines. If Bylayer is accepted as the default, the color will be assigned once each layer has been defined. Typically, most offices set colors using the Bylayer option rather than assigning colors to individual objects.

LAYERS

The use of layers in a CAD drawing is similar to overlay drafting procedures of manual design. Overlay drawing will have a base drawing, such as the valve assembly seen in Figure 7–21, on one sheet of Mylar®, and have overlay sheets of notes and dimensions on separate sheets of Mylar®, as seen in Figure 7–22. Depending on who needed what, different sheets could be attached to the base and various prints made. Layers allow all the flexibility with none of the alignment problems of overlay drafting procedures.

Until now you have been drawing on the 0 layer. AutoCAD will allow you to create different layers, assign each layer a different linetype and a different color, allow some layers to be made invisible, and allow differences to be made in how each layer is reproduced.

Entering the LAYER Command

The LAYER command can be started by selecting the Layer icon from the Object Properties toolbar, or by selecting Layer... from the Format pull-down menu. Options for layer may be entered by keyboard by entering **LA** [enter], **LAYER** [enter], or **DDLMODES** [enter] at the command prompt. Each will display the dialog box shown in Figure 7–23. Options for controlling layers can also be accessed from the command prompt and will be introduced later in this chapter.

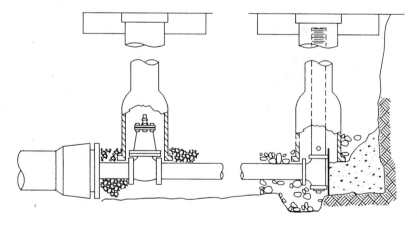

FIGURE 7–21 The use of layers is similar to placing information on several sheets of paper in overlay drafting. This drawing shows the basic drawing with no supplemental information provided. (Courtesy Department of Environmental Services, City of Gresham, OR)

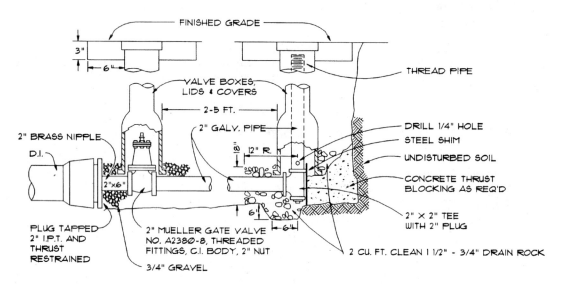

NOTES:

1. VALVE BOX SHALL BE PER STANDARD DETAIL NO. 414 OR 415.

2. VALVE BOX TO BE ASPHALT ENCASED AS SHOWN, IF NOT IN PAVED AREA.

3. BLOW-OFF UNIT SHALL BE GRAVEL BACK FILLED AND COMPACTED AS SHOWN.

FIGURE 7-22 Supplemental information added to the base drawing. (Courtesy Department of Environmental Services, City of Gresham, OR)

Creating a New Layer

Once in the Layer and Linetype Properties dialog box choose the New button from the upper right hand corner. This will add the highlighted name of Layer 1 in the name edit box. By clicking on Layer 1, you will be allowed to assign a new layer name by keyboard. Press the select button of the mouse to accept the new layer name. In Figure 7–24a, layer 1 is assigned the name of COOLSTUFF. Layer names can be up to 31 characters long and contain letters, digits, and special characters such as **$, -,** or _. No blank spaces can be used in the name. Guidelines for naming layers will be discussed in Chapter 15. For now keep the names short, simple, and descriptive using names that will identify the contents. Once the name has been supplied, picking the OK button will save the layer and return the drawing area. Just as with Linetype, all you've done is to create a new layer. The current layer is still the default layer of **0**.

Altering Layer Qualities. Reenter the Layer Properties dialog box, select the COOLSTUFF layer and then pick the Current: button. COOLSTUFF will now be

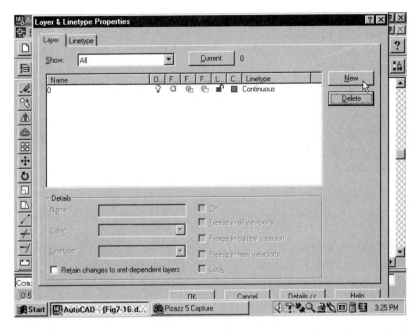

FIGURE 7–23 Picking the Layer icon from the Object Properties toolbar will display the Layer and Linetype Properties dialog box. The box can be used to create and control the properties of each drawing layer.

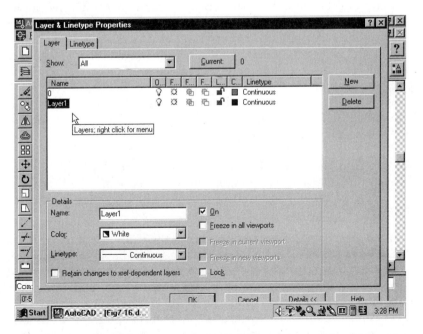

FIGURE 7–24a Selecting the New button will display a layer title of Layer 1.

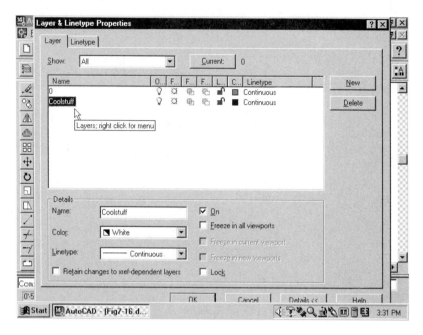

FIGURE 7–24b The selected layer name of Figure 7-24a can be altered to reflect the desired layer name.

displayed in the Details portion of the dialog box. A linetype can be assigned to all objects created, when this layer is the current layer, by picking the current Linetype display. This will produce the menu shown in Figure 7–25a. Only linetypes that have been loaded into the drawing will be displayed. Picking one of the linetypes will use that linetype for all drawing entities created using that layer. Picking the Color edit box will produce the menu shown in Figure 7–25b. Once the desired qualities of a layer have been assigned, pick the Current button if you want to use that layer as you return to the drawing editor. Selecting the OK button will close the dialog box and return you to the drawing screen.

Controlling the Use of Layers. Six icons are displayed by each layer name. These icons represent various controls for the use of each layer. The light bulb icon is a toggle for the ON/OFF setting of a layer. The drawing shown in Figure 7–21 and 7–22 is controlled by toggling the TEXT layer ON/OFF.

The Sun icon can also be used to control the visibility of a layer. The second icon is a toggle between Thaw/Freeze. When the sun is displayed, the contents of that layer will be displayed. Picking the sun icon will change it to an icicle, and the contents of the layer are said to be frozen. Items on a frozen layer will not be displayed. The benefits of using the ON/OFF and Thaw/FREEZE options as well as each of the other layer controls, will be discussed in Chapter 15. To remove the contents of a layer from the screen display, select the desired layer name and then pick the OFF or FREEZE icon and the pick the OK button.

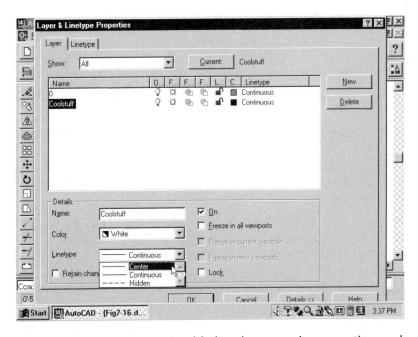

FIGURE 7–25a Once a layer name is added to the menu, its properties can be adjusted. To set the linetype of the new layer, pick one of the linetypes from the Linetype menu found in the Details box.

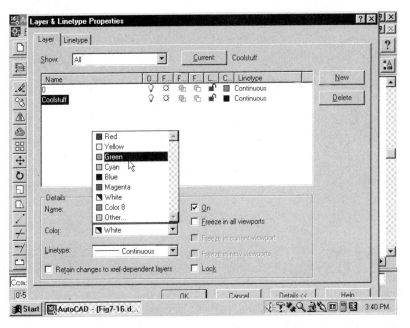

FIGURE 7–25b The color of the new layer can be altered by picking the Color icon from the Details box.

AN INTRODUCTION TO PRINTING

Once you've created a drawing, you'll need some method of passing the contents of the file onto those who will put it to use. This might include sending the drawing electronically to another firm, or sending the drawing by disk or on some other storage medium. For many it also means making a hard copy such as a paper print or plot. No matter how technology advances, for most CADD operators, a paper copy is still the most dependable method to check drawing. This chapter will introduce the process of making a paper print using the system printer configured using your Windows platform. Chapter 24 will provide additional information on choosing additional printers, configuring a plotter, and controlling the many variables associated with plotting.

The print process can be started by picking the Print icon from the left end of the Object Properties toolbar, by picking Print... from the File pull-down menu, or by entering **PRINT** [enter] or **PLOT** [enter] at the command prompt. Each selection method will produce the Print / Plot Configuration dialog box shown in Figure 7–26. The dialog box contains six major areas that must be considered to produce a print.

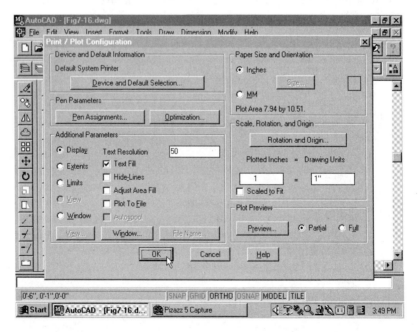

FIGURE 7–26 Selecting the Plot icon from the Object Properties toolbar, picking Plot from the File pull-down menu, or typing **PRINT** [enter] at the command line will display the Print / Plot Configuration dialog box. The dialog box can be used to control each aspect of creating a print.

Device and Default Selection

The Device and Selection... button displays the current printing device as well as allowing new printing and plotting devices to be added and configured. Printer selections can also be controlled by selecting the Printer Tab from Preferences of the Tools pull-down menu. For the current discussion, the default printer will be used.

Pen Parameters

The pen parameters box contains the Pen Assignment... and Optimization... boxes. Optimization will be introduced in Chapter 24. Picking the Pen Assignment... button will display the dialog box shown in Figure 7–27. The box can be used to control the pen parameter for plotters with multiple pens. Depending on the type of printer used, the box can also be used to control the width of lines. To alter a line width, select the line to be changed. In Figure 7–27, color 1 is selected to be changed from it's current width of 0.010. Enter the desired value, and then pick the OK button. Although a printer does not have a physical pen, settings can be provided to alter the width and color of the line. Figure 7–28 shows an example of various line weights that will result as the pen width is varied. The color of the line setting can effect the line quality of the print. To alter a value of a pen, pick the desired pen number and then provide the desired width, followed by picking the OK button.

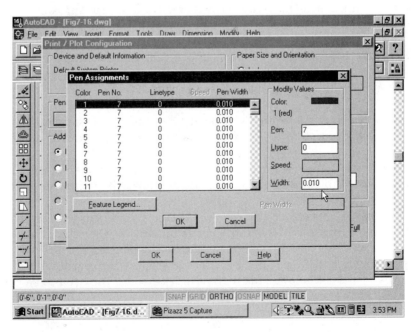

FIGURE 7–27 Although a printer has no physical pens, many printers allow the widths of lines to be altered by changing the value for the pen width.

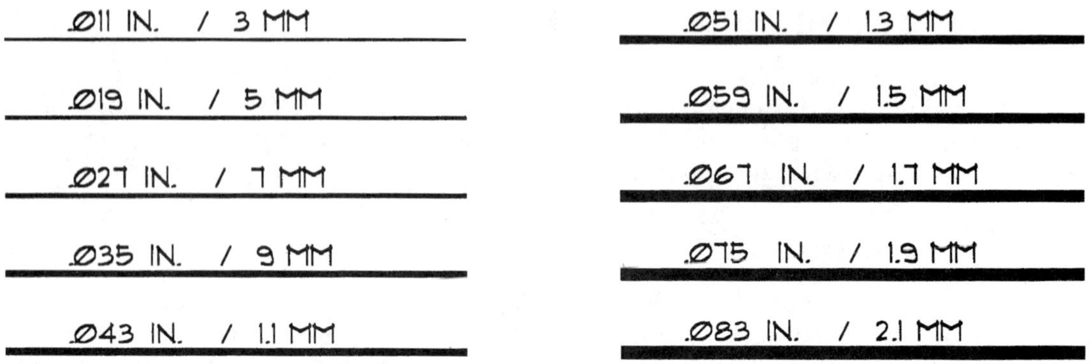

.011 IN. / .3 MM	.051 IN. / 1.3 MM
.019 IN. / .5 MM	.059 IN. / 1.5 MM
.027 IN. / .7 MM	.067 IN. / 1.7 MM
.035 IN. / .9 MM	.075 IN. / 1.9 MM
.043 IN. / 1.1 MM	.083 IN. / 2.1 MM

FIGURE 7–28 The results of various pen width settings on a printed drawing.

The Additional Parameters Selection Box

This selection box provides options for describing what portion of the current drawing will be printed, options for describing how specific entities will be handled during the printing process, and a box for choosing how the file will be stored. Because the entire drawing often does not need to be printed, five options are given to decide what portion of the drawing will be plotted.

Display.   The default setting is to print only the material that is currently displayed on the screen. If you have zoomed into a drawing, only that portion of the drawing will be plotted.

Extents.   Choosing this option will print the current drawing with the extent of the drawing entities as the maximum limits that will be displayed in the plot. Using this option may eliminate entities at the perimeter of the drawing.

Limits.   Using this option will print all of the drawing entities that lie within the current drawing limits that were defined during the drawing setup.

View.   This option is currently not active. Methods of creating views will be introduced in Chapter 10 and the use of this option will be introduced in Chapter 24

Window.   This option will allow a specific area of a drawing to be defined for plotting. Initially the Window radio box is inactive. By selecting the Window... button, a Window Selection dialog box similar to the one in Figure 7–29 will be displayed. The window can be specified by entering the drawing coordinates that will define the corners of the window. This method works well if the area to be printed is not currently on the screen. Typically the Pick option will be used. Selecting this option will remove the dialog box from the screen and the command line will show the prompt:

First corner:
Other corner:

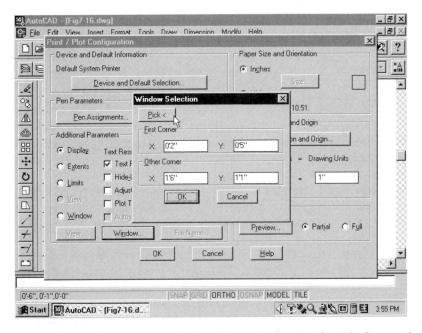

FIGURE 7–29 When objects are selected to be printed using the Window option, the size of the window can be set by providing the coordinates of the window corners, or by picking the window corners with the cursor. Selecting the Pick < button will remove the menu and return the drawing to the screen. Once the corners of the select window are located, the dialog box will be displayed again so that other print options can be set.

Once the window is defined, the Window Selection box will be redisplayed, allowing the coordinates to be confirmed or edited. Press the OK button if you're satisfied with the selected window, or the Pick button if you would like to redefine the area of the drawing to be printed. Once approved, the Plot Configuration dialog box will be returned to the screen. The Window radio button will be activated, allowing other options to be set. Each option will be discussed in Chapter 24.

Paper Size and Orientation Selection Box

This selection box allows the printing units and orientation to be specified. Most construction projects will use the default setting of Inches for the plotting standard. The two other features of this box are the Icon box, and the Size... box.

The box on the right side of the selection box is the orientation icon and not a selection box. The current orientation of the box indicates the drawing will be printed in the normal orientation for construction drawing. This setting is referred to as "Landscape." When the print is rotated so that it is read from the titleblock end, the print is referred to as a Portrait print. A portrait print is more likely to be used with laser printers. The orientation of the drawing to the paper can be adjusted by picking the appropriate radio button in the Size... dialog box. The Size... button will be inactive if a printer is configured as the current plotting device. (See Chapter 24 for using a plotter.)

Scale, Rotation, and Origin Selection Box

The Plot Configuration dialog box controls settings for the Scale, Rotation, and Origin of plot. The current printing scale is indicated beneath the Rotation selection box.

Rotation and Origin . . . Picking this selection box will display a dialog box similar to the display seen in Figure 7–30. The Plot Rotation radio buttons will allow for rotating the plot in a clockwise direction on the paper.

Plot Origin. The default setting for plotting origin is X = 0 – Y = 0. Printers typically specify the origin from the upper right-hand corner. Specifying various locations for the origin can be a useful technique for altering the drawing location on the paper. By selecting the X or Y origin box, new values can be entered. Entering an X value of 12 and a Y value as 10 would place the new origin 12 units to the right and 10 units above the original plotting origin.

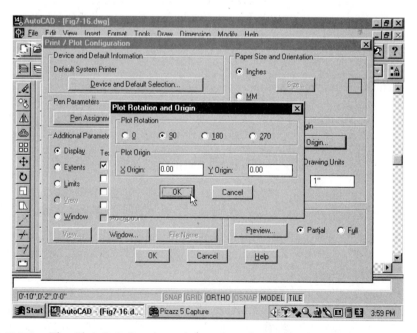

FIGURE 7–30. The Plot Rotation box determines how the drawing will be placed on paper. Plots can be rotated in 90° increments in a clockwise direction.

Selecting a Scale. In the default setting, when a print is made it will be reduced in size to fit on the current printer paper. This is indicated by a check in the Scaled to Fit box. Unless you'll be printing a small detail, most architectural drawings can't be printed to scale on a printer. Rarely could you get the floor plan for a structure to fit on "A" size material. You can, however, print the entire floor plan on "A" size material if the scale is reduced. The Scaled to Fit option reduces the selected material to be printed to an appropriate size to fill the current paper. You can print portions of a drawing to scale by selecting the desired scale, and using the window selection method. To print a drawing to a scale, pick the Scaled to Fit button to deactivate this feature. As the check is removed, you'll notice the values in the scale box are changed to display 1 = 1. To print a detail at a scale of 3/4"=1'-0" enter 3/4 or .75 in the box on the left and enter 12 or 1' in the box on the right. Figure 7–31a and 7–31b compare the differences in scaling options.

The Print Preview Selection Box

The final dialog box to be displayed for printing is accessed from the Plot Preview box. Selecting Preview allows the outcome of the current plot settings prior to the actual plot. Two options are available for previewing the print—Partial or Full Preview.

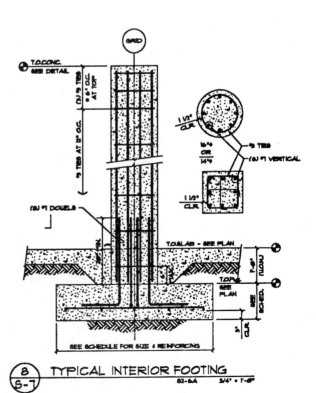

FIGURE 7–31a A detail printed using the Plot to Fit option. Most professionals use drawing printed on using the Fit option for check prints prior to plotting the drawing to scale.

TYPICAL INTERIOR FOOTING

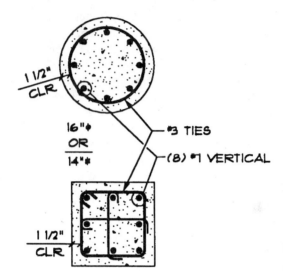

1 1/2" CLR

16"∅ OR 14"∅

#3 TIES

(8) #7 VERTICAL

1 1/2" CLR

FIGURE 7–31b A portion of the detail from Figure 7–31a printed using the scale of 3/4"=1'-0".

Partial Preview. The default setting of a preview is partial. Selecting the Preview... button with Partial active will produce a display similar to the one seen in Figure 7–32. The paper size is represented by a red line and the limits of the effective printing area are specified on the screen by a blue line. If either of the two sizes matches, a dashed red and blue line will be used to represent the plot. A warning also will be given of any problems that may be encountered during the plot. Common warnings displayed in the warning box include:

> Effective area too small to display
> Origin forced effective area off display
> Plotting area exceeds plotter maximum

A small triangular Rotation Icon is displayed in the lower left-hand corner to represent a 0 rotation. As the rotation is set to 90 degrees, the icon is moved to the upper left-hand corner, to the upper right-hand corner for 180 degrees, and to the lower right-hand corner for 270-degree rotation.

Full Preview. Choosing this option prior to picking the Preview... box will produce a graphic display of the drawing as it will appear when printed. Figure 7–33 shows an example of a full preview. An outline of the paper size will be drawn on the screen, which displays the portion of the drawing to be printed. The cursor is changed to show a magnifying glass with a + and − sign beside it. This represents the ZOOM command and allows the image to be printed to be enlarged or reduced. By pressing and holding the mouse pick button the size of the image can be altered. Moving the mouse to the top of the screen enlarges the image, and moving the image to the bottom of the screen reduces the size of the image. Pressing the right mouse button will display the menu shown in Figure 7–34 allowing other options to be selected. Picking the Pan option allows the image to be shifted within the limits of the paper, but

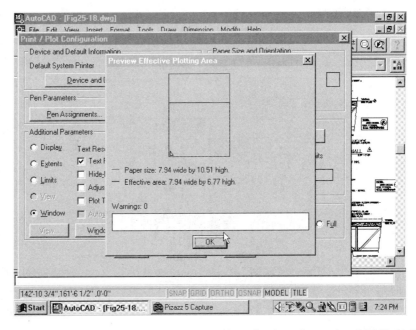

FIGURE 7–32 Athe detail from Figure 7–3a printed using the scale of 3/4"=1'-0".

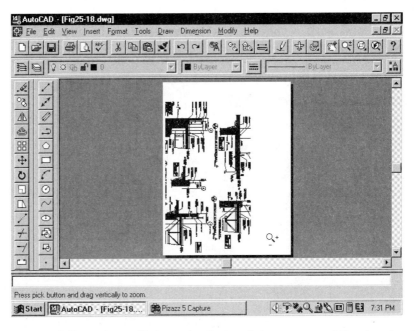

FIGURE 7–33 A full preview will show the limits of the paper and the drawing to be plotted.

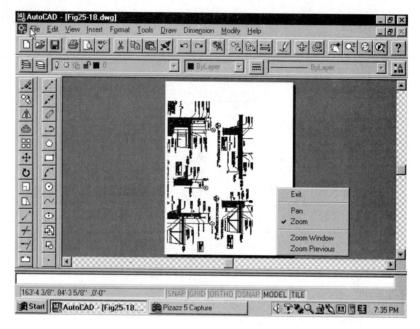

FIGURE 7–34 Pressing the right mouse button while using a full preview will display this menu. The options can be used to control the display to be printed.

the magnification of the image is not altered. Each of the options will be further discussed in Chapter 24. To continue with the print process pick the Exit option to remove the menu.

Producing the Print

Seven areas of the Plot Configuration dialog box have been examined. If you change your mind on the need for a print, select the Cancel box; the dialog box will be terminated, and the drawing will be redisplayed. If you are satisfied with the printing parameters that have been established, select the OK box. The dialog box will be removed from the screen and the print will be produced. Figure 7–35a and 7–35b show the results of the selected plot settings.

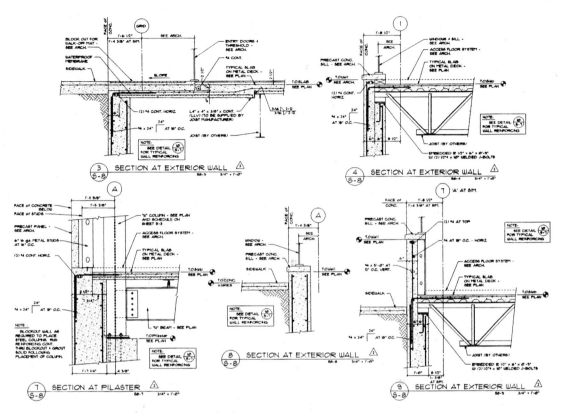

FIGURE 7–35a The results of the preview shown in Figure 7–32. (Courtesy Van Domelen / Looijenga / McGarrigle / Knauf Consulting Engineers

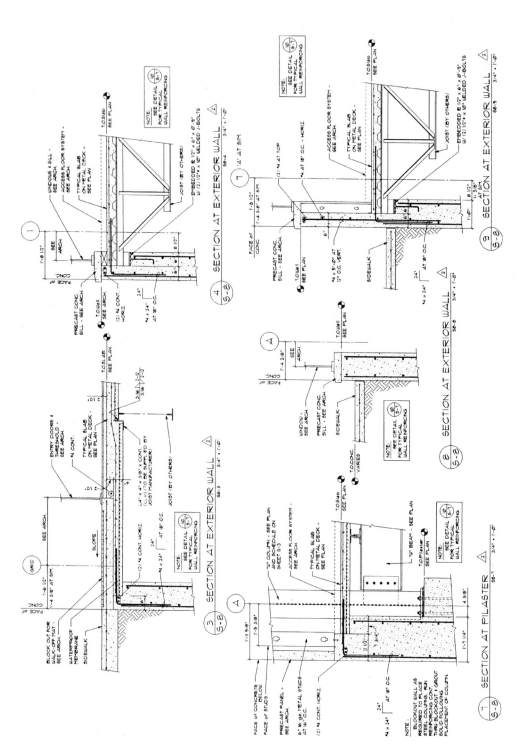

FIGURE 7–35b The results of the preview shown in Figure 7–33. The End Preview button was selected and the 90° rotation was picked prior to selecting the OK button to produce the print. (Courtesy Van Domelen / Looijenga / McGarrigle / Knaff Consulting Engineers)

CHAPTER 7 EXERCISES

Use the disk that was formatted in Chapter 4 to save the following exercises. Set limits, units, snap, and grid to best suit each project.

E-7-1. Use the following coordinates to lay out the required shape. Save as E7-1.

Starting at point 7.75,7.95

> @–2,0
> @0,–2
> @–1.50, 1.75
> @–.96,0
> @1.62<137°
> @.46<227°
> @1.22<310½°
> @–.92,0
> @1.62,–.50
>   2.17,2.91
> @3.5,0
> @1.62,1.62
> @1.75<308
> @3.32<80.73°
> back to the point of beginning

E-7-2. A parcel of land is to be developed. Use the following metes and bounds supplied by the surveyor and draw the site plan and the indicated utility easement. Set limits to 200′,150′. Use a phantom line to represent the property line and use a center line to represent the easement. Select different colors and layers for each linetype.

> SITE. Beginning at a point to be described as the northwesterly corner, 100.00′ due south, thence 37.52 N89° 37′E, thence 96.85′ N58°40′52″E, thence 57.25′ N3°50′40″W, and then back to the true point of beginning.
>
> EASEMENT. The easement is approximately an inverted "L"-shaped easement. The long leg of the "L" is 10 feet wide; the short leg is 15′ wide. Lay out the easement beginning at a point that lies on the southerly property line, 20′-0″ from the southwesterly corner of the property. Thence northerly 70′-0″ along the center line of said easement to a point at N7°32′15″W, thence a 15′-wide easement laying along a center line extending to the westerly property line at N50°55′04″W.

Save the drawing on a floppy disk in a directory labeled EXCER with the filename E7-2 SITE.

E-7-3. Use the dimensions on the floor plan below and draw the outline for the perimeter of the structure. No footings or beams are to be drawn. Save the drawing in a directory labeled EXCER with the filename E7-4 PLAN.

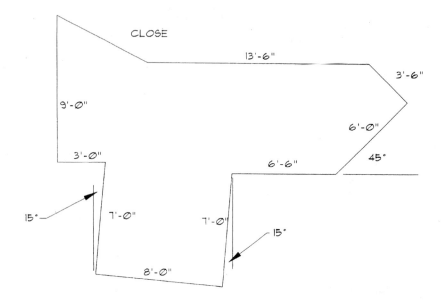

E-7-4. Create a templet drawing containing appropriate values for an architectural drawing. Include settings for each of the options discussed in Chapter 4. Set limits and paper sizes assuming 24" × 36" plotting paper. Load linetypes for lines that might be found on a plan view. Create a layer titled FLRWALLS. Save the file as a templet drawing.

E-7-5. Use the templet drawing created in exercise 7–4 and the attached drawing to draw the roof line of the structure. Assume all hips and valleys are drawn at 45° angle. Save the drawing as E-7-5.

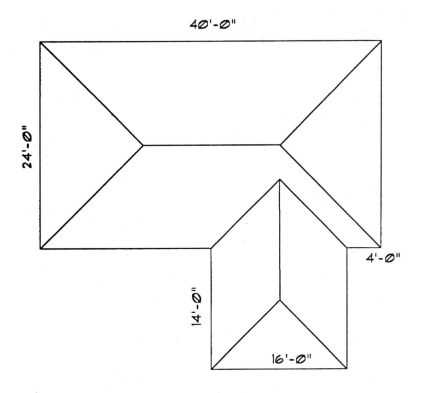

E-7-6. Use the attached drawing to draw a section of a one-story footing for a wood floor system. Assume the bottom of the footing extends 12" into the grade, and the top of the stem wall extends 8" above grade. Save the drawing as E-7-6.

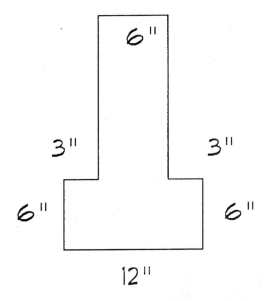

E-7-7. Use the attached drawing to draw a section of a one-story footing for an on-grade concrete floor system. Save the drawing as E-7-7.

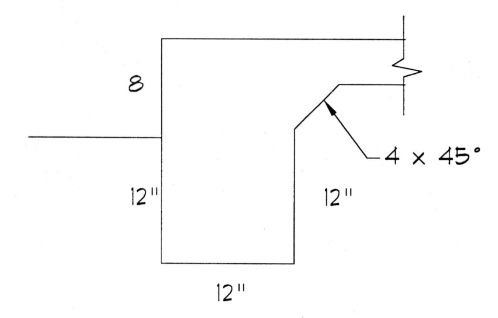

8

12"

12"

12"

4 x 45°

E-7-8. Use the attached drawing to complete a side view of a steel beam resting on a steel plate supported on a steel column. Assume the beam to be 18" deep, and show the thickness of each plate as 7/8" thick. Provide a 10"-long support plate, and an 18"-high stiffener plate at the end of the beam. The support column is a TS 6 × 6. Save the drawing as E-7-8.

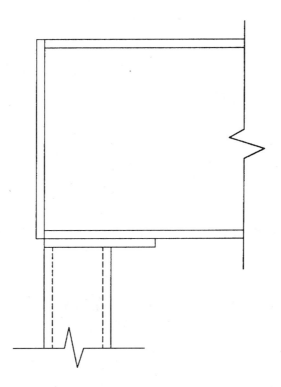

E-7-9. Use the attached drawing to complete a section through a 6"-wide concrete panel resting on a concrete footing 24" wide × 12" deep. Support the wall panel on 1" deep grout with beveled edges. Show a 3" × 6" deep pocket in the footing for reinforcing. Save the drawing as E-7-9.

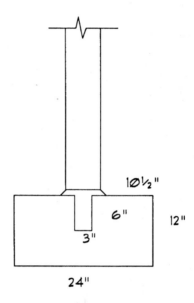

E-7-10. Use the attached drawing to complete a section view of a floor truss intersecting a 6"-wide concrete tilt-up wall. Assume the truss to be 24" deep with 2" deep top and bottom.

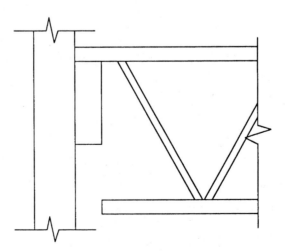

CHAPTER 7 QUIZ

1. How can the LINE command be accessed from the pull-down menu?

   _____

   _____

2. What are the results of selecting "1 Line" from the screen menu?

   _____

   _____

3. What will the L optionof the ERASE command do?

   _____

   _____

   _____

4. List five methods for accessing the LINE command.

   _____

   _____

   _____

5. What steps are required to erase a line?

   _____

   _____

6. If a line has been erased by mistake how can it be restored?

   _____

   _____

7. What effect does ERASE have on the cursor?

   _____

   _____

8. List four methods of starting the ERASE command.

   _____

   _____

9. Describe the difference between OOPS and UNDO.

_____

_____

10. List nine types of lines often found on construction drawings.

_____

_____

11. Should the BLIPMODE be ON or OFF if you do not want markers placed for each drawing entity?

_____

_____

12. What option should be used if you're working on a drawing with many line segments?

_____

_____

13. What effect will **C** [enter] have on a line segment?

_____

_____

14. What command will allow the line pattern to be altered?

_____

_____

15. How does the Snap setting affect the use of the LINE command?

_____

_____

16. What type of coordinate entry is best suited for drawing E-7-3?

_____

_____

17. You've just loaded five linetypes, and have entered the LINE command. Only continuous lines are being drawn? List two possible problems.

_____

_____

18. What is the difference between assigning a color to a linetype or a layer?

19. List four options for entering the Layer command.

20. What is the current layer name, and how can a new layer be named?

21. What are the limitations to assigning layer names?

22. List four options for controlling the visibility of a layer.

23. How does the Pen Parameters affect a print made on a device with no pens?

24. Explain the difference between printing using the Display and Window options.

25. Describe the options for a print preview.

• • • • • • • • • • • • •

CHAPTER

8

DRAWING GEOMETRIC SHAPES

• •

L ines and points may be the base of all drawings, but it's difficult to imagine a large engineering or architectural project without curved features. This chapter will introduce you to drawing circles, arcs, ellipses, polygons, and donuts.

CIRCLES

Circles are found in a variety of projects throughout the construction world. Round windows, concrete piers and columns, steel tubing, and conduit lines—circular features will be prevalent throughout your drawings. Figure 8–1 uses circles to represent concrete piers for a structure. The CIRCLE command may be entered by keyboard by picking the Circle icon from the Draw toolbar, by keyboard, or by picking Circle from the Draw pull-down menu. Each can be seen in Figure 8–2. To draw a circle, pick the Circle icon from the Draw toolbar, type **C** [enter], or type **CIRCLE** [enter] at the command prompt. This will display the prompt:

Command: **C** [enter]
3P/2P/TTR/<Centerpoint>:

As you enter the command you'll notice that the prompt changes to:

CIRCLE 3P/2P/TTR/<Centerpoint>:

AutoCAD is waiting for you to pick a centerpoint for the origin point of the circle. Once the center point is selected, you're prompted for the edge of a circle. The process can be seen in Figure 8–3.

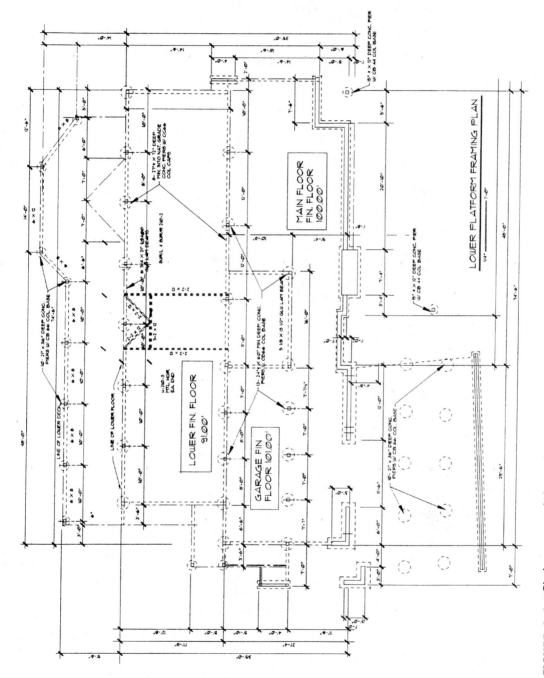

FIGURE 8-1 Circles are used throughout architecture to represent many building components. (Courtesy Residential Designs)

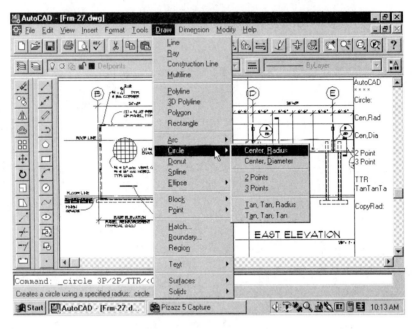

FIGURE 8–2 Circles can be drawn by picking the Circle icon from the Draw toolbar, by typing **C** [enter] or **CIRCLE** [enter] at the command prompt, or by picking Circle from the Draw pull-down menu.

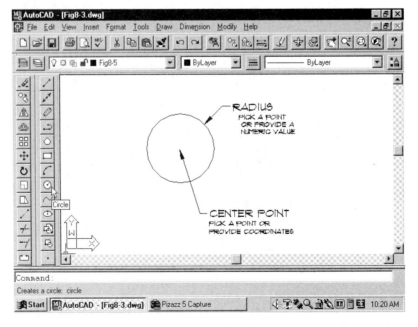

FIGURE 8–3 The size of a circle can be controlled by entering a value for the radius or the diameter.

Circles may be drawn by entering seven different combinations of information about the circle. Each category of information is listed in the pull-down menu shown in Figure 8–2. The DONUT command, which is listed in this menu, produces a circle with varied line width and will be discussed later in this chapter.

Notice that once the centerpoint is selected, a circle will appear on the screen, but the size will be altered as the mouse is moved. AutoCAD refers to this as *drag* or *rubberband*. Moving the mouse and changing the circle size refers to *dragging* the circle into position. This drag image is typically very helpful in determining the final locations of lines and other geometric shapes. If you find it distracting, it can be disabled by typing DRAGMODE at the command prompt.

AutoCAD provides three different options for controlling the display. Each can be found by typing **DRAGMODE** at the command prompt. This will produce the prompt:

Command: **DRAGMODE** [enter]
ON/OFF/Auto <Auto>:

Auto is the current default setting. Using Auto (A) will use drag automatically on each command that supports the drag mode.

Entering the ON option will allow drag to be used, but only at your discretion. Set the dragmode to ON and draw a circle. A line will extend from the Center point to the circle edge, but the circle is not seen until the radius is entered. Typing **DRAG** [enter] at the radius prompt reveals the limits of the circle for this command sequence. Entering the OFF option deactivates DRAG so that all attempts to use DRAG are denied. To reactivate, enter DRAGMODE and choose ON or A.

Center and Radius

The default method of drawing a circle is to enter the centerpoint and a radius. As the CIRCLE is started, the centerpoint may be entered by typing coordinates or by picking a center point with the pointing device. In Figure 8–4 three circles have been drawn, and a fourth has been dragged into position. Notice that the rubberband line extends from the chosen centerpoint to the radius indicated by the cross-hairs. By moving the cross-hairs, the size of the circle can be changed. By entering a numeric value and [enter], a circle will be drawn using the desired radius.

Center and Diameter

Sizes of circular features such as piers, pipes, and windows typically are given as a diameter. At the circle prompt enter **D** [enter] to switch the prompt from radius to diameter. The command sequence is:

Command: **C** [enter]
CIRCLE 3P/2P/TTR/<Centerpoint>: (*Select any point.*)
Diameter/<Radius>: **D** [enter]

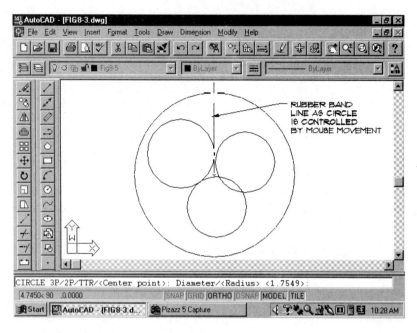

FIGURE 8–4 Rather than entering a specific size, use the rubberband line to see the effect moving the cursor will have on the circle to be drawn.

Diameter: (*Enter a diameter value or move the cross-hair to the desired location.*)
Command:

Once the circle center is selected, the center will drag across the screen but will not line up with the cross-hairs. Figure 8–5 shows an example of a circle being created using the diameter. Notice that the length of the line extending from the centerpoint is the diameter of the circle to be drawn. The line will disappear once the diameter is picked. The diameter entered will become the default setting for the next circle drawn.

Two-Point Circles

Drawing a two-point circle can be useful if the diameter is known, but the center-point is not. The two-point circle is drawn by picking a point and then picking a point on the opposite side of the circle as seen in Figure 8–6. The command sequence is:

Command: **C** [enter]
CIRCLE 3P/2P/TTR/<Centerpoint>: **2P** [enter]
First point on diameter: (*Enter coordinates or pick a point.*)
Second point on diameter: (*Enter coordinates or pick a point.*)
Command:

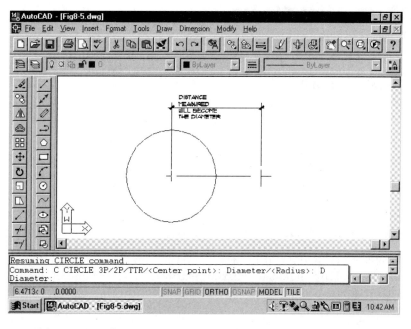

FIGURE 8–5 When using the Diameter option, the distance that the cursor is moved from the centerpoint will become the diameter of the circle.

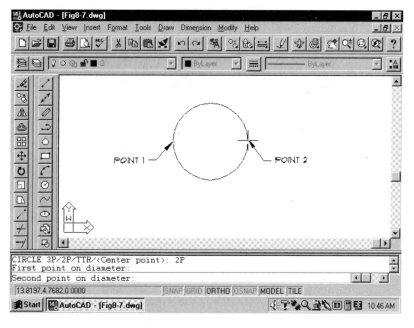

FIGURE 8–6 A circle can be drawn by specifying the endpoints of a line that would be the diamter.

Three-Point Circles

If certain points of a circle are known, a circle can be drawn by entering any three points. Figure 8–7 shows an example of a three-point circle. The points may be picked with the cross-hairs, or coordinates may be entered. The process for drawing a three-point circle is:

Command: **C** ⏎
CIRCLE 3P/2P/TTR/<Centerpoint>: **3P** ⏎
First point: (*Enter coordinates or pick a point.*)
Second point: (*Enter coordinates or pick a point.*)
Third point: (*Enter coordinates or pick a point.*)
Command:

Tangent, Tangent, and Radius

An object is tangent if it intersects another object at one and only one point. A circle may be drawn tangent to existing features by selecting two features that the circle will be tangent to, and then specifying a radius. This can be seen in Figure 8–8. The command sequence is:

Command: **C** ⏎
CIRCLE 3P/2P/TTR/<Centerpoint>: ⏎ **TTR** ⏎

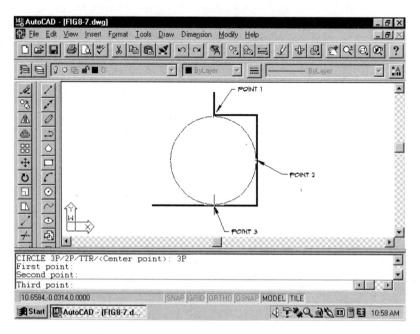

FIGURE 8–7 The size of a circle can be determined by specifying three tangent points.

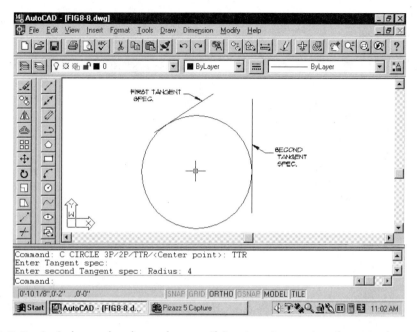

FIGURE 8–8 A circle can be drawn by specifying two tangent surfaces and a radius.

Enter Tangent spec: (*Select a line or circle.*)
Enter second Tangent spec: (*Select a line or circle.*)
Radius: (*Enter a numeric value.*) [enter]
Command:

In Figure 8–9 each line was selected as a tangent point and a radius was entered. Because the radius is smaller than the distance between the lines, the first line does not touch the circle. If the line were extended, it would be tangent to circle.

Tangent, Tangent, Tangent

What sounds like the name for a great Hollywood drafting movie is actually a CIR-CLE option for drawing a circle which is tangent to three objects. This option can only be accessed through Tan, Tan, Tan of the pull-down and screen menus. The command sequence would be:

Command: _circle 3P/2P/TTR/<Centerpoint>: _3p First point: _tan to (*Pick the first object the circle is to be tangent to.*)
Second point: _tan to (*Pick the second object the circle is to be tangent to.*)
Third point: _tan to (*Pick the third object the circle is to be tangent to.*)
Command:

Figure 8–10 shows the development of a circle tangent to three other circles.

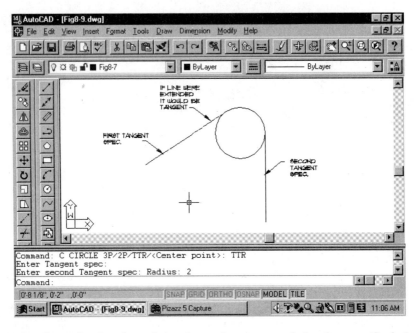

FIGURE 8–9 If a radius is selected that is not large enough for the specified circle to touch each line, a circle is drawn tangent to one line and tangent to where the other line would be if it were to be extended.

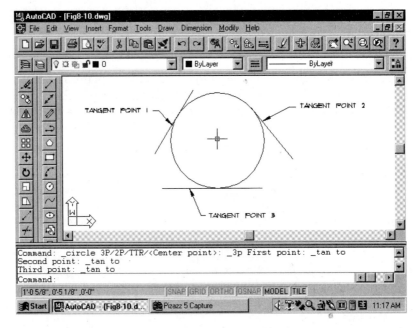

FIGURE 8–10 Tan, Tan, Tan: can be used to draw circles that are tangent to three surfaces.

Copy Rad

The Copy Rad option allows you to draw a circle with the same radius as an existing circle or arc. This is helpful if you wish to add a matching circular feature to a drawing, but have forgotten the original size. The Copy Rad option is only available from the screen menu. See Chapter 3 to activate the screen menu. The command sequence would be:

> Command: _circle 3P/2P/TTR/<Centerpoint>: *(Pick Copy Rad from the screen menu.)*
> New value for CIRCLERAD <1.000>: '_cal>> Expression: rad
> \>> Select circle, arc or polyline segment for RAD function: *(Pick the existing circle that you wish to duplicate.)*
> Command: **C** [enter]
> CIRCLE 3P/2P/TTR/<Centerpoint>: *(Pick desired centerpoint.)*
> Diameter/<radius> <1.000> [enter]
> Command:

Figure 8–11 shows the development of a circle using the Copy Rad option of the CIRCLE command. In addition to drawing circles, picking this command also activates the geometry calculator. This aspect of the option will be discussed in Chapter 17.

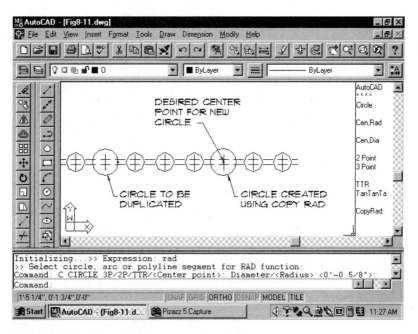

FIGURE 8–11 When Copy Rad is selected from the screen menu a circle with the exact same radius as an existing circle or arc can be drawn.

Circle@

This option allows a circle to be drawn using the last endpoint of the last line drawn for the centerpoint of a new circle. The option is best used for drawing concentric circles associated with pipes and circular columns. The command sequence would be:

Command: **C** [enter]
CIRCLE 3P/2P/TTR/<Centerpoint>: **@** [enter]
Diameter/<radius> <1.000>(*Pick desired diameter.*)
Command:

DRAWING ARCS

Arcs are used where a full circle is not required. Curved components are found throughout the structure seen in Figure 8–12. Arcs may be drawn by typing **ARC** [enter] or **A** [enter] on the command line which will produce the prompt:

Command: **A** [enter]
ARC Center/<Start point>:

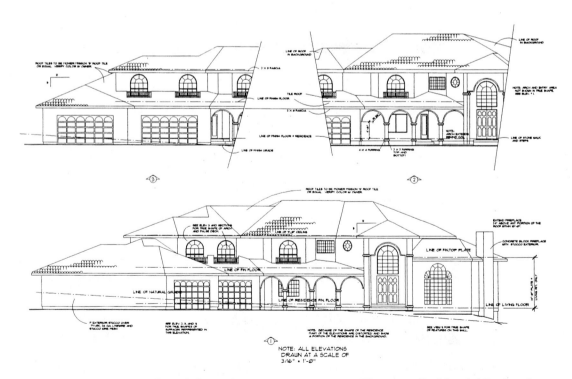

FIGURE 8–12 Arcs are found throughout many structures. (Courtesy Residential Designs)

The command can also be started by picking ARC from the Draw toolbar or by picking Arc from the Draw pull-down menu. The pull-down menu can be seen in Figure 8–13. Eleven different options are available for drawing arcs, depending on what information is available or on personal preference. The default setting is three-point arc.

With the exception of the three-point arc, each arc drawing method may be selected by typing a letter and [enter] or the space bar. Letters and the function of the arc they represent are:

> A: included Angle
>
> C: Center
>
> D: starting Direction
>
> E: Endpoint
>
> L: Length of chord
>
> R: Radius

Each of the terms used to describe an arc can be seen in Figure 8–14. Notice that center describes the center of the circle of which the arc is part, not the center of the arc.

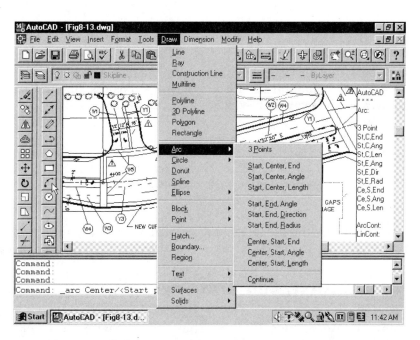

FIGURE 8–13   The Arc pull-down menu. (Courtesy Lee Engineering, Inc., Oregon City, OR)

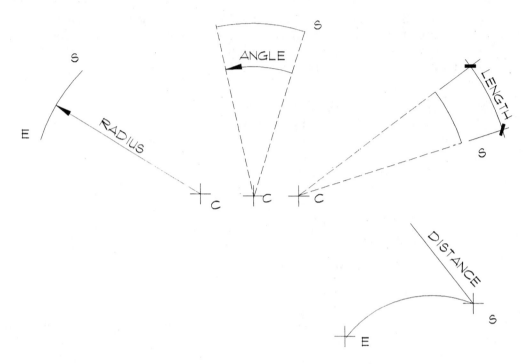

FIGURE 8–14 Terms used to describe arcs are Start, Center, End, Angle, Length, Radius, and Distance.

Three-Point Arcs

Drawing a 3-POINT arc is similar to drawing a 3P circle and can be seen in Figure 8–15. The first and third points are the arc end points; the second point is any point on the arc. The command sequence for a three-point arc is:

> Command: **A** [enter]
> ARC Center/Start point>: *(Select any point or enter coordinate.)*
> Center/End/ <second point>: *(Select second point.)*
> Endpoint: *(Select the arc's endpoint or enter coordinates.)*
> Command:

Notice that once the second point is selected, the arc will drag across the screen with the cross-hairs. The arc will continue to change size until the endpoint is picked.

Start, Center, End Arcs

When the ends and center of an arch are known, the St, C, End arc can be used. The stair railing in Figure 8–16 used a Start, Center, End arc. The command sequence is:

> Command: **A** [enter]
> Arc Center/<Start point>: *(Select any point or enter coordinates.)*

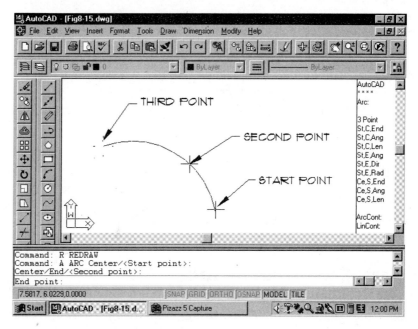

FIGURE 8–15 Drawing a three-point arc is similar to drawing a three-point circle.

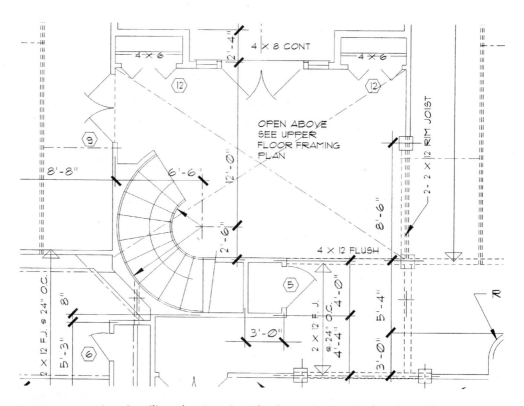

FIGURE 8–16 A stair railing drawn using the Start, Center, End option. (Courtesy Residential Designs)

233

Center/End<second point>: **C** [enter]
Center: (*Select any point or enter coordinates.*)
Angle/Length of Chord/<endpoint>: (*Select the arc's endpoint or enter coordinates.*)
Command:

As the last point is entered, the arc will be drawn and the command prompt will be returned. The process can be seen in Figure 8–17. Notice that as the last point is entered, the arc may not end at the point entered. The endpoint that was specified is used only to determine the angle at which the arc will end. The actual radius was determined by the start and Center points.

Start, Center, Angle Arcs

Often when doing road and utility layouts on a site plan it is necessary to draw an arc using lengths or angles. Drawing an arc using the Start, Center point and the included Angle is ideal for this type of work. When entering the angle, if a positive value is used, the angle will be drawn in a counterclockwise direction. If a negative value is used, the angle will be made in a clockwise direction. The effects of positive and negative values can be seen in Figure 8–18.

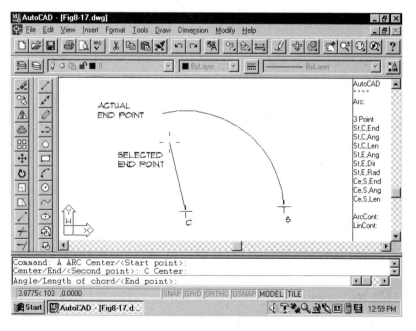

FIGURE 8–17 The selected endpoint is used to determine the angle of the specified circle. The actual size of the circle is determined by the distance from the Start to the Centerpoint.

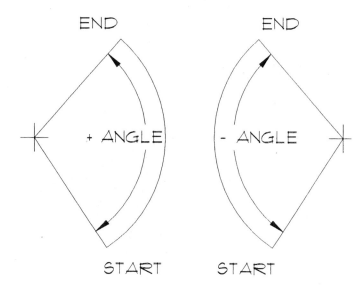

FIGURE 8–18 The direction of layout is controlled by entering a positive (counter-clockwise) or negative (clockwise) angle when drawing arcs with the Start, Center, Arc option.

The command sequence for a Start, Center, Angle arc is:

Command: **A** [enter]
ARC Center/Start point>: (*Select any point or enter coordinates.*)
Center/End/<second point>: **C** [enter]
Center: (*Select any point or enter coordinates.*)
Angle/Length of chord/<endpoint>: (*Select the desired angle or
    pick point.*) **30** [enter]
Command:

When the desired angle is entered, the arc will be drawn and the command line will return.

Start, Center, Length Arcs

A chord is the straight line connecting the endpoints of an arc. Subdivision maps and other drawings relating to land often include arcs that require use of the chord length. Figure 8–19 shows an example of arcs that could be drawn using Start, Center, and Length of chord. The arc will be drawn by entering the start and center-points and then supplying the chord length. A positive or negative value must be entered with the length. Figure 8–20 shows the difference between a chord of +3 or –3". The command sequence for an Start, Center, Length arc is:

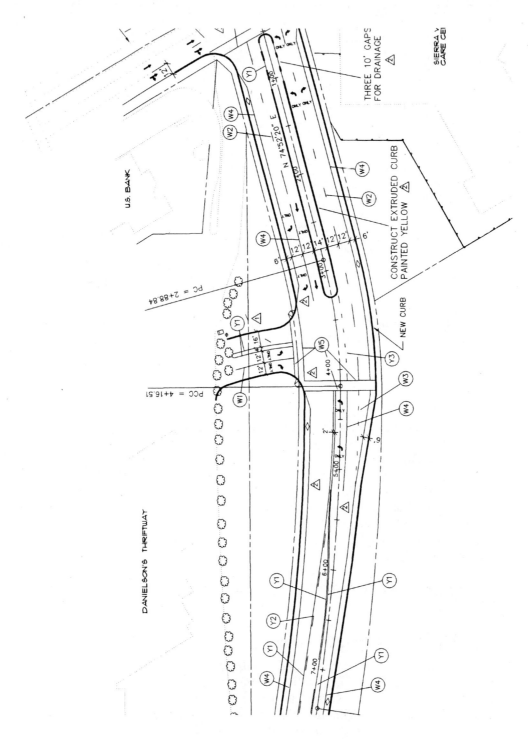

FIGURE 8-19 The Start, Center, Arc option is useful when drawing parcels of land. (Courtesy Lee Engineering, Inc., Oregon City, OR)

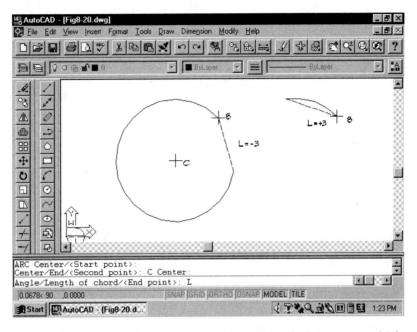

FIGURE 8–20 Entering a positive or negative chord length determines which portion of the arc will be drawn using the Start, Center, Length ARC option.

Command: **A** [enter]
ARC Center/<Start point>: (*Select any point or enter coordinates.*)
Center/End/<Second point>: **C** [enter]
Center: (*Select any point or enter coordinates.*)
Angle/Length of chord/<endpoint>: **L** [enter]
Length of chord:(*Select desired length.*) **3** [enter]
Command:

When a length is specified or a point is picked, the arc will be drawn and the command line will return.

Start, End, Angle Arcs

An arc drawn using Start, End, and Included Angle is drawn counterclockwise from the start point to the endpoint. If a negative angle is entered, the angle will be drawn clockwise. The results of coordinate entry can be seen in Figure 8–21. The command sequence is:

Command: **A** [enter]
ARC Center/<Start point>: (*Select any point or enter coordinates.*)
Center/End/<Second point>: **E** [enter]
Endpoint: (*Select the desired endpoint or enter coordinates.*)

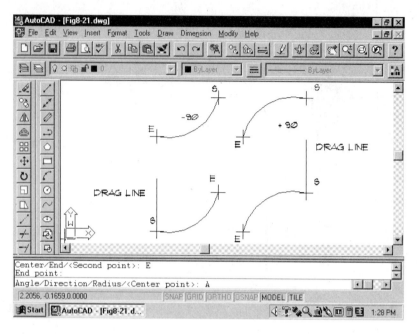

FIGURE 8–21 Entering a positive or negative angle determines which direction an arc will be drawn using the Start, End, Angle ARC option.

Angle/Direction/Radius/<Centerpoint>: **A** [enter]
Included angle: (*Select desired angle.*)
Command:

When a length is specified or a point is picked, the arc will be drawn and the command line will return.

Start, End, Direction Arcs

The Start, End, Direction option allows the direction of the arc to be picked. Major or minor arcs may be drawn in any orientation to the start point. Effects can be seen in Figure 8–22. The command sequence is:

Command: **A** [enter]
Arc Center/Start point>: (*Select any point or enter coordinates.*)
Center/End/<second point>: **E** [enter]
Endpoint: (*Select any point or enter coordinate.*)
Angle/Direction/Radius<Centerpoint>: **D** [enter]
Direction from start point: (*Select a point.*)
Command:

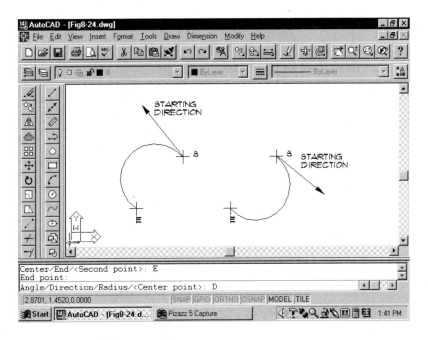

FIGURE 8–22 The Start, End, Direction Arc option allows the direction of the arc to be selected.

When a length is specified or a point is picked, the arc will be drawn and the command line will return.

Start, End, Radius Arcs

Start, End, Radius arcs can only be drawn counterclockwise. Because the variables can produce different options, the arc is drawn counterclockwise from the Start point. Entering a positive or negative value for the radius will determine if the major or minor arc will be drawn. The effects of the value can be seen in Figure 8–23. The command sequence is:

Command: **A** ⏎
Arc Center/Start point>: (*Select any point or enter coordinates.*)
Center/End/<second point>: **E** ⏎
Endpoint: (*Select any point or enter coordinate.*)
Angle/Direction/Radius<Centerpoint>: **R** ⏎
Radius: (*Select the desired radius or enter a point.*)
Command:

When a length is specified or a point is picked, the arc will be drawn and the command line will return (see Figure 8–24).

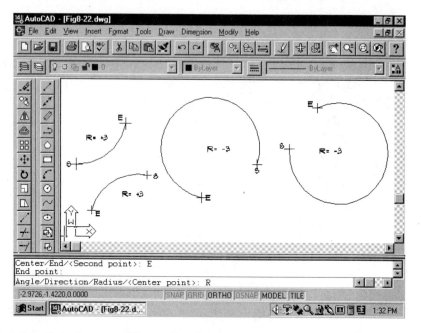

FIGURE 8–23 Entering a positive radius determines which portion of the arc will be drawn using the Start, End, Radius ARC option. The direction will always be counter-clockwise.

Center, Start, End Arcs

The Center, Start, End option is similar to the Start, Center, End option. The effects of each command can be seen in Figure 8–25. The command sequence is:

> Command: **A** enter
> ARC Center/<Start point>: **C** enter
> Center: (*Select any point or enter coordinates for the Centerpoint.*)
> Start point: (*Select any point or enter coordinates to start the arc.*)
> Angle/Length of chord/<endpoint>: (*Select desired endpoint.*)
> Command:

When a length is specified or a point is picked, the arc will be drawn and the command line will return.

Center, Start, Angle Arcs

The Center, Start, Angle option is similar to the Start, Center, Angle option. The effects of each command can be seen in Figure 8–26. The command sequence for this arc is:

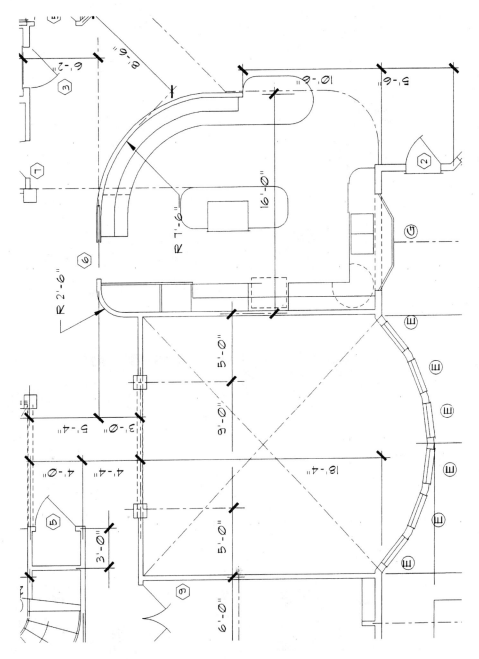

FIGURE 8–24 Walls drawn using the Start, End, Radius arc option. (Courtesy Residential Designs)

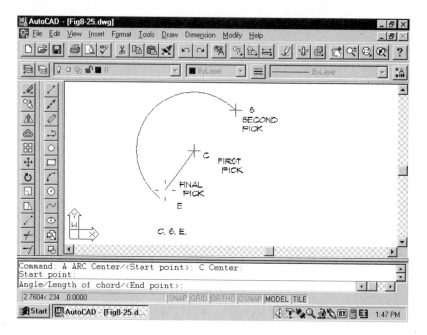

FIGURE 8–25   The Center, Start, End ARC option is similar to the Start, Center, End option.

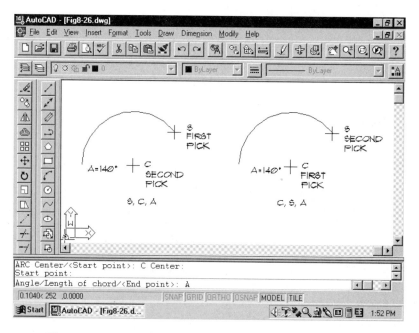

FIGURE 8–26   The Center, Start, Angle ARC option produces similar results to the Start, Center, Angle option.

Command: **A** [enter]
ARC Center/<Start point>: **C** [enter]
Center: (*Select any point or enter coordinates for the centerpoint.*)
Start point: (*Select any point or enter coordinates to start the arc.*)
Angle/Length of chord/<Endpoint>: **A** [enter]
Included angle: (*Select desired angle.*)
Command:

When a length is specified or a point is picked, the arc will be drawn and the command line will return.

Center, Start, Length Arcs

The Center, Start, Length of chord option is similar to the Start, Center, Length option. The effects of each command can be seen in Figure 8–27. The command sequence is:

Command: **A** [enter]
ARC Center/<Start point>: **C** [enter]
Center: (*Select any point or enter coordinates for the centerpoint.*)
Start point: (*Select any point or enter coordinates for the start point.*)
Angle/Length of chord/<Endpoint>: **L** [enter]

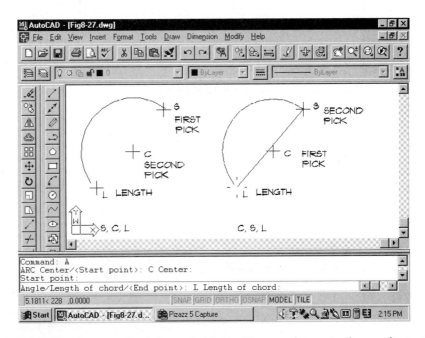

FIGURE 8–27 The Center, Start, Length ARC option produces similar results to the Start, Center, Length option.

Length of chord: (*Select desired length.*)
Command:

When a length is specified or a point is picked, the arc will be drawn and the command line will return.

Continue:

Similar to the MULTIPLE LINE command, an arc may be continued from the previous arc. The ARC command may be continued by pressing the [enter] key or the space bar, or by picking Continue from a menu in lieu of picking the first arc entry point. By continuing from a previous arc, the current arc's start point and direction are taken from the endpoint and ending direction of the last arc. Figure 8–28 shows an example of four continuous arcs. The command sequence is:

ARC 1

Command: **A** [enter]
Center/<Start point>: (*Select any point or enter coordinates.*)
Center/End<second point>: (*Select any point or enter coordinates.*)
Endpoint: (*Select the arc's endpoint or enter coordinates.*)

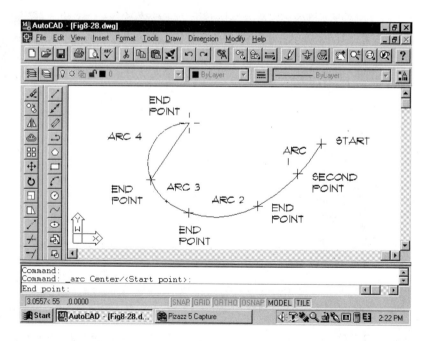

FIGURE 8–28   The Continue: option of ARC allows for multiple arcs to be drawn.

ARC 2

Command: [enter] (*To resume arc from last endpoint.*)
ARC Center/<Start point>: (*Select desired arc point.*)

ARC 3

Command: [enter]
ARC Center/<Start point>: [enter] (*To resume arc from last endpoint.*)
Endpoint: (*Select the arc's endpoint or enter coordinates.*)
Command:

ARC 4

Command: [enter]
ARC Center/<Start point>: [enter] (*To resume arc from last endpoint.*)
Endpoint: (*Select the arc's endpoint or enter coordinates.*)
Command:

The Continue option also can be used to extend an arc tangent to the endpoint of a line. Figure 8–29 shows an example of continuous arcs drawn from a straight line. The command sequence is:

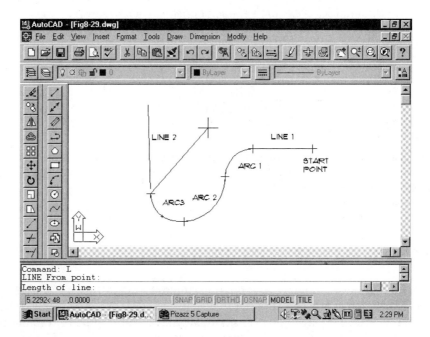

FIGURE 8–29 Continue: can be used to extend an arc tangent to the last line segment.

LINE 1

> Command: **L** [enter]
> From point: (*Enter point or coordinate.*)
> To point: (*Enter point or coordinate.*)
> To point: [enter]
> Command:

ARC 1

> Command: **A** [enter]
> Center/<Start point>: [enter] (*This will start at the end of the last line drawn.*)
> Endpoint: (*Select the arc's endpoint or enter coordinates.*)

ARC 2

> Command: [enter]
> ARC Center/<Start point>: [enter] (*To resume arc from endpoint of last arc.*)
> Endpoint: (*Select the arc's endpoint or enter coordinates.*)

ARC 3

> Command: [enter]
> ARC Center/<Start point>: [enter] (*To resume arc from endpoint of last arc.*)
> Endpoint: (*Select the arc's endpoint or enter coordinates.*)

LINE 2

> Command: **L** [enter]
> From point: [enter]
> To point: (*Enter point or coordinate.*)
> To point: [enter] (*To end drawing sequence.*)

Notice in Figure 8–29 that line 2 was drawn tangent to arc 3 because Continue: was used. To place a line that continues from an arc but is not tangent to the arc will require the arc endpoint to be selected as the LINE "From point" using the methods described in Chapter 9.

USING THE ELLIPSE COMMAND

Although not a major part of engineering drawing, ellipses sometimes are used for construction symbols. Ellipses are also used to show round objects that are not being seen in true orthographic projection, such as a round pipe that is not perpendicular to the viewing plane of the drawing. To enter the ELLIPSE command, pick the Ellipse icon from the Draw toolbar, type **EL** [enter] or **ELLIPSE** [enter] at the command prompt, or select Ellipse from the Draw pull-down menu. Menu options for ELLIPSE can be seen in Figure 8–30. Figure 8–31 shows the major and minor axis that will be referred to in laying out an ellipse. To draw an ellipse at the command prompt, type:

Command: **EL** enter
Arc/ Center/<Axis endpoint 1>:(*Select a point or enter coordinates for the endpoint.*)
Axis endpoint 2: (*Select a point or enter coordinates for the endpoint.*)
<Other axis distance>/Rotation: (*Select a point or enter coordinates for the endpoint.*)
Command:

An ellipse is now shown on the screen between the two selected points. The process can be seen in Figure 8–32. AutoCAD is waiting for you to enter the axis or rotation. If the angle is not known, the cross-hairs can be moved until the ellipse represents the desired shape.

If you know the ellipse represents a pipe that is at a 60° angle to the viewing plane, enter **R** enter. This will produce the prompt:

Command: **EL** enter
Arc/ Center/<Axis endpoint 1>: (*Select a point or enter coordinates for the endpoint.*)

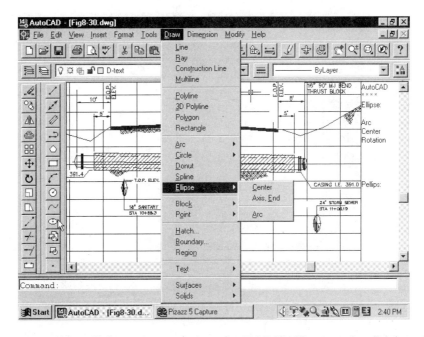

FIGURE 8–30 The pull-down menu options for ELLIPSE. (Courtesy Lee Engineering, Inc., Oregon City, OR)

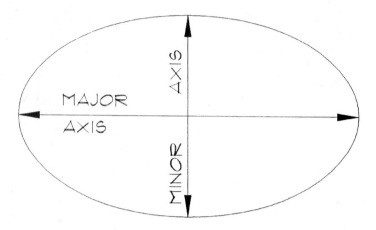

FIGURE 8–31 An ellipse is defined by the major and minor axis.

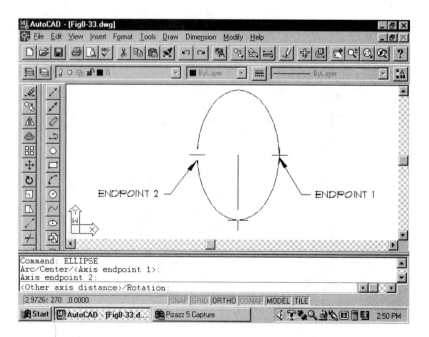

FIGURE 8–32 Once the two endpoints are specified, the ellipse will rubberband as the cursor is moved.

Axis endpoint 2: (*Select a point or enter coordinates for the endpoint.*)
<Other axis distance>/Rotation: **R** [enter]
Rotation around major axis: (*Enter desired angle of rotation.*) **60** [enter]
Command:

and will produce the ellipse shown in Figure 8–33. Figure 8–34 shows the effect of rotating a circular pipe at 10° increments. An ellipse with a 90° rotation cannot be drawn using the ELLIPSE command since it would produce a single straight line.

An ellipse also can be drawn by locating the centerpoint and then specifying the end-points of each axis. This process can be seen in Figure 8–35. The command process is:

Command: **EL** [enter]
Arc/ Center/<Axis endpoint 1>: **C** [enter]
Center of ellipse: (*Select a point or enter coordinates for the centerpoint.*)
Axis endpoint: (*Select a point or enter coordinates for the endpoint.*)
<Other axis distance>/Rotation: (*Select a point or enter coordinates.*)
Command:

When the second axis endpoint is entered the command prompt is returned and the ellipse is drawn on the screen.

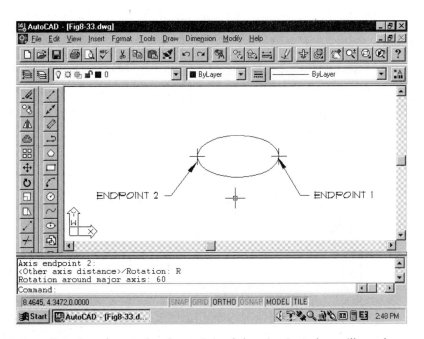

FIGURE 8–33   Entering the angle of rotation of the viewing plan will produce an ellipse. Using a 60 degree rotation produces the ellipse shown.

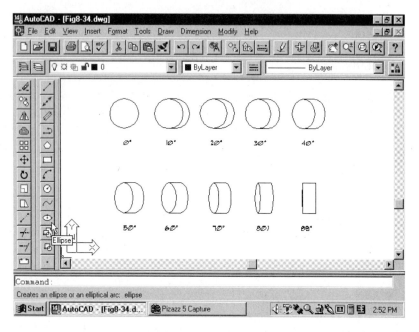

FIGURE 8–34 The effects of rotating a circular object in 10° increments.

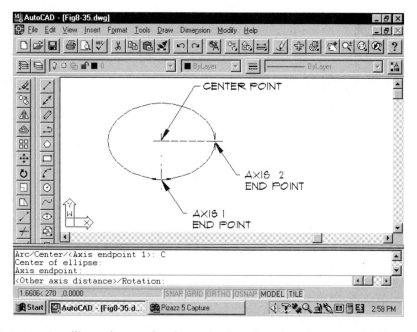

FIGURE 8–35 An ellipse also can be drawn by locating the centerpoint followed by an endpoint for the major and minor axis.

DRAWING POLYGONS

A polygon is any geometric shape lying on one plane, bound by three or more straight lines. Triangles, squares, pentagons, hexagons, and octagons are used as symbols and material shapes in construction drawings. AutoCAD draws polygons with equal length lines and angles having between three and 1,024 sides. To draw a polygon, you will be asked to choose the number of sides and whether the object is to be inscribed in or circumscribed around a circle. The effects of a triangle inscribed in and circumscribed around 2.5" circle can be seen in Figure 8–36.

The polygon command can be accessed by picking the Polygon icon from the Draw toolbar, by typing **POL** ⏎ or **POLYGON** ⏎ at the command prompt, or by picking Polygon from the Draw pull-down menu. Figure 8–37 shows the pull-down menu. Typing **POL** ⏎ at the command prompt will produce the following prompts:

> Command: **POL** ⏎
> Number of sides <4>: (*Select the desired number of sides.*) **8** ⏎
> Edge/<Center of polygon>: (*Select a point or enter coordinates for the Center-point.*)
> Inscribed in circle/ Circumscribed about circle (I/C) <I>: (*Accept I or select C depending on need.*) ⏎

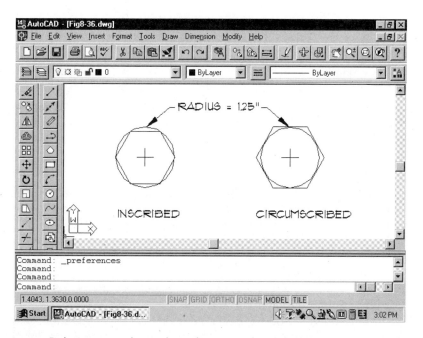

FIGURE 8–36 Polygons are drawn based on a circle. Once the number of sides has been specified, you will be asked to specify whether the polygon will be placed inside or outside the circle.

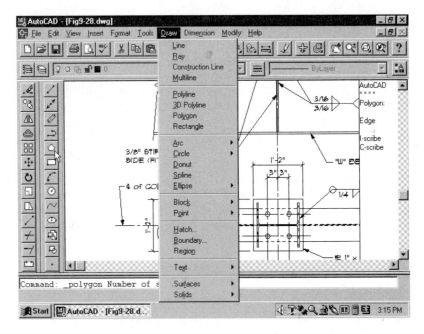

FIGURE 8–37 The Pull-down menu for POLYGON. (Courtesy Lee Engineering, Inc., Oregon City, OR)

Radius of circle: (*Enter desired radius.*) **2.75** [enter]
Command:

The results of the sequence can be seen in Figure 8–38. The setting of ORTHO will affect how the polygon will be placed. When Ortho is ON, the edges of the polygon will be parallel or perpendicular to the drawing screen. When Ortho is OFF, the edges of the polygon will rotate based on the movement of the mouse. To activate ORTHO, double-click the Ortho button in the Task bar with the mouse select button.

A polygon also can be drawn based on the location of one edge rather than the location of the centerpoint. Once the number of sides has been provided, you will be asked for an edge or centerpoint. By entering **E** [enter] at the prompt, a polygon similar to the one in Figure 8–39 can be drawn. The command sequence is:

Command: **POL** [enter]
Number of sides <4>: (*Select the desired number of sides.*) **6** [enter]
Edge/<Center of polygon>: **E** [enter]
First endpoint of edge: (*Select a point or enter coordinates for the first endpoint.*)
Second endpoint of edge: (*Select a point or enter coordinates for the second endpoint.*)
Command:

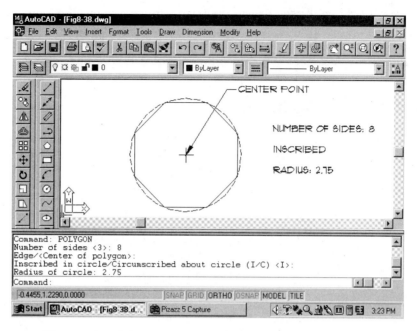

FIGURE 8–38 The results of drawing an inscribed octagon with a 2.75" radius.

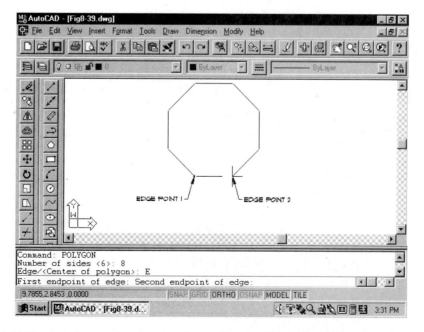

FIGURE 8–39 A polygon also can be drawn by specifying the points that define one edge.

As the first point is entered, a polygon will be displayed which will vary in size as the mouse is moved to the second point. As the second point is entered, the command prompt will be returned and the polygon will be drawn. The setting of ORTHO will affect the placement of a polygon using the Edge option.

DRAWING RECTANGLES

A square can be drawn using the POLYGON command. A rectangle is drawn by selecting the Rectangle icon from the Draw toolbar, by entering **REC** [enter] at the command prompt, or by selecting Rectangle from the Draw pull-down menu. The command sequence is:

Command: **REC** [enter]
Chamfer/Elevation/Fillet/Thickness/Width/<First corner>: (*Select an option or pick a corner or the rectangle with the mouse.*)
Other corner: (*Select the second corner of the rectangle.*)
Command:

As the cursor is moved between the first and second corner, a rectangle will be displayed on the screen allowing the exact size to be determined. The sequence can be seen in Figure 8–40. Selecting either the Chamfer or Fillet option allows the corners of the rectangle to be altered. Selecting the Width option allows the width of the line

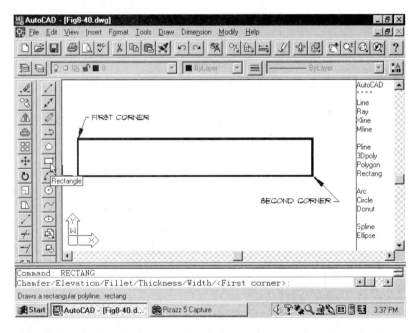

FIGURE 8–40 Rectangular shapes can be drawn as one entity using the Rectangle option of the pull-down menu or by typing **REC** [enter] at the command line.

to be altered. The other options only effect a rectangle drawn in 3D and will not be discussed in this chapter.

Chamfer

Drawing a rectangle with the Chamfer option active will place a chamfer on each corner of the rectangle. Once the Chamfer option is selected, prompts will be given for the first and second chamfer distances. If the distances are equal, the result will be as shown in Figure 8–41. Providing different values will produce a rectangle similar to the example on the right in Figure 8–41. The command sequence to draw a rectangle with a chamfer is:

Command: **REC** [enter]
Chamfer/Elevation/Fillet/Thickness/Width/<First corner>: **C** [enter]
First chamfer distance for rectangles <0'-0">: *(Enter the desired size.)* **1** [enter]
Second chamfer distance for rectangles <0'-1">: [enter]
Chamfer/Elevation/Fillet/Thickness/Width/<First corner>: *(Select a corner of the rectangle with the mouse.)*
Other corner: *(Select the second corner of the rectangle.)*
Command:

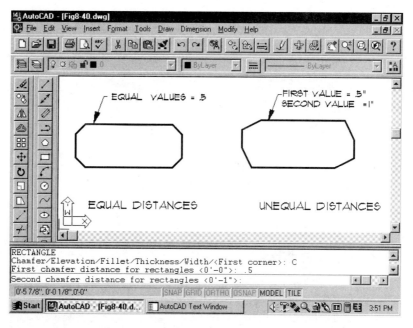

FIGURE 8–41 The Chamfer option of the RECTANGLE command can be used to provide a chamfer at each corner of a rectangle. The left option shows distance 1 and 2 with a value of .5". The right option shows distance 1 with a value of .5" and distance 2 with a value of 1".

Fillet

Drawing a rectangle with the Fillet option active will place a fillet on each corner of the rectangle. Once the Fillet option is selected, a prompt will be given to provide a radius for the fillet. Figure 8–42 shows an example of a rectangle with filleted corners. The command sequence to draw a rectangle with a fillet is:

Command: **REC** (enter)
Chamfer/Elevation/Fillet/Thickness/Width/<First corner>: **F** (enter)
Fillet radius for rectangles <0'-1">: *(Enter the desired radius)* **.5** (enter)
Chamfer/Elevation/Fillet/Thickness/Width/<First corner>: *(Select a corner of the rectangle with the mouse.)*
Other corner: *(Select the second corner of the rectangle.)*
Command:

Notice that the default radius value is <0'-1">. The value that was provided for the chamfer will become the default value for all future chamfer and fillets until a new value is provided.

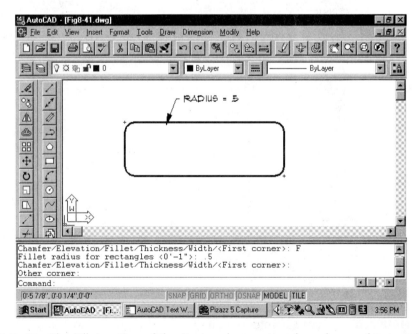

FIGURE 8–42 The Fillet option of the Rectangle command can be used to provide a fillet at each corner of a rectangle. The value used for the radius will remain constant for future use until it is altered.

Width

Selecting the Width option allows the line thickness to be altered. Because the fillet option was just used, setting the thickness for a rectangle will result in a rectangle with 2" radius corners to be drawn with a thick line. To draw a rectangle with square corners, enter the RECTANGLE command and alter the value for Fillet to 0, and then set the thickness, or vise versa. The command sequence is:

Command: **REC** ⏎
Chamfer/Elevation/Fillet/Thickness/Width/<First corner>: **F** ⏎
Fillet radius for rectangles <0'-1">: *(Enter the desired radius)* **0** ⏎
Chamfer/Elevation/Fillet/Thickness/Width/<First corner>: **W** ⏎
Width for rectangles <0'-0">: **.25** ⏎
Chamfer/Elevation/Fillet/Thickness/Width/<First corner>(*Select a corner of the rectangle with the mouse*).
Other corner: (*Select the second corner of the rectangle.*)
Command:

The results of this command sequence can be seen in Figure 8–43. The width will remain the current width for all future rectangles until the value is altered. Methods of providing thickness to other entities will be discussed in Chapter 13 as Polylines are introduced.

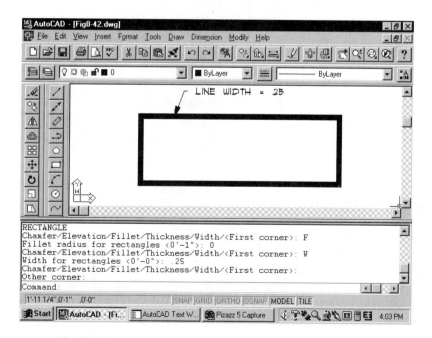

FIGURE 8–43   The Width option can be used to assign a line width to the RECTANGLE command.

DRAWING DONUTS

Sorry if you've started thinking about something covered with chocolate or coconut. A donut in AutoCAD is a filled circle or ring. The command can be accessed by typing **DO** [enter], **DONUT** [enter] or **DOUGHNUT** [enter] at the command prompt, or by selecting Donut from the Draw pull-down menu. Each can be seen in Figure 8–44.

Three different options are available for drawing donuts: a ring, a filled circle that is solid black, or a circle filled with lines. Each can be seen in Figure 8–45. The fill material is determined by the status of the FILL command. Typing **FILL** [enter] at the command prompt will allow for a choice to be made for DONUT and several other options. By entering OFF, filled objects would resemble the bottom circle in Figure 8–45. With FILL ON, filled objects will resemble the middle circle in Figure 8–45.

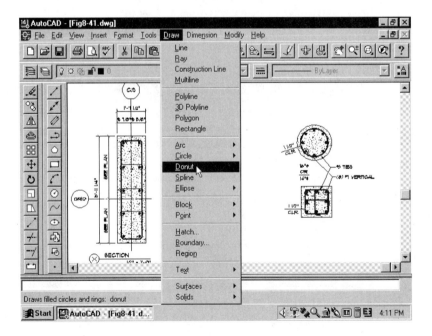

FIGURE 8–44 Donuts may be drawn using the pull-down menu or from the comand prompt.

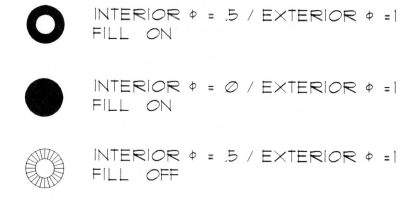

FIGURE 8–45 Three options are available for drawing a donut.

CHAPTER 8 EXERCISES

Save the following exercises on your floppy disk.

E-8-1. Draw a 3" diameter circle, a circle with a 1" radius, and a 1" diameter circle that is tangent to two 1" long perpendicular lines. Save the drawing as E-8-1.

E-8-2. Draw two parallel lines that are 1" long and 1" apart, and provide an arc at each end of the lines to form a slotted hole. Save the drawing as E-8-2.

E-8-3. Draw a 4" line. Use the continuous option to place a 1"- diameter circle at each end of the line. Save the drawing as E-8-3.

E-8-4. Draw a box formed by three 1" long lines. Draw an arc to close the polygon. Save as drawing E-8-4.

E-8-5. Draw an arc with a start point that is 2" to the right of the endpoint with an included angle of 232°. Save as drawing E-8-5.

E-8-6. Draw an arc with an endpoint that is @2,2 from the centerpoint with an included angle of 45°. Save as drawing E-8-6.

E-8-7. Draw an ellipse with a 3" long axis at a 42° rotation. Save as drawing E-8-7.

E-8-8. Draw an ellipse with a centerpoint that is 2.578 inches from the end of the major axis and 1" from the minor axis. Draw a four-sided polygon using the center of the ellipse as the center of the polygon. Make two edges of the square touch the ellipse. Save as drawing E-8-8.

E-8-9. Draw the following polygons:

a.	3 sides	inscribed	2″ diameter
b.	3 sides	circumscribed	2″ diameter
c.	4 sides	inscribed	.5″ radius
d.	4 sides	circumscribed	.5″ radius
e.	6 sides	inscribed	.75″ radius
f.	6 sides	circumscribed	.75″ radius
g.	8 sides	inscribed	1.25″ radius
h.	8 sides	circumscribed	1.25″ radius

Save the drawing as E-8-9.

E-8-10. Draw a donut with an inside diameter of .25″, outside diameter of .75″, and fill set at on. Draw another donut with an inside diameter of .0″ and an outside diameter of .375″ with fill set at off. Draw a donut with an inside diameter of 0 and an outside diameter of .5″ with Fill set to On. Save the drawing as E-8-10.

E-8-11. Draw a rectangle with filleted corners using a 2″radius. Save the drawing as E-8-11.

E-8-12. Draw a rectangle with chamfered corners using a 1″ distance for both settings. Save the drawing as E-8-12.

E-8-13. Draw a rectangle with chamfered corners using a 1.5″ distance for the first distance and a 2.5 value for the second setting. Assign a line width of .065 and save the drawing as E- 8-13.

E-8-14. Open a new drawing and draw at least three connected combinations of lines and arcs. Save the Drawing as E-8-14.

E-8-15. Draw two perpendicular lines that are two inches long and form a corner. Draw a 3″- diameter circle that is tangent to each of the lines. Draw three lines that are not connected and are neither parallel nor perpendicular to each other. Draw a circle that is tangent to all three line segments. Save this drawing as E-8-15.

CHAPTER 8 QUIZ

1. What is the maximum number of sides that can be drawn for a polygon? _____

2. When using the ARC command, what do the following letters represent?

St _____

Ce _____

Len _____

E _____

Ang _____

R _____

D _____

3. What is the default method for drawing arcs? _____

4. What is the command sequence to draw three continuous arcs?

_____

_____

_____

5. What command determines whether the interior of a doughnut will be black or hatched with lines?

_____

6. What is the difference between inscribed and circumscribed and what are the effects on a polygon?

_____

_____

7. What four components can be used to construct an ellipse?

_____

_____

8. When would TTR be used for drawing a circle?

_____

_____

9. What are the options available for drawing circles?

_____

_____

10. What command controls the rubberband effect of geometric shapes?

_____

_____

11. What options must be entered before a donut can be drawn?

_____

_____

12. What is the largest angle rotation that can be used with an ellipse? _____

13. What shape would a polygon with 500 sides appear as? _____

14. How does ORTHO affect the drawing of a polygon?

    _____

    _____

15. What process is used to set the program to drag objects only at your request?

    _____

    _____

    _____

9

DRAWING TOOLS

· ·

In Chapter 4 you discovered how to use the SNAP command to make the cross-hairs move at a specific interval. As you begin to combine geometric shapes into drawings it is often helpful to join the shapes at specific locations. Object snap (OSNAP) will allow a line or other entity to be snapped to a specific point. For instance, OSNAP will allow a line to be started exactly at the midpoint of an existing line, with absolutely no guessing. OSNAP has two modes of operation, which include single-point override and running. Each is used by picking an object with an aperture box.

APERTURE BOX

The normal cursor for the drawing area is a cross hair surrounded by a box. If you enter a drawing command such as LINE, the box is removed from the cross hairs as you pick the From: and To: points. As OSNAP is activated, a target box is added to the cross-hairs, as seen in Figure 9–1. The shape of the aperture will vary depending on the mode being used. Various shapes will be introduced later in this chapter. Any object that lies within the target or aperture box is subject to the OSNAP command. Although objects may be added or deleted from the target box, this is time consuming. To obtain higher accuracy in entity selection, the size of the aperture box can be adjusted. To adjust the size of the aperture box, at the command prompt type:

Command: **APERTURE** ⏎
Object snap target height (1-50 pixels) <10>:

A numeric value can now be entered. The larger the number selected, the larger the box will be. Figure 9–2 shows the effect of changing the value. The value entered will

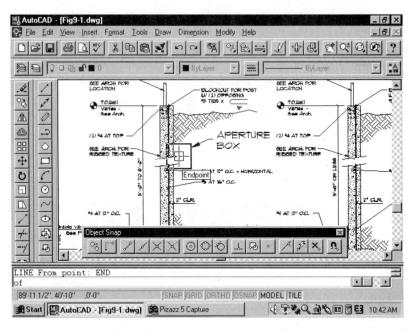

FIGURE 9–1 As OSNAP is activated, an aperture box is added to the cross-hairs. The aperture box is used for selecting existing drawing entities to be used as a selection point for a new line, circle, arc, or ellipse. (Courtesy Van Domelen/Looijenga/McGarrigle/Knauf Consulting Engineers.)

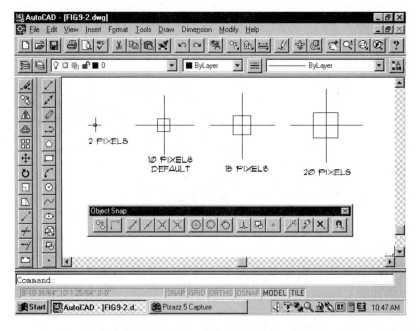

FIGURE 9–2 As the value for the height of the target box is altered, the height is changed.

now be the default for all future OSNAP commands until a new value is entered. Changing the default for the OSNAP aperture box will not affect the size of the pick box used for erase.

The size of the aperture box also can be adjusted by typing **SETVAR** [enter] at the command prompt. This will produce the prompt:

Command: **SETVAR** [enter]
Variable name or ?: **APERTURE** [enter]
New value for APERTURE <10>:

If the question mark is selected, a list of variables will be listed.

The size of the aperture box can also be set by using the Running Object Snap dialog box shown in Figure 9–3. The dialog box is displayed by picking the Object Snap Setting icon from the Object Snap toolbar, by picking Running Object Snap from Toolbars... in the View pull-down menu, or by typing **OSNAP** [enter] at the command prompt. Picking and holding the slide bar allows the size of the aperture box to be altered. The size of the aperture box is altered in the display box as the slide box is moved.

OSNAP

Joining objects by eye is difficult and should rarely be done. Objects that appear to be tangent may overlap when the ZOOM command is used. (See Figure 9–4.) Perfect intersections can be achieved by using the OSNAP options. OSNAP can be used to select a start or endpoint for connecting lines, so that rather than hoping you're close, exact points can be determined. The desired OSNAP can be activated by selecting the OSNAP icon from the Object Snap toolbar from the Osnap icon in the standard toolbar. The toolbar can be activated by picking Object Snap from Toolbars... of the View pull-down menu. Figure 9-5 shows a listing of each Object Snap mode found in the Object Snap toolbar and the aperture that will be displayed with each mode. The Running Object Snap dialog box allows for picking the desired object snap mode for continuous use and for changing the size of the aperture box. Running OSNAP will be discussed later in this chapter. Although four methods have been given to access OBJECT SNAP, the easiest and quickest way is often by the toolbar or the keyboard. No matter how they are accessed, using the OSNAP options will save time and greatly improve your accuracy.

Once OSNAP is activated, the aperture box is used to select objects to snap to. Move the cross-hairs so that the desired object lies within the aperture box, then press the pick button. If more than one object lies within the aperture box, the closest object will be selected. Two options are available if no objects are found; these will be discussed after we explore the OSNAP modes.

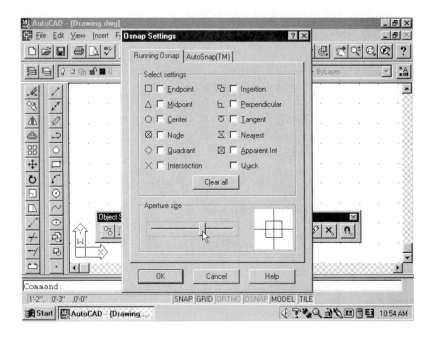

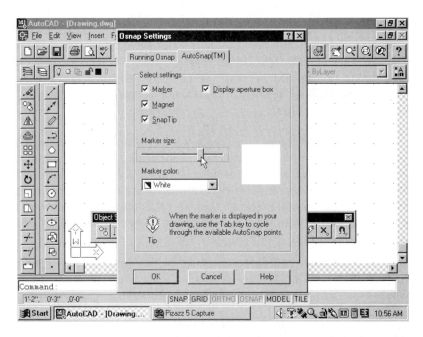

FIGURE 9–3 Options for OSNAP and the aperture box size can be adjusted by using the Running Object Snap dialog box. The box can be access by typing **OSNAP** [enter] at the command prompt.

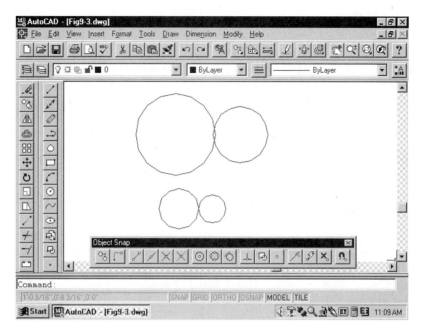

FIGURE 9–4 Objects such as the lower circles that are placed by eye often appear to be perfectly placed. As the scale is changed, or the drawing is Zoomed, circles that appeared tangent are crossing.

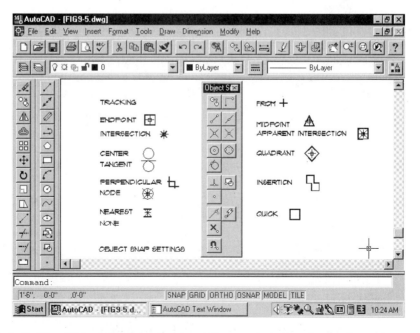

FIGURE 9–5 The Object Snap toolbar and the aperture box for each mode.

OSNAP MODES

OSNAP tools include ENDpoint, MIDpoint, INTersection, APParent Intersection, CENter, QUAdrant, TANgent, PERpendicular, INSertion, NODe, NEArest, QUIck, From, and None. These options can be used for drawing lines that connect at certain points to existing lines, circles, arcs, or ellipses. Osnap snap options also can be used for COPY, MOVE, and INSERT commands, which will be discussed in later chapters. Notice that as the options were listed, the first three letters of some of the options were written in capitol letters. Typing these letters at the prompt for start point or endpoint will activate the desired object snap option.

Endpoint

This option allows an arc, line, or circle to be drawn to or from the endpoint of a previous line or arc by snapping to the desired endpoint. The process is started by picking the Endpoint icon from the Object Snap toolbar when you are prompted for a From point. To add a line that extends between the endpoints of two existing lines, the command sequence would be:

Command: **L** [enter]
From point: (*Select the Endpoint icon*)

Now move the cursor near the end of the desired line that you would like the new line to connect to. The cursor does not have to touch the end of the line but must be between the midpoint and endpoint so the correct end of the line is selected. As the cursor touches a line, the aperture box is displayed around the nearest endpoint . See Figure 9–6. Press the mouse pick button, and the new line segment will be connected to the selected endpoint. Pick the Endpoint icon again at the prompt for the To point: to connect the new line to the end of the second line. The new line will automatically end at the selected end of the existing line segment.

The endpoint can also be selected at the keyboard by entering **END** [enter] as the "From point" of the desired command. Move the aperture box and pick the desired line. To draw a line from the endpoint of an existing line to the end-point of a second line use the following command sequence:

Command: **L** [enter]
LINE From point: **END** [enter]
of (*Select desired line.*)
To point: **END** [enter]
of (*Select desired line.*)
To point: [enter]
Command:

The command sequence can be seen in Figure 9–7.

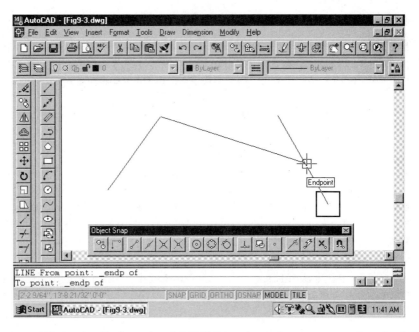

FIGURE 9–6 When the End mode of OSNAP is selected, an aperture box is placed on the end point of the selected line to indicate the selected From or To point.

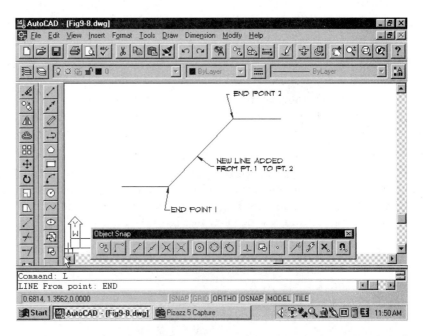

FIGURE 9–7 To draw a line from the ENDpoint of one line to the ENDpoint of another, pickup the Endpoint icon or type **END** [enter] when the "From point" and the "To point" prompts of the LINE command are displayed.

Midpoint

MIDpoint allows for the selection of a line or arc at its midpoint by picking the Midpoint icon in response to the From point: prompt. To add a line that extends between the middle of two existing lines, the command sequence would be:

Command: **L** (enter)
From point: (*Select the Midpoint icon*)

Now move the cursor near the desired line that you would like the new line to connect to. As the cursor touches a line, the aperture box is displayed around the nearest endpoint. Press the pick button and the new line segment will be connected to the selected midpoint of the existing line. Pick the Midpoint icon as the To point: is requested at the prompt and the new line will automatically end at the selected midpoint of the existing line segment.

The midpoint can also be selected by entering **MID** as the "From" or "To" point, an arc or line can extend from or to the midpoint of a line or arc. To draw a line from the midpoint of an existing line to the midpoint of a second line use the following command sequence:

Command: **L** (enter)
LINE From point: **MID** (enter)
of (*Select desired line.*)
To point: **MID** (enter)
of (*Select desired line.*)
To point: (enter)
Command:

The command sequence can be seen in Figure 9–8.

Intersection

As its name implies, INTersection allows a line or arc to be extended to or from the intersection of any lines, arcs, or circles. An example of the use of the intersection option can be seen in Figure 9–9. The command sequence is similar to other OSNAP command entries and can be seen in Figure 9–10. By picking the intersection icon from the Object Snap toolbar or by entering **INT** (enter) as a "From" or "To" point the desired intersection can be snapped to. The command sequence is:

Command: **L** (enter)
LINE From point: **INT** (enter)
To point: (*Select desired to point.*)
To point: (enter)
Command:

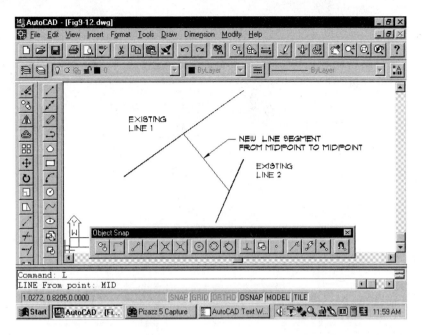

FIGURE 9–8 The MIDpoint option can be used to draw a line from the midpoints of one line to another by picking the Midpoint icon or by typing **MID** ⏎ for the "From point" and the "To point" prompts of the LINE command.

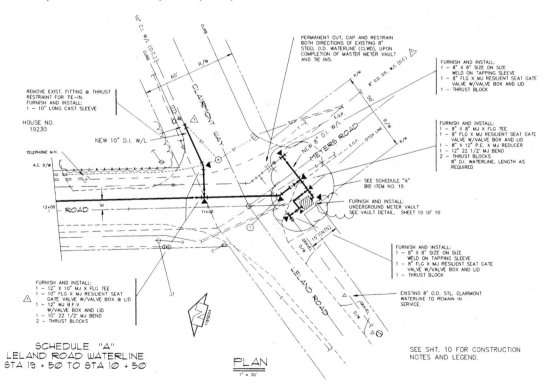

FIGURE 9–9 The INTersection option of OSNAP allows lines and arcs to be connected to the intersection of other geometric shapes. (Courtesy Lee Engineering, Inc., Oregon City OR)

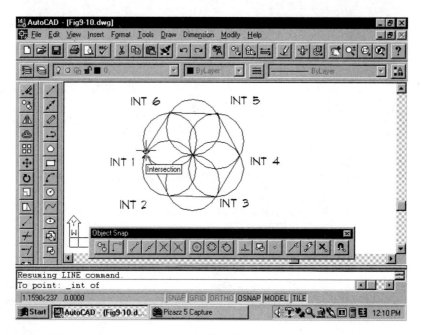

FIGURE 9–10 The Intersection option of OSNAP is used to draw a line from the intersection of each circle.

Apparent Intersection

This option finds the apparent intersection, or where two lines would intersect if one or both are extended. To find the apparent intersection pick the Apparent Intersection icon when prompted for the From point. Using the keyboard, enter **APP** [enter] at the From or To prompt. Because the line will be generated from where two lines will intersect, two From points will need to be selected. The prompt will read From point *of / and* to remind you to select two lines. The new line will be placed where the selected lines will merge. From the command prompt use the following command sequence:

Command: **L** [enter]
LINE From point: **APP** [enter]
of (*Select desired line.*)
and (*Select desired line.*)
To point: (*Select desired To point.*)
Command:

Picking [enter] will extend the line. The command sequence can be seen in Figure 9–11. Chapter 12 will present two alternatives to this option.

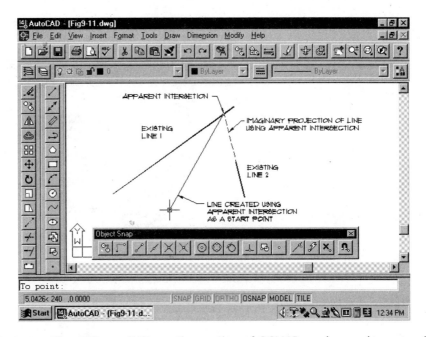

FIGURE 9–11   The APParen INTersection option of OSNAP can be used to extend a line between two lines that would intersect. The line was started by using the END option to select the end of line 2, picking the icon from the toolbar, or typing **APP** ⏎ for the "To point," and using the aperture box to select line 1 as the line to intersect. The endpoint of the original line must be selected to display the new line segment.

Center

The CENter option can be used any time an object needs to be drawn based on the center of a circle. A line or arc often needs to be drawn from the center of an existing circle, or several circles may need to be drawn around one centerpoint, as seen in Figure 9–12. This drawing could be easily drawn using the CENter option of OSNAP.

A line or arc may be extended from the centerpoint of a circle by picking the Center icon from the Object Snap toolbar. The Centerpoint option can also be selected by keyboard using the following command sequence:

> Command: **L** ⏎
> LINE From point: **CEN** ⏎
> of: *(Select a circle.)*
> To point: *(Select the desired endpoint.)*
> To point: ⏎
> Command:

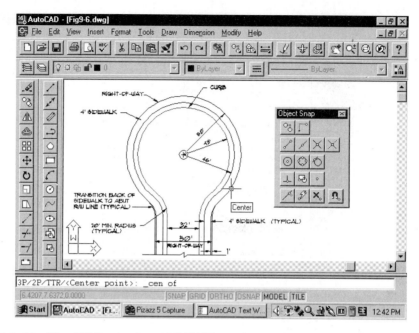

FIGURE 9–12 The CENter option of OSNAP can be used whenever a drawing entity needs to be placed based on the center of an existing circle. (Courtesy Lee Engineering.)

Touch any part of the circle and a rubberband line now extends from the exact center of the circle until a "To point" is selected. This process can be seen in Figure 9–13.

Quadrant

This OSNAP option will allow a line or arc to be snapped to the 0-, 90-, 180-, and 270-degree positions of a circle or arc. The command is activated by picking the Quadrant icon from the toolbar or by entering **QUA** ⏎ as the "From" or "To" point. Place the aperture box on the circle or arc near the quadrant and the new entity will automatically be joined at the quadrant, as seen in Figure 9–14. To draw a line from the quadrant point of an existing circle, use the following command sequence:

> Command: **L** ⏎
> LINE From point: **QUA** ⏎
> of *(Select desired circle.)*
> To point: *(Select desired to point.)*
> To point: ⏎
> Command:

Tangent

This OSNAP mode will allow a line to be snapped tangent to a circle or an arc. This very common combination of geometric shapes can be seen in Figure 9–15. This

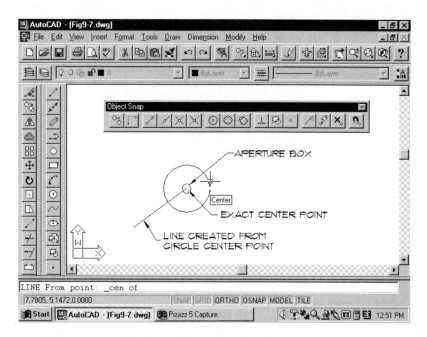

FIGURE 9–13 To extend a line from the center of a circle, enter the LINE command. When prompted for a LINE From point:, pick the Center icon or type **CEN** ⏎ at the command prompt. This will produce the aperture box which can be used to select the desired circle. When you place the aperture box anywhere on the circle and pick the circle, AutoCAD will automatically use the centerpoint of that circle as the From point: for a line.

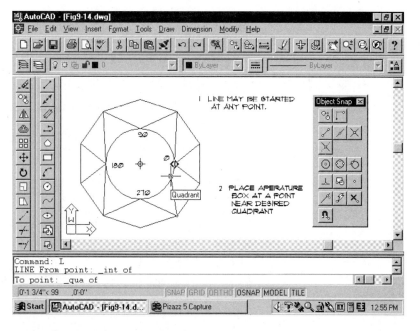

FIGURE 9–14 The QUAdrant option of OSNAP can be used to attach a line or arc to the quadrants of an existing circle.

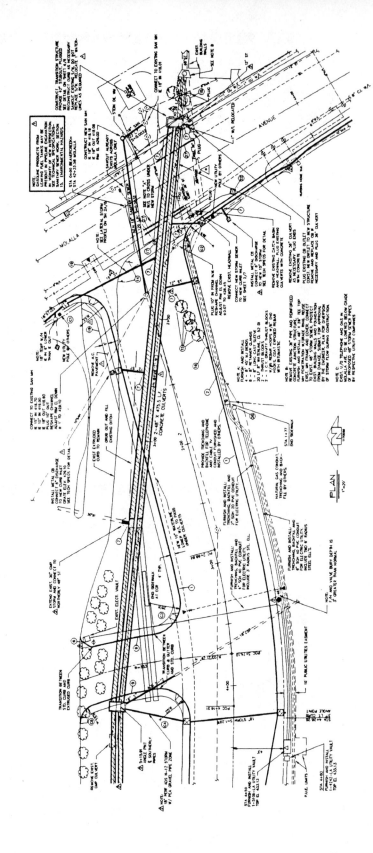

FIGURE 9–15 A common drawing requirement is to draw tangent surfaces. Tangent OSNAP will allow a line or arc to be drawn tangent to an existing arc or circle. (Courtesy Lee Engineering, Inc., Oregon City, OR)

mode is started by selecting the Tangent icon from the Object Snap toolbar or by typing **TAN** [enter] at the "From" or "To" point prompts. Figure 9–16 shows an example of how Tangent can be used. To draw a line tangent to an existing circle use the following command sequence:

Command: **L** [enter]
LINE From point: *(Select desired point.)*
To point: **TAN** [enter]
to *(Select desired arc or circle.)*
To point: [enter]
Command:

To draw a line tangent to an existing circle to another point use the following command sequence:

Command: **L** [enter]
LINE From point: **TAN** [enter]
to *(Select desired point.)*
To point: [enter]
Command:

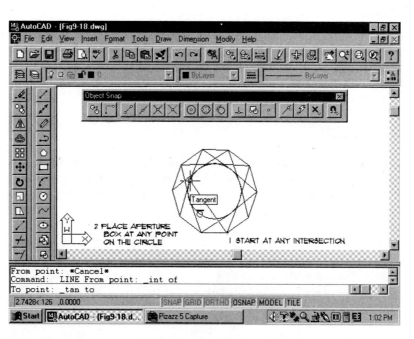

FIGURE 9–16 Tangent is started by selecting the Tangent icon or by entering **TAN** [enter] for the "From" or "To point" of the new line or arc.

Perpendicular

As seen in Figure 9–17, perpendicular lines are a basis of most construction draw-ings. Previous OSNAP options could be used to bring a line or an arc toward or away from an object. PERpendicular is used to draw a line that is perpendicular to an existing line. The process can be seen in Figure 9–18. To draw a line from any point perpendicular to an existing line pick the Perpendicular icon when prompted for a From or To point:. Type **PER** [enter] at the command prompt and use the following command sequence:

> Command: **L** [enter]
> LINE From point: *(Select desired point.)*
> To point: **PER** [enter]
> to *(Select desired line.)*
> To point: [enter]
> Command:

Insertion

This command will snap to the insertion point of a Shape, Text, Attribute, or Block. Meaningless stuff now, but combined with the information in Chapters 21 and 22, it will allow information to be inserted to an exact point.

Node

This option can be used to snap to a point. By entering Node, lines or arcs can be extended from or to predetermined points. Methods of drawing a point will be intro-duced in Chapter 14. This could be used, for instance, to lay out points on a route survey. To activate this OSNAP option, pick the Node icon or type **NOD** [enter] at either the "From" or "To" point.

Nearest

This option can be used to snap to the circle, line, or arc that is nearest to the crosshairs. To activate this OSNAP option, pick the Nearest icon or type **NEA** [enter] at either the "From" or "To" point. To draw a line from any point of an existing line use the following command sequence:

> Command: **L** [enter]
> LINE From point: **NEA** [enter]
> of *(Select desired line.)*
> To point: *(Select desired to point.)*
> To point: [enter]
> Command:

The command sequence can be seen in Figure 9–19.

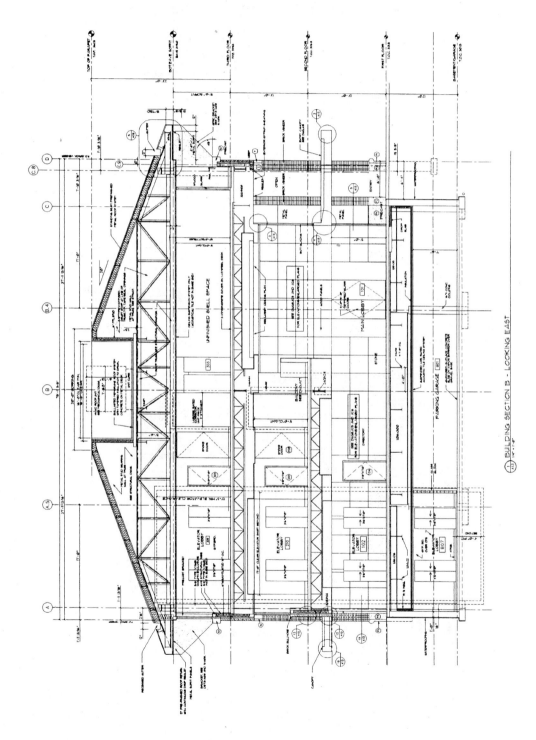

FIGURE 9-17 The perpendicular option of OSNAP can be used to draw a line perpendicular to another surface. This can be useful if the surface is not true vertical or horizontal. Courtesy Peck, Smiley, Ettlin Architects.

279

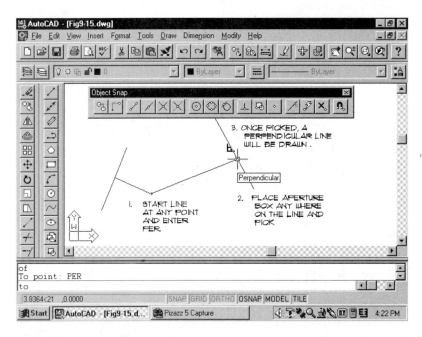

FIGURE 9–18 The Perpendicular option can be used to project a new line so that it will be at a 90° angle to an existing line.

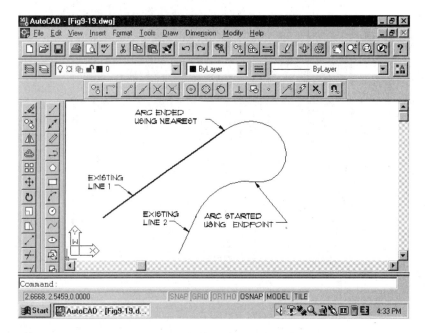

FIGURE 9–19 The Nearest option can be used to select the nearest drawing entity for a "From point" or "To point" when placing lines, arcs, circles, ellipses, and other geometric shapes.

Quick

Although it may seem that objects are selected very quickly, AutoCAD must search each entity of your drawing to find the designated objects of the aperture box. This process can be simplified by using QUICK. Quick is used in conjunction with another OSNAP option and allows for AutoCAD to stop searching as soon as it finds one object with at least one point of the specified type. OSNAP searches for all objects crossing the target and selects the closest potential snap points. If two or more objects that match the OSNAP request are in the aperture box, QUICK mode will choose the first one it sees. On a complex drawing, this may not always give the desired results. The command sequence to use QUICK and draw a line from one line endpoint to the midpoint of another line is:

> Command: **LINE** [enter]
> LINE From point: **QUI,END** [enter]
> To point: **QUI, MID** [enter]
> To point: [enter]
> Command:

From

The From point selection of OSNAP can be used to establish a point for a line or other drawing entity based on a known base point. The base point can be located by selecting a point with the cursor, or by providing polar or relative coordinates for the point. The From option could be useful on site related drawings when objects are to be located based on a specific point.

Figure 9–20 shows the use of FROM for locating a circle from the endpoint of an existing line. The command can be entered by picking the From icon from the Object Snap toolbar or by entering **FROM** [enter] at the command prompt. The command sequence is:

> Command: **C** [enter]
> 3P/2P/TTR/<Centerpoint>: **FROM** [enter]
> Base point: **END** [enter]
> of: (*Select desired line endpoint.*)
> of<offset>: (*Select location or provide coordinates from base point.*) **1.5** [enter]
> Diameter/<Radius>: (*Pick radius point or provide coordinates.*)
> Command:

Tracking

The Tracking mode can be used to find the center of a rectangle. The option functions similar to the Apparant Intersection mode. By picking the midpoint on the bottom side of the rectangle, and then picking the midpoint of one of the sides, a line can be drawn form the imaginary intersection of these two lines. Tracking can be started by selecting the Tracking icon from the Object Snap toolbar or by entering **TK** [enter],

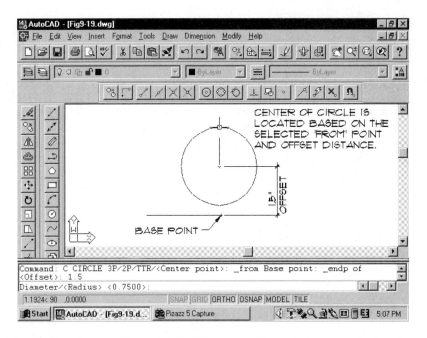

FIGURE 9–20 The From setting of OSNAP allows a line, circle, or arc to be drawn from a base point by entering **FROM** [enter] for one of the selection points for the drawing entity.

TRACK [enter], or **TRACKING** [enter] at the command prompt. The command sequence to draw a line from the center of a rectangle is:

Command: **L** [enter]
LINE From point: **TK** [enter]
First tracking point: **MID** [enter]
of (*Pick the bottom edge of the rectangle.*)
Next point (*Press ENTER to end tracking.*): **MID** [enter]
of (*Press ENTER to end tracking.*) (*Pick the edge of the rectangle.*)
Next point (*Press ENTER to end tracking.*): [enter]
To point: (*Select the desired endpoint of the line.*)
Command:

As the tracking sequence is ended, the rubberband line will now be centered in the rectangle and the prompt for a To point: is given. The command sequence can be seen in Figure 9–21.

RUNNING OSNAP

OSNAP has been used up to this point in its override mode. The Osnap mode was picked by icon or keyboard for one specific point, and then the option was ended. It

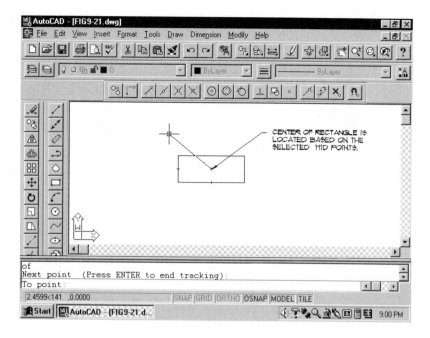

FIGURE 9–21 Tracking can be used to determine the center point of a rectangle. After picking the Tracking option, select the midpoint of the bottom and one side of the rectangle.

may be necessary to draw several lines to various endpoints of existing lines. Although you could continue to pick END each time you draw a line, there is a faster way. OSNAP selection can be set to be in effect until the command is no longer needed. Running OSNAP can be set using the Running Osnap folder of the Osnap Setting dialog box. The dialog box can be displayed by selecting the Object Snap Settings icon from the Object Snap toolbar, by entering **OS** (enter) or **OSNAP** (enter) at the command prompt, or by selecting Object Snap Settings... from the Tools pull-down menu. Once the Running Osnap tab is displayed, place a check in the mode to be activated, and then press the OK button. If Center is selected, a centerpoint will automatically be selected if a circle is selected for a From or To point each time a line or arc is drawn. Figure 9–22 shows an example where the Center option is used to place columns in the center of each pier, and lines to represent beams that start at the center of each column. The selection can be ended or altered by returning to the dialog box and altering the active modes.

Running OSNAP can also be set by typing **–OS** (enter) at the command prompt. The sequence to activate the Center Running OSNAP is:

Command: **–OS** (enter)
Object snap modes: **CEN** (enter)
Command:

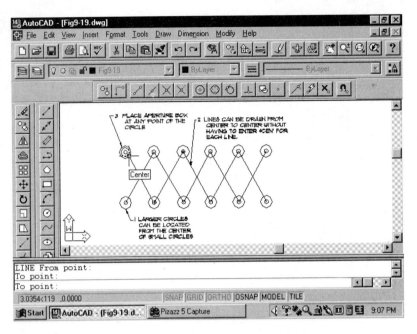

FIGURE 9–22 An OSNAP can be used repeatedly by picking multiple boxes from the Setting dialog box or by typing **–OS** enter at the command prompt and specifying the desired OSNAP mode. The selected mode can be deactivated by entering **OFF** enter as the current mode or double-clicking the ORTHO mode.

Running OSNAP can also be set to combine multiple options. This would be useful, for instance, if several lines need to be constructed from the INTersection of one object to the CENter of other objects, as seen in Figure 9–23. Multiple options are activated by picking the desired options from the Object Settings dialog box. There is no limit to the number of boxes that can be selected. To set multiple settings from the keyboard, the command sequence is:

> Command: **–OS** enter
> Object snap modes: **INT,CEN** enter
> Command: **LINE** enter

Now the running OSNAP mode is set to draw lines from intersections to center-points. A temporary override of running OSNAP can be achieved by entering **NONE** enter at the OSNAP prompt or picking NONE from the toolbar. Running OSNAP can be deactivated by double-clicking the OSNAP button or by typing **OFF** enter for the "Object Snap mode."

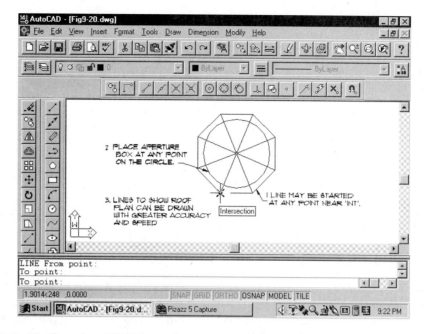

FIGURE 9–23 Running OSNAP can be set to combine multiple options for From and To points.

CHAPTER 9 EXERCISES

E-9-1. Draw an equilateral triangle inscribed in a 4" diameter circle. Use OSNAP midpoint to form a second triangle inside the first. Draw a third triangle inside the second. Save the drawing as E-9-1.

E-9-2. Draw the base to a chimney in plan view that is 60" × 32" with 8" wide walls. Draw an 8" wide support wall in the center of the chimney. Save as drawing E-9-2.

E-9-3. Use the drawing on the next page to draw a 1" diameter circle, a 2.25" diameter circle, a four-sided polygon, and a triangle. Complete the drawing using the CEN, MID, INT, and TAN OSNAP options.

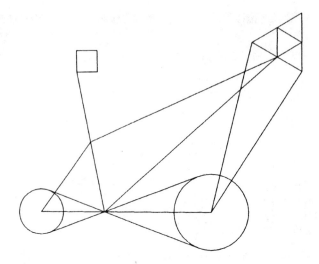

E-9-4. Set the limits and units to draw a 100′ × 75′ lot. Assume north to be at the top of the page. The north and south property lines will be 100′ long. Five trees are located on the property. Each is located from the southwest corner with coordinates and diameters listed as follows:

Coordinates	Trunk diameter	Branch diameter
a. 15′,12′-9″	9″ diameter	6′ diameter
b. 30′,17′	13″ diameter	7.5′ diameter
c. 42′-6″, 19′-3″	18″ diameter	9.5′ diameter
d. 50′, 28′-2″	15″ diameter	8.5′ diameter
e. 62′-0″,5′-3″	26″ diameter	14′ diameter

Locate each tree with a point. Use an "x" to locate each tree. Use the appropriate OSNAP options to draw the diameter of the trunk and the branch structure around each tree center. Draw a line from the southwest property corner through the center of the trees to the southeast corner. Draw another line from the southeast property corner to the northern limits and tangent to the branch diameter of each tree and ending at the southwest corner. Save the drawing as E-9-4.

E-9-5. Use the drawing below as a guide to draw a 4' × 4' window with a half-round window above. Draw all window dividers as one-half-inch wide. Save the drawing as E-9-5.

E-9-6. Using the attached drawing as a guide, open the ARCHBASE templet and draw a 32" × 21" double sink. Show the outline of the sink with rounded corners using a 2" radius. Divide the sink into two equal portions. Draw a 2"-diameter circle in the center of the right portion of the sink. Save the drawing as E-9-6SNK.

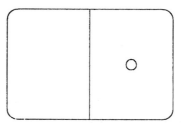

E-9-7. Using the attached drawing as a guide, open the ARCHBASE templet and draw a rectangle to represent a 60" × 32" tub/shower. Draw a rectangle with 3" radius rounded corners at one end and a full radius at the other end. Draw a triangle at the squared end to represent a shower head. Save the drawing as E-9-7TUB.

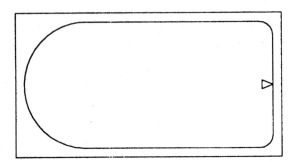

E-9-8. Using the attached drawing as a guide, open the ARCHBASE templet and draw a 48" wide × 24" deep 0' clearance fireplace. Draw the face of the fireplace as 8" wide on each side, with the interior edges 20" long, set at a 15 degree angle. Save the drawing as E-9-8FIRE.

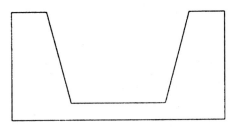

E-9-9. Open the ARCHBASE templet and draw a 42" × 36" rectangle using continuous lines. Draw diagonal lines from opposite corners using hidden lines and show a 3"-diameter circle at the intersection of the interior crossing lines. Save the drawing as E-9-9SHOW.

E-9-10. Draw a plan view of a 10' long × 24" wide bath room vanity. The vanity is to have a 30" wide knee space in the center that should be represented by hidden lines. Place a 19" × 16" oval sink centered on each side of the knee space. Save the drawing as E-9-10VAN.

CHAPTER 9 QUIZ

1. What command is used to change the size of the aperture box?

   _____

   _____

2. What is the size range for the aperture box? _____

3. Type the command sequence to provide running snap to an end point. _____

4. List three osnap options for connecting a line to a circle.

   _____

   _____

5. What can be done to save time when using OSNAP on a huge drawing?

   _____

   _____

6. List three methods to stop the running OSNAP mode.

   _____

   _____

7. What steps would be needed to draw a perpendicular line to the midpoint of an existing line?

   _____

   _____

8. List the letters that are used to activate the OSNAP mode used to connect geometric shapes.

   _____

   _____

9. How can you get the dialog box for OSNAP to be displayed?

   _____

   _____

10. When using quadrant, at what degrees will lines be attached to a circle?

   _____

   _____

   _____

10

DRAWING DISPLAY OPTIONS

In the last chapter you were introduced to drawing more accurately by snapping to specific points of an object. Your drawing accuracy also can be improved by changing the size of the view of the drawing and by keeping the number of drawing blips to a minimum. Cleaning the screen can be accomplished by controlling the BLIPMODE, or using the REDRAW or REGENeration commands. The drawing view and the view size can be adjusted by the ZOOM, VIEW, PAN, AERIAL VIEW and VIEWPORTS commands. Each of these commands can be used while another command is in progress to aid viewing. Changing the view resolution can adjust the quality of the view and the speed at which the views are changed.

BLIPMODE

The BLIPMODE command controls the generation of blips. BLIPMODE was introduced in Chapter 7. In the default mode, no blips or markers will be made as lines, arcs, circles, or other drawing entities are created. These markers may be helpful for some commands, such as CIRCLE or ARC, to help you visualize the starting points. The setting may be changed throughout the life of the drawing. To change the display, type **BLIPMODE** [enter] at the command prompt.

Command: **BLIPMODE** [enter]
ON/OFF <ON>:

With BLIPMODE set in the ON default, markers are displayed on the screen each time a point is designated. No markers are displayed when the OFF option is selected. Blips also can be controlled by using the SETVAR command, although more commands are required. The process is:

Command: **SETVAR** [enter]
SETVAR Variable name or ? **BLIPMODE:** [enter]
New value for BLIPMODE <0>: (*Enter 0 or 1.*) **1** [enter]

For most uses, BLIPMODE can be left ON and removed as desired. When ON is selected, each time an entity is drawn, a blip is created on the screen to mark the From, To, or Center point. Throughout a drawing session, these blips can accumulate and be quite distracting , as seen in Figure 10–1. Blips can be removed by toggling GRID between ON/OFF with the F7 key or by picking BLIPS from the Drawing Aids dialog box shown in Figure 10–2. The dialog box is accessed by typing **DR** [enter] or **DDRMODES** [enter] at the command prompt or by picking Drawing Aids... from the Tools pull-down menu. The commands REDRAW, REGEN, ZOOM, and PAN will also remove blips.

REDRAW

The REDRAW command removes the current screen display momentarily, and then replaces the drawing. As the drawing is replaced, blips are removed from the draw-

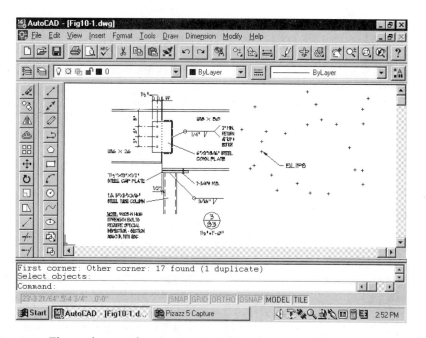

FIGURE 10–1 Throughout a drawing session, blips can be added to the screen to mark the location of a point. Blips will be added if BLIPMODE is set to ON.

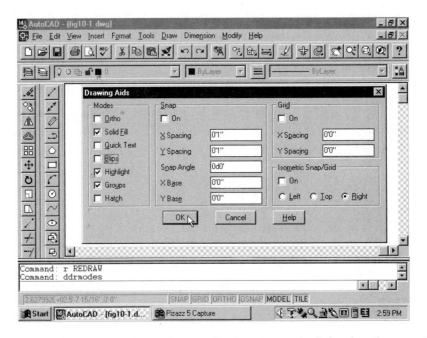

FIGURE 10-2 Blips can be adjusted using the Drawing Aids dialog box by entering **DR** [enter] or **DDRMODES** [enter] at the command prompt.

ing editor. Lines that may have been partially erased while editing other entities also will be restored. The command is begun by typing **R** [enter] or **REDRAW** [enter] at the command prompt. The same results can be achieved by picking Redraw View from the View pull-down menu, or by picking the Redraw All View icon from the Standard toolbar. REDRAW may be used while another command is in progress if it is entered at any nontext prompt. On drawings similar in size to the exercises that you've been doing, REDRAW will seem almost instantaneous. On larger drawings, such as a subdivision map or floor plan, REDRAW can take slightly longer. To terminate a REDRAW, press the ESC key. This may be helpful if your next command after the redraw is a ZOOM, which automatically cleans the screen.

REGEN

Similar to REDRAW is the slightly slower REGEN command. One of the truly great advances of R14 is it's ability to do quick REGENs of large drawings. AutoCAD clears the drawing screen and then regenerates the entire drawing. AutoCAD recalculates the current screen display with 14-place accuracy, keeping track of each pixel in the X and Y direction. If you have zoomed in on a specific part of a drawing, REGEN is regenerating even the part of the drawing that cannot be seen. Later in this chapter you'll learn how to create viewports. Using viewports will speed up REGENs since only the current viewport is regenerated. A screen regeneration can be started by picking Regen from the View pull-down menu or by entering **REGEN** ⏎ at the command prompt. To terminate a REGEN, press the ESC key.

REGENALL

Later in this chapter you'll learn how to divide the drawing screen into different views so that specific areas of the drawing can be viewed quickly. When multiple views are created, the REGEN command will only affect the current drawing view. All other viewports will remain unaffected. The REGENALL command will regenerate all of the views, so that blip in each view will be removed and reindex the drawing database to increase objection selection. The command can be accessed by selecting Regen All from the View pull-down menu or by entering **REA** ⏎ or **REGENALL** ⏎ at the command prompt. The significance of the command will be discussed later in this chapter.

REGENAUTO

Just as you can save a drawing automatically at a specified time interval, AutoCAD allows the regeneration time to be set. On a small drawing, an automatic regeneration may not even be noticed. As the drawing size increases, an automatic regeneration may prove a frustration. This aspect of REGEN can only be entered by keyboard. The command sequence is:

Command: **REGENAUTO** ⏎
ON / OFF <ON>:(*Enter On or Off, or press* ⏎)

If On option is retained, AutoCAD will preform automatic regenerations when needed when altering the screen display in commands such as ZOOM. When Off is the current mode and a regeneration is needed as the screen is altered, AutoCAD will display the prompt:

About to Regen--proceed <Y> Enter Y or N, or press ENTER

If you accept the default, the drawing regeneration will proceed allowing the ZOOM to take place. If the NO option is selected, the ZOOM will not be completed. Because of AutoCAD's increased ability to handle huge amounts of information quickly and easily, you really can leave this option in the "That's nice" file of your mind and explore other viewing options

ZOOM

Similar to the zoom lens of a camera, AutoCAD's ZOOM command has a potential zoom ratio of 10 trillion to one. "Zooming in" on a drawing, can magnify a portion of a drawing to allow for better visual quality. Zoom can be used in the middle of another command sequence, which can greatly improve accuracy on detailed drawings. As the size of an object is increased, the area of the drawing that can be seen is reduced. This can be seen in Figures 10–3 and 10–4. The opposite holds true when

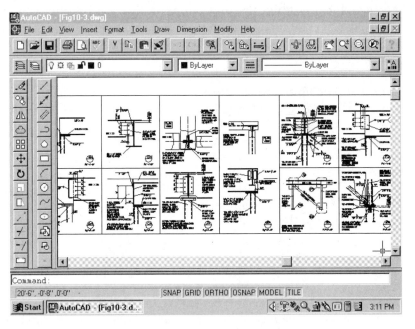

FIGURE 10–3 As the drawing limits are enlarged, many features of the drawing may become unreadable. (Courtesy StructureForm Masters, Inc.)

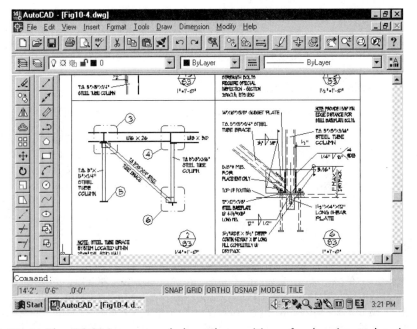

FIGURE 10–4 The ZOOM command alters the position of a drawing to be viewed, giving the sense that the portion of the drawing being viewed has been enlarged. (Courtesy SctructureForm Masters, Inc.)

you reduce the apparent size of an object, more of the drawing can be seen. It is important to keep in mind that the actual size of the object is not changing, only the magnification.

ZOOM can be activated by picking Zoom from the View pull-down menu. This will produce the selections seen in Figure 10–5a. Zoom can also be accessed by selecting one of three different icons in the Standard toolbar. These icons include the Zoom Realtime, Zoom Flyout, and Zoom Previous and will be discussed throughout the chapter. Figure 10–5b shows the Zoom flyout menu found in the Standard toolbar. The command sequence can also be started by typing **Z** [enter] or **ZOOM** [enter] at the command prompt. This will produce the prompt:

Command: **Z** [enter]
All/Center/Dynamic/Extents/Previous/Scale (X/XP)/Window/<Realtime>:

Each of the options listed at the command prompt and in the pull-down menu can be selected from the Zoom flyout icon in the Standard toolbar.

The REALTIME Command

Real time zooming means that as you move the cursor, you see the results. The REALTIME command is entered by selecting the Realtime icon from the Standard toolbar, by entering **RTZOOM** [enter] at the command prompt and then pressing the

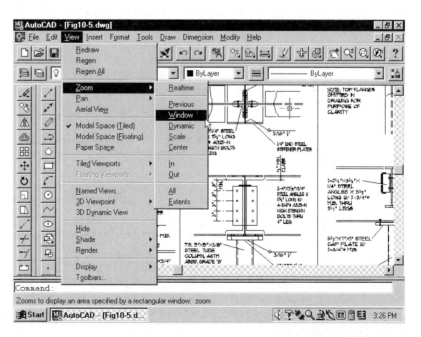

FIGURE 10–5a The ZOOM command can be selected by typing **Z** [enter] or **ZOOM** [enter] at the command prompt or by selecting the command from the View pull-down menu.

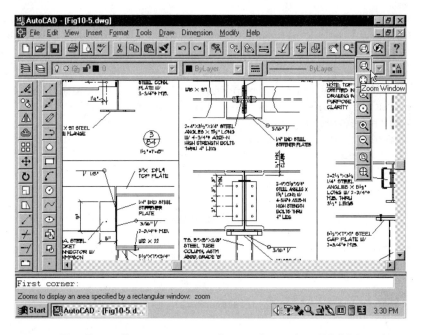

FIGURE 10–5b   The Zoom flyout menu can be used to select ZOOM options.

right mouse button to activate the command. The command can also be accessed by entering **Z** [enter] at the command prompt and the pressing the [enter] key to activate the command. As the command is activated, the cursor will change to a magnifying glass with a + and – symbol. The command prompt now displays the prompt:

All/Center/Dynamic/Extents/Previous/Scale (X/XP)/Window/<Realtime>:
(Press Esc or [enter] to exit, or right-click to activate pop-up menu.)

Move the mouse around the screen and nothing happens. As you press and hold the pick button, and move the mouse toward the top of the screen, the magnification of the drawings is enlarged. Placing the magnifying glass in the middle of the screen and moving to the top will enlarge the display 100%. Pressing and holding the mouse select button and moving from the middle to the bottom of the screen will decrease the magnification 100%. Placing the magnifying glass at the bottom of the screen and moving to the top will enlarge the display 200%. Moving from the top to the bottom of the screen will decrease the magnification 200%. Each command can be repeated indefinitely, until the desired magnification is achieved. If you have trouble making Realtime Zoom alter the drawing size, remember that you must hold down the right mouse button as the mouse is moved. The command is ended by pressing the ESC or [enter] key or by pressing the right mouse button. Pressing the right mouse button will produce the menu shown in Figure 10–5c. Picking Exit will end ZOOM and return the command prompt. The menu also allows other viewing options to be selected. By altering between the Pan and the Zoom option, a drawing can be

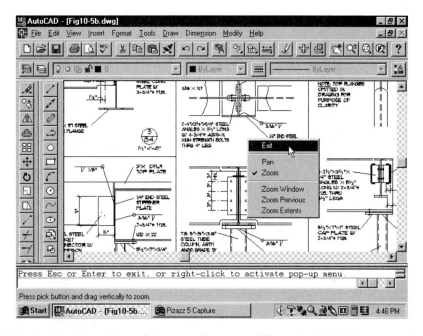

FIGURE 10–5c Pressing the right mouse button while in the ZOOM command will produce the the zoom menu.

enlarged, keeping the desired portion on the screen. Each of these options will be discussed throughout the balance of this chapter.

Zoom Window

The Window option of ZOOM is a common option for zooming into a drawing. This option lets you select the area you wish to view by providing two opposite corner locations that will form the viewing window. The size of the view box can be selected by entering coordinates or by using the mouse to select locations on the screen. The command can be accessed by picking the Zoom Windows icon from the Flyout icon of the Standard toolbar or by entering **Z** at the keyboard. For point selection, the command process is:

> Command: **Z** (enter)
> All/Center/Dynamic/Extents/Previous/Scale (X/XP)/Window/<Realtime>: **W** (enter)
> First corner: (*Pick a corner.*)
> Other corner: (*Pick a corner.*)

This process can be seen in Figure 10-6. Objects within the view box will be redisplayed on the screen as the new display. The process is similar when entering coordinates with a mouse. Instead of entering coordinates, you select the first corner of the zoom display by moving the cursor to the desired location. Once a corner is

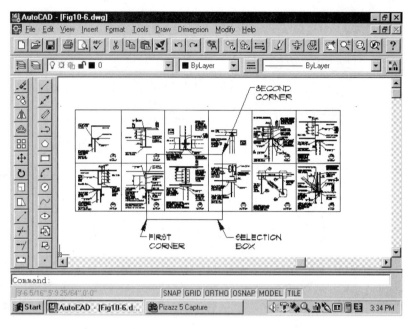

FIGURE 10–6 The ZOOM window option allows the display area to be selected by specifying opposite corners of the "window." (Courtesy StructureForm Masters, Inc.)

picked, the cross-hairs will switch to a box that is enlarged or reduced as the mouse is moved across the drawing area. Figure 10–7 shows the results of the selection made in Figure 10–6. Entering **Z** [enter], **W** [enter]**,** or picking the Zoom Window icon are the most efficient methods of starting the command.

Transparent Zoom

One of the really great features of ZOOM is that it is transparent, meaning it can be used in the middle of another command. If you start to draw a line, and realize that you're straining to see the line, the screen image can be magnified using ZOOM in the middle of a LINE command sequence. To use a transparent command type an **'** (apostrophe) prior to the command. The command sequence for using the window option of ZOOM in the middle of the LINE command would be:

Command: **L** [enter]
LINE From point: **'Z** [enter]
All/Center/Dynamic/Extents/Previous/Scale (X/XP)/Window/<Realtime>:**W** [enter]
>>First corner: *(Pick desired start point.)*
>>Other corner: *(Pick opposite corner to be enlarged.)*
Resuming LINE command.
From point: *(Pick desired start point.)*
To point: *(Pick desired to point.)*

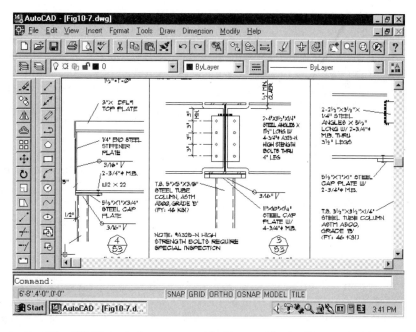

FIGURE 10–7 The ZOOM window created in Figure 10–6 will now display the selected area. (Courtesy StructureForm Masters, Inc.)

To point: [enter]
Command:

ZOOM could be used at any point during the LINE or any other command when a prompt is given. In the LINE command, the 'ZOOM command could have been started at the prompt for "From point" or at the "To point" prompt. Enter the LINE, CIRCLE or ARC commands and then use 'ZOOM to become familiar with this option of zoom. Your eyes and back will love you as you zoom more and squint less.

Zoom All

This option will allow for an entire drawing to be viewed, based on the drawing limits. If the drawing extends beyond the drawing limits, the screen display will show the entire drawing. Zoom All should be used if you are zoomed into a small area of a drawing such as Figure 10–4. Using the All option returns all of the drawing to the screen, so that the display resembles Figure 10–3. Zoom All can be selected from the Zoom flyout icon menu in the Standard toolbar or by entering **Z** [enter], **A** [enter] at the command prompt. The command sequence is:

Command: **Z** [enter]
All/Center/Dynamic/Extents/Previous/Scale (X/XP)/Window/<Realtime>: **A** [enter]

Zoom Center

This option allows a zoom to be specified by its desired centerpoint. Once a center-point is selected, a prompt will be given for the height of the display. If the current default is maintained, the drawing is redisplayed based on the new centerpoint, but the magnification is not changed. If a smaller value for the height is selected, the display magnification will be increased. If a larger value for the height is selected, the display magnification will be reduced. The command sequence is:

Command: **Z** [enter]
All/Center/Dynamic/Extents/Previous/Scale (X/XP)/Window/<Realtime>: **C** [enter]
Center point: (*Enter coordinates or pick a point.*)
Magnification or Height < 17'-4">: **10'** [enter]
Command:

Zoom Dynamic

This option will allow specific areas of the drawing to be selected for ZOOM viewing. The command will seem complicated at first, but once you master it, you'll be able to select any portion of the drawing for the next view without requiring a REGEN. The importance of this may seem lost on you as you compare a dynamic to a window zoom working on the exercises. As you start to work on larger projects using a couple of million bytes of information, a Dynamic ZOOM will be well worth the effort because of the savings in time of not having to regenerate a drawing. The original view box can be enlarged or reduced and moved about the drawing screen to select the exact portion of the drawing to be zoomed. The command sequence is:

Command: **Z** [enter]
All/Center/Dynamic/Extents/Previous/Scale (X/XP)/Window/<Realtime>: **D** [enter]

The screen will be cleared, then the original view will be returned surrounded by three color-coded boxes. Each box can be seen in Figure 10–8 and represents a different viewing area.

The first box that will be displayed is the Pan box. The Pan box allows you to drag the drawing across the screen to determine which portion of the drawing you would like to view. (See Figure 10–8.) Once the desired portion of the drawing is shown on the screen display, press the mouse pick button and the Zooming window is displayed. The view box represents the current viewport. Viewports will be discussed later in this chapter. For now, you only have one viewport. The view box allows the size of the current viewport to be adjusted, either by shrinking or enlarging, or by moving it at its current size around the screen. Adjust the size of the window and press [enter] . Press the pick button if you need to readjust the portion of the drawing to be displayed. Sound complicated? It's not, but it may not be worth the effort either. The Dynamic Zoom was a great method of moving through a drawing, until Realtime Zoom and Pan were added in Release 13. With these two options added to

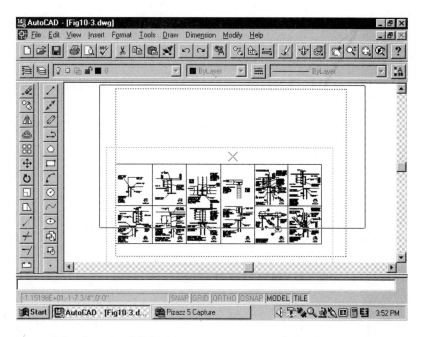

FIGURE 10–8 Selecting the ZOOM Dynamic option will display the original drawing area as well as three view boxes.

the Standard toolbar in Release 14, Zoom Dynamic has almost been rendered obsolete.

Zoom Extents

This option will display the entire drawing based on the size of the drawing and not the size of the limits that were used for Zoom All. If a small object is being examined with large drawing limits, Extents will enlarge the object to fill the drawing screen. This can be quite helpful during the initial stages of a drawing. Zoom Extents is also helpful if the extent of the drawing exceeds the drawing limits, because the option allows the entire drawing to be examined. The command sequence is:

Command: **Z** 🄴
All/Center/Dynamic/Extents/Previous/Scale (X/XP)/Window/<Realtime>: **E** 🄴

Zoom Previous

Each of the other ZOOM options are used to enlarge an area of a drawing for easier viewing. The Previous option will allow you to go backward, reducing the magnification so that more of the drawing area can be seen. When you use this option, the drawing displayed prior to the ZOOM will be returned to the screen. A maximum of

10 previous views can be restored using the Previous option. The command sequence for Zoom Previous is:

Command: **Z** `enter`
All/Center/Dynamic/Extents/Previous/Scale (X/XP)/Window/<Realtime>: **P** `enter`

Zoom Scale

The default selection for the ZOOM command is Scale or magnification factor (X). The scale may be set to be relative to the full view or to a current view.

Scale Relative to Full View.   When you enter a number for the scale factor, the entire drawing will be affected. A scale factor of 1 will display the current size. A scale factor of 3 will make objects appear three times as large. By entering a number smaller than 1, you will decrease the size of the object. If you enter .5, the object will appear half as big as the full display. The command sequence is:

Command: **Z** `enter`
All/Center/Dynamic/Extents/Previous/Scale (X/XP)/Window/<Realtime>: **S** `enter`
Enter scale factor: **.5** `enter`
Command:

Scale Relative to Current View.   If you enter a numeric value followed by an "X," the scale of the zoom will be relative to the current view rather than the entire drawing. The command sequence is:

Command: **Z** `enter`
All/Center/Dynamic/Extents/Previous/Scale (X/XP)/Window/<Realtime>: **S** `enter`
Enter scale factor: **1.5X** `enter`

Scale Relative to Paper-Space Units.   The XP scaling will be covered in Chapter 23. AutoCAD thinks of space as model and paper space. 2D and 3D drawings are typically drawn using model space. Paper space is used to plot views of the model that have been created. Zoom XP can be used to scale each space relative to paper space units. This will be helpful, for instance, in printing drawings on the same sheet that are drawn at different scales. This topic will be covered in more detail in Chapter 23.

Zoom In

Zoom In automatically scales the current drawing screen display by a scale factor of 2. Selecting this Zoom option produces the same result as picking a scale factor of 2X from the command line. The Zoom In option is only available as an icon in the Zoom Flyout menu in the Standard toolbar and from the Zoom listing in the View pull-down menu.

Zoom Out

Zoom Out automatically enlarges the drawing area that is displayed on the screen by a factor of 2. The drawing is, in effect reduced by a scale factor of .5X. The Zoom Out option is only available as an icon in the Zoom Flyout Standard toolbar and from the Zoom listing in the View pull-down menu.

PAN

The PAN command can best be visualized by thinking of your drawing on a large sheet of paper that is rolled up. As you pan across your drawing, it's just as if you were unrolling the drawing to view a different portion. The drawing magnification is not changed, just the portion of the drawing that is displayed. This can be helpful when you've zoomed in on a drawing similar to the one in Figure 10–4 and are looking at a specific detail. Rather than having to ZOOM All, and then ZOOM Window to a new detail, PAN can be used to bring in the desired detail. The PAN command is accessed by selecting Pan from the View pull-down menu or by selecting the Pan Realtime icon from the Standard toolbar. The submenu for PAN can be seen in Figure 10–9.

PAN REALTIME

The Pan Realtime command is similar to Zoom Realtime. The command is entered at the command prompt by entering **RTPAN** [enter] or by selecting Realtime from PAN of the View pull-down menu. The easiest method of entering the command is to pick the Pan Realtime icon in the Standard toolbar. No matter the access method, as the command is activated, the cursor will change to a hand cursor. The command prompt now displays the prompt:

Press Esc or [enter] to exit, or right-click to activate pop-up menu.

Move the cursor to the desired pick point and press and hold the pick button. This point now becomes the displacement point of the Pan Command. As the cursor is moved and the mouse pick butting is being pressed, the screen display is altered. Panning can be repeated indefinitely, until the desired display is achieved. The command is ended by pressing the ESC or [enter] key. Figures 10–10 and 10–11 compares the effects of using the Pan Command.

Combining Realtime Zoom and Pan. By entering either Pan Realtime or Zoom Realtime using the appropriate icon in the Standard toolbar, you can easily toggle between the two options and have a convenient method of moving between different portions of a large drawing. To experiment, open a drawing and draw several lines and circles. Press the Zoom Realtime icon and use the mouse to enlarge the drawing. Now press the Pan Realtime icon and the cursor will change to the hand and allow the enlarged view to be panned. Remember that by pressing the right mouse button, a pop-up menu will be displayed that allows you to toggle between the Zoom and Pan commands.

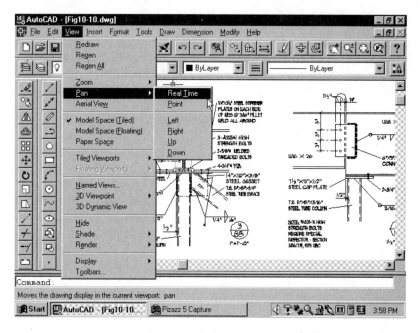

FIGURE 10–9 The PAN command is accessed by typing **P** [enter] at the command prompt, by picking the Pan icon from the Standard toolbar, or by selecting PAN from the View pull-down menu.

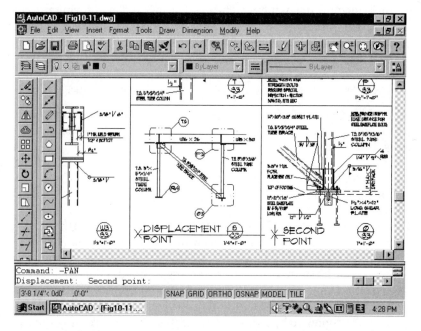

FIGURE 10–10 PAN allows a drawing to be "scrolled" across the screen without requiring a REGEN or REDRAW. (Courtesy StructureForm Masters, Inc.)

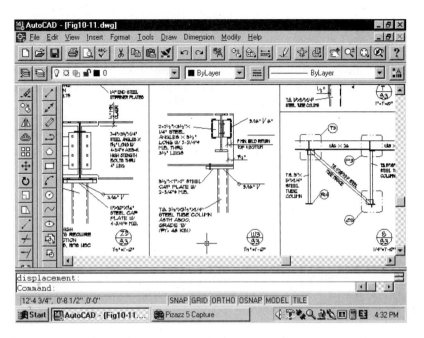

FIGURE 10-11 The effects of the PAN that was started in Figure 10–10. (Courtesy StructureForm Masters, Inc.)

Entering Pan from the Keyboard. The PAN command can also be entered by typing –P [enter] or –PAN [enter] at the command prompt. Be sure to enter the minus sign or the command will function just as it did when the icon is picked

Once entered, the command will prompt you to pick a point on the drawing screen and then ask, in effect, where would you like to move that point. Points may be entered by providing coordinates from the keyboard or by picking a point on the screen. The drawing can be panned in greater increments by entering coordinates. Figure 10–10 shows the PAN process. The results can be seen in Figure 10–11. The command sequence is:

Command: **–PAN** [enter]
Displacement: (*Enter point or coordinates of point to be dragged.*)
Second point: (*Enter point or coordinates to bring first point to.*)

For most cases, the Pan Realtime option is much quicker to enter and easier to use. The –PAN option is there to make users of older AutoCAD versions feel at home.

VIEW

An alternative to ZOOM and PAN is the use of prenamed drawing views. The VIEW command allows for views or viewports to be named and saved so that they can be restored later. By default AutoCAD begins each drawing as one viewport. As the limits are set, you are deciding how big the viewport is. The screen can be divided into smaller viewports for easy movement between drawings. On a drawing similar to Figure 10–3, each detail box could be named as a viewport so that each could be revised easily. By naming each detail as a view, you have in effect predefined the areas to be viewed. Instead of using ZOOM or PAN to select a view, you could select a desired VIEW to be displayed. Like other commands, the VIEW command may seem like a waste of time on small exercises. The VIEW command offers great flexibility for quickly viewing areas of a large, detailed drawing if you need to constantly move from one place to another. VIEW can be accessed by picking the Named View icon in the Standard toolbar, or by picking the Named View icon from the View Point toolbox. It is also found by selecting, Named Views... from the View pull-down menu which will produce the View Control dialog box shown in Figure 10–12. The View Control dialog box can also be produced by typing **V** [enter] or **DDVIEW** [enter] at the command prompt. Each method will produce the View Control dialog box shown in Figure 10–12. Figure 10–13 shows the sheet of details that have been used throughout this chapter and the view names that are used. The command can also be started by typing **VIEW** [enter] at the command prompt. The command format for VIEW is:

Command: **VIEW** [enter]
?/Delete/Restore/Save/Window: (*Select one option.*) [enter]

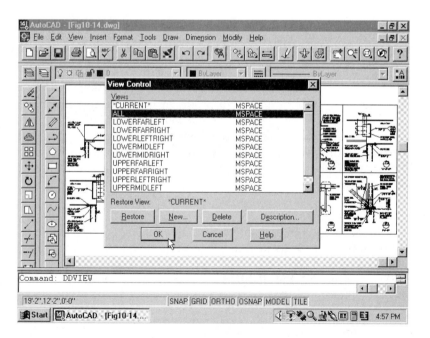

FIGURE 10–12 The View Control dialog box provides a list of named views within a drawing file. The dialog box can be accessed by picking Named Views... from the View pull-down menu or by typing **V**[enter] or **DDVIEW** [enter] at the command prompt.

The SAVE and WINDOW options each allows for creating a view.

Save

The Save option saves the current screen display using a name that you supply. View titles may be up to 31 characters, including letters, numbers, and special characters such as $, –, _. Views are often named by their location in the job or by their specific function. If four rows of details will be drawn, each row could be given a letter and each column of details could be given a number to produce an easy grid to work with. Each room of a floor plan can also be turned into a view for easy movement through a project. To save the detail in the upper left corner of Figure 10–13 as a view, the ZOOM command was used so that the detail was displayed on the screen. With the desired view selected, the SAVE option of the VIEW command was used. The command sequence is:

Command: **VIEW** [enter]
?/Delete/Restore/Save/Window: **S** [enter]
View name to save: **UPPERFARLEFT** [enter]

A view can also be selected by using the View Control dialog box. Select the portion of the screen you want saved as a view and then type **V** [enter] or **DDVIEW** [enter] at the command prompt. Selecting the New... button will produce the dialog box shown in

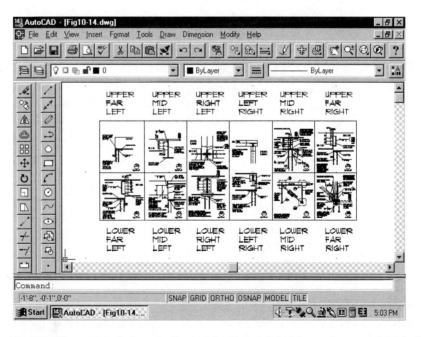

FIGURE 10–13 The named views reflected in Figure 10–12 can be seen in this figure. (Courtesy StructureForm Masters, Inc.)

Figure 10–14. Type the desired view name into the select box, and pick the Save View button.

Window

This option will allow a portion of the current screen display to be saved as a view. The proposed view is selected with a window, exactly as was done with the ZOOM command. The command sequence is:

Command: **VIEW** [enter]
?/Delete/Restore/Save/Window: **W** [enter]
View name to save: **UPPERMIDLEFT** [enter]
First corner: (*Select a point to start the window.*)
Second corner: (*Select a point to form the opposite window corner.*)

? (The Question Mark Option)

This option will display a list of named views contained in the drawing file. The command sequence is:

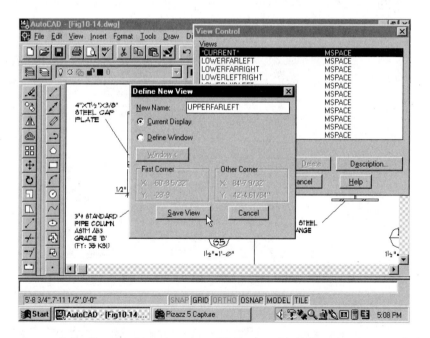

FIGURE 10–14 Selecting the New button will display the Define New View dialog box which can be used to name views. A view is saved by displaying the desired view in the screen and then typing the view name in the New Name box and then picking the Save button.

Command: **VIEW** [enter]
?/Delete/Restore/Save/Window: **?** [enter]
View(s) to list<*>: [enter]

If you enter [enter], each saved view will be listed. Choosing the **?** option for the views in Figure 10–13 would produce the following listing:

LOWERFARLEFT	MSPACE
LOWERFARRIGHT	MSPACE
LOWERLEFTRIGHT	MSPACE
LOWERMIDLEFT	MSPACE
LOWERMIDRIGHT	MSPACE
UPPERFARLEFT	MSPACE
UPPERFARRIGHT	MSPACE
UPPERLEFTRIGHT	MSPACE
UPPERMIDLEFT	MSPACE
UPPERMIDRIGHT	MSPACE

Notice that each listing in Figure 10–12 is followed by the word MSPACE which represents model space. An alternative is PSPACE which represents paper space. Each will be explained in Chapter 23.

Delete

This option will allow one or more views to be removed from the list of saved views. If more than one name is to be removed, put a comma between each view name. The command sequence is:

Command: **VIEW** [enter]
?/Delete/Restore/Save/Window: **D** [enter]
View name(s) to Delete: (*Enter view name.*) [enter]

Views can also be deleted from the drawing base by using the View Control dialog box. To delete a view, highlight the desired view or views and pick the Delete button. The information on the drawing will remain unchanged, but that portion of the drawing will not be a named view.

Restore

This will redisplay a saved view as the current view. The command process is:

Command: **VIEW** [enter]
?/Delete/Restore/Save/Window: **R** [enter]
View name to restore: (*Enter name.*) [enter]

The View Control dialog box can also be used to easily change between views. Type **DDVIEW** at the command prompt and highlight the desired view you want displayed. Once the view is selected, pick the Restore button followed by the OK button. This will remove the dialog box from the screen and alter the view that is displayed.

View Description

Picking the Description... button of the View Control dialog box will display View Description dialog box which can be seen in Figure 10–15. This box provides information on the width and height of the selected view. It also contains information that will be needed to produce 3D drawings.

VIEWPORTS

In addition to being able to name your own views, AutoCAD also provides you with prearranged views called VIEWPORTS or VPORTS. Viewports allow you to divide the screen into multiple images so that you have multiple zoom images of different parts of one drawing file. This could allow you to zoom in on a beam to column connection in one viewport, a column to footing detail in another viewport and a display of the a whole sheet of details in a third viewport as shown in Figure 10–16. Up to 16 viewports can be visible at one time, but work can be done in only one viewport at a time. This multiple use of enlarged viewports allows for drawing and editing within the enlarged portions of the drawing without having to ZOOM in and out to move from one detail to another. It is important to remember that even though you

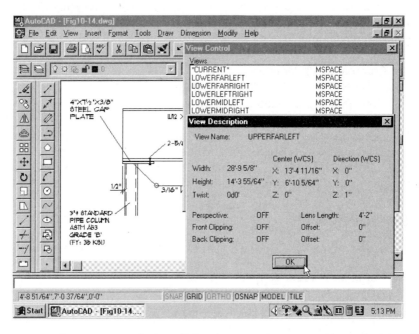

FIGURE 10–15 The View Description dialog box can be accessed to provide information about named views.

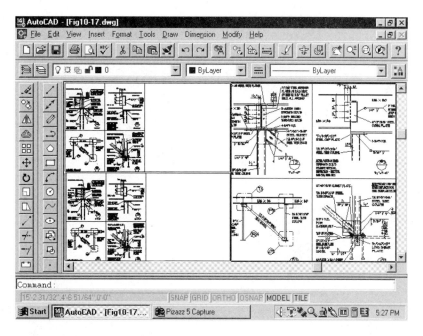

FIGURE 10–16 Viewports can be established to be able to view two or more parts of the drawing at the same time. (Courtesy StructureForm Masters, Inc.)

are seeing multiple images on the screen, that you're working with only data base. An object that is drawn or edited in one viewport will also affect the other viewports.

Viewports Commands

Before viewports can be established, the TILEMODE system variable must be set to 1. This can be done by double-clicking both the MODEL and TILE buttons in the Status bar, by entering **TILEMODE** [enter] at the command prompt, or by selecting the Model Space(Tiled) from the View pull-down menu. Selecting any one of these options will activate the Tiled Viewports menu in the View pull-down menu. This will produce the menu seen in Figure 10–17. The command can also be started by typing **VPORTS** [enter] or **VIEWPORTS** [enter] at the command prompt. This will produce the following display:

Command: **VPORTS** [enter]
Save/Restore/Delete/Join/Single/?/2/<3>/4: [enter]

Accept the default value of **3** for now by pressing the [enter] key. This will produce the prompt:

Horizontal/Vertical/Above/Below/Left/<Right>: [enter]

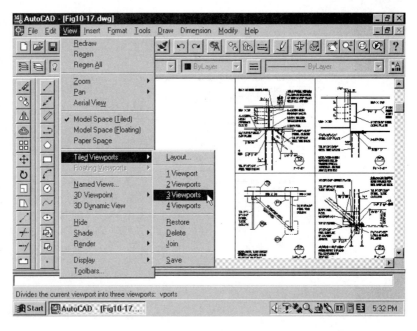

FIGURE 10–17 The Tile Viewports menu can be selected from the View pull-down menu.

Accepting the default of Right will produce the display seen in Figure 10–18. By accepting the defaults for viewports, AutoCAD now displays three images of your drawing. Possible arrangements for two, three and four viewports can be seen in Figure 10–19. The Tiled Viewport Layout dialog box is displayed by selecting Layout... from the Tiled Viewports listing in the View pull-down menu. Rather than using the keyboard selection of 3 viewports, the display shown in Figure 10–18 could have been established by picking the indicated icon shown in Figure 10–19, or by picking Three right: from the menu. Multiple viewports can be ether tiled or untiled. Tiled viewports are arranged in nice, neat patterns similar to floor tile where each tile aligns with another tile. Untiled viewports can overlap or be separated by space.

Notice in Figure 10–19, that the viewport on the right side is surrounded by a bold black line. This viewport also contains the cross-hairs. As you move the cursor around the screen, you'll notice that the cross-hairs only work in the drawing on the right side. This view is called the active viewport. You can draw or edit in this viewport just as in any other drawing. As the cursor crosses the center of the drawing into the viewport on the left, the cross-hairs change to an arrow.

Notice that by comparing Figure 10–18 with Figure 10–20, information related to the steel column was removed from the lower left corner of the upper left viewport, which also removed the information from the data base, and the total view. Although the total detail is unreadable, you can see that information is missing. The fact that the information is unreadable is the point. Viewports make multiple enlarged views of one drawing which are readable for easy drawing or editing while leaving a view

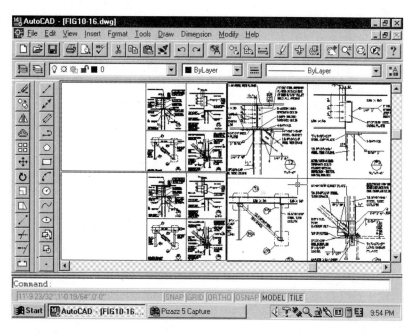

FIGURE 10–18 Using the default values for viewport, three views will be created. The viewport on the right is active by default. (Courtesy StructureForm Masters, Inc.)

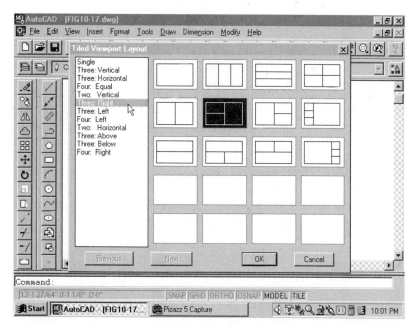

FIGURE 10–19 Viewports can be placed in several different arrangements.

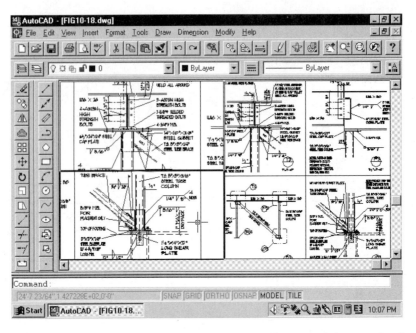

FIGURE 10–20 The viewports created in Figure 10–18 have each been edited to show this display. By making the upper left viewport active, the upper right corner was enlarged using the ZOOM command to display the beam-beam-column detail. The strut-column-plate detail was enlarged by making the lower viewport active and then enlarging the lower right hand detail. (Courtesy StructureForm Masters, Inc.)

of the overall drawing for perspective. Viewports and its options have several other advanced uses which will be discussed in Chapter 23.

To make a different viewport active, move the cursor to the desired viewport and click. To convert a drawing similar to Figure 10–18 to a drawing similar to Figure 10–20, change between active viewports and use the ZOOM command. Remember once active, all drawing and editing commands function normally. First make the upper left viewport active. Then select the detail in the upper right quadrant for a ZOOM window and enlarge to fill the viewport. Then make viewport on the lower left active and select the detail in the lower right quadrant for a ZOOM using the Window option. With these details enlarged, you can make changes to each without having to zoom from one detail to another.

AERIAL VIEW

Aerial View is an extremely useful method of viewing different portions of a large drawing file. An overall view of the entire drawing is displayed in a small window which is overlaid over the drawing screen. This is like picture-in-picture of a TV.

Portions of the drawing can be selected for viewing from this small window and redisplayed while still being able to see the entire drawing in the view window. This is especially helpful when working on a very large file. Many CAD users often lose perspective of where they are in the drawing file when they have done several ZOOMs into the drawing. The use of Aerial View provides that perspective by displaying the overall drawing in the window.

Accessing the Aerial View Window

The Aerial View window is displayed by picking the Aerial View icon from the Standard toolbar, by selecting Aerial View from the View pull-down menu, or by typing **AV** [enter] or **DSVIEWER** [enter] at the command prompt. Each method will display the window seen in Figure 10–21. The Aerial View display has many features similar to other windows including the minimize/maximize buttons and the pull-down bar menu. The size and location of the Aerial View display can be altered using either the corner or side stretching methods, and the scroll can be used to pan the display seen in the aerial view at its current magnification. Other controls of the display box include pull-down menus, and control button.

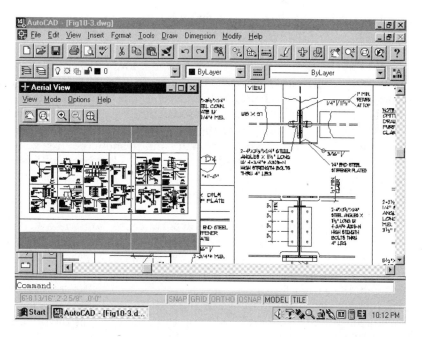

FIGURE 10–21 Aerial View provides an option for moving between views of a drawing. The drawing extents are shown in the Aerial View Window box.

PULL-DOWN MENUS

Three pull-down menus of Aerial include View, Mode and Options.

View

The options for view include Zoom In, Zoom Out, and Global. If the entire drawing is displayed as Aerial View is entered, the Zoom In/Out options will not be active. These options are used to change the magnification of the Aerial View.

Zoom In. The Zoom In option of Aerial is the same as the option found in the Zoom option of the View pull-down menu. The option increases the magnification of the image displayed in the Aerial view window.

Zoom Out. The Zoom Out option of Aerial is the same as the option found in the Zoom option of the View pull-down menu. The option decreases the magnification of the image in the Aerial view window.

Global. Selecting this option will display the entire view in the Aerial view window.

Mode

The Mode pull-down menu offers the options of Pan and Zoom. Each option can also be selected from the View pull-down menu.

Pan. This option allows the display to be altered from side to side or top to bottom without changing the drawing magnification. Pan can also be completed using the scroll bars. The Aerial View provides a point of reference by showing the entire structure. By selecting the Pan option, a reference box will be displayed in the Aerial View to pan the drawing. As the reference box is moved in the aerial view, the portion of the drawing displayed in the drawing portion of the screen is altered. Move the reference box, and the drawing will snap to show the area contained in the reference box. By pressing and holding the mouse select button, the drawing will drag in realtime to reflect the movement of the reference box. Figure 10–22 shows an example of the effect of the Pan option.

Zoom. This option changes the viewports center and view size. The Zoom mode changes the current view by allowing two points to be selected that will become the new corners of the current view. Display the Aerial View and pick the Zoom option. With the Zoom option active, pick and hold the mouse select button. As the button is picked, the corner for the Zoom window is selected. While still pressing the select button, move the mouse to select the opposite corner of the window. As the mouse is moved, the window size is altered and the area and size of the drawing is altered. Figure 10–23 shows the effect of using the zoom option.

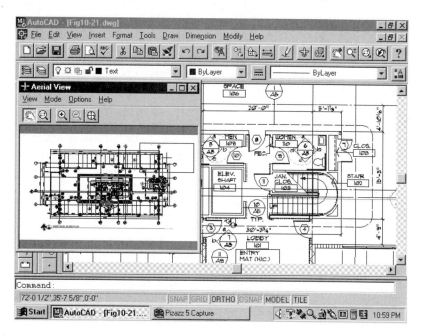

FIGURE 10–22 With the Pan option active, by moving the reference box, the information shown in the box will be displayed in the drawing editor. The current screen display is surrounded by a bold black border in the aerial view.

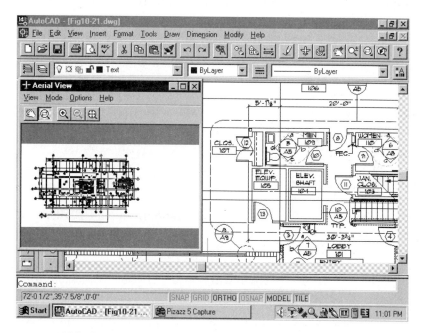

FIGURE 10–23 With the Zoom option active, a view window can be created by pressing and holding the mouse pick button. As the mouse is moved, the size of the window is altered. Once the window is selected, releasing the select button will alter the drawing editor to reflect the contents of the viewing window.

Options. The Options menu includes the options of Auto Viewport, Dynamic Update, Locator Magnification... and Display Statistics....

Auto Viewport. This option is a toggle switch between ON/OFF. When a check is displayed, the setting is ON. In the ON mode, switching to a different tiled viewport will automatically cause the new viewport to be displayed in the aerial view window. In the OFF mode, a new viewport will not be displayed.

Dynamic Update. This option is a toggle switch between ON/OFF. When a check is displayed, the setting is ON. In the ON mode, the window will update the display after each change in the drawing.

Real Time Zoom. This option controls how the window will be updated. In the default setting, the window will be zoomed in realtime. By clicking this option, the view will be enlarged in Zoom rather than in Real Time Zoom.

Control Buttons

Below the menu bar are five icons that can be used to control the aerial view display. From left to right these buttons include Pan, Zoom, Locate, Zoom In, Zoom out, and Global. The icon produces the same effect as the option found in the menu bar.

Using the Aerial Window

As you move the cursor into the Aerial View window it turns into cross-hairs. Picking two points in the view window with the cross-hairs will now display the selected area on the drawing as a zoom. Selecting two new points redisplays the selected area on the drawing screen at the same scale as the first display. The process can be seen in Figure 10–22 and 10–23.

VIEW RESOLUTION

The speed of each ZOOM or PAN can be enhanced if a REDRAW is used rather than a REGEN. The speed of all view changes also can be influenced by adjusting the view resolution. Resolution refers to the amount of detail that is represented when arcs and circles are drawn. The higher the resolution, the more lines that are used and the smoother the arc or circle will appear. Higher resolution also means that more time is needed for zooms and pans.

The resolution of arcs and circles is controlled by the VIEWRES command. The command sequence is:

Command: **VIEWRES** [enter]
Do you want fast zooms? <y> (*a Y or N is required.*) [enter]
Enter circle zoom percent (1–20000) <100>:

Fast Zooms

In setting VIEWRES, the first decision to be made is for fast zooms. Responding **N** enter will cause all ZOOM, PAN, and VIEW restorations to be done by regenerations. Responding **Y** enter will keep a large virtual screen and execute most ZOOM, PAN, and VIEW restorations at REDRAW speed. Zooming to the limits of a drawing may require a regeneration even though you have selected fast zooms.

Circle Percent

This selection will allow for controlling how circles and arcs will be represented on the screen. Circles and arcs are drawn using many short, straight line segments. AutoCAD computes the optimum number of sectors required for each arc or circle to appear at the given zoom magnification. CIRCLE PERCENT controls the computation required to gain speed at the cost of accuracy. The default value of 100 allows AutoCAD to run its computations without alteration. As a circle is zoomed in on, its smoothness will begin to disappear.

As the value is increased, more line segments will be added to the circle. As a circle is zoomed in on, it will still retain its smoothness. As the value is decreased, fewer line segments will be used to draw the circle. Each line segment can be seen without zooming in.

CHAPTER 10 EXERCISES

E-10-1. Load drawing E-8-9. Use the zoom command to make the triangles the current view. Save this view as TRIANGLES. Zoom into each of the remaining shapes, and make each shape a separate saved view. Save the entire drawing as E-10-1. Write a list of each view names in the space provided.

_____

_____

_____

E-10-2. Using drawing E-10-1 as a base, make the view containing the squares the current view. Draw a square in the center of one of the existing squares using midpoint to form a diagonal square. Repeat this process four additional times, using zoom as required. Inside the final square, draw a circle that is tangent to the smallest square. Save this view with the name TINY and save the entire exercise as E-10-2.

E-10-3. Load E-9-1 and enlarge the drawing to twice its current size. Save this view as VIEW2X. Zoom out so that the edges of the largest triangle touch the edge of the screen. Name this view with the name of the command required to perform the zoom. Set the drawings so that no screen markers will be produced and save the drawing as E-10-3. Write the name of the views created and the commands used in the space below.

VIEWS   COMMANDS

_____   _____

_____   _____

E-10-4. Load exercise E-9-3 and use it as a base for this exercise. Change the limits of this drawing to 36,18 and set the screen display so that the total content of the limits will be displayed. The original view is now displayed:

_____   _____

_____   _____

Use ZOOM dynamic to make the object's original view fill the screen. Save the drawing as E-10-4. Write the steps that were required to adjust the views.

_____   _____

_____   _____

E-10-5. Start a new drawing and set the limits to 96' × 144' with a grid of 24" and a snap of 6". Set units to measure in architectural, with fractions set at 16. Set angles to be measured in degrees/minutes/seconds with two-place accuracy. Set the view resolution to 150. Divide the drawing into the following seven views: All, upper left, upper cen, upper right, low right, low cen, and low left. Save view All as the current view. Save the drawing as a drawing templet. This drawing can be used as a future base for drawings done at ¼" = 1'-0" scale.

E-10-6. Open a file of your choice and establish a minimum of three Views. Provide names that quickly describe area captured in the viewport. Save the file as E-10-6.

E-10-7. Open a file of your choice and establish a minimum of three viewports. Enlarge the information in one of the viewports and shrink the information in a second. Save the file as E-10-7.

E-10-8. Open a file of your choice and establish an aerial view window. Use the window to zoom a portion of the drawing and then save the file as E-10-8.

CHAPTER 10 QUIZ

1. What command controls the use of screen markers and how should it be set to provide markers?

_____

_____

_____

2. List four methods of removing screen markers other than the command used in question 1.

_____

_____

_____

_____

3. Compare and explain the difference in VIEWRES settings between 50 and 500 and the default value.

_____

_____

_____

_____

4. Explain the difference between REGEN and REDRAW.

_____

_____

_____

5. When is paper space typically used?

_____

_____

_____

6. Describe the difference between ZOOM ALL and EXTENTS.

_____

_____

_____

7. What commands will be needed to make a drawing exactly twice as large as the current screen display?

_____

_____

_____

8. What option is used to scale space relative to paper space units?

_____

_____

9. Describe the meaning of "displacement point" and "second point."

_____

_____

_____

10. List and define the options for VIEW.

_____

_____

_____

_____

_____

_____

11. List the steps to enlarge a portion of a drawing using the zoom command.

_____

_____

12. You're tired of seeing markers left every time you edit a drawing. How can you end this frustration?

_____

_____

_____

13. List six options for accessing the ZOOM command.

_____

_____

_____

_____

14. Describe the process which allows the ZOOM command to be used in the middle of another command?

_____

_____

15. What term is used to describe a command that can be used while another command is being executed?

16. What command would you use if you wanted to create a picture-in-picture of an AutoCAD drawing?

17. When will an hourglass be displayed in the drawing area and what does it mean?

18. Describe the difference in effects between using an icon or menu in the Aerial View.

19. What must be typed at the command prompt to start Aerial view?

20. What must be typed at the command prompt to access the View control Dialog box?

21. How many viewports are active at any one time?

22. Sketch the results of selecting the following options of VPORTS. 3 viewports, above; 3 viewports, below; 3 viewports, left; 3 viewports, right; and 4 viewports, vertical.

23. Describe how to reduce the size of the aerial view window.

_____

_____

_____

24. Use the help command to find and describe the difference between tiled and untiled viewports.

_____

_____

25. Other than keeping you up late at night reading, what's the point of all of the options presented in this chapter?

_____

_____

_____

SECTION 3

ENHANCING DRAWINGS

Selecting and Modifying Drawing Entities

Modifying and Selecting Drawing Objects

Polylines

Supplemental Drawing Commands

Linetypes, Layers, and Colors

Solids, Bhatch, and Hatching

Inquiry Commands

······························

CHAPTER

_____

11

SELECTING AND MODIFYING
DRAWING ENTITIES

·······························

B y now you've tackled the basic components of a drawing. In this chapter you'll learn how to modify your drawing. This will include expanding your knowledge of the ERASE command and exploring four commands that can be used to alter drawings. These commands include COPY, MIRROR, OFFSET and ARRAY. Each will reproduce all or parts of a drawing. A second group of editing commands for refining drawings, and two new methods of sorting objects will be explored in the next chapter.

SELECTING OBJECTS FOR EDITING
WITH THE PICKBOX

To this point, only one entity at a time has been selected to be edited (ERASE). One or more entities have been selected using the pick box that is controlled by the mouse. As the prompt for the edit command is entered, the cross-hairs will change to an object selection target and the entity to be edited can be selected. Any object that is included in the target box will be affected by the command. The size of the box can be altered to enlarge or reduce the size of the selection area by using the PICKBOX system variable. Typing **PICKBOX** [enter] at the command prompt will produce a prompt of:

Command: **PICKBOX** [enter]
New value for PICKBOX <3>:

AutoCAD is now waiting for you to enter a numeric value to represent the size of the pickbox. If the size of the pickbox is altered it will effect the accuracy of your selections. As the size of the box is enlarged, information that is not desired in the selection set may be added. With the box too small, you may have trouble selecting information.

The size of the target box can also be selected by picking Selection... from the Tools pull-down menu, or by typing **SE** (enter) or **DDSELECT** (enter) at the command prompt. Each will produce the Objection Selection Settings dialog box shown in Figure 11–1. The slide bar can be used to control the size of the Pickbox. Other aspects of this dialog box will be discussed later in this chapter.

SELECTING OBJECTS FOR EDITING

In addition to selecting objects to edit using the pick box, AutoCAD provides the options of Single, A Point, Window, Box, Crossing, Wpolygon, Cpolygon, Fence, All, Last, Add, Remove, Previous, and Multiple to select an object for editing.

As different methods of selecting objects for editing are discussed, each method will be introduced using the ERASE command. Keep in mind, these options are in no way limited to one command. These options can be used anytime a selection needs to be

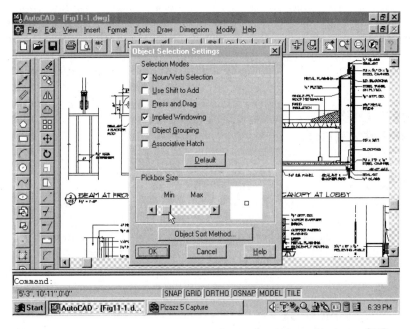

FIGURE 11–1 The size of the pickbox can be adjusted by typing **PICKBOX** (enter) or **DDSELECT** (enter) at the command prompt. If the box is too large, inappropriate objects may be included into the selection set. With the box too small, it may be difficult to include the desired objects.

made for any command using any of the selection methods for the desired editing command. Picking an icon from a toolbar or entering the command at the keyboard are generally the fastest methods. No matter the method used to access the command, a prompt will be displayed asking you to:

Select objects:

Each of the following selection methods can be entered at the Select objects: prompt for the editing command.

Single

When an editing command such as ERASE is entered, once an object is selected, you are prompted to select other objects to be edited.

The command sequence for ERASE is:

Command: **E** ⏎
ERASE
Select objects: *(AutoCAD is now waiting for you to select an object to be erased. Select an object. The selected object is highlighted as you pick it with the mouse.)*
Select object: *(This prompt will be displayed indefinitely until the* ⏎ *key is pressed.)* ⏎

You have the option to alter the selection set method so that only one option is selected and the command will be terminated. This can be done by typing **SI** ⏎ for *single* at the Select Objects: prompt. The command sequence is:

Command: **E** ⏎
Select objects: **SI** ⏎
Select objects: *(Select object to be edited, in this case erased. As the entity is picked, the command will terminate.)*
Command:

A Point

Objects to be edited can be selected by picking any point of the object. Points are selected by placing the target box across the desired entity box and pressing the pick button. Entities also may be selected by entering coordinates for the location of the target. When selecting objects, be careful not to pick objects at their intersections.

Window

Rather than selecting objects to edit one at a time, a group of entities may be selected by surrounding them completely in a window by responding **W** ⏎ at Select objects prompt. This procedure was used in Chapter 10 with the ZOOM command, and is shown in Figure 11–2. Any object that lies entirely inside the window will be edited.

Any object that is partially inside the window will not be edited. If the selection set in Figure 11–2a is accepted, the text that is within the selection box will be removed, and the leader lines which pass through the right side of the box will be not be affected by the edit. The command sequence to ERASE with a window is:

Command: **E** [enter]
Select objects: **W** [enter]
First corner: *(Select point for first corner.)*
Other corner: *(As the first point is selected, a window will drag across the screen until a second point is selected.)*
Select objects: *(Selected objects will be highlighted. You now have an opportunity to alter the selection set. The selection set can be altered by adding additional objects using another window, or by using the single, add, or remove selection methods. If no changes need to be made, press the enter key. Making changes to the selection set will be introduced in this chapter.)* [enter]

The result of the selection set can be sen in Figure 11–2b

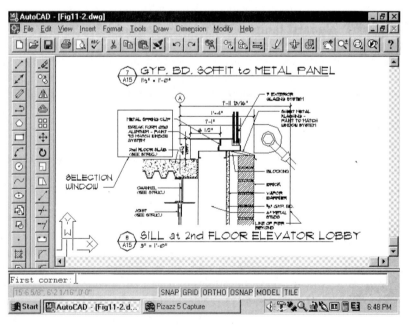

FIGURE 11–2a If the Window option is used for selecting objects to be edited, all objects that lie entirely in the window will be edited. In this selection set, the text in the window will be removed, but the leader lines that cross the right side of the selection window will not be included in the selection set.
(Courtesy Peck, Smiley, Ettlin Architects)

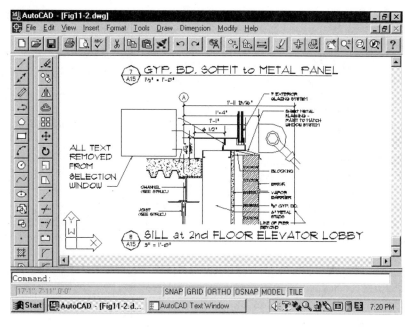

FIGURE 11–2b The results of the ERASE Window option started in Figure 11–2a.

Box

Using a box to select an object is similar to selecting objects using a window. When prompted for the object to be erased, press the mouse select button. This will turn the pick box into the corner of the selection box. As the mouse is moved, the window will be formed. Any items the window crosses will be selected. When you're satisfied with the size of the selection box, press the select button to accept the objects. Any objects crossed by the box will be highlighted for editing. To remove the objects, press [enter].

Crossing

Choosing an object to edit with the C option is similar to the Window option. Groups of entities may be selected by surrounding them partially in a window by typing **C** [enter] at the Select objects: prompt. The command process will be the same as the Window option. Any objects that lie entirely in the window will be edited. Any objects that are partially inside the window also will be edited. If the selection set in Figure 11–2a was selected using the C (Crossing) option and accepted for editing, both the text and the leader lines would be changed by the specified editing command. See Figure 11–3.

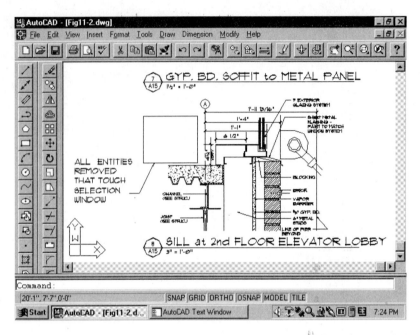

FIGURE 11–3 The results of the ERASE Crossing option started in Figure 11–2a.

WPolygon

The WP option is especially useful for selecting an irregular-shaped group of objects to edit. The option is similar to the window option, but you are allowed to select points to form a polygon around objects to be edited. The polygon may be any shape except a shape that crosses itself, such as a figure eight. The command sequence is started once the editing option has been entered and you are prompted to select objects:

> Command: **E** [enter] (*Any editing command may be entered.*)
> Select objects: **WP** [enter]
> First polygon point: (*Enter a point.*)
> Undo/<endpoint of line>: (*Enter a point.*)

Points to define the polygon may be entered by using a pointing device or entered as coordinates, as seen in Figure 11–4. The polygon is closed by entering [enter].

> Undo/<endpoint of line>: [enter]
> Select objects: [enter]
> Command:

All objects lying totally inside the polygon will be edited.

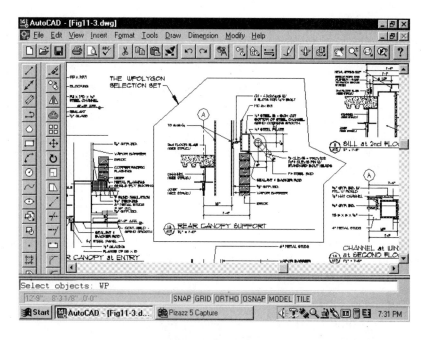

FIGURE 11-4 The WP (Wpolygon) option can be used as an alternative to "C" to select objects in an irregular-shaped group. Points for the polygon may be entered by picking points or entering coordinates. (Courtesy Peck, Smiley, Ettlin Architects)

CPolygon

Objects can also be edited by entering **CP** [enter] at the Select objects: prompt. This option will be similar to the effects of combining Crossing and WPolygon in that it will edit all objects that are in or cross the polygon. The command sequence is:

Command: **E** [enter] (*Any editing command may be entered.*)
Select objects: **CP** [enter]
First polygon point: (*Enter a point.*)
Undo/<endpoint of line>: (*Enter a point.*)
Undo/<endpoint of line>: [enter]
Select objects: [enter]
Command:

All objects lying inside or touching the polygon will be edited.

Fence

Entities can be selected for editing by Fence. This option is similar to the WPolygon and CPolygon, except that it does not close the polygon. The Fence may be a series of line segments or may form a polygon that can cross itself. This allows long rows or columns to be edited. As seen in Figure 11–5, the selection Fence only edits items that intersect or cross the fence. Fence points may be entered with a pointing device or by entering coordinates. The command sequence is:

Command: **E** enter (*Any editing command may be entered.*)
Select options: **F** enter
First fence point: (*Enter a point.*)
Undo/<endpoint of line>: (*Enter a point.*)
Undo/<endpoint of line>: enter (*Selecting enter will highlight selected objects.*)
Select objects: enter (*Selecting enter will execute the editing command.*)
Command:

All

When this option is used, all entities that are current will be affected. In Chapter 7, you learned how to freeze certain layers so that they cannot be seen. All objects except those on frozen layers will be edited.

Command: **E** enter (*Any editing command may be entered.*)
Select objects: **ALL** enter (*Items will be highlighted.*)
Select objects: enter (*Items will be removed.*)
Command:

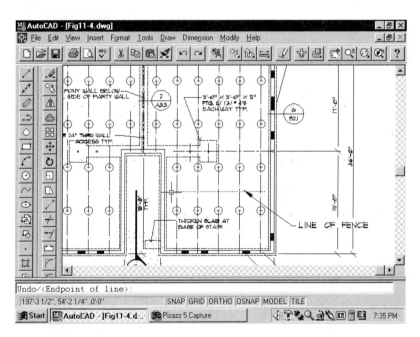

FIGURE 11–5 The Fence option allows straight lines to be used to include objects into the selection set. Any objects that are touched by the Fence will be included in the selection set. (Courtesy of Scott R. Beck, Architect)

Last

The LAST (L) option allows the most recently created object to be selected for editing.

> Command: **E** [enter] (*Any editing command may be entered.*)
> Select options: **L** [enter]
> Select options: [enter]
> Command:

Add

When selections are made using windows, not all intended objects for editing may be selected. ADD will allow for additional objects to be added to the selection set.

> Command: **E** [enter] (*Any editing command may be entered.*)
> Select objects: **W** (*window*) [enter]
> First corner: (*pick*) second corner: (*pick*) 20 found (*Quantity will vary with each selection.*)
> Select objects: **A** (*add*) [enter]
> Select objects: [enter]
> Command:

Objects may now be added to the selection set individually or with any other selection method. When you are satisfied with the selection set, press the pick button to edit the drawing.

Remove

This option will allow for objects to be removed from the selection set. Once objects are selected for editing, entering **R** [enter] will allow entities to be taken out of the selected entities to edit. The command sequence is:

> Command: **E** [enter] (*Any editing command may be entered.*)
> Select objects: **C** (*crossing window*) [enter]
> First corner: (*pick*) second corner: (*pick*) 30 found (*Quantity will vary with each selection.*)
> Select options: **R** (*remove*) [enter]
> Select options: [enter]
> Command:

Objects may now be removed from the selection set individually or with any other selection method. When you are satisfied with the selection set, press the pick button to edit the drawing.

Previous

Entering P at the select objects: prompt selects the last entity drawn to be edited. The command sequence is:

Command: **E** [enter] (*Any editing command may be entered.*)
Select objects: **P** (*previous*) [enter]

Multiple

This option for choosing the selection group is for when drawing entities are stacked over each other. To this point, all of your drawings have been created on one layer. In Chapter 7 you learned to place parts of your drawing on different levels. The Multiple option will allow you to edit material on different levels at one time.

SELECTING ENTITIES WITH A DIALOG BOX

Entities can also be added to the selection for editing set by typing **SE** [enter] or **DDSE-LECT** [enter] or by picking Selection.... from the Tools pull-down menu. Either will produce the dialog box shown in Figure 11–1.

Entity Selection Settings

The Object Selection Settings dialog box offers six options for selecting entities for editing with the dialog box. Notice that in Figure 11–1 two boxes have an **X** in the box indicating the default active settings for selecting objects. Any combination of selection methods may be used.

Noun/Verb.   Up to this point, as you have edited an entity, you've used a process AutoCAD refers to as Verb/Noun. You've picked the verb (ERASE) and then selected the object to be edited (the NOUN). AutoCAD also allows you to reverse the process. This box activates a new selection method that allows the selection of the object to be entered first, and then the selection of the editing command. The command prompt is:

Command: (*Pick object to edit using any method.*)
Command: (*Enter any of the listed edit commands.*)

With Noun/Verb active, any entity can be selected merely by selecting it with the pickbox. Enter a drawing and select a line. Notice the line is highlighted and a box appears in the middle and at each end of the line, but the command prompt has not changed.

Entering **E** [enter] will erase the selected line.

The Noun/Verb selection can be used with the following commands to edit entities:

ARRAY	DDCHROP	MIRROR	WBLOCK
BLOCK	DVIEW	MOVE	EXPLODE
CHANGE	ERASE	ROTATE	
CHPROP	HATCH	SCALE	
COPY	LIST	STRETCH	

Some commands require that the edit command be selected before the entities can be selected. These commands include:

BREAK	DIVIDE	FILLET	OFFSET
CHAMFER	EXTEND	MEASURE	TRIM

If you try to edit these commands, they will ignore any selection set created prior to starting the command.

Use Shift to Add. This selection method controls how entities are added to an existing selection set. Open a new drawing and draw a few lines. Type **SE** ⟨enter⟩ or **DDSELECT** ⟨enter⟩ and toggle Noun/Verb to OFF (no X) and Use Shift to Add to ON (X). Select OK and return to the drawing. Select an object to erase and it will be highlighted. Select a second object and the first object no longer will be highlighted. To retain the first object, hold the shift key down while the second object is selected. Both objects will now be edited. If Use Shift To Add is not enabled, objects are added to the selection by picking them individually or using a selection window. To remove objects from the selection set, press SHIFT while selecting the object. Shift to Add can also be toggled ON/OFF by using the PICKADD variable at the command prompt of 0 (active) or 1 (inactive).

Press and Drag. This control box determines the method used for drawing a selection window. With the option active, the selection window can be made with one button selection rather than two. The selection window is drawn by pressing and holding the pick button and dragging the cursor diagonally. The window is completed by releasing the pick button when the window reaches its desired size. When this option is not selected, objects may be selected for editing using the normal window option. Press and Drag can also be toggled ON/OFF by using the PICKDRAG variable at the command prompt of 1 (active) or 0 (inactive).

Implied Windowing. Making this option current will create a selection window automatically when the Select object prompt is displayed by picking the first corner point. A prompt will then be given requesting the other corner. If you draw the window from left to right, the window selects objects that lie entirely within the window. If the window is drawn from right to left, all objects in and touching the window will be edited. With the option off, the window must be set for each edit. Implied windowing can also be toggled ON/OFF by using the PICKAUTO variable at the command prompt of 1 (active) or 0 (inactive).

Object Grouping. This option determines if grouped objects will be recognized as individual objects or as a group. In the default setting of ON, grouped entities function as a group. When OFF, grouped entities can be edited individually. Grouping of objects will be discussed later in this chapter.

Associative Hatch. This option affects how HATCH patterns are edited and will be discussed in Chapter 16.

Default. Selecting the Default button will restore each of the six selection modes to their original setting.

Object Sort Method...

Selecting this button will display the menu seen in Figure 11–6. This menu can be used to control the order for display and plotting of drawing entities. Plotting will be introduced in Chapter 23.

Object Selection. With this option active, as window selections are created, they will be displayed in the order that they were placed in the drawing database.

Object Snap. With this option active, as OSNAP is used, (See Chapter 9) object snap modes will be selected based on the order they were placed in the database. If

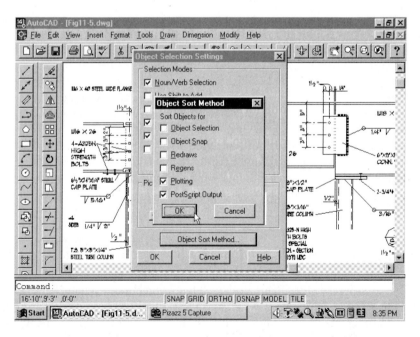

FIGURE 11–6 Selecting the Object Sort Method... button will display a new dialog box which can be used to determine the order in which objects in the selection set are displayed.

you try and SNAP to the end of two lines that are in the pickbox, the first line drawn will be picked and the second will be ignored for the selection set.

Redraws. With this option active, objects will be redisplayed in the order they were drawn when the REDRAW command is used.

Regens. With this option active, objects will be regenerated in the order they were drawn when the REGEN command is used.

Plotting. With this option active, objects will be plotted in the order they were drawn when the PLOT command is used. Plotting will be further discussed in Chapter 23.

PostScript Output. PostScript is a page description language which is used in the desktop publishing industry. Files are controlled by PSOUT. With this option active, PSOUT will process information in the order that it was placed in the database. This option will not be further discussed in this text but can be researched in AutoCAD's HELP file.

Don't panic. Think of AutoCAD as a really custom, super complete, $10,000 stereo sound system. It can do ANYTHING and all you want to do is listen to your favorite CD. Bit by bit though, you'll get hooked, read the manual, and start adjusting knobs. You'll crank the base and adjust the tweeters. AutoCAD is similar. All of the edits you'll be introduced to can be done without the dialog boxes. Once you feel more comfortable, try them. It's easy to learn one way to do something, and then not experiment with other methods. You'll really benefit by exploring. Don't be afraid to crank the bass and test the options and settings. If things get really spooky, exit the drawing, reboot the computer, and start all over again. Just make sure to make back-up copies of your work before you go exploring. It might also help to take a few notes as you go along. If you stumble into some really neat stuff, you may want to be able to do it a second time, and save the settings to a templet drawing so you can use them repeatedly.

MODIFYING A DRAWING USING ERASE

The ERASE command, which was introduced in Chapter 7, allows you to remove unwanted objects (mistakes) from a drawing. Since you're familiar with ERASE by now, the command will serve as a means of using some of the selection methods just covered. The selection prompts Point, Multiple, Window, Crossing, WPolygon, CPolygon, Fence, All, Last, Add, Remove, and Previous should start to seem more meaningful. Figure 11–7 shows a detail marker that has been selected for editing by entering "a Point." The selected entity is highlighted, while AutoCAD waits for you to select other objects. When the selection process is complete, press [enter] to remove the object.

The Window option can be seen in Figure 11–8. Notice that at the Select object prompt a "W" was entered rather than WINDOW. Remember, an object must be

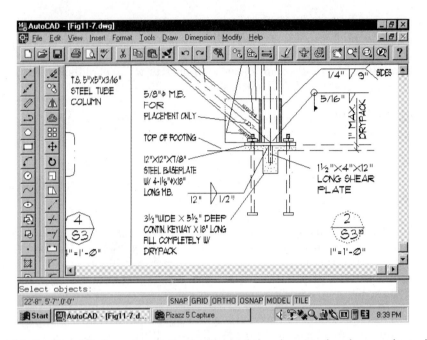

FIGURE 11–7 The detail marker at the bottom of the drawing has been selected for ERASE. (Courtesy StructureForm Masters, Inc.)

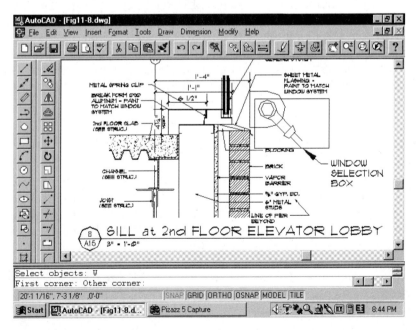

FIGURE 11–8 Using the W option to form a selection set for the ERASE command. (Courtesy Peck, Smiley, Ettlin Architects)

entirely inside the window to be edited. Figure 11–9 shows the objects to be erased highlighted. If you would like to ADD or REMOVE objects from the selection window, enter the corresponding letter at the Select objects prompt. Figure 11–10 shows the effects of the Window ERASE with no objects added or removed.

If you type **C** [enter] at the Select object: prompt, objects that are in or cross the window are selected to be erased, as seen in Figure 11–11. The results of a Crossing window selection can be seen in Figure 11–12. Because the engineer would like to keep the beams and remove the column, the REMOVE option was used to restore the selected lines of the beams. Figure 11–13 shows the selection set, and Figure 11–14 shows the results of making this selection. This drawing could now be used as the basis to form a new detail of two intersecting beams of different sizes with no support column.

MODIFYING A DRAWING BY ADDING OBJECTS

The COPY, MIRROR, OFFSET and ARRAY commands can each be used to modify a drawing by adding objects to the drawing base. Objects to be added using one of these commands can be selected using any of the selection methods introduced earlier in this chapter.

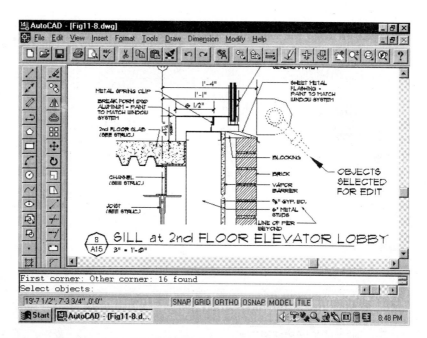

FIGURE 11–9 Results of the W (Window) selection set are now displayed to confirm the choice of objects to be included in the selection set. Notice that the number of objects included in the set is displayed above the command prompt. (Courtesy Peck, Smiley, Ettlin Architects)

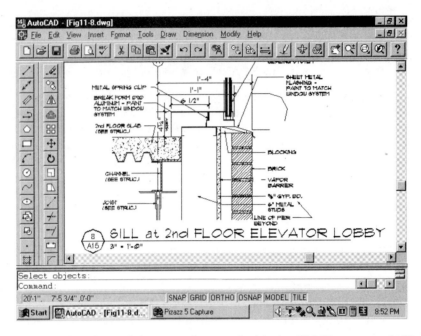

FIGURE 11–10. The effect of the W option used with the ERASE command. (Courtesy Peck, Smiley, Ettlin Architects)

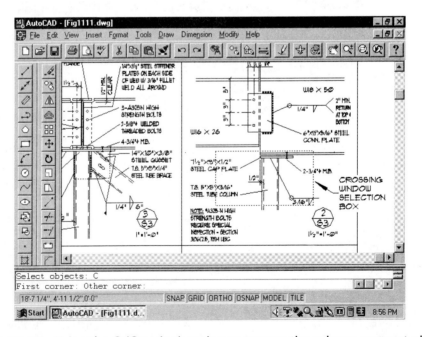

FIGURE 11–11 Using the C (Crossing) option to remove the column, support plate and related information. Any entity touched by the window will be edited. (Courtesy StructureForm Masters, Inc.)

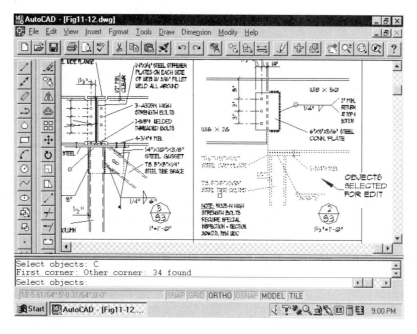

FIGURE 11–12 Results of the C (Crossing Window) selection set are now displayed to confirm the choice of objects to be included in the selection set. (Courtesy Structure-Form Masters, Inc.)

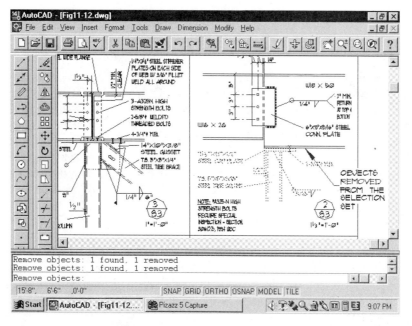

FIGURE 11–13 If you would like to include more objects in the set, enter **A** [enter] at the Select objects: prompt. Entering **R** [enter] at the prompt will remove objects from the set. To accept the set, press the [enter] key. (Courtesy StructureForm Masters, Inc.)

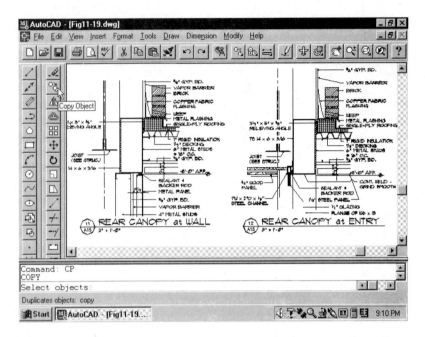

FIGURE 11–14 The COPY command can be used to duplicate common features. The command can be typed, picked from the Modify toolbar, or selected from the Modify pull-down menu. (Courtesy Peck, Smiley, Ettlin Architects)

COPY

Whenever multiple objects need to be drawn, the COPY command is a convenient method of reproducing objects. ARRAY, BLOCK, and WBLOCK are also efficient means of reproducing objects and will be covered in the coming chapters. The COPY command can be started by typing **CP** [enter] or **COPY** [enter] at the command prompt, by selecting Copy from the Modify pull-down menu, or by selecting the Copy icon from the Modify toolbar. Figure 11–14 shows an example of a drawing created using the copy command. The drawing on the left was drawn first and copied to produce the drawing on the right. Once it was copied, the text was edited slightly to produce the detail. The copy process is similar to the MOVE command, but the original objects are left intact. The command sequence is:

> Command: **CP** [enter]
> Select objects: (*Pick objects by desired selection method.*) [enter]
> <Base point or displacement>/Multiple: (*Select base point or enter coordinates.*)
> Second point of displacement: (*Select a point to relocate the base point to.*)
> Command:

The copy process can be seen in Figure 11–15a through 11–15d.

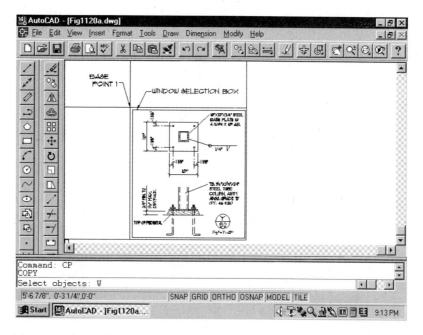

FIGURE 11–15a The COPY command is started by defining the selection set. (Courtesy StructureForm Masters, Inc.)

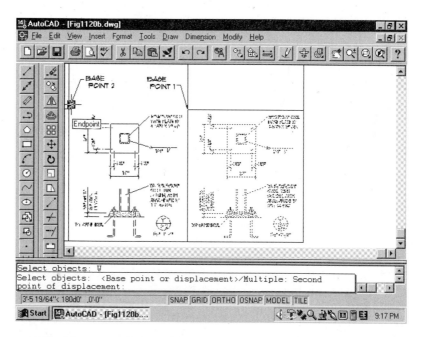

FIGURE 11–15b The second step of the COPY command is to specify a base point and then specify a new location for the base point. (Courtesy StructureForm Masters, Inc.)

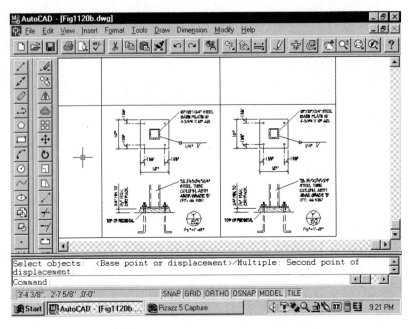

FIGURE 11–15c Once the [enter] key is pressed for the new location, the objects in the selection set will be duplicated. (Courtesy StructureForm Masters, Inc.)

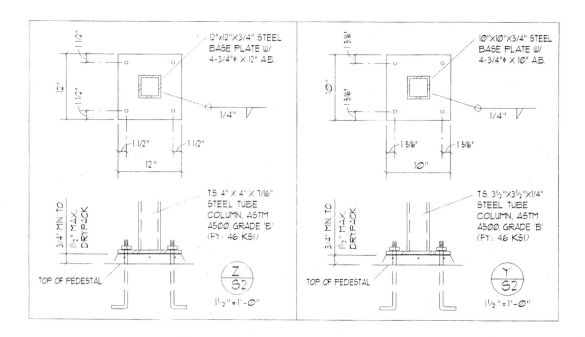

FIGURE 11–15d Once copied, the detail can be edited to provide information about a similar column-plate intersection. Notice that the size of the plate and column, and the bolt locations have been edited. (Courtesy StructureForm Masters, Inc.)

Multiple Copies. It's not uncommon to need to reproduce multiple copies of an object throughout a drawing. Multiple copies of an object or group of objects may be made once the selection set is picked. Notice in the copy command sequence that the alternative to the first point of displacement is multiple. When you enter **M** [enter] at the prompt, the selection group will be placed at an unlimited amount of locations. The command sequence is:

> Command: **CP** [enter]
> Select objects: (*Pick objects by desired selection method.*) [enter]
> <Base point or displacement>/Multiple: **M** [enter]
> Base point: (*Select base point or enter coordinates.*)
> Second point of displacement: (*Select a point to relocate the base point to.*)
> Second point of displacement: (*Select a point to relocate the base point to.*) [enter]

The process can be continued until all of the desired copies are drawn. To end the sequence enter [enter] for the Second displacement: prompt.

MIRROR

Objects in a drawing often need to be reversed. Something as simple as reversing a door swing, or as complex as flopping a floor plan for an apartment unit floor plan as shown in Figure 11–16 can be done using the MIRROR command. The command also can be used any time objects are symmetrical, such as a fireplace or double doors. When you draw half the object, you can create the other half by using MIRROR. The MIRROR command can be started by typing **MI** [enter] or **MIRROR** [enter] at the command prompt, by selecting Mirror from the Modify pull-down menu, or by selecting the Mirror icon from the Modify toolbar.

Once objects are selected to be mirrored (flopped), you will be asked to describe a "mirror line." The process can be seen in Figure 11–17. Think of the mirror line as a fold line between the old and the new objects to be mirrored. As you draw the first point of the mirror, drag the selection set into position and alter as the mouse is moved. The use of ORTHO and Object Snaps can greatly aid in the placement of the mirrored image. The new placement can be seen in Figure 11–18.

You'll be given the option of keeping or discarding the original object prior to making the MIRROR permanent. Your choice will depend on the use of the drawing. If the object is being completed by MIRROR, don't delete the old object. The final result will look like Figure 11–19. The command sequence will be:

> Command: **MI** [enter]
> Select objects: (*Choose objects using the desired selection method or combination of methods.*)
> Select objects: [enter]
> First point of mirror line: (*Select a point or enter coordinates.*)
> Second point: (*Select a point or enter coordinates.*)
> Delete old objects? <N> [enter]
> Command:

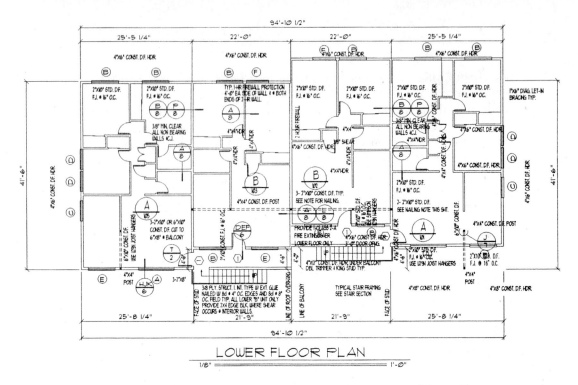

FIGURE 11–16 The MIRROR command can be used to flop existing objects. The command can be selected from the Mirror icon from the Modify toolbar or from the Modify pull-down menu or entered by keyboard. (Courtesy StructureForm Masters, Inc.)

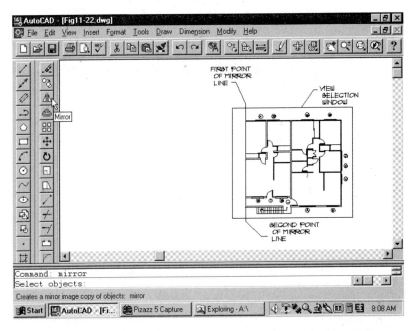

FIGURE 11–17 Once the selection set has been defined, two points will be required to define the mirror line. (Courtesy StructureForm Masters, Inc.)

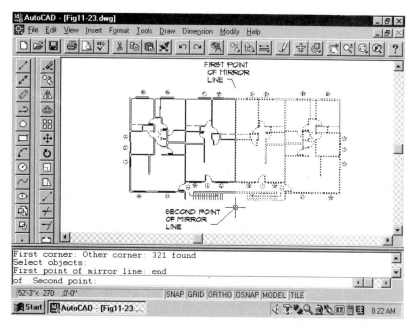

FIGURE 11–18 As the second point is entered, the selection set will be mirrored into position. (Courtesy StructureForm Masters, Inc.)

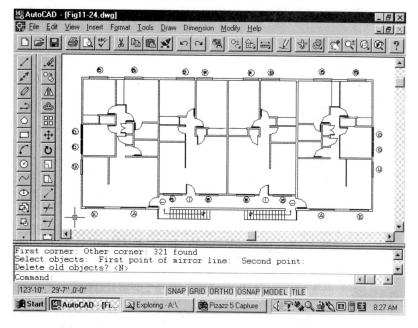

FIGURE 11–19 Before the MIRROR command is complete, an option will be given to "Delete old objects?" By responding **N** [enter] the selection set will be mirrored and the original objects will be retained. (Courtesy StructureForm Masters, Inc.)

If the object is being flopped, you will want to remove the old object as in Figure 11–20. The command prompts would be similar to those just noted until the last line is reached.

Delete old objects? <N> **Y** [enter]

Typing **Y** [enter] for yes will now delete the object that was mirrored and display the mirrored copy of the original object.

One of the drawbacks of the MIRROR command can be text. Although you will not add text to a drawing until Chapter 18, an additional feature of MIRROR will be helpful soon. As AutoCAD is currently set, if you were to mirror a drawing with text, the text would be upside-down. (See Figure 11–21). To mirror the drawing without flopping the text, adjust the MIRRTEXT value using SETVAR to zero. With MIRR-TEXT set to zero, the drawing will be flopped, but the text will remain readable, as seen in Figure 11–22. This will be covered again in Chapter 18, but you might want to set the value in your drawing template now so that when you do have text, you won't need the UNDO or OOPS commands. The command sequence is:

Command: **MIRRTEXT** [enter]
New value for MIRRTEXT <1>: **0** [enter]
Command:

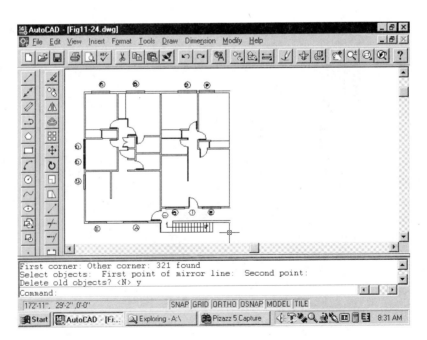

FIGURE 11–20 By responding **Y** to the "Delete old objects?" the selection set will be mirrored and the original objects will be removed. (Courtesy StructureForm Masters, Inc.)

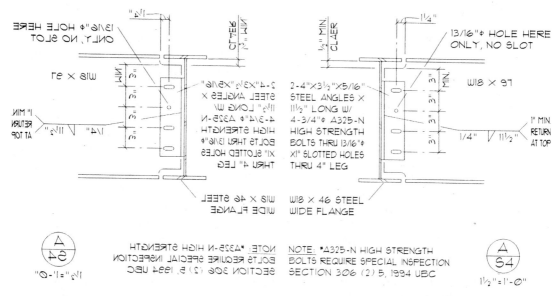

FIGURE 11–21 The default setting of the MIRROR command also will reverse text. (Courtesy StructureForm Masters, Inc.)

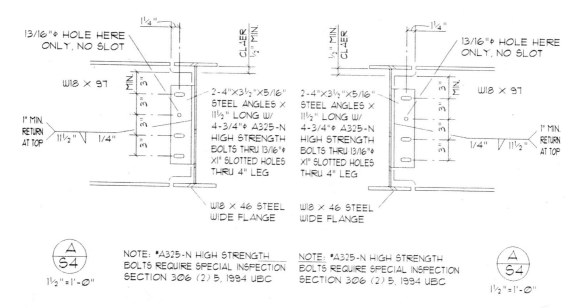

FIGURE 11–22 By setting the MIRRTEXT variable set at 0, text will also be mirrored. (Courtesy StructureForm Masters, Inc.)

OFFSET

The OFFSET command allows an entity to be copied and relocated parallel to and at a specific distance from the original object. The uses of this command are almost endless. A circle can be drawn and offset to represent a conduit or sewer pipe. Given a horizontal and vertical line, an entire rectangular structure can be designed by off-setting lines, and then cleaned up using other editing commands presented in Chapter 12.

OFFSET can be used in a manner similar to blueline layout of manual drafting. Lines can be drawn in the exact location without giving thought to their exact length. Once all of the boxes are lightly drawn to outline the desired rooms, the desired lines are darkened. Use OFFSET with the same mentality. Get the location right without worrying that the line may be too long, then edit the length with FIL-LET, STRETCH or TRIM. Each will be introduced in Chapter 12. Figures 11–23a and 11-23b show the use of OFFSET to lay out a simple floor plan.

The command process begins by typing **O** [enter] or **OFFSET** [enter] at the command prompt, by selecting the Offset icon from the Modify toolbar or by picking Offset from the Modify pull-down menu. The sequence is:

Command: **O** [enter]
Offset distance or Through (through>: **1'6** [enter]

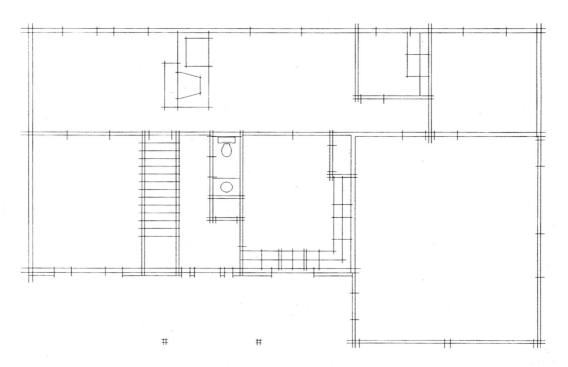

FIGURE 11–23a Using one horizontal and vertical line, the OFFSET command can be used for drawing parallel lines a specified distance apart to create a floor plan.

A distance may be entered by providing a numeric value or by entering the letter **T** [enter]. When the distance is entered, a new prompt will be displayed that asks:

Select the object to offset:

A Line, Arc, Circle, or Polygon may be selected to be offset. Once selected, the object will be highlighted and the prompt will read:

Side to offset:

Move the pointing device slightly to the desired side and press the pick button. Pressing the pick button will offset the object and redisplay the prompt:

Select object to offset:

The process may be continued indefinitely with the offset distance that was entered originally remaining the default. The OFFSET process is terminated by [enter]. Once the process is terminated, a new default distance may be entered or a new command sequence may be started. The command sequence can be seen in Figure 11–24.

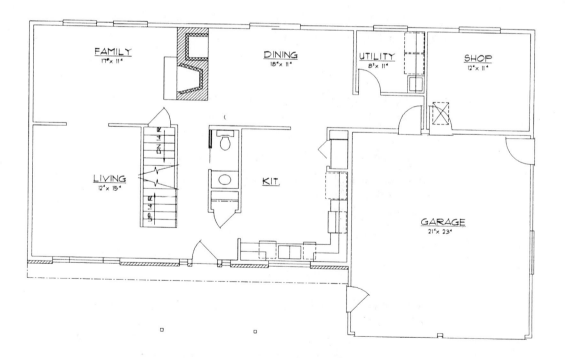

MAIN FLOOR PLAN
SCALE : ¼" = 1'-0"

FIGURE 11–23b Once the basic lines have been put in the correct position using OFFSET, other editing commands can be used to finish the drawing.

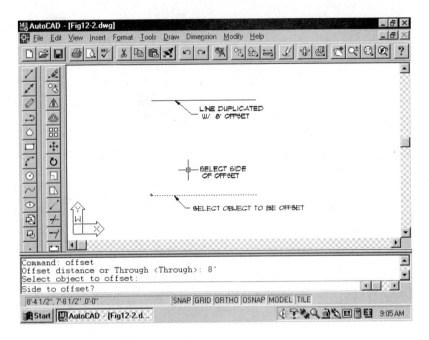

FIGURE 11-24 The OFFSET command is completed by selecting the distance to be offset, the object to be offset, and on which side of the original the offset will occur.

Entering **T** [enter] at the original prompt allows a point to be entered with a pointing device, and the object to be offset will pass through the indicated point. The command sequence is:

Command: **O** [enter]
Offset distance or Through (through>: **T** [enter]
Select object to offset: (*Pick an entity.*)
Through point: (*Move cursor to the desired point and pick.*)
Select object to offset: [enter]
Command:

The process can be seen in Figure 11–25.

Practical Uses for OFFSET. Open the ARCHBASE templet. Now draw a horizontal and a vertical line which intersect. Refer to the floor plan in Figure 11–23b. Many clients come to a designer with a floor plan that they like, along with a list of a thousand changes that they would like to make. The OFFSET command can be useful in the layout of a floor plan. Assume the two lines that have been drawn to represent the outer edges of the walls in the upper right-hand corner of the plan shown in Figure 11–23b. Offset each line 6" in toward the inside of the house. Since the room in Figure 11–23b is 12'-4" wide, offset the vertical line you just created a

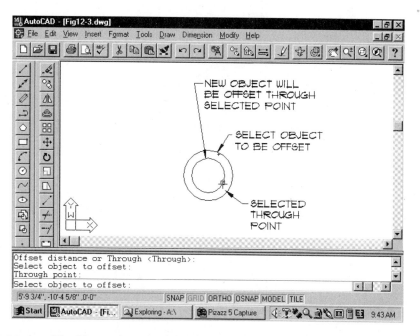

FIGURE 11–25 The Through option of OFFSET allows the offset distance to be specified using the pointing device.

distance of 12'-4" to the left. Offset this same line 4" to represent the wall thickness. Offset this line 8'-2" and then again 4". This will provide the wall on the left side of the Utility room.

Now go back to the inner horizontal line that you drew. Offset this line 11'-4" toward the bottom of the drawing. Offset this line 6" toward the bottom of the drawing. These two lines will represent the wall between the shop and the garage. To place the wall between the utility room and the hall, offset the inner garage wall 42". Offset this new line 4". The entire process is shown in Figure 11–26. For now save this crude drawing as FLOOR11.

ARRAY

Until now, if you have wanted to reproduce a drawing element, you used the COPY command. The ARRAY command will also allow for multiple copies of an object or group of objects to be reproduced such as the beam, post, and columns of a post-and-beam foundation. The command also provides several added features not available with COPY. ARRAY reproduces objects in rectangular or circular (polar) patterns, allows the object to be rotated as it is reproduced, and allows for easy control of the spacing of the object during the ARRAY. Figures 11–27a and 11-27b show examples of a rectangular and a polar array. The command can be accessed from the Modify pull-down menu by selecting one of the Array icons from the Copy icon from the Mod-

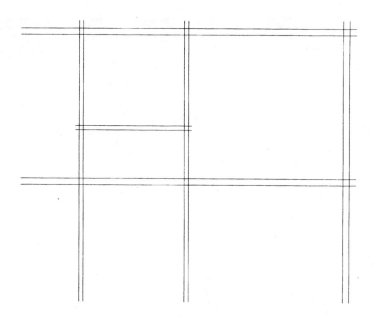

FIGURE 11–26 By drawing two perpendicular lines, and using the OFFSET command, a floor plan can easily be drawn.

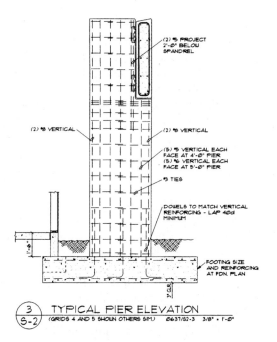

(2) #5 PROJECT 2'-0" BELOW SPANDREL

(2) #8 VERTICAL

(2) #8 VERTICAL

(5) #5 VERTICAL EACH FACE AT 4'-0" PIER
(5) #6 VERTICAL EACH FACE AT 5'-0" PIER

#3 TIES

DOWELS TO MATCH VERTICAL REINFORCING - LAP 40d MINIMUM

FOOTING SIZE AND REINFORCING AT FDN. PLAN

3"—

3 TYPICAL PIER ELEVATION
S-2 (GRIDS 4 AND 5 SHOWN OTHERS SIM.) 0637/82-3 3/8" = 1'-0"

FIGURE 11–27a The steel reinforcement was placed using the ARRAY command. (Courtesy Van Domelen/Looijenga/McGarrigle/Knaff Consulting Engineers)

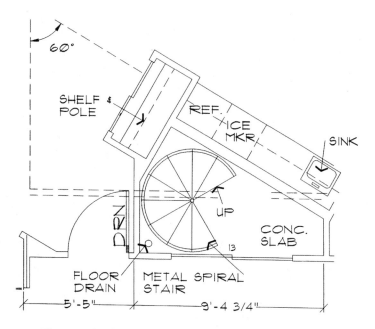

FIGURE 11–27b The treads of this stair were placed using the polar option of the ARRAY command. (Courtesy Piercy & Barclay Designers, Inc.)

ify toolbar, or by typing **AR** [enter] or **ARRRAY** [enter] at the command prompt. This command should not be confused with the DISARRAY option, which results from poor planning or procrastination.

RECTANGULAR ARRAY

Once the command has been accessed, objects to be arrayed may be selected by any of AutoCAD's select features. Once selected for ARRAY, the command prompt will ask for the type of display. The prompt is:

> Command: **AR** [enter]
> Select objects: (*Pick desired objects.*)
> Select objects: [enter]
> Rectangular or Polar array (<R>/P): [enter]

with R being the default. A rectangular array will prove suitable for construction projects which are arranged in rows or columns, such as placement of posts or columns for a floor or foundation plan, or for drawing sheets of plywood for a roof-framing plan. Four specifications must be provided to perform a rectangular array. These include the number of rows, number of columns, distance between rows, and distance between columns. The effects of each can be seen in Figure 11–28.

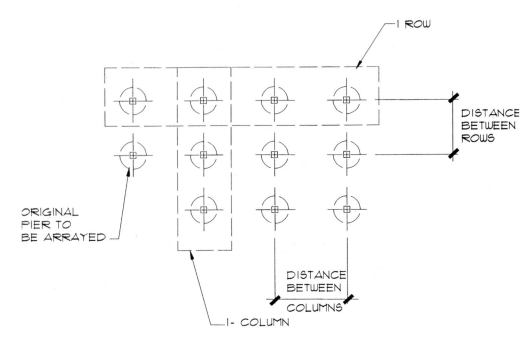

FIGURE 11–28 The number of rows and columns and the spacing of each must be specified to use a rectangular ARRAY.

Rows

The first prompt to respond to for a rectangular array will be:

Number of rows (---) <1>:

Any whole number may be entered for the value. In defining the quantity of rows, you are defining the quantity of horizontal rows to be created. Figure 11–29 shows an example of an array with two, three and four rows.

Columns

Specifying the number of columns will dictate the quantity of vertical columns to be used. Any whole number may be entered for the value. If one was used for the row value, one may not be used for the column value. If it is, AutoCAD will respond with the prompt:

One-element array, nothing to do.

and return the command prompt. Figure 11–30 shows the effect of arrays with two, three, and four columns.

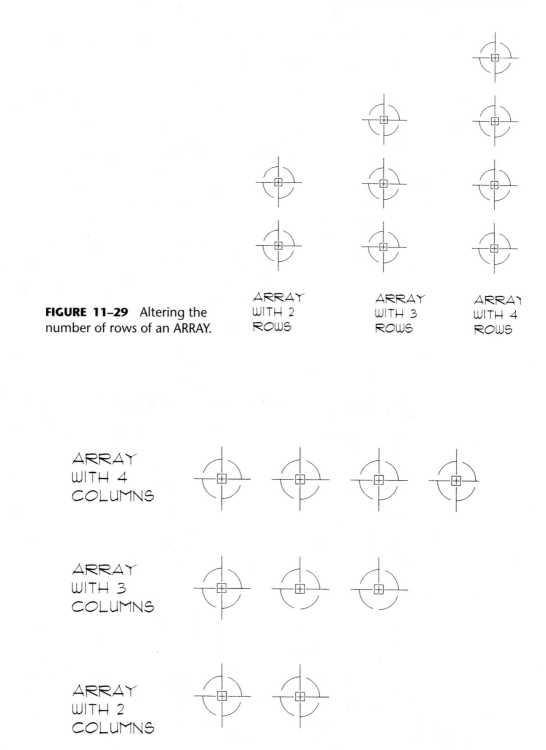

FIGURE 11-29 Altering the number of rows of an ARRAY.

ARRAY WITH 2 ROWS

ARRAY WITH 3 ROWS

ARRAY WITH 4 ROWS

ARRAY WITH 4 COLUMNS

ARRAY WITH 3 COLUMNS

ARRAY WITH 2 COLUMNS

FIGURE 11-30 Altering the number of columns in an ARRAY.

Distance Between Rows

The third prompt for a rectangular array is:

Unit cell or distance between rows (---):

This prompt allows the distance to be set for row spacing, also called the unit cell. The unit cell distance includes the size of the object to be arrayed. If you are arranging 16" wide chairs in rows, and would like 16" between rows, the spacing would require a 32" unit cell. This is shown in Figure 11–31. If you are arranging sheets of plywood, which typically are laid edge to edge, the unit spacing would be the spacing of the unit to be arrayed, in this case 48" for a sheet of plywood. This layout can be seen in Figure 11–32.

You also can set the distance between rows by picking the unit cell as two opposite points of a rectangle. The rectangle will provide information for the row spacing. If more than one column is to be arrayed, the rectangle will provide information for both the row and the column distances. The unit cell spacing can be seen in Figure 11–33.

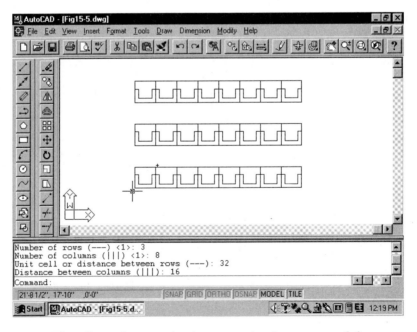

FIGURE 11–31 The effect of row and column spacing in an array of three rows and eight columns

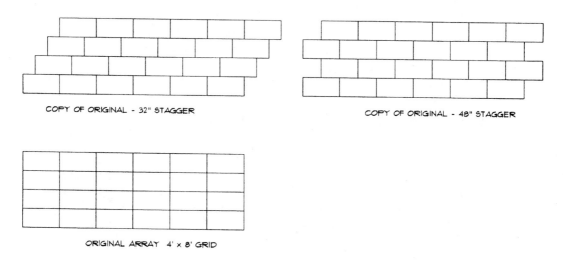

COPY OF ORIGINAL - 32" STAGGER

COPY OF ORIGINAL - 48" STAGGER

ORIGINAL ARRAY 4' x 8' GRID

FIGURE 11–32 The original array pattern at the top was created using four rows and six columns. The pattern at the bottom was altered by moving two rows.

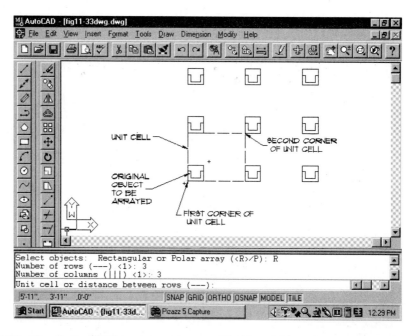

FIGURE 11–33 Unit cell distance can be specified by picking the two opposite corners of a window.

Distance Between Columns

The final prompt for a rectangular array is to provide the distance between columns. The same guidelines for spacing rows apply to spacing columns. The prompt is:

Distance between columns (| | |):

Positive and Negative Values

As values are provided for rows and columns, they may be either positive or negative, or a combination of both. Using positive values for both will array the objects up and to the right of the original object, as seen in Figure 11–34. If a negative value is used for the row distance, new objects will be added below the original. If negative values are used for the column distance, new objects will be added to the left of the original object, as seen in Figure 11–35. Figures 11–36 and 11–37 show the effects of combining values.

Rotated Rectangular Arrays

The array pattern is normally based on a horizontal or vertical baseline. The baseline is based on the current SNAP rotation value. By using the SNAP Rotate command discussed in Chapter 4, you can create a rotated array. By rotating the

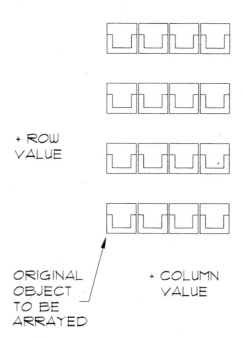

+ ROW
VALUE

ORIGINAL
OBJECT
TO BE
ARRAYED

+ COLUMN
VALUE

FIGURE 11–34 An ARRAY pattern based on entering two positive distances.

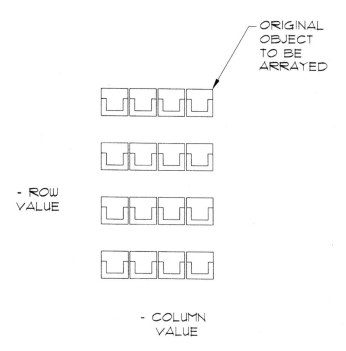

FIGURE 11–35 An ARRAY pattern based on entering two negative values.

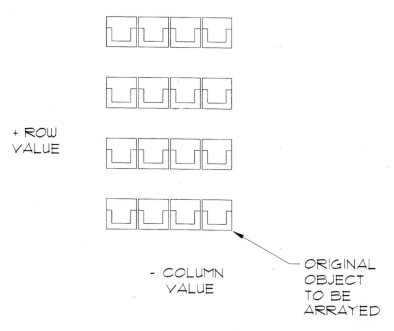

FIGURE 11–36 An ARRAY based on a positive row value with a negative column value.

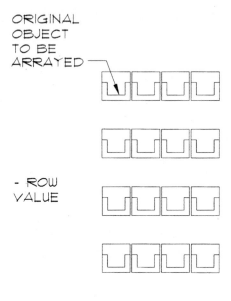

ORIGINAL
OBJECT
TO BE
ARRAYED

- ROW
VALUE

+ COLUMN
VALUE

FIGURE 11–37 An ARRAY based on a negative row value with a positive column value.

horizontal and vertical baselines, objects may be arrayed along the new baseline. The command sequence is:

Command: **SN** [enter]
Snap spacing or ON/OFF/Aspect/ Rotate/ Style <0′-1″>: **R** [enter]
Base Point <0′-0″,0′-0″>: (*Select desired base point.*)
Rotation Angle <0>: **45** [enter]

With the baseline rotated to 45°, the cross-hairs also will be rotated and any objects that are arrayed will be located on the current cross-hairs, as seen in Figure 11–38.

POLAR ARRAY

A polar array is a circular layout of objects based on a centerpoint. The process is started once the objects to be arrayed have been selected. The command sequence is:

Command: **AR** [enter]
Select objects: (*Pick desired objects.*)
Select objects: [enter]
Rectangular or Polar array (<R>/P): **P** [enter]
Base/<Specify center point of array>: (*Pick desired centerpoint.*)

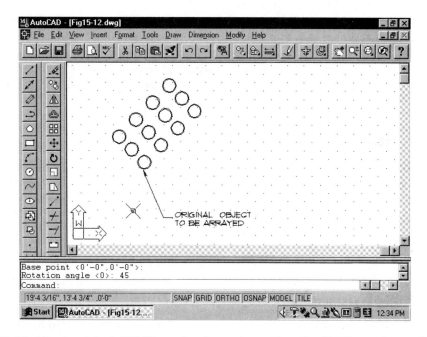

FIGURE 11–38   A rotated ARRAY can be drawn by setting the SNAP spacing to ROTATE and providing the desired angle.

The centerpoint is the point around with selected objects will be arrayed. Three options are now presented to set parameters for the polar array. Any two of the three may be used. The parameters include:

Number of items in the array
Angle to fill (+ = CCW, – = CW) <360>:
The angle between objects in the array

Number of Items

The number of items prompt allows you to select how many items, including the original, will be displayed around the centerpoint. If a 0 is entered, the other two prompts will be given. If a positive value is entered, one of the two remaining prompts will be given.

Angle to Fill

This prompt controls the limits of the display around the centerpoint. The prompt is:

Angle to fill (+ = CCW, – = CW)<360>:

The default of 360° will locate the specified number of items around a full circle. A number smaller than 360° will array the indicated number of objects around a selected centerpoint in a specified degree range. The pattern will start with the original object and spread the balance of the pattern from its location. A positive value will produce a counterclockwise rotation. A negative value will cause a clockwise rotation. Examples of each pattern can be seen in Figure 11–39. Entering a 0 will be accepted if the number of items has been provided. The location will now be arrayed by the final prompt requesting the angle between items.

Angle Between Items

The third prompt that may be used for a polar array is to specify the angle between items. If the number of items and the angle to fill prompt have been specified, this prompt will not be given. The number of items to be rotated combined with the angle distance between each item often is used to locate bolts or other similar features for drawing connectors (see Figure 11–40). Materials also can be arrayed over a specified degree range and intervals, such as showing pipes that intersect a cooling unit over a 75° area at each 15°, as seen in Figure 11–41.

Rotating Arrayed Objects

The final prompt for a polar array asks if the selected item to be arrayed should be rotated as it is arrayed. The default is yes, which will rotate objects as they are arrayed around the centerpoint. If an N is entered at the prompt, the objects will not

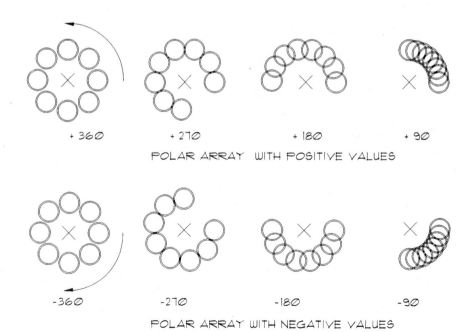

FIGURE 11-39 The effect of adjusting the "Angle to be filled" option for a polar array.

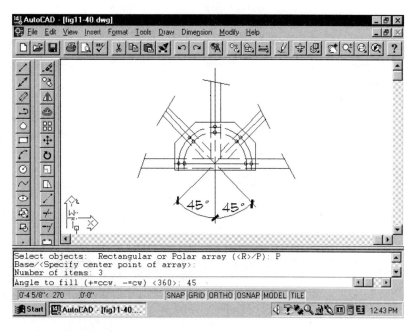

FIGURE 11–40 Specifying the number of items to be rotated and the "Angle between items" is a common method for defining an array pattern.

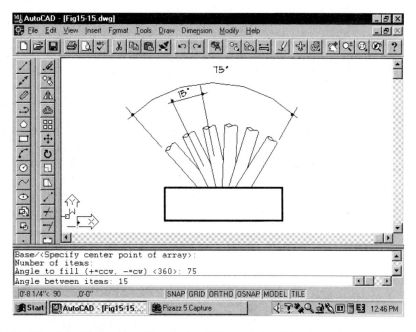

FIGURE 11–41 Specifying the "Angle to fill" and the "Angle between items" is also used to define the array pattern.

be rotated and will appear similar to Figure 11–42. The entire command sequence for a polar array of 50 segments around a 250° arc is:

Command: **AR** [enter]
Select objects: (*Pick desired objects.*)
Select objects: [enter]
Rectangular or Polar array (<R>/P): **P** [enter]
Base /< Specify center point of array >: (*Select desired centerpoint.*)
Number of items: **5** [enter]
Angle to fill (+ = CCW,– = CW)<360>: **250** [enter]
Rotate items as they are copied? <Y>: [enter]
Command:

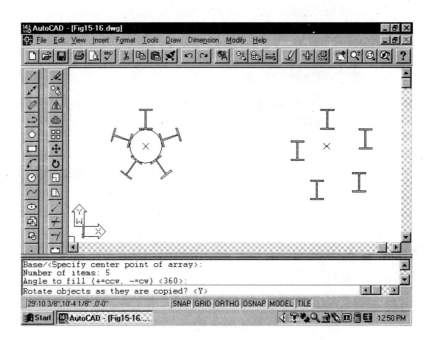

FIGURE 11–42 Objects can be rotated as they are arrayed by adjusting the "Rotate objects as they are copied?" option.

CHAPTER 11 EXERCISES

E-11-1. Start a new drawing and draw a horizontal line. Draw polygons with three, four, five, six, and eight sides with the centerline of each polygon on the horizontal line. Draw all polygons inscribed in a circle with a .5" radius. Make copies of the polygons with three, five, and eight sides and place them directly below their counterparts. Make copies of the four- and six-sided polygons and place them in a third row directly below their counterparts. Mirror the entire drawing with a horizontal mirror line, and delete the old objects. Save the drawing as E-11-1.

E-11-2. Open your existing file E-9-5. Copy the window so that a total of three pairs of windows are drawn exactly 4" apart. Align the bottoms of the windows. Save the drawing as E-11-2.

E-11-3. Draw half of a W 8" × 28" (I-shaped) beam. The beam has a total height of 8.28". The top and bottom flange are .25" thick, with a total width of 5.25". The web is a total of .28" thick. Use the mirror command to complete the drawing. Save the drawing as E113BM.

E-11-4. Using the following figure as a guide, draw an exterior swinging door that is 36" wide with a 12" wide sidelight. Use the mirror command to draw a pair of doors with double sidelights. Save the drawing as E114DOOR.

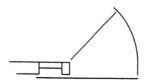

E-11-5. Using the following figure as a guide, draw a steel connector plate using the sketch below. The plate is 15½" × 10". Draw the bolts as ¾" diameter hexagons, 1½" down and 1½" in from the top and ends. Bolts are 3" o.c. Each end of the plate is symmetrical. Draw a 2½" × 8" strap centered in the bottom of the plate. Provide two bolts with similar spacing as the plate. Save the drawing as E-11-5.

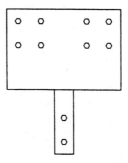

E-11-6. Use the sketch below as a guide and lay out the outline of the apartment floor plan. Draw the exterior walls as 6" wide. Use this unit plan and lay out four different building layouts with a minimum of four units per building. Save the drawing as APARTFLR.

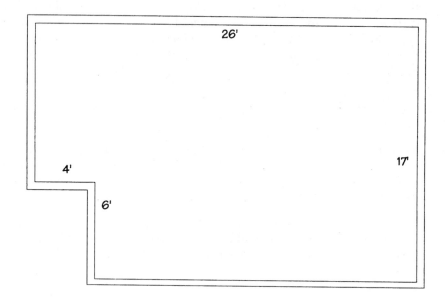

E-11-7. Draw a post-and-beam foundation plan for a 24' × 16' structure. Exterior walls will be 8" thick. Piers are 18" diameter at 8' o.c. along a 4" wide beam. Repeat the beam and piers at 4' o.c. Save as PB-FND.

E-11-8. Start a new drawing that can be used as a base for shingle siding in elevation. Draw several lines at 10" o.c. approximately 36" long. Draw shingles similar to the sketch with widths that vary from 4" through 12". Copy this row of shingles to form a minimum of five rows of shingles. Stagger each row with the rows above and below it so that the shingle seams do not line up. Save the drawings as SHINGLE.

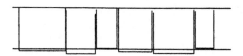

E-11-9. Extend the limits as needed to draw a 4" square centered in an 15" diameter circle. Draw a 8" and a 12" diameter circle with the same centerpoint as the 15" circle. Make three additional copies of this object and move each so that the four 15" circles are tangent. Save the drawing as E-11-9.

E-11-10. Open drawing E-9-5. Your client thinks he would like a grid in each half, but isn't really sure. Copy the drawing and draw a grid in one of the windows. Provide one vertical and three horizontal dividers. Mirror the grid into the other. Save the drawing as E-11-10.

E-11-11. Draw a 60' wide × 44' deep structure. Use 8"-wide walls. No doors or windows need to be shown. Create a layer to represent the suspended ceiling and draw the plan for a suspended ceiling using 24" × 48" panels. Make a second copy of the plan and show the panels laid out in the opposite direction. Save the drawing as E-11-11.

E-11-12. Use the base drawing started in Figure 11-11 and draw a roof framing plan. Freeze the suspended ceiling information and create a new layer to represent roof trusses at 32" o.c. Use the attached drawing as a guide to show a portion of the roof with 4' × 8' plywood. over the trusses. Save the drawing as E-11-12.

E-11-13. Draw a column that is 10'-0" tall and 12" wide. Show a 20" × 4" and a 16" × 4" corbeling 4" down from the top and 4" up from the bottom of the column. Save the drawing as E-11-13.

E-11-14. A circular structure is to be built on a hillside. Use the drawing below as a guide and draw the pier plan. Save the drawing as E-11-14.

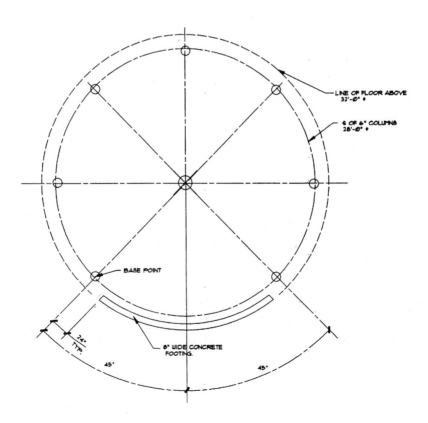

E-11-15. Draw a plan view of a 40-foot diameter water reservoir. The reservoir will be supported on (12)-6" diameter × ⅜" steel columns. The columns will be equally spaced, on a 36-foot diameter circle centered on an 18" diameter × ¾" thick vertical inlet pipe in the center of the reservoir. Assume north to be at the top of the drawing, and locate a 12" diameter × ⅜" thick vertical outlet pipe at the outer edge of the tank, at 37° west of true north.

CHAPTER 11 QUIZ

1. Describe the process to draw a wall that is 6" wide.

2. Several copies of an object need to be made. What option is available so that the COPY command does not have to be repeated?

3. You have selected several objects for editing when you realize you have accidentally selected an entity that you do not want changed. What can you do so that you do not have to start the edit sequence over? _____

4. How will UNDO affect objects that have just been moved?

5. List three methods suitable for selecting irregular-shaped objects for editing.

6. When you use the Fence option, what objects will it select for editing?

7. What option will allow you to select objects and then decide how to edit them?

8. What command will display the dialog box that controls selection methods?

9. List the commands and sequence required to complete Exercise E-11-9.

10. Explain when the Through option of OFFSET could be used.

11. Open the ERASE command using each method and describe the advantages or disadvantages.

12. Experiment with the mirror command. Describe the effects of ORTHO on the mirror line.

13. Mirror is being used to complete a symmetrical object. What options are available and how do they affect the mirror sequence?

14. Where is Erase listed in the pull-down menu, and what options are listed?

15. Describe how to change the size of the pick box.

16. What is the effect of Base point on the COPY command, and how should it be selected?

17. List ways that the COPY command could be used to lay out a floor plan.

18. What is the benefit of having so many different methods of selecting objects for editing?

19. What would be the drawback to enlarging the pickbox to twice its normal size? To half it's normal size?

20. Explain the difference between the normal setting for an editing command and the SINGLE option.

21. Explain the difference between Noun/Verb and Verb/Noun selection.

22. Explain the meaning and the effect of the following letters for selection groups. W, C, F, WP, CP, R, A, and P.

23. What effect does Press and drag have on selection?

24. List the effects of the PICKADD, PICKAUTO, and PICKDRAG variables.

25. How can OBJECT SNAP be used with editing commands?

26. Explain the difference between the two types of ARRAY patterns.

_____

_____

_____

27. Twelve steel columns need to be drawn 22 feet on center in one horizontal line. What will the array sequence be?

_____

_____

28. Three horizontal lines of 10 steel columns need to be drawn 18 feet on center. The horizontal lines of columns will be 32 feet apart. What will the array sequence be?

_____

_____

29. Use the drawing for Exercise 11–14 and give the command sequence to array the supports.

_____

_____

30. Using an illustration as an example, define a unit cell.

31. How could an array pattern for objects aligned at 45° from horizontal be drawn?

_____

_____

32. Ten rows and three columns are to be arrayed below and to the left of the original object at 60" o.c. How can this be done?

_____

_____

33. Describe how to enter information to produce a clockwise rotation.

_____

_____

34. Give the combination of alternatives for describing a polar array.

35. If a positive column distance and a negative row distance are entered, where will the objects be arrayed in relation to the original object?

12

MODIFYING AND SELECTING DRAWING OBJECTS

· ·

In Chapter 11 you were introduced to editing commands that can affect an entire drawing. The commands discussed in this chapter can be used to modify part of an object within a drawing. These commands include MOVE, ROTATE, ALIGN, SCALE, STRETCH, LENGTHEN, TRIM, EXTEND, BREAK, CHAMFER and FILLET. Mastering these commands will greatly increase your drawing ability. All of the commands can be entered by keyboard, selected from the Modify toolbar, or picked from the Modify pull-down menu. In addition to these editing commands, the DIVIDE and MEASURE commands will be introduced. These two commands can be helpful in deciding the placement of repetitive objects. The final commands to be introduced will be GRIPS, SELECT and GROUPS which are methods of selecting objects for editing.

MODIFYING DRAWINGS BY ALTERING THE POSITION AND SIZE

The Modify menu offers five options for modifying a drawing by changing the position of an object or by changing the size of the object. These commands include MOVE, ROTATE, SCALE, STRETCH, and LENGTHEN.

MOVE

Have you ever started a manual drawing project, only to find out that after all of your careful planning, because of client changes, the project no longer fits on the paper? This common problem led to such solutions as splicing drawings into new sheets or using sepias, and Mylar® copies. AutoCAD has a better solution: move the object. The MOVE command allows one or more entities to be selected and moved from their current location to a new one. The command can be accessed by picking the Move icon from the Modify toolbar, by entering **M** [enter] or **MOVE** [enter] at the command prompt, or by picking Move from the Modify pull-down menu. Figure 12–1a and 1b compare the effects of the MOVE command.

The command sequence for MOVE is:

> Command: **M** [enter]
> Select objects: (*Use the desired selection method.*)
> Select objects: [enter]
> Base point of displacement: (*Enter point or coordinates.*)
> Second point of displacement: (*Enter point or coordinates.*)
> Command:

The process can be seen in Figure 12–2a through c. Use any selection means that best suits your needs to select the objects to be moved. The Base point of displacement: prompt is asking for a reference point for moving the selection set. Any point may be selected, but it may help you to visualize the results of moving by picking a point in the drawing such as an intersection, or endpoint. This is an excellent time to make use of OSNAP.

The Second point of displacement: prompt is asking where you would like to move the first point to. Each point may be entered by picking a point with a pointing device or by entering coordinates. SNAP, ORTHO, OSNAP, and ZOOM can be very helpful for aligning related objects as they are moved.

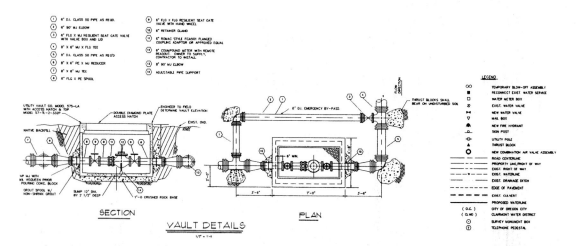

FIGURE 12–1a The MOVE command can be used to alter the layout of a drawing. (Courtesy Lee Engineering, Inc., Oregon City OR)

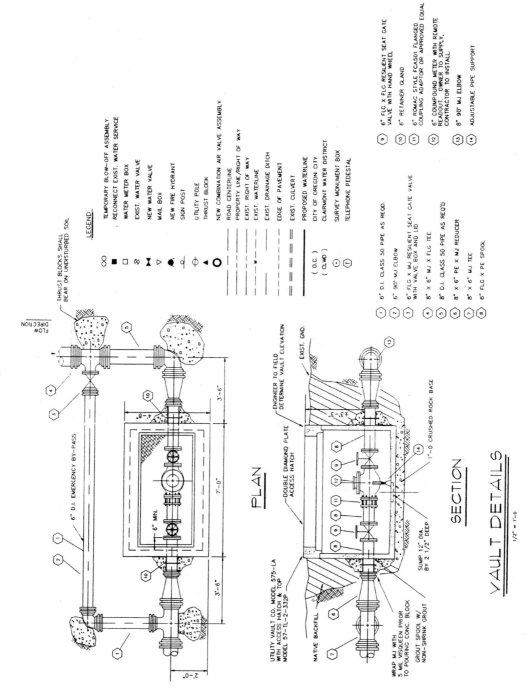

LEGEND

○○ TEMPORARY BLOW-OFF ASSEMBLY
. RECONNECT EXIST. WATER SERVICE
■ WATER METER BOX
□ EXIST. WATER BOX
⊗ EXIST. WATER VALVE
▷◁ NEW WATER VALVE
▷ MAIL BOX
◉ NEW FIRE HYDRANT
♀ SIGN POST
♦ UTILITY POLE
▲ THRUST BLOCK
○ NEW COMBINATION AIR VALVE ASSEMBLY

ROAD CENTERLINE
PROPERTY LINE/RIGHT OF WAY
EXIST. RIGHT OF WAY
EXIST. WATERLINE
──w── EXIST. DRAINAGE DITCH
EDGE OF PAVEMENT
──▷── EXIST. CULVERT
PROPOSED WATERLINE
(O.C.) CITY OF OREGON CITY
(CLWD) CLAIRMONT WATER DISTRICT
⊙ SURVEY MONUMENT BOX
① TELEPHONE PEDESTAL

① 6" D.I. CLASS 50 PIPE AS REQD.
② 6" 90° MJ ELBOW
③ 6" FLG x MJ RESILIENT SEAT GATE VALVE WITH VALVE BOX AND LID
④ 8" x 6" MJ x FLG TEE
⑤ 8" D.I. CLASS 50 PIPE AS REQ'D
⑥ 8" x 6" PE x MJ REDUCER
⑦ 8" x 6" MJ TEE
⑧ 6" FLG x PE SPOOL

⑨ 6" FLG X FLG RESILIENT SEAT GATE VALVE WITH HAND WHEEL
⑩ 6" RETAINER GLAND
⑪ 6" ROMAC STYLE FCA501 FLANGED COUPLING ADAPTOR OR APPROVED EQUAL
⑫ 6" COMPOUND METER WITH REMOTE READOUT. OWNER TO SUPPLY, CONTRACTOR TO INSTALL.
⑬ 8" 90° MJ ELBOW
⑭ ADJUSTABLE PIPE SUPPORT

THRUST BLOCKS SHALL BEAR ON UNDISTURBED SOIL

FLOW DIRECTION

6" D.I. EMERGENCY BY-PASS

6" MIN.

3'-6"

7'-0"

3'-6"

2'-0"

PLAN

UTILITY VAULT CO. MODEL 575-LA WITH ACCESS HATCH & TOP MODEL 57-7L-2-332P

DOUBLE DIAMOND PLATE ACCESS HATCH

NATIVE BACKFILL

WRAP MJ WITH 5 MIL VISQUEEN PRIOR TO POURING CONC. BLOCK

GROUT SPOOL W/ NON-SHRINK GROUT

ENGINEER TO FIELD DETERMINE VAULT ELEVATION

EXIST. GND.

SUMP 12" DIA. BY 2 1/2" DEEP

1"-0 CRUSHED ROCK BASE

SECTION

VAULT DETAILS
1/2" = 1'-0

FIGURE 12–1b The MOVE command was used to place the plan view of Figure 12–1a above the section drawing. (Courtesy Lee Engineering, Inc., Oregon City OR)

379

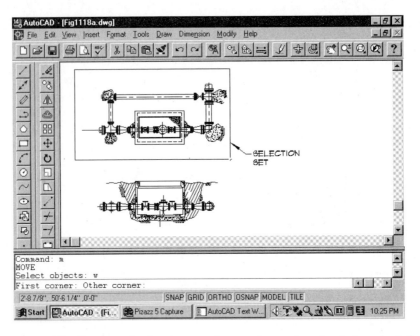

FIGURE 12–2a The first step in the process for moving a drawing is to select the objects to be moved. (Courtesy Lee Engineering, Inc. Oregon City, OR)

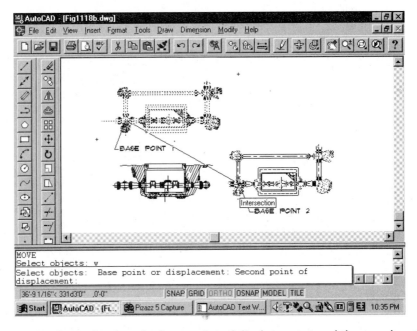

FIGURE 12–2b Second, select the base point of displacement and the new location for this point. (Courtesy Lee Engieering, Inc., Oregon City, OR)

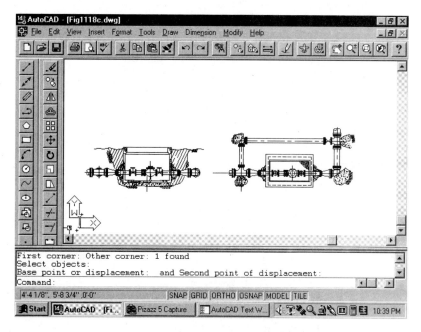

FIGURE 12–2c Third, as the second point of displacement is selected, the MOVE process is complete. (Courtesy Lee Engineering, Inc., Oregon City, OR)

ROTATE

In Chapter 11 you learned how to rotate the grid to lay out a portion of a drawing that may not be oriented to horizontal or vertical. In a similar fashion, you can draw an object or group of objects and then rotate them around a base point to change their orientation. The command can be started by selecting Rotate icon from the Modify toolbar by entering **RO** [enter] or **ROTATE** [enter] at the command prompt or by selecting Rotate from the Modify pull-down menu. Figure 12–3 shows an example of a window that was drawn for a horizontal wall on a floor plan. The window was then copied and rotated 90° to be used in a vertical wall.

To complete the command, you will need to select objects to be rotated, a base point to rotate the object around, and the desired degree of rotation. Entering a positive rotation angle rotates the selected object in a counterclockwise direction. Entering a negative rotation angle produces a clockwise rotation. The base point may be anywhere on the drawing screen, but selecting one corner of the object as the base point may ease visualization of the rotation. The command sequence is:

> Command: **RO** [enter]
> Select objects: (*Use any selection method.*)
> Select objects: [enter]
> Base point: (*Select any point.*)
> <Rotation angle>/Reference: (*Enter angle or drag into position.*)
> Command:

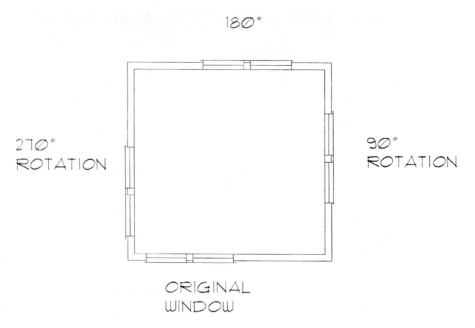

180°

270°
ROTATION

90°
ROTATION

ORIGINAL
WINDOW

FIGURE 12–3 The ROTATE command will allow objects to be drawn and rotated into a new position.

Notice that instead of entering a precise angle, the selected objects may be dragged into the desired position as the mouse is moved. If ORTHO is toggled ON, the objects will be snapped at 90° intervals. With ORTHO off, the objects can be dragged into any desired angle. When you enter an angle, you can rotate objects to exact locations based on the current location. The process can be seen in Figure 12–4.

Objects also can be rotated based on absolute rotation angles. Suppose you've drawn the top chord of a truss at 22½° for a 5/12 pitch. The owners change their mind and desire a 6/12 pitch (27½°). You could rotate the existing top chord 5° and hope you get it close enough, or use the Reference option and place the roof exactly where it should be. This uses the following command sequence:

Command: **RO** ⏎
Select objects: (*Use any selection method.*)
Select objects: ⏎
Base point: (*Select any point.*)
<Rotation angle>/Reference: **R** ⏎
Reference angle: **22.5** ⏎
New angle 27.5 ⏎
Command:

The process can be seen in Figures 12–5a, 12–5b, and 12–5c.

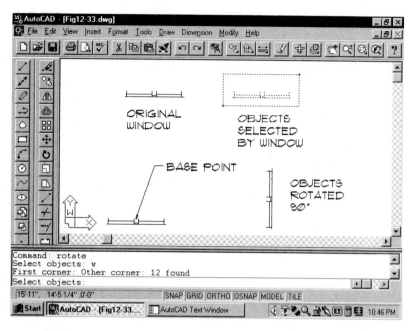

FIGURE 12–4 Objects are rotated by defining the selection set, specifying a base point for the rotation, and providing the rotation angle.

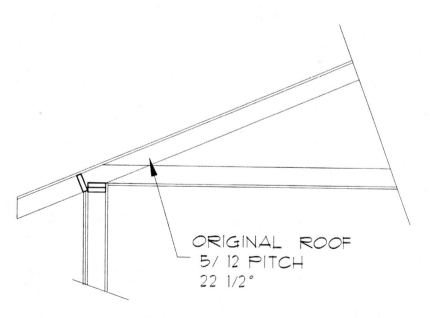

FIGURE 12–5a The pitch of a roof can be altered by use of the ROTATE command.

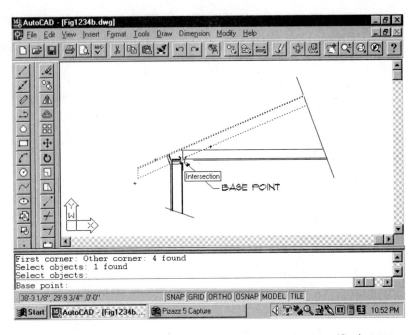

FIGURE 12–5b The selection set defined and the base point specified.

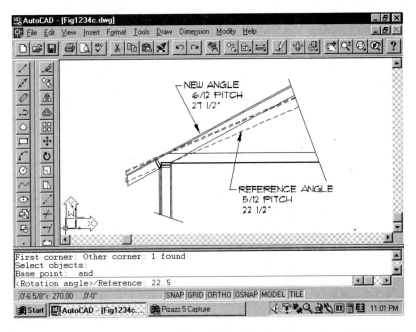

FIGURE 12–5c With the reference angle specified (existing angle of 22.5) the new angle of 27.5 is specified and the roof is rotated.

ALIGN

Think of the ALIGN command as a combination of the MOVE and ROTATE commands. Although primarily a command for 3D drawing, it works well when an object or group of objects need to be moved and rotated. The command can be started by typing **AL** [enter] or **ALIGN** [enter] at the command prompt, or selected from the 3D Operation of the Modify pull-down menu. The command will prompt you for two source points and two destination points. The source points will be used to define points on the original object in its original position. The destination points represent the locations where the object will be placed. The command sequence to move and rotate a kitchen from one area of a floor plan to another is:

Command: **AL** [enter]
Select object: *(Select object using any selection method.)*
Specify 1 st source point: *(Select point with mouse or enter coordinates.)*
Specify 1st destination point: *(Select point with mouse or enter coordinates.)*
Specify 2nd source point: *(Select point with mouse or enter coordinates.)*
Specify 2nd destination point: *(Select point with mouse or enter coordinates.)*
Specify 3 rd source point or <continue>: [enter]
Scale objects to alignment points? [Yes/No] <No>: [enter]

As the enter key is selected the object will be moved and rotated. The selection process can be seen in Figure 12–6. The results of the command can be seen in Figure 12–7. If **Y** [enter] is entered at the last prompt, the objects being rotated will be reduced to match the distance indicated by the first and second destination points. Figure 12–8 shows the results of scaling the objects to meet the alignment points.

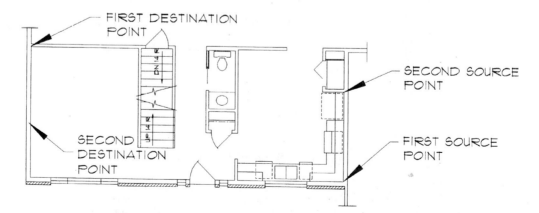

FIGURE 12–6 The ALIGN command allows portions of a drawing to be moved and rotated. The command will prompt you to provide source points, and then ask where you would like that point moved to.

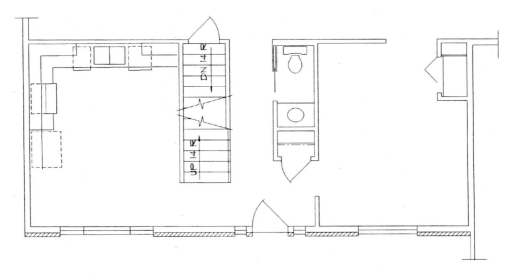

FIGURE 12–7 By using the destination points shown in Figure 12–6, the kitchen was moved and rotated to a new position.

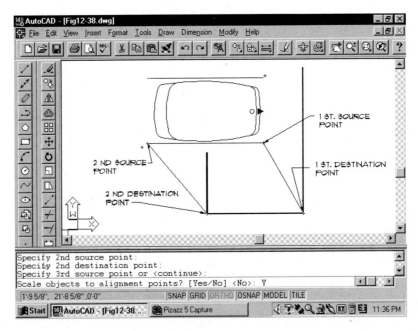

FIGURE 12–8a By selecting the Scale objects to alignment points option, an object or group of objects can be moved and scaled to match the selected destination points.

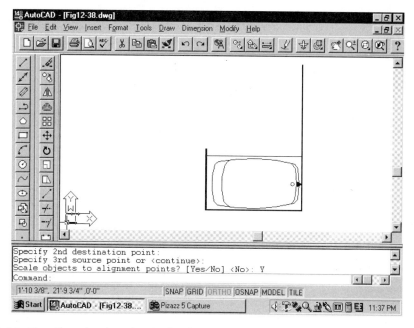

FIGURE 12–8b By selecting the Scale objects to alignment points option, the 6' tub is scaled to fit in a 5' space.

SCALE

Your drawings to this point have been drawn at full scale. If you needed to draw a plot of land that was 100 feet square, you set the limits and units accordingly, zoomed out to the limits, and drew the plot of land. The whole process was done at real scale with the size of the zoom used to control the viewing screen. The SCALE command will allow you to change the size of an existing object or an entire drawing, as seen in Figure 12–9. The command can be started by selecting the Scale icon from the Modify toolbar, by entering **SC** (enter) or **SCALE** (enter) at the command prompt, or by selecting Scale from the Modify pull-down menu.

The command process is:

> Command: **SC** (enter)
> Select objects: (*Select objects using any selection method.*)
> Select object: (enter)
> Base point: (*Pick base point.*)
> <Scale factor>/Reference:

Scale Factor. The default for entering the effect of scaling is to enter a numeric value. To enlarge an object to twice its existing size, enter **2** (enter) at the prompt.

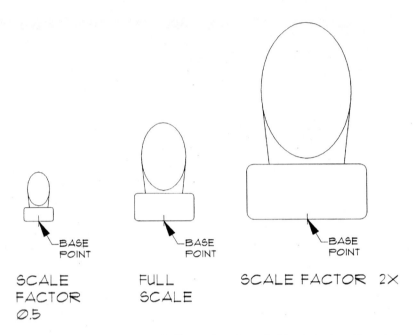

BASE
POINT

BASE
POINT

BASE
POINT

SCALE
FACTOR
0.5

FULL
SCALE

SCALE FACTOR 2X

FIGURE 12–9 The SCALE command allows the size of objects to be altered. In this case, the toilet on the left is half the size of the one on the right.

Using whole numbers will enlarge the selected objects by that factor. Enter **3** [enter] at the prompt to enlarge the objects to three times their existing size. Entering a fraction such as **.5** [enter] will reduce the selected object to half of its original size.

Reference Factor

This option will allow an object to be enlarged by an absolute length rather than a relative scale. For instance, a tub that is 5 feet long can be scaled to 6 feet long. Rather than figuring the proportion of enlargement, enter **R** [enter] at the command prompt. The process is:

Command: **SC** [enter]
Select objects: (*Select objects using any selection method.*)
Select object: [enter]
Base point: (*Pick base point.*)
<Scale Factor>/Reference: **R** [enter]
Reference length <1>: **5'** [enter]
New length: **6'** [enter]

The process can be seen in Figure 12–10.

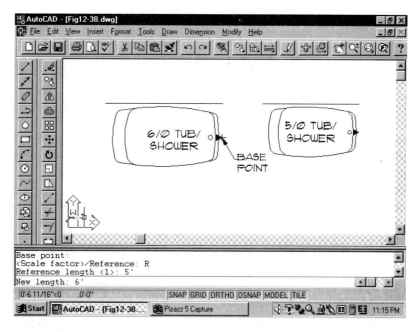

FIGURE 12–10   Entering a reference factor allows an object to be scaled by an absolute length.

STRETCH

The STRETCH command allows you to elongate or shrink drawing entities. This can be especially helpful when editing the shape and size of a floor plan to meet the changing needs of a client, as seen in Figure 12–11. Drawings made with Lines, Arcs, Traces, Solids (Chapter 16), and Polylines (Chapter 13) can all be stretched. The command can be started by selecting the Stretch icon from the Modify toolbar, by typing **S** [enter] or **STRETCH** [enter] at the command prompt, or by picking Stretch from the Modify pull-down menu. The command sequence is:

Command: **S** [enter]
Select objects to stretch by crossing-window or -polygon…
Select objects: *(Select object to be stretched using a window, crossing window, or a polygon options.)* **C** [enter]
First corner: *(Pick a point for the corner of the window.)*
Other corner: *(Pick the second point for the corner so that the object to be stretched is included in the window.)*
Select objects: [enter]

The selection window only needs to cross a portion of the desired object since the default selection is Crossing. Objects can be added to the selection set by any of the typical methods described in Chapter 11. When the Window option is used to pick

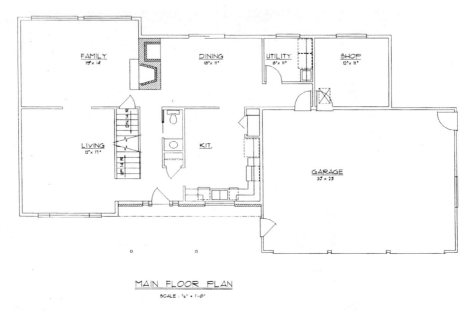

MAIN FLOOR PLAN
SCALE : ¼" = 1'-0"

FIGURE 12–11 The STRETCH command can be used to elongate an entity or group of entities. Compare the garage size with the garage in Figure 11–23b.

the selection set for objects to be stretched, the desired effects are not always achieved. If too little of the group is picked, nothing will happen. If the entire object is selected, STRETCH will function like the MOVE command and move the entire object. The Crossing option can be used to consistently achieve the desired results. Once the selection set is complete, the prompt will display:

> Base point or displacement: *(Select a point to serve as a reference. Object snap modes work great for selecting a point.)* **END** enter
> End of *(Select end point of line to be stretched.)*
> Second point of displacement: *(Select the desired location that you would like to stretch the base point to.)*
> Command:

When DRAGMODE is toggled to ON, the rubberband effect will be displayed as the object is moved between the base and new points. The process can be seen in Figures 12–12a and 12–12b.

Not only can Stretch be used to lengthen objects, it also can be used to stretch some objects and move others, as seen in Figures 12–13a, 12–13b, and 12–13c.

LENGTHEN

The LENGTHEN command is similar to the STRETCH command. STRETCH can be used to edit the length of groups of objects and closed polygons. LENGTHEN can

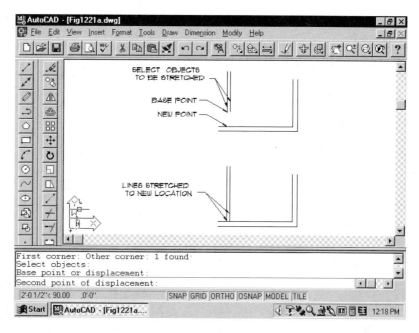

FIGURE 12–12a The Window option is used to form the selection set of objects that are to be stretched. Once selected, prompts will be displayed for the basepoint of the original objects and the new position they are to be stretched to.

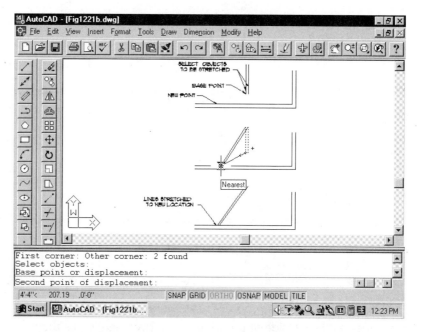

FIGURE 12–12b In addition to adding length to line segments, the STRETCH command can be used to alter the angle of the selected lines.

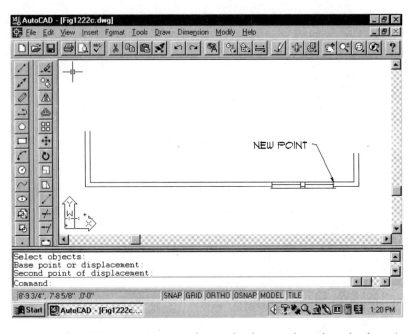

FIGURE 12–13a The STRETCH command can also be used to alter the location of objects.

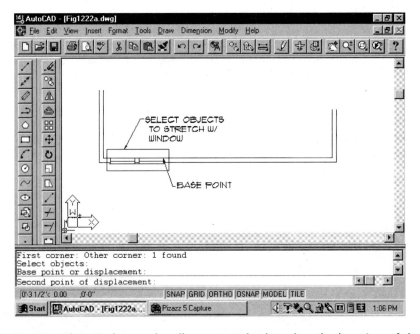

FIGURE 12–13b The window and walls are stretched to alter the location of the window. Lines on the left side of the window are stretched, while the lines on the right side are shortened.

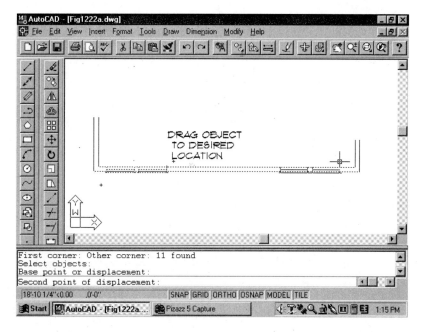

FIGURE 12–13c The completed STRETCH command. This is much more efficient than moving the window, stretching the left lines and trimming the right lines.

be used to edit the length of objects but not closed polygons. Like other edit commands, the LENGTHEN command can be started by selecting the Lengthen icon from the Modify toolbar, by typing **LEN** ⏎ or **LENGTHEN** ⏎ at the command prompt, or by picking Lengthen from the Modify pull-down menu. The command sequence is:

> Command: **LEN** ⏎
> DElta/Percent/Total/DYnamic/<Select object>: *(Select an object to edit.)*
> Current Length: 10'-0"*(Length will vary.)*
> DElta/Percent/Total/DYnamic/<Select object>: **DE** ⏎

Entering **DE** ⏎ at the Select object: prompt will produce the prompt:

> Angle/<Enter delta length (0'-0")>: *(Enter the desired length.)* **5'** ⏎
> <Select object to change>/Undo: *(Select the object again.)*
> <Select object to change>/Undo: ⏎
> Command:

The command sequence can be seen in Figure 12–14. Notice that when the DElta was entered at the Select prompt: a choice could be made between angle or length. The default is to provide a length. Entering a positive number lengthens the line. Providing a negative response shortens the line.

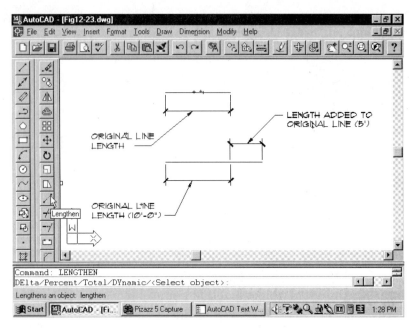

FIGURE 12–14 The DElta option of the LENGTHEN command can be used to lengthen or shorten a line.

The Angle option can be used to change the included angle of an arc. The command sequence would be:

Command: **LEN** [enter]
DElta/Percent/Total/DYnamic/<Select object>: **DE** [enter]
Angle/<Enter delta length (0'-0")>: **A** [enter]
Enter delta angle<00>: **60** [enter]
<Select object to change>/Undo: *(Select desired arc to be edited.)*
<Select object to change>/Undo: [enter]
Command:

The effects of this command sequence can be seen in Figure 12–15.

Using the Percent Option of LENGTHEN. The Percent option allows the length of a line or arc to be altered by a percentage of the existing length. The existing line length represents 100%. To lengthen the line, provide a value greater than 100%. Providing a percentage of less than 100% will shorten the line. The command sequence to lengthen a 2" line to a 3" line using the Percent option is:

Command: **LEN** [enter]
DElta/Percent/Total/DYnamic/<Select object>: **P** [enter]
Enter percent length <100.0000>: **150** [enter]

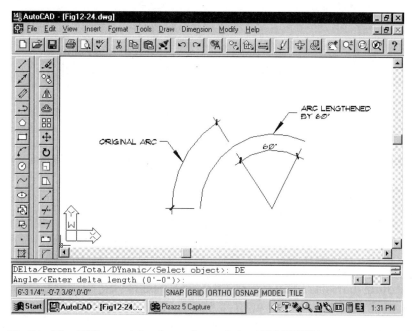

FIGURE 12–15 The DElta and Angle options of the LENGTHEN command can be used to lengthen or shorten a angle.

<Select object to change>/Undo: *(Select desired line to be edited.)*
<Select object to change>/Undo: enter
Command:

The effects of the Percent option can be seen in Figure 12–16. When used with an arc, the angle of the arc will be edited based on a percentage of the total angle of the specified arc.

Using the Total Option of LENGTHEN. The Total option allows the length of a line or angle of an arc to be edited by providing the desired ending result. The command sequence to alter an existing line to a 3″ line would be:

Command: **LEN** enter
DElta/Percent/Total/DYnamic/<Select object>: **T** enter
Angle/<Enter total length (0'-0")**35'** enter
<Select object to change>/Undo: *(Select desired line to be edited.)*
<Select object to change>/Undo: enter
Command:

The results of this sequence can be seen in Figure 12–17. When first prompted for Select object: you may prefer to select an object before choosing the Total option. When a line is selected first, the length of the existing line will be displayed.

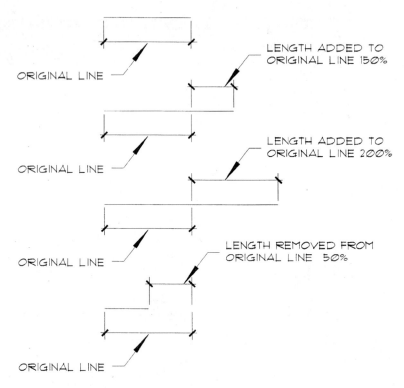

FIGURE 12–16 The lengths of lines and arcs can be altered by entering a percentage of the original object. Percentages less than 100% will shorten the line, and those greater than 100% will lengthen.

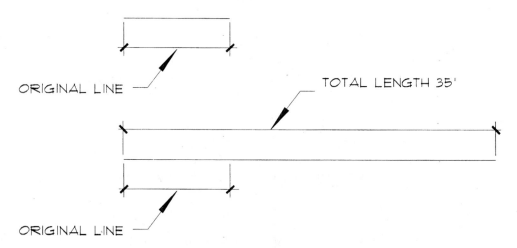

FIGURE 12–17 The Total option of LENGTHEN allows a line or angle of an arc to be lengthened based on the desired total length or angle.

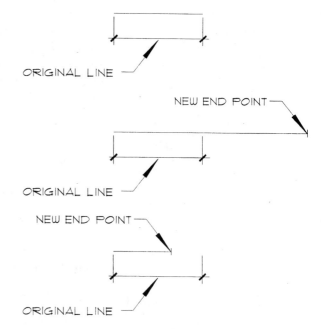

FIGURE 12–18 The DYnamic option of LENGTHEN allows a line or angle of an arc to be dragged to the desired length by selecting a new ending point.

Using the DYnamic Option of LENGTHEN. The DYnamic option allows the selected line or angle to be dragged to the required length.

Command: **LEN** [enter]
DElta/Percent/Total/DYnamic/<Select object>: **DY** [enter]
Specify new end point.
<Select object to change>/Undo: *(Select desired line to be edited.)*
<Select object to change>/Undo: [enter]
Command:

The results of the Dynamic option can be seen in Figure 12–18.

MODIFYING OBJECTS BY ALTERING LINES

To this point, you've learned to modify objects by making additional copies and by altering the position or size of an object. The commands contained in this portion of the Modif menu will allow you to modify a drawing by altering intersections of lines, circles or arc, or by removing a portion of an object. The commands include TRIM, EXTEND, BREAK, CHAMFER, and FILLET.

TRIM

The TRIM command is located in the Modify pull-down menu and in the Modify toolbar. The command can also be started by typing **TR** [enter] or **TRIM** [enter] at the command prompt. The command allows you to edit Lines, Arcs, Circles, or Polylines using a cutting edge. Polylines will be introduced in Chapter 13. Any portion of the selected object extending past the designated cutting edge will be removed. Lines, Arcs, Circles, and Polylines may be used as cutting edges. TRIM can be used to clean the corners of intersecting walls in plan view, as seen in Figure 12–19, or to modify the size of entities. The command sequence is:

Command: **TR** [enter]
Select cutting edges: (Projmode = UCS, Edgemode = No extend)
Select objects: *(Select the cutting edge.)* 1 found
Select objects: [enter]
<Select object to trim>/Project/Edge/Undo: *(Select object to be trimmed.)*
<Select object to trim>/Project/Edge/Undo: *(Select another object to be trimmed*
        or press enter when finished.) [enter]
Command:

The command will continue until [enter] is selected. The process can be seen in Figure 12–20. Notice that an object can be selected to be a cutting edge and it can also be an object to be trimmed.

Using the TRIM Edge Option.   The Edge option of TRIM can be used to trim two objects that do not intersect, but would if one of the objects were longer. AutoCAD

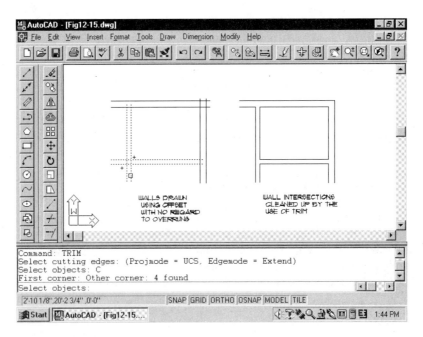

FIGURE 12–19 The TRIM command can be used to eliminate line overruns created by using OFFSET.

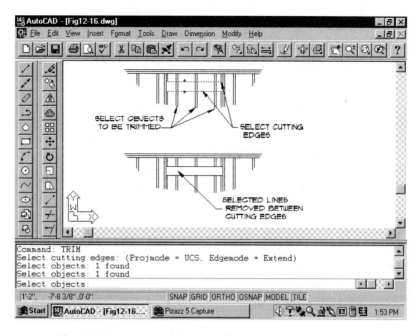

FIGURE 12–20 The TRIM command is started by selecting one or several "Cutting edges." Any segment extending past the selected cutting edge will be trimmed.

refers to this as an *implied intersection*. To use the Edge option, two objects must be selected. Once these edges have been selected, enter **E** [enter] when prompted to select another object. This will now display the option of:

Extend/No extend <No extend>:

When Extend is selected, AutoCAD checks the cutting edge object to verify if it will intersect when extended. If it would touch, the implied intersection is used to trim the second object. This option is shown in Figure 12–21. The command sequence is:

Command: **TR** [enter]
Select cutting edges: (Projmode = UCS, Edgemode = No extend)
Select objects: *(Select the cutting edge.)*
Select objects: *(Select a second cutting edge.)* 1 found
Select objects: [enter]
<Select object to trim>/Project/Edge/Undo: **E** [enter]
Extend/No extend <No extend>: **E** [enter]
<Select object to trim>/Project/Edge/Undo: *(Select object to be trimmed.)*
<Select object to trim>/Project/Edge/Undo: *(Select another object to be trimmed.)*
<Select object to trim>/Project/Edge/Undo: [enter]
Command:

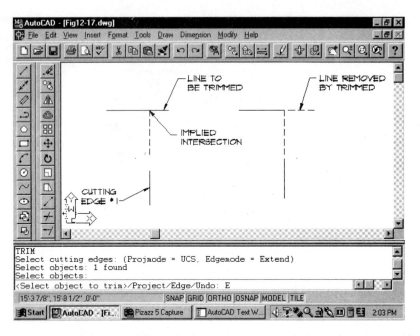

FIGURE 12–21 Lines that would touch if one were extended can be trimmed by selecting the Edge and Extend options.

Using the TRIM Undo option. Occasionally you may trim an object and not be pleased with the results. The TRIM Undo option will allow an unsatisfactory TRIM to be undone while still in the command sequence. The command sequence is:

> Command: **TR** ⏎
> Select cutting edges: (Projmode = UCS, Edgemode = No extend)
> Select objects: *(Select the cutting edge.)*
> Select objects: ⏎
> <Select object to trim>/Project/Edge/Undo: *(Select object to be trimmed.)*
> <Select object to trim>/Project/Edge/Undo: **U** ⏎
> Command has been completely undone.
> <Select object to trim>/Project/Edge/Undo: *(Select object to be trimmed.)*
> <Select object to trim>/Project/Edge/Undo: ⏎
> Command:

Using the TRIM Project Option. The TRIM Project option is a 3D option and will not be discussed in this text. Refer to the HELP menu for further information.

Practical use of TRIM. Open the FLOOR12 file that was started earlier in this chapter. By combining the use of TRIM and FILLET this partial floor plan can be completed so that it more closely resembles the drawing in Figure 11–23b.

EXTEND

The EXTEND command can be started by selecting Extend from the Modify pull-down menu, by selecting the Modify icon from the Trim flyout menu in the Modify toolbar, or by typing **EX** [enter] or **EXTEND** [enter]. The command allows an entity to be lengthened to an exact boundary point. The command will ask you to designate boundary edges, and then select objects to be extended to that edge. The boundary edge may be a Line, Arc, Circle, or Polyline. The process can be seen in Figure 12–22. The command sequence is:

Command: **EX** [enter]
Select boundary edges: (Projmode = USC, Edgemode = No extend)
Select objects: *(Select object to be used as boundary.)*
Select object: 1 found
Select object: [enter]
<Select object to extend>/project/Edge/Undo: *(Select object to be extended.)*
<Select object to extend>/project/Edge/Undo: [enter]
Command:

As the object to be extended is selected, the line will be extended to the object that was selected as the boundary edge.

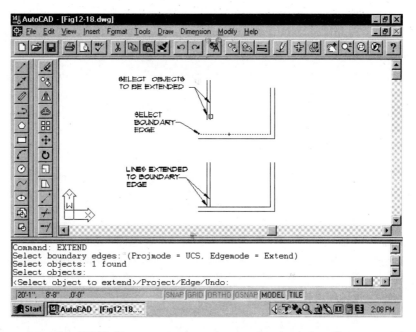

FIGURE 12–22 The EXTEND command can be used to lengthen lines to a selected "Boundary edge."

Using the EXTEND Edge Option. The EXTEND Edge option is similar to the Edge option presented with the TRIM command. Edge will allow lines that do not touch but have an implied intersection to be extended. The command sequence is:

Command: **EX** [enter]
Select boundary edges: (Projmode = USC, Edgemode = No extend)
Select objects: *(Select object to be used as boundary.)*
Select object: 1 found
Select object: [enter]
<Select object to extend>/project/Edge/Undo: **E** [enter]
Extend/No extend <Extend>: [enter]
<Select object to extend>/project/Edge/Undo: *(Select object to be extended.)*
<Select object to extend>/project/Edge/Undo: [enter]
Command:

The results of the command can be seen in Figure 12–23.

Using the EXTEND Undo Option. The EXTEND Undo option allows a sequence that has not produced satisfactory results to be undone while still in the command sequence.

Using the EXTEND Project Option. This mode is used for 3D drawings and will not be discussed in this text. Use the HELP option for further information on this subject.

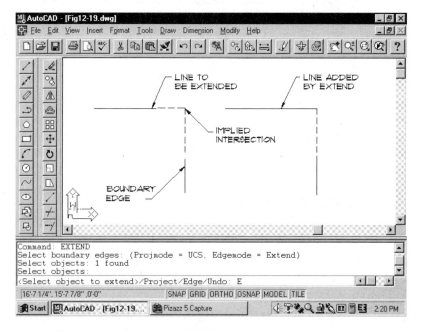

FIGURE 12–23 Lines which would touch if both were extended can be extended by selecting the Edge option.

BREAK

The BREAK command can be used to remove a portion of a Line, Circle, Arc, Polyline, or Trace. Polylines will be introduced in Chapter 13 and Trace will be introduced in Chapter 14. The BREAK command can be selected by picking the Break icon from the Modify toolbar, by typing **BR** [enter] or **BREAK** [enter] at the command prompt or by picking Break from the Modify pull-down menu. BREAK can be completed using two different methods, but their effects are similar.

Select Object 2nd Point. This option allows you to remove a segment of an entity by using two selection points. The first selection point picks the segment to be broken and specifies where the break will begin. The second point specifies where the BREAK will end. The command is:

Command: **BR** [enter]
BREAK Select object: *(Pick an object to break.)*
Enter second point (or F for first point): *(Pick a second point.)*
Command.

Once the second point is picked, the portion of the object from the first to the second point will be removed and the command prompt will be returned. The sequence can be seen in Figure 12–24. If a circle is being edited, the break occurs in a counterclockwise direction from the first point to the second point, as seen in Figure 12–25. Enter

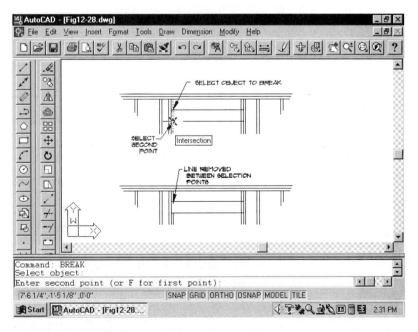

FIGURE 12–24 The BREAK First option can be used to remove a segment from a specified line. Specified points are used to define which part of the line will be removed.

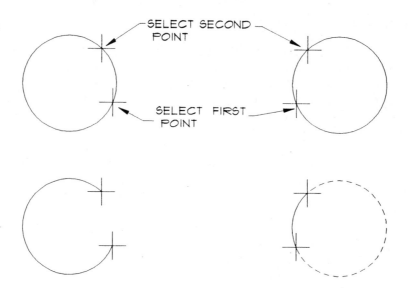

FIGURE 12–25 When a break is specified for a circle, it will occur in a counterclockwise direction.

F [enter] when the second point is requested to allow the procedure to follow the Select Object Two Points method.

Select Object Two Points. This option allows you to remove a segment of an entity by using three selection points. The first request picks the segment to be broken. Once the segment has been defined for editing, the first selection point specifies where the break will begin and the second selection point specifies where the BREAK will end. The command sequence is:

> Command: **BR** [enter]
> BREAK Select object: *(Pick an object to break.)*
> Enter second point (or F for first point): **F** [enter]
> Enter first point: *(Pick a point to start the break.)*
> Enter second point: *(Pick a point to end the break.)*
> Command:

The command sequence can be seen in Figure 12–26.

Break @. This option will break a segment, but no apparent change will take place. BREAK @ will divide an entity into two portions, but no space will be left between the portions. The @ option is useful to erase only a portion of an object by using BREAK@ and then using the ERASE command. The command is started by using one of the methods to start the BREAK command. The command sequence is:

> Command: **BR** [enter]
> Select objects: *(pick a break point.)*

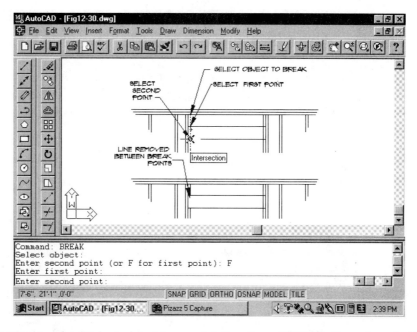

FIGURE 12–26 The "Select object two points" option of BREAK allows for selecting the line to be broken, then specifying the segment to be removed.

Enter second point (or F for first point): @ [enter]
Command:

The sequence is shown in Figure 12–27.

CHAMFER

A chamfer is an angled edge formed between two intersecting surfaces, typical of what might be found on the edge of the cabinet counter shown in Figure 12–28. The CHAMFER command trims two intersecting lines and provides a mitered corner. You'll be asked to select two lines and two distances, which will form a new line segment between the two existing line segments. The Chamfer command can be started by selecting the Chamfer icon from the Modify toolbar, by typing **CHA** [enter] or **CHAMFER** [enter] at the command prompt, or by selecting Chamfer from the Modify pull-down menu. The command sequence to set the default values to 0" is:

Command: **CHA** [enter]
(TRIM mode) Current chamfer Dist1 = 0'-0", Dist2 = 0'-0"
Polyline/Distance/Angle/Trim/Method/<Select first line>: *(Pick first line.)*
Select second line: *(Select second line.)*
Command:

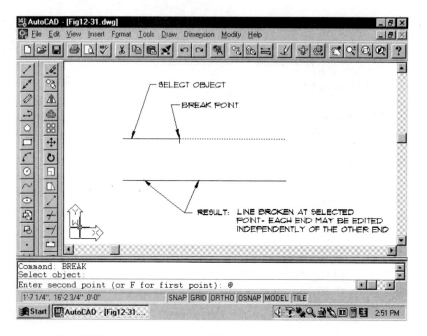

FIGURE 12–27 The BREAK @ command will break a line into two segments, but no visible change will occur.

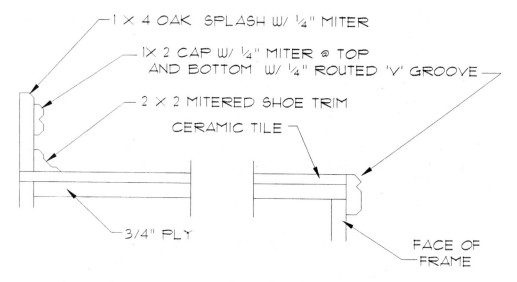

FIGURE 12–28 The CHAMFER command can be used to miter a corner.

Because no chamfer distance has been entered, the process and results would resemble Figure 12–29.

To provide a mitered corner, the sequence would be similar. For the following example, a distance of 36" will be used, although any numeric value can be used as long as the value is no longer than that of the line being chamfered. If the value exceeds the length, the message:

> Distance is too large
> *Invalid*

will be displayed and the command terminated. The command sequence is:

> Command: **CHA** (enter)
> (TRIM mode) Current chamfer Dist1 = 0'-0", Dist2 = 0'-0"
> Polyline/Distance/Angle/Trim/Method/<Select first line>: **D** (enter)
> Chamfer distance <0'-0">: **36** (enter)
> Enter second chamfer distance <3'-0">: (enter)
> Command:

The first value will become the default for the second value. When the distances are equal, the results will produce a 45° miter. When the second value or the (enter) key is entered, the command prompt will be returned. Pressing the (enter) key again will allow the chamfer to be drawn.

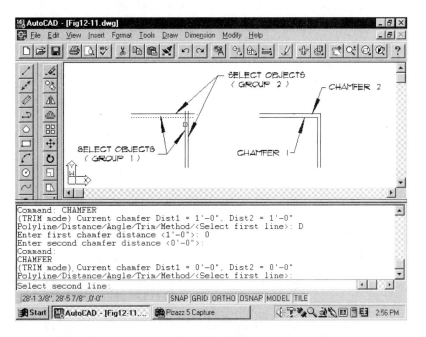

FIGURE 12–29 With the value set at 0, CHAMFER can be used to trim corners.

Command: ⏎
Polyline/Distance/Angle/Trim/Method/<Select first line>: *(Select first line.)*
Select second line: *(Select second line.)*
Command:

As the second line is selected, the chamfer will be performed and the command prompt returned. The effects of the command sequence can be seen in Figure 12–30. Figure 12–31 shows an example of varied distance values. CHAMFER can be used to square or chamfer two intersecting lines, as well as to square or chamfer two lines that do not touch.

Selecting the CHAMFER Angle. CHAMFER also allows the chamfer to be created using one distance and an angle. This option is started by entering **A** ⏎ when prompted for the first line. The command sequence to place a 30 degree angle 4" from a corner would be:

Command: **CHA** ⏎
(TRIM mode) Current chamfer Dist1 = 0'-0", Dist2 = 0'-0"
Polyline/Distance/Angle/Trim/Method/<Select first line>: **A** ⏎
Enter chamfer length on the first line <0'-0">: **4** ⏎
Enter chamfer angle from the first line <0'-0">: **30** ⏎

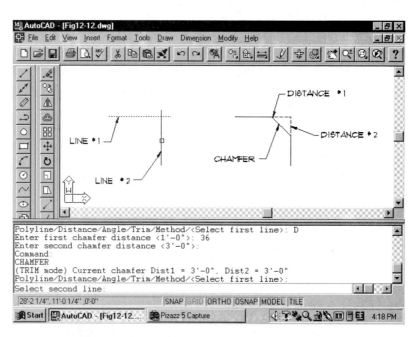

FIGURE 12–30 To miter a corner, prompts will be displayed for the "first distance," and the "second distance." Once provided, a new segment will be provided between the two points.

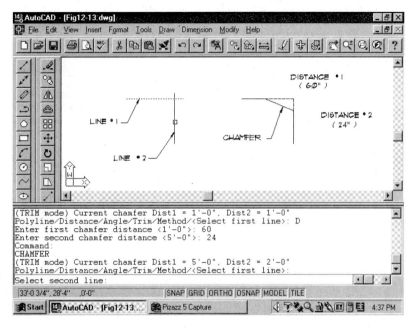

FIGURE 12–31 Providing different lengths for the chamfer distances will affect the angle of the chamfer.

```
Command: [enter]
CHAMFER
(TRIM mode) Current chamfer Length = 0'-4", Angle = 2'-6"
Polyline/Distance/Angle/Trim/Method/<Select first line>: (Select first line.)
Select second line: (Select second line.)
Command:
```

The process can be seen in Figure 12–32.

Selecting the CHAMFER Trim Mode. The common method of using the CHAMFER command is to create the chamfer between two lines and remove a portion of the lines where the chamfer is created. Using the Method option allows for a choice of removing or not removing the lines by the chamfer. The command sequence is:

```
Command: CHA [enter]
(TRIM mode) Current chamfer Dist1 = 0'-0", Dist2 = 0'-0"
Polyline/Distance/Angle/Trim/Method/<Select first line>: T [enter]
Trim/No trim <Trim>: N [enter]
Polyline/Distance/Angle/Trim/Method/<Select first line>: (Select first line.)
Select second line: (Select second line.)
Command: [enter]
CHAMFER
```

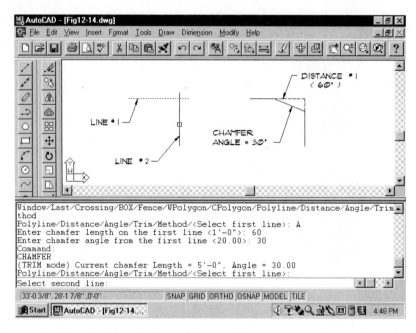

FIGURE 12–32 Selecting the Angle option provides a prompt for one length and an angle. AutoCAD converts the angle to a length.

Selecting the CHAMFER Method. When a chamfer is placed using the Distance or Angle options, the values are kept as the defaults until you change them. Using the Method option, the values will not affect each other and you can switch between using either the Distance or Angle. The command sequence is:

Command: **CHA** [enter]
(TRIM mode) Current chamfer Dist1 = 0'-0", Dist2 = 0'-0"
Polyline/Distance/Angle/Trim/Method/<Select first line>: **M** [enter]
Distance/Angle <Angle>: **D** [enter]
Polyline/Distance/Angle/Trim/Method/<Select first line>: *(Select first line.)*
Select second line: *(Select second line.)*
Command:

FILLET

You may be most familiar with a fillet as it relates to welding; it earns its name because the weld fills in a corner. The FILLET command of AutoCAD will connect two lines to form a corner, or allow two lines, arcs, or circles to be fitted together smoothly in rounded corners. The command can be started by selecting the Fillet icon from the Modify toolbar, by entering **F** [enter] or **FILLET** [enter] at the command prompt, or by selecting Fillet from the Modify pull-down menu. The command sequence is:

Command: **F** enter
(TRIM mode) Current fillet radius = 0'-0"
Polyline/Radius/Trim/<Select first object>:

The default is set for one of the most common uses of FILLET, forming a corner with no radius. Once the first object is picked, select a second object at the prompt:

Select second object:

As the second entity is selected, the corners will be closed and the command prompt will be returned. The fillet process can be seen in Figure 12–33. If the Method option of CHAMFER is set to No Trim, the results of the FILLET will not be evident. Be sure the Method option of CHAMFER is set to Trim. The FILLET command can also be used to form a corner between two lines that do not touch.

For the command to function with crossing lines as you intend, you need to be careful which portion of the lines you select. Draw two more intersecting lines similar to those in Figure 12–33. Enter the fillet command and pick the short ends of the lines.

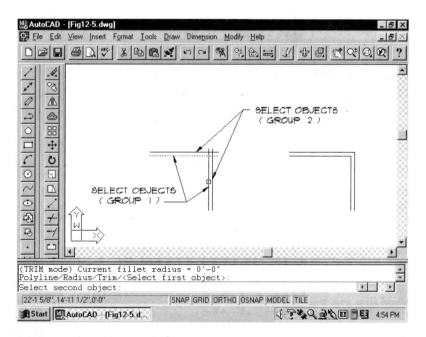

FIGURE 12–33 With the value of the Radius set to 0, the FILLET command can be used to square corners of crossing lines.

Once the FILLET command is completed, only the small portion of the lines will remain as shown in Figure 12–34.

By entering **R** [enter] at the first prompt, a radius may be entered for rounded corners. To transform two existing perpendicular lines into a 60" radius corner, the command sequence is:

Command: **F** [enter]
(TRIM mode) Current fillet radius = 0'-0"
Polyline/Radius/Trim/<Select first object>: **R** [enter]
Enter fillet radius <0'-0">: **60** (or **5'**) [enter]
Command:

Once the radius is entered, the command prompt is returned. Press [enter] to reenter the command. If you continue through the command sequence, you will draw a 60" radius. The balance of the command sequence is:

Command: [enter]
FILLET
(TRIM mode) Current fillet radius = 5'-0"
Polyline/Radius/Trim/<Select first object>: (*Select object.*)
Select second object: (*Select object.*)
Command:

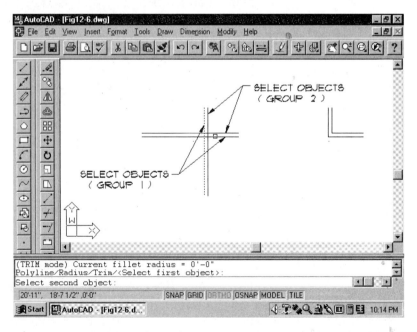

FIGURE 12–34 Care must be taken when selecting objects to be edited by the FILLET command. Selecting the short ends of the lines removes that portion of the line.

As the second object is selected, the radius will be formed between the two selected lines, as seen in Figure 12–35. The portion of the object that is picked will affect how the FILLET will be drawn. Examples of selections and their results can be seen in Figure 12–36. Don't speed past the command prompt. If you've entered a radius value that exceeds the length of the lines, no change will be made to the drawing, and the following prompt will be displayed:

Radius is too large
Invalid
Command:

Before you start thinking that you've really messed up, and only have five seconds before your drawing file self destructs, relax! Remember, rarely are you going to damage a drawing file. Keep exploring. You just need to pick a smaller radius value to make the command work.

Preset Fillet Radius. By entering **SET** ⏎ or **SETVAR** ⏎ at the command prompt, you can enter a default fillet radius. The command process to set the default value at 24" is:

Command: **SET** ⏎
Variable name or ?: **FILLETRAD** ⏎
New value for FILLETRAD <0'-0">: **24** ⏎

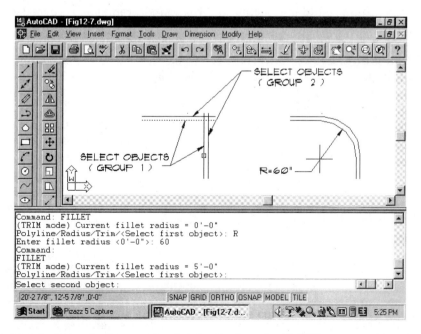

FIGURE 12–35 By assigning a value for the Radius, the FILLET command can be used to round square corners.

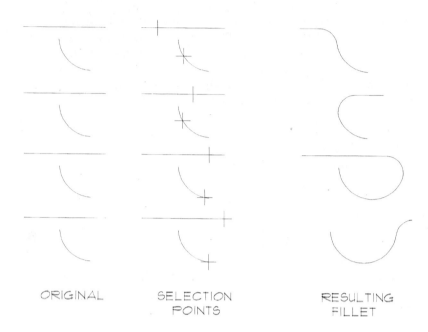

ORIGINAL   SELECTION   RESULTING
                POINTS   FILLET

FIGURE 12–36 Selecting different portions of the segment will affect the results of the FILLET.

Once a new value is selected, the command prompt is returned, and the value is stored as a system variable.

Combining OFFSET and FILLET.   Open the FLOOR11 drawing that you started in chapter 11. Since you've seen the finished floor plan in Figure 11–23b it's not too hard to visualize the lines in Figure 11–26 representing the rooms of the finished plan. Using a combination of OFFSET and FILLET will make it much easier to visualize. Figure 12–37 shows two examples of how FILLET can be used to clean up the floor plan. Notice that if FILLET is used at Fillet 3, the lower portion of the line will be removed and need to be redrawn. This would be a good opportunity to use TRIM instead of FILLET.

SELECTING OBJECTS WITH GRIPS

In Chapter 11 you were introduced to several methods of selecting objects for editing. The GRIP command will provide an additional method of selecting objects for editing as well as a method of combining several of the editing procedures introduced in this chapter.

GRIPS is similar to moving a heavy object by its handles. You can move the object without using the handles, but they make life so much easier. Grips are handles placed on objects for easy manipulation, similar to OSNAP points with STRETCH,

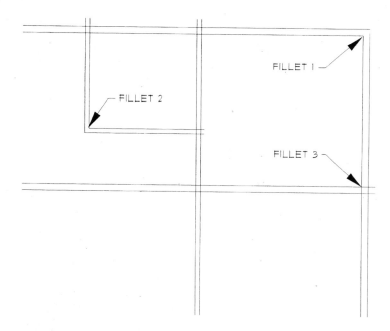

FIGURE 12–37 The FILLET command was used to clean up the corners at locations 1 and 2. If FILLET had been used at location 3, the lines that will be used to form the garage would have been removed.

MOVE, ROTATE, SCALE, and MIRROR. With the AUTOEDIT system variable toggled to ON (1), selected entities can be edited using GRIPS. AUTOEDIT will display grip boxes at various entity-specific positions, as seen in Figure 12–38.

The default setting for GRIPS is ON. Grips can be toggled On/Off by typing **GRIPS** [enter] at the command prompt or by picking Grips...from the Tools pull-down menu and setting the value to <1>. Typing **GR** [enter] or **DDGRIPS** [enter] at the command prompt will produce the dialog box seen in Figure 12–39 and will allow adjustment of the grips. Picking "Enable Grips" will allow grips to be displayed when an edit or inquiry command is given. When the cursor is moved over a grip, the cursor will snap to the grip for editing. This will save time over the use of an OSNAP location.

Using Grips

Enter AutoCAD and draw several lines, arcs and circles. Move the cursor so that it is over one of the lines and press the pick button. Three boxes or *grips* are now displayed. Move the cursor to the center grip, and press the pick button. The grip has changed from a hollow box to a solid square. This grip is now said to be *hot*. Notice that the command prompt has also changed to display the following prompt:

```
** STRETCH **
<Stretch to point>/Base point/Copy/Undo/eXit:
```

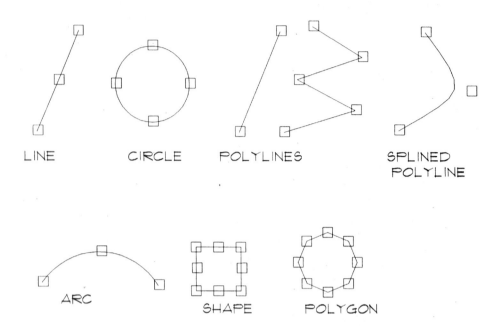

FIGURE 12–38 With the AUTOEDIT system variable toggled to ON, GRIPS will be displayed for each geometric shape.

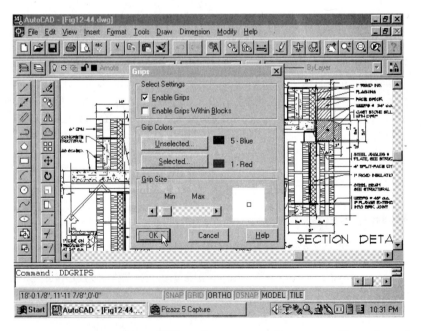

FIGURE 12–39 Typing **DDGRIPS** [enter] at the command prompt will produce the Grips dialog box. (Courtesy G. Williamson Archer, A.I.A. of Archer & Archer, P.A.)

We'll come back to the prompt later. With the center grip hot, and the prompt reflecting STRETCH, move the cursor slightly. As the cursor moves so does the line. Pressing the pick button will move the line to this new location. The hot grip in the center of the line functions like the MOVE command. Select the same line with the cursor and make one of the end grips hot. Now the grip at the other end remains a stationary point, and as the hot grip is moved, the selected line is stretched. The effect of STRETCH on a line can be seen in Figure 12–40. GRIPS function in a similar manner with arc and circles. Figure 12–41 shows the effects of stretching each. If you decide you don't want to edit a selected line, press ESC to deactivate the grips.

Grip Options

Earlier you discovered that each command displayed in grips has several options. These options include:

Base point. Typing **B** [enter] at the prompt allows the hot box to be disregarded, and a new base point to be selected.

Copy. Typing **C** [enter] at the prompt allows one or more copies of the selected object to be made.

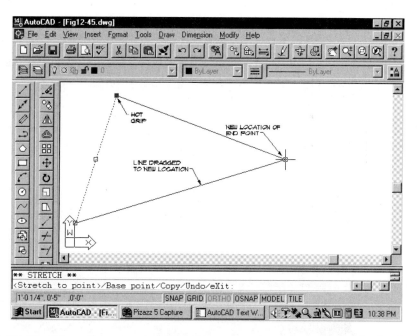

FIGURE 12–40 Move the cursor so that it is over a drawing entity to display its grips. Moving to one of the grips and pressing the pick button will make that grip HOT. The entity can now be edited by STRETCH, MOVE, ROTATE, SCALE, and MIRROR.

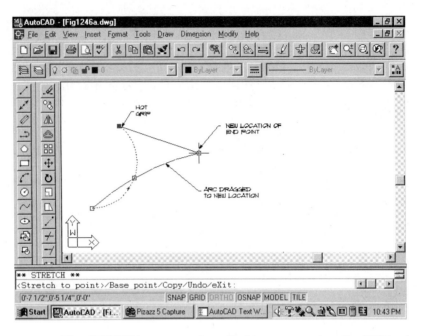

FIGURE 12–41a The STRETCH command applied to an arc using the HOT grip.

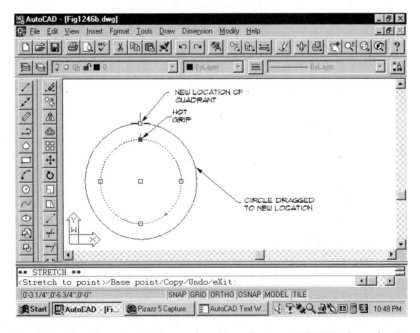

FIGURE 12–41b The size of a circle can be altered using STRETCH by selecting a HOT grip and a new location.

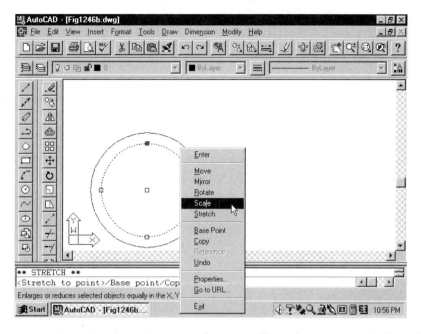

FIGURE 12–42   Pressing the right mouse button will produce this menu that allows the Grip edit option to be altered.

eXit.   Typing **X** [enter] at the prompt terminates the GRIP edit and returns you to the command prompt.

Undo.   Typing **U** [enter] at the prompt will undo the previous selection.

Changing Grip Options

Once a grip box as been selected and made hot, pressing the right mouse button will produce the menu shown in Figure 12–42. Any one of the options can be selected to alter the function to be preformed by the GRIP.

SELECTING OBJECTS FOR MULTIPLE EDITS

Often an object or group of objects may need to be edited with several editing commands. If you were arranging units in a multi-building apartment complex, a unit may need to be mirrored, moved and rotated to form a new building. Up to this point you would have to select the objects to be mirrored, use the MIRROR command, select the objects to be moved, use the MOVE command, select the objects to be rotated, and use the ROTATE command. It works, but you'll never impress the boss as worthy of the big bucks. The SELECT command will allow an object or group of objects to be preselected and used for one or more editing functions. The command sequence to mirror, is:

Command: **SELECT** [enter]
Select objects: *(Select objects using any of the selection methods.)* **W** [enter]
First corner: Other corner: 100 found
Select objects: [enter]
Command:

The command prompt is returned but no apparent changes have occurred. The objects which were selected have been grouped as a set and will be affected by any editing command. Enter the MIRROR command and when prompted to select objects, type **P** [enter]. This will automatically select the objects which were just placed in the set.

Type MIRROR at the command prompt. The sequence to mirror, move and rotate the objects is:

Command: **MI** [enter]
Select objects: **P** [enter]
100 objects found
Select objects: [enter]
First point of mirror line: *(Select first point.)*
Second point: *(Select second point.)*
Delete old objects? <N>**Y** [enter]

Command: **M** [enter]
Select objects: **P** [enter]
100 objects found
Select objects: [enter]
Base point or displacement: *(Select base point.)*
Second point of displacement: *(Select new location for base point.)*

Command: **RO** [enter]
Select objects: **P** [enter]
100 objects found
Select objects: [enter]
Base point:*(Select base point.)*
<Rotation angle>/Reference: **45** [enter]
Command:

The group that was created using SELECT will remain with the drawing file throughout the life of the file. The GROUP command allows similar functions, but with more power.

NAMING OBJECT GROUPS

SELECT allows a group to be created. The GROUP command allows multiple groups to be created, named and saved. For instance, the apartment unit that was just edited could be saved as a group called Base. It could be mirrored, and both units saved as a new group called Units. A group that is part of another group is referred to as *nested*. The Units group could also be mirrored and saved as a third group called BLDG1. The GROUP command can be started by typing **G** [enter] or **GROUP** [enter] at the command prompt, by picking Grouped Objects... from the Tools pull-down menu. Each will produce the Object grouping dialog box shown in Figure 12–42.

Naming a Group

In Figure 12–43 you can see that this drawing has no named groups. To create a group, enter the desired name at the keyboard. As you type, the name will be entered into the Group Name: box of the Group Identification area. Names can be up to 31 characters long and include number, letters or the characters **-**, **_**, or **$**. Two separate words can't be used, but compound words can be joined by a dash or underline to create a name such as **base_unit**. As with the naming of drawing files, be sure to use names that will be easily recognized and that will clearly describe the contents of the group.

Description: A description can be created to help you remember the contents of a specific group. To create a description, move the arrow to the Description: box and press the pick button. A description of up to 64 characters including spaces can be created.

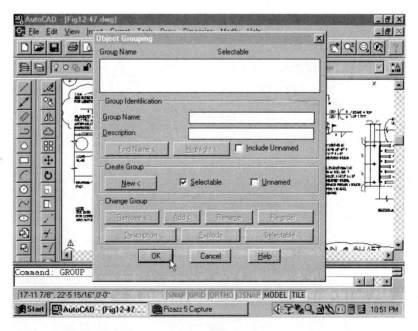

FIGURE 12–43 Typing **G** [enter] at the command line will display the object grouping dialog box. Group can be used to place drawing entities into groups which can then be edited. (Courtesy KPFF Engineers.)

New <. Once the desired name and description have been entered, select the New < button. This will remove the dialog box from the screen, and prompt you to Select objects:. Objects to be included into the group can be selected using any of the selection methods presented in Chapter 11. When the selection process is complete, the dialog box will be redisplayed and the group name will be displayed in the Group Name box. As the group is created, the word "yes" is now displayed beside the group name. If a group is selectable, picking one object in the group picks the entire group.

Once the group is created, the other area of the dialog box can be used to alter or use the group. These options include:

Group Identification Box

Key element of this box can be seen in Figure 12–44. Group Name: and Description: were covered in the preceding section, Naming a Group.

Find Name <. Selecting this button removes the dialog box from the screen and provides the prompt

Pick a member of a group.

When an object is selected, a Group Member List box similar to the display shown in Figure 12–45 is displayed. This box shows all groups that the selected object is a

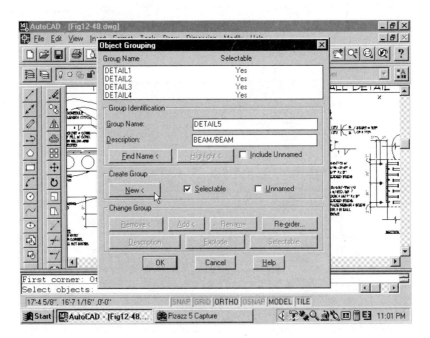

FIGURE 12–44 Once groups are identified, selecting one of the groups will activate each portion of the dialog box. (Courtesy KPFF Engineers.)

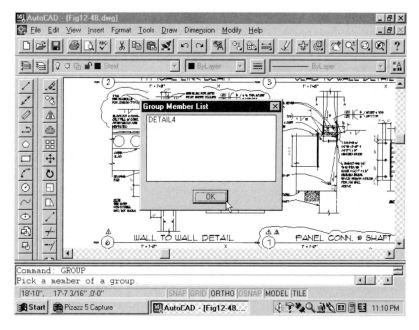

FIGURE 12–45 Selecting Find Name < will display the group member List box. Pick an object on the drawing and the group that it belongs to will be displayed. (Courtesy KPFF Engineers.)

member of. Selecting the **OK** button returns the Object Grouping dialog box. Selecting an unnamed group, and the Add< box allows a name to be provided to an unnamed group.

Highlight <.   This box is not active as the dialog box is displayed. Moving the cursor to one of the named groups and pressing the pick button activates the button. Pressing the Highlight button removes the dialog box from the screen and shows the named group in dashed lines. This option is especially helpful on complicated drawings with multiple groups. Pressing the Continue button restores the Object Grouping dialog box.

Include Unnamed.   Activating this box will display unnamed groups in the Group Name box. Unnamed groups can be renamed using the Rename option of Change Group.

Create Group Box

The key element of this box can be seen in Figure 12–44. The New< box was covered in the section, Naming a Group. Choices include:

Selectable.   Activating this box determines the status of named groups. Remember that selecting one object in a *named* group selects the entire group.

Unnamed. If this button is active, AutoCAD will provide names for unnamed groups.

Create Change Box

Key element of this box can be seen in Figure 12–44. All elements except Re-Order… are inactive as the dialog box is displayed. Highlighting one of the named groups will activate all of the boxes in the Change Group area.

Add<. This option allows drawing entities to be added to an existing group. Once picked, you will be prompted to select objects. If the selected object is part of an existing group, the group name and description boxes will display the current status of the object. The selection is completed by picking the OK button.

Description. This box allows the Description: to be updated and saved. To alter a description, move the arrow to the cursor box and type the desired material to be added. When complete, picking the Description button updates the description and saves it for future reference.

Explode. The Explode box is used to remove the selected group from the effects of GROUP select. The objects that were in the group will not be altered, only the ability to select those objects as a group. Groups are exploded by highlighting the group to be removed in the Group Name box, and then pressing the Explode button.

Remove<. Selection of the Remove box allows individual objects to be removed from the grouping. Highlighting the desired group to be edited and picking Remove will display the Remove objects: prompt. The dialog box will be removed and the objects in the group will be highlighted. Objects can be selected for removal using any of the selection methods.

Rename. This option can be used to rename an existing group. Highlight the name of an existing group so that the name and description are displayed in the Group Identification area. Edit the name as desired and pick Rename to make the new name the current name.

Re-order… . Selecting this box will produce the dialog box shown in Figure 12–46. Notice that the lower portion of the box includes three listings with numbered entries. Objects in a group are numbered in the same order that they were selected as the group was created. Re-ordering is typically used in CAD-CAM applications of manufacturing and is not often a part of the construction drawings. Options include:

Group Name. Displays the name of the group to be reordered.

Description. This selection displays the description of the selected group.

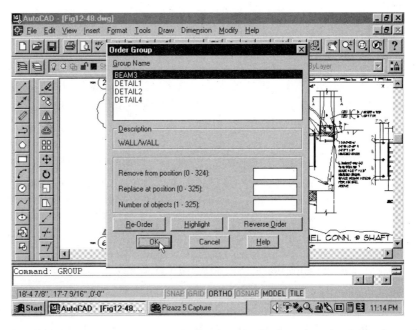

FIGURE 12–46 The Re-order...Group dialog box is displayed as the Re-order... button is selected. Re-ordering allows the order in which entities are selected for certain functions. It is generally not used with construction drawings.

Remove from position (0–x): For each listing, x will vary depending on the size of the objects in the group. This selection specifies the current numbered position of the entity to be reordered.

Replace at position (0–x): This selection specifies the new position the object will be moved to.

Number of objects (1–x): This selection list the range of numbers to be re-ordered.

CHAPTER 12 EXERCISES

E-12-1. Open a templet and draw a 38' long × 18½" deep floor truss. Top and bottom chords will be made out of two 3 × 4s with 1" diameter aluminum webs at approximately 45° equally spaced throughout the truss. Save the drawing as E-12-1.

E-12-2. Draw four sets of perpendicular crossing lines. On one set of lines, use the fillet command to square the corners. On the second set of lines, provide a 12" fillet. On the third set of lines use a 12" chamfer. On the fourth set of lines use a 12" and a 24" chamfer.

E-12-3. Draw an 8" wide concrete block retaining wall extending 8' above a 4" thick concrete floor slab. Show the wall extending 8" above the finish grade. Thicken the slab to 8" thick at the edge resting on the footing. Use a 16" wide × 12" deep concrete footing, and show a 4" diameter drain on the soil side of the wall. Show a 2 × 6 top plate supporting 2 × 8 floor joists flush with the soil side of the wall. Reinforce the wall with #5 diameter @ 12" o.c. each way, 2" clear of the interior face. Use PLINES for concrete, grade, sectioned wood, and steel. Adjust the widths to achieve good contrast. Save the drawing as E-12-3.

E-12-4. Open drawing E-11-4. Copy and rotate the pair of doors to provide doors for walls at 45°, 90°, 135°, 180°, and 270° rotations. Save the drawing as E-12-4.

E-12-5. Open drawing E-11-5 and make a copy of the column cap. Adjust the base strap so that it is 3" longer and 1" wider. Add one additional bolt. Each end of the side plate must be 2" longer (4" total) and 1½" higher. Save the drawing as E-12-5.

E-12-6. Open drawing E-11-7. Enlarge the foundation plan 8 feet in each direction. Add required beams and piers to meet the original criteria. Place 4' × 4' diagonal corners in the exterior walls. Save the drawing as E-12-6.

E-12-7. Open drawing E-9-1 and make three additional copies. In the original drawing, provide a 6" long line from each intersection and perpendicular to the opposite side of the triangle. Rotate the new triangles and place a point of each new triangle tangent to the existing triangle and perpendicular to the 6" lines. Draw a 10" diameter circle with a center point in the center of the original triangle. Trim all elements that extend beyond the circle. When complete, your drawing should resemble the drawing on the next page. Save the drawing as E-12-7.

E-12-8. Start a new drawing and draw a metal chimney cap that extends a total of 12" above a 4' wide wood chimney chase. Trim the chase with 3" wide trim and show 6" wide horizontal siding. The top of the chimney cap is 14" wide × 1". The main portion of the cap is 12" × 7". The chimney is 10" wide × 4" high. Use MIRROR OFFSET, TRIM, and FILLET to complete the drawing. When complete, your drawing should resemble the drawing below. Save as E-12-8.

E-12-9. Start a new drawing and draw a section of a standard one-story foundation similar to the sketch below. Copy and edit the drawing to the proper size for a two-story footing. Save the drawing as Figure E-12-9.

E-12-10. Open a new drawing and draw one half of a roof truss with a 5/12 pitch with a 26' span. All chords will be made of 6" wide material; all webs will be made from 4" material. Use a 24" overhang. The webs should be spaced equally at the center of the bottom chord. As a minimum, use OFFSET, OSNAP, and TRIM. Use the mirror command to complete the drawing. Save the drawing as E-12–10.

E-12-11. Use the floor plan shown in Figure 11–23b and draw the walls and cabinets for this residence. No doors, windows, plumbing fixtures or appliances need to be drawn at this time. Assume:

- All exterior heated walls and toilet walls are 6" thick.
- All exterior unheated walls, interior walls are 4" thick.
- Assume the fireplace is 3'-4" wide. Determine the length by allowing for a 36" hallway.
- Assume the stairway, the hall between the stair and bathroom and between the stair and the front wall to be 42" wide.
- Assume the bathroom is 36" wide.
- Although door and windows symbols should not be drawn at this time, provide the opening in the wall for each door.
- Provide the outline of all cabinets. Assume 24" wide base and 12" wide upper cabinets. Do not draw appliances.

Save this drawing as EFLOOR.

E-12-12. Open drawing EFLOOR. Assume North to be at the top of the drawing so that the front door is presently facing south. Make all changes so that the current wall between the family and living rooms remains straight for the entire length of the house. Alter the width of the wall as needed as it changes from interior/exterior/toilet wall.

- Add 2'-0" to the family room in the north/south (n/s) direction.
- Add 3'-6" to the living room in the east/west (e/w) direction. Add 18" in the n/s direction.
- Enlarge the bathroom to 42" wide. (+6" e/w)
- Remove 24" e/w from the shop.
- Remove the garage doors on the south face of the garage and place them on the east wall. Remove the window from the east wall and place (2)- 6'-0" wide windows on the south wall. In addition to the 2'-8" personnel door on the east wall, enlarge the garage to accommodate (3) 8'-0" wide doors with a 12" minimum space. Provide 24" space on each end of the garage between the door and the end of the wall. Save the drawing as E12FLOOR.

E-12-13. Draw five concentric circles with the smallest diameter of 6" and the other sizes increasing in 3" increments. Remove a 1" wide strip going across each of the circles in both the vertical and horizontal directions. The center of the strip to be removed should be at the center of each quadrant. Rotate the remaining portions of the second and forth circles (starting at the inside) so that the remaining portions of these circles have been revolved 45° while remaining concentric. Save the drawing as E-12-13.

E-12-14. Use EFLOOR as a base. Draw the plan view of a simple tree. Draw the trees in groups of two and three with each tree in the group to have a different size. Place at least four groups of trees around the house. Design walkways, patios and a driveway.

E-12-15. Draw concentric shapes of 3, 4, 5, 6, 7, and 8 sides which are tangent to one circle. Make copies of this object which are ¼ size, ½ size, full size, 2× normal size and 90% of the normal size. Save the drawing as E-12-15.

E-12-16. Open drawing E-11-4 and make two copies of the door. Scale one door to be 2'-6" wide and the other door to be 2'-8" wide. Leave the window beside the door as is on each option. Once the sizes are altered, make copies so that double doors are available for each option. A total of 4 door arrangements should be available when complete. Save the doors as E-12-16.

E-12-17. Open Drawing E-11-10 and make two additional copies of the window. Enlarge one window to 5'-0" wide. Reduce the other window to 75% of the original size. Save the drawing as E-12-17.

E-12-18. Draw a U-shaped kitchen with a 30" × 48" island. Provide a 32" × 21" sink, a 24" × 24" dishwasher, a 34" × 24" refrigerator, and 27" × 27" oven in the kitchen, with a 30" × 22" cook top in the island. Create a second copy of the kitchen which will be flopped and set at a 45° angle to the first kitchen. Save the drawing as E-12-18.

E-12-19. Using the drawing below as a guide, lay out the outline of the office floor plan. Use 8" wide exterior walls. Copy the plan. Save the drawing as E-12-19.

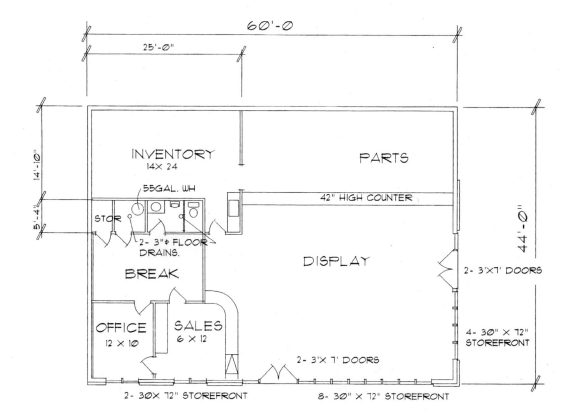

E-12-20. Draw a 3" diameter circle, a 6" long line, and a polygon with 8 sides which will fit in a 5.75" circle. Save this portion of the drawings as E-12-20a. Make enough copies of each object and the use each option of each editing function of GRIPS to edit each object. Save the drawing as E-12-20b.

E-12-21. Use E-12-20a as a templet drawing to complete this assignment. Make a second copy of these objects. Use SELECT to create a group consisting of the two circles and two polygons. Use MIRROR, ROTATE, and SCALE to edit the group. Save the drawing as E-12-21.

E-12-22. Use Drawing E-12-20a as a templet drawing to complete this assignment. Make a second copy of these objects and then mirror both groups so that you have a total of four circles, lines and polygons. Use GROUP to name and save at least 6 different groups. Save the file as E-12-22.

E-12-23. Using the drawings to the right as a guide, redraw the plan view of the manhole cover showing 16-½" wide support flanges equally spaced around the plate. Save the drawing as E-12-23.

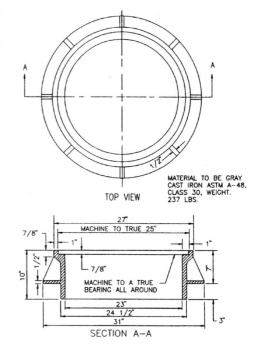

E-12-24. Using the drawing below as a guide, draw and array a pattern to represent Spanish clay tile roofing. Save the drawing as E-12-24.

E-12-25. Use a reference guide to determine the size of a W8 × 35 steel piling. Draw a plan for a 38 foot × 76 foot structure showing three rows of piling. Each of the two exterior rows will be inset 8" from the exterior face of the structure. The interior row of pilings will be 18 feet in from the front (the 76 foot long surface) of the structure. Pilings will be placed at 19 feet o.c. equally spaced from each exterior wall. Save the drawing as E-12-25.

CHAPTER 12 QUIZ

1. Explain the uses of the BREAK command.

2. Describe two different uses of the FILLET command.

3. List the command sequence for ROTATE.

4. Explain the difference between MOVE and COPY.

5. List the command sequence to EXTEND a line to meet another line and sketch and label the process in the space provided.

6. Two perpendicular lines cross at a corner. List four methods to edit the lines to form a noncrossing corner.

7. A line has been drawn short of its required destination. How can the error be corrected?

8. What two options are available for offsetting an object?

9. In the CHAMFER command, what is the relationship between the distances and the selection of lines?

10. You are preparing to MOVE some objects so you type "P" at the Select objects prompt. What will be moved?

11. What SETVAR controls a preset FILLET value? _____

12. How can an object be rotated in a clockwise direction?

13. A door is drawn at a 45° angle. It should be 49°. How can this error be corrected?

14. Six objects need to be moved. As you select the objects, an extra entity is included in the selection set, and an intended object is not included. Describe the command process and prompts to move these objects.

15. List and describe the three break options.

16. Describe how each of the editing commands presented in this chapter could be used in drawing a floor plan.

17. What will happen if you try to place a 60" radius fillet in two intersecting lines that are 48" long?

18. Explain how CHAMFER will affect two lines that do not touch if they are selected for the objects of this command.

19. What will be the effect of the Trim mode of CHAMFER on two intersecting lines?

20. What will be the effect of the TRIM edge option on two lines that do not intersect?

21. How will the Edge option of EXTEND affect two lines?

22. How will window and crossing affect objects selected for STRETCH?

_____

_____

_____

23. What are the options that are given if the DElta option of LENGTHEN is selected?

_____

_____

_____

24. A 6" long line is selected to be edited using LENGTHEN. What will be the result of entering 50%? 150%? 200%?

_____

_____

_____

25. How many selection points will be required when breaking an arc using BREAK@?

_____

_____

_____

26. Provide an example of how the ALIGN command could save time.

_____

_____

_____

27. List the command used to create E-12-9.

_____

_____

_____

28. What must be done to display GRIPS on a line?

_____

_____

_____

29. What five editing functions can be executed after a grip is hot?

30. What is a *hot* grip and how do you get one?

31. You're exploring options and select Grips B. What will the B do and what can you do if you don't want this option?

32. What advantage does SELECT have over GRIPS?

33. Describe the process to select and name a group of objects using GROUP.

34. What is a nested group?

35. What is a *selectable* group?

36. You've named a group, but come to realize the name does not describe the drawing well? How do you fix the problem?

_____

_____

_____

37. Save a group of objects without providing a name. How does AutoCAD keep track of this group?

_____

_____

_____

38. You're exploring again and use the Explode option of Group. What effect will this have on your computer? What effect will it have on the current drawing? What effect will it have on the selected group?

_____

_____

_____

39. How do you make the options in the Change Group box of Object grouping active?

_____

_____

_____

40. Use the help menu and determine what variables affect grips and what are the options and their effects?

_____

_____

_____

_____

CHAPTER

13

POLYLINES

• •

One of the best ways to get objects to stand out on a complicated drawing is to vary the line thickness. This can be done easily with the PLINE command, which represents polylines. In plan views, walls typically are drawn thicker than all other types of lines. On site drawings, polylines are often used to represent contour elevations of specific intervals such as every five or ten feet. Polylines are used throughout structural drawings to represent various materials. Sections and details often use several thicknesses of lines to help distinguish between concrete, steel, and wood that has been cut by the cutting plane, and materials that lie beyond the cutting plane. Figure 13–1 shows examples of polylines used to represent steel reinforcing in concrete details.

In addition to being able to control the width, Polyline has several other unique features. When the line is drawn using the LINE command, each segment is a separate entity. An irregular-shaped object can be drawn with PLINE, yet each individual line is part of the whole. Notice that in Figure 13–2 the horizontal line, the arc, and the vertical line forming the L-shaped rebar has been selected to be erased. Because these features were drawn using the PLINE command they function as one entity. This feature is an important editing feature when the HATCH command is used. Features of polylines to be explored include the PLINE command, changing widths, PLINE arcs, undoing a PLINE, editing a PLINE, exploding a PLINE, PEDIT, and editing a vertex.

PLINE COMMAND

The PLINE command can be accessed by picking the Pline icon from the Draw toolbar, by typing **PL** [enter] or **PLINE** [enter] at the command prompt or by selecting Polyline from the Draw pull-down menu.

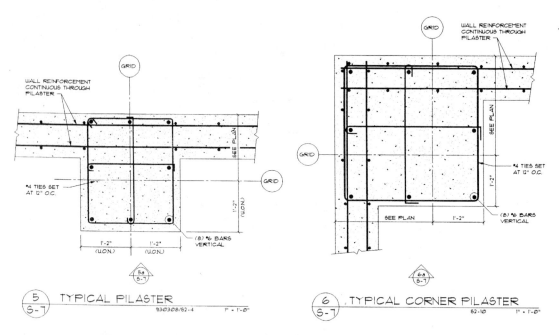

FIGURE 13–1. The PLINE command can be used to distinguish between various materials. Notice that the steel is drawn with a polyline and can be easily separated from the edge of the concrete. (Courtesy Van Domelen / Looijenga / McGarrigle / Knauf Consulting Engineers)

The command sequence is:

Command: **PL** [enter]
From point: (*Pick any point.*)

Once a beginning point has been selected, a new prompt will be displayed:

From point:
Current width is 0'-0"
Arc/ Close/ Halfwidth/ Length/ Undo/ Width/<Endpoint of line>:

This prompt will continue to be displayed each time a new endpoint is entered. Each new line segment will continue from the end of the preceding line. The command will continue until the [enter] key is pressed.

CHANGING THE WIDTH

Enter a new drawing and draw a few polylines. Once you've drawn a polyline with several different drawing segments, pick one of the segments to be erased. During editing is when you can really appreciate a polyline. Picking one segment of a line

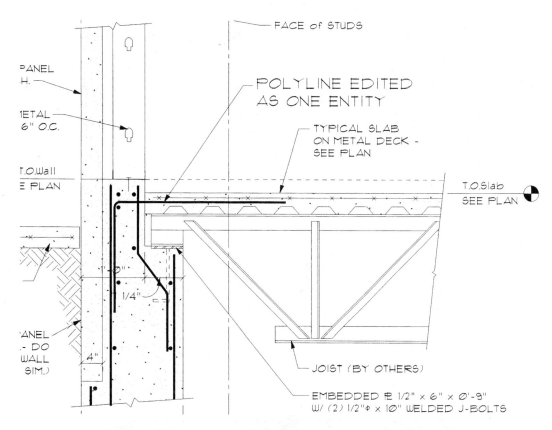

FACE of STUDS

PANEL
H.

POLYLINE EDITED
AS ONE ENTITY

1ETAL
6" O.C.

TYPICAL SLAB
ON METAL DECK -
SEE PLAN

T.O.Wall
E PLAN

T.O.Slab
SEE PLAN

1/4"

'ANEL
.- DO
WALL
SIM.)

4"

JOIST (BY OTHERS)

EMBEDDED ₽ 1/2" × 6" × 0'-9"
W/ (2) 1/2"⏀ × 10" WELDED J-BOLTS

FIGURE 13–2. Selecting the L-shaped rebar which was drawn using PLINE at any point will select the horizontal line, the arc, and the vertical line as one entity. (Courtesy Van Domelen / Looijenga / McGarrigle / Knauf Consulting Engineers)

selects the entire polyline for editing. A line can be made up of hundreds of segments, but if one is selected, they are all selected. With the line width set to "0," polylines look just like any other line. The 0 width works great if you want to outline an irregular-shaped area and find the area or the perimeter where line width is not important. To change the width of a line, start the PLINE sequence and select a start point. As the option prompt is displayed, enter a **W** [enter]. This will allow information for a starting and ending width to be entered. The command sequence to draw a ⅛" wide line is:

Command: **PL** [enter]
From point: (*Pick any point.*)
Current width is 0'-0"
Arc/ Close/ Halfwidth/ Length/ Undo/ Width/<Endpoint of line>: **W** [enter]
Starting width < 0'-0">: **.125** [enter]
Ending width < <0'-0 ⅛">: [enter]

AutoCAD rounds the option to the nearest fraction depending on how the UNITS have been set. If you enter a fraction and the default returns as 0, exit PLINE and enter UNITS and set the fractional values as desired. For the examples used in this chapter the units are set to architectural and the denominator for fractions is set to ⅛" (.125). Notice that the Ending width prompt shows ⅛" as a default. By responding [enter] at the prompt, the option prompt will return allowing for selection of the end-point. With the width selected, the command sequence is:

> Arc/ Close/ Halfwidth/ Length/ Undo/ Width/<Endpoint of line>: (*Pick desired endpoint.*)

With the starting and ending width the same, a line with uniform thickness will be drawn, as shown in Figure 13–3.

Drawing Tapered Lines

PLINE width allows the width of a Polyline to be changed for each segment or from one end of a segment to the other. A tapered polyline with a starting width of ⅛" and an ending width of ¼" can be drawn by using the following command sequence:

> Command: **PL** [enter]
> From point: (*Pick any point.*)
> Current width is 0'-0"

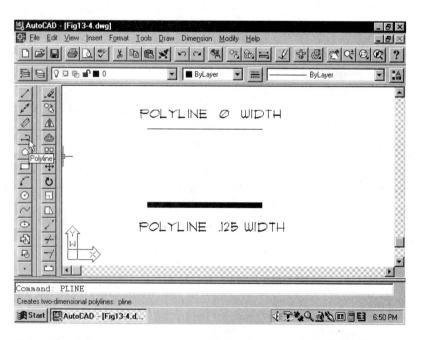

FIGURE 13–3. If polyline is drawn with a width value of 0, the pline will appear as any other line segment.

Arc/ Close/ Halfwidth/ Length/ Undo/ Width/<Endpoint of line>: **W** [enter]
Starting width < 0'-0">: **.125** [enter]
Ending width < <0'-0⅛">: **.25** [enter]
Arc/ Close/ Halfwidth/ Length/ Undo/ Width/<Endpoint of line>: (*Pick end-point.*)
Arc/Close/Halfwidth/Length/Undo/Width/<Endpoint of line>: [enter]
Command:

As the endpoint is entered, the line will be drawn and the option prompt will be returned to allow continuing a polyline. Notice that if the command is continued, the default for the next line segment is the ending width. Figure 13–4 shows an example of varied line width.

Choosing a Line Width

When you experiment with PLINE, the width selection is not terribly important. As you begin to work on more involved projects that will be printed or plotted, the width selection is critical. One of the considerations in selecting a line width is the scale at which the drawing eventually will be printed. A .25" PLINE will look huge on the screen when you've zoomed in on a feature, but if printed or plotted at ⅜" = 1'-0" will appear as a 0 width line on a drawing plotted at 1" = 30'-0".

It's also important to realize that line width can be controlled in the plotting process. This will be covered in Chapter 15 as layers and colors are discussed, and in Chapter

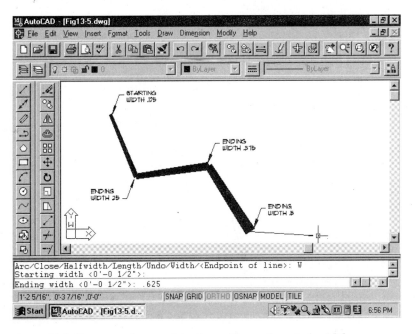

FIGURE 13–4. Polylines can be used to draw lines of varied widths.

24 as you explore plotting options. For now, the key is to realize that pline width can be easily altered. Chapter 14 will include an introduction to TRACE, which also can be used to achieve wider line widths.

Halfwidth Option

This option allows the width of a polyline to be set from the center of the polyline to an edge and is very similar to setting the width. It's like describing the size of a circle by the radius instead of the diameter. The command sequence is:

> Command: **PL** [enter]
> From point: (*Pick any point.*)
> Current width is 0'- ¼"
> Arc/ Close/ Halfwidth/ Length/ Undo/ Width/<Endpoint of line>: **H** [enter]
> Starting halfwidth < 0'-0 ⅛">: [enter]
> Ending halfwidth < <0'-0 ⅛">: [enter]
> Arc/ Close/ Halfwidth/ Length/ Undo/ Width/<Endpoint of line>: (*Pick end-point.*)
> Command:

Length Option

This option will allow a polyline to be drawn to a specified length and parallel to the last polyline drawn. The new line will extend in the same direction as previous lines. To change the direction in which the line is drawn, enter a negative length. This process can be seen in Figure 13–5. If the last segment was an arc, a line drawn with L will be tangent to the arc.

> Command: **PL** [enter]
> From point: (*Pick any point.*)
> Current width is 0'-¼"
> Arc/ Close/ Halfwidth/ Length/ Undo/ Width/<Endpoint of line>: **L** [enter]
> Length of line: (*Enter desired length.*) **2** [enter]
> Arc/ Close/ Halfwidth/ Length/ Undo/ Width/<Endpoint of line>: [enter]
> Command:

PLINE ARCS

Arc is the first option listed in the PLINE menu. When you select ARC by entering **A** [enter] at the prompt, AutoCAD enters the ARC submenu and displays the prompt:

> Command: **PL** [enter]
> From point: (*Pick any point.*)
> Current width is 0'-¼"
> Arc/ Close/ Halfwidth/ Length/ Undo/ Width/<Endpoint of line>: **A** [enter]

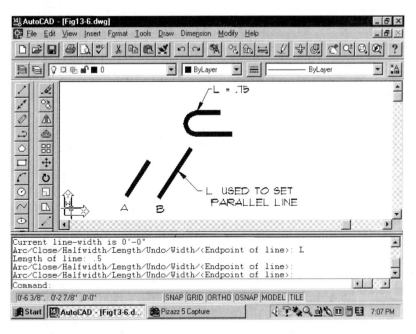

FIGURE 13–5. The L option of PLINE will draw a pline parallel to another pline at a specified length.

Angle/ CEnter/ CLose/ Direction/ Halfwidth/ Line/ Radius /Second pt/ Undo/
    Width/<Endpoint of arc>: (*Pick any point.*)
<Endpoint of arc>: [enter]
Command:

The Arc option will continue to prompt for arc endpoints until the [enter] key is selected.

As the submenu is displayed, an arc is started at the first point that was picked, and is dragged to the current location of the cursor. Several of the options for ARC are the same as the options for straight line PLINE. The Width option is similar, with the width for an ARC ranging from 0 up to the diameter of the arc.

Continuous Pline Arcs

If an arc is started from a previous PLINE segment, the new arc will be tangent to the existing line by default. The arc will continue in the same direction as the most recent PLINE segment. An example of how the arc will be constructed can be seen in Figure 13–6. An arc can also be constructed using the Angle, CEnter, CLose, Direction, Line, Radius, and Second pt methods.

FIGURE 13–6. Arcs can be constructed of plines by using the Arc option of PLINE. The Arc options can now be seen on the command line.

Pline Angles

This option will allow the start point, center, and angle that the arc is to span (included angle). By default, the angle will be drawn counterclockwise. By entering a negative value, the angle will be drawn clockwise. The command process to draw an arc with a 45° included angle is:

> Command: **PL** ⏎
> From point: *(Select desired starting point.)*
> Current line-width is 0'-⅛" Arc/Close/Halfwidth/Length/Undo/Width/<Endpoint of line>: **A** ⏎
> Angle/CEnter/CLose/Direction/Halfwidth/Line/Radius/Second pt/Undo/Width/<Endpoint of arc>: **A** ⏎
> Included angle: **45** ⏎
> Center/Radius/<Endpoint>: *(Select desired endpoint.)*
> <Endpoint of arc>: ⏎
> Command:

As the endpoint is picked, the center is rotated as needed, based on the two end-points. This process can be seen in Figure 13–7.

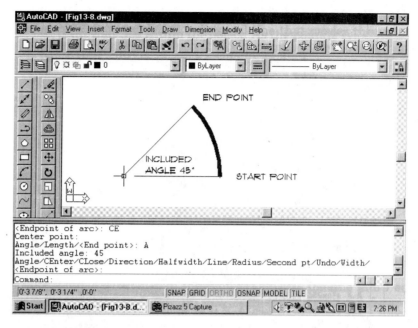

FIGURE 13–7. Selecting the Angle option of Arc will allow an arc to be drawn based on the provide angle.

Arc Center

An arc can be drawn by specifying the endpoint and the centerpoint. When you enter **CE** [enter] at the last prompt, you will be allowed to define the "first endpoint" and the "centerpoint" of the arc. The first steps of the command sequence are identical to other ARC commands. The command sequence for drawing an ARC by defining the centerpoint is:

Command: **PL** [enter]
From point: *(Select desired starting point.)*
Current line-width is 0'-⅛" Arc/Close/Halfwidth/Length/Undo/Width/<End-point of line>: **A** [enter]
Angle/CEnter/CLose/Direction/Halfwidth/Line/Radius/Second pt/Undo/Width/ <Endpoint of arc>: **CE** [enter]
Centerpoint: *(Select desired centerpoint.)*
Angle/Length/<Endpoint>: **A** [enter]
Included angle: **45** [enter]
<Endpoint of arc>: [enter]
Command:

Arc Radius

Entering **R** [enter] will allow for specifying the first endpoint and the radius to be selected. The command sequence is:

> Command: **PL** [enter]
> From point: *(Select desired starting point.)*
> Current line-width is 0'-⅛" Arc/Close/Halfwidth/Length/Undo/Width/<Endpoint of line>: **A** [enter]
> Angle/CEnter/CLose/Direction/Halfwidth/Line/Radius/Second pt/Undo/Width/<Endpoint of arc>: **R** [enter]
> Radius:*(Select desired radius.)* **3.25** [enter]
> Angle/<Endpoint>: **A** [enter]
> Included angle: **45** [enter]
> Direction of chord<235>: [enter]
> <Endpoint of arc>: *(Select desired endpoint or angle.)* [enter]
> Command:

By entering an angle for the chord (the line from endpoints of the arc) the arc will be drawn and the prompt will be redisplayed for continuing the arc command. Then a new arc will be rubberbanded on the screen. Press [enter] to stop the command or enter the letter representing the desired option to continue.

Arc Direction

The Direction option allows for control of the bearing of the ARC chord. The default for PLINE is to draw the arc tangent to the preceding segment. The Direction option will allow you to override this action and specify a starting direction for the arc that is not tangent to the previous line. The sequence is:

> Command: **PL** [enter]
> From point: *(Select desired starting point.)*
> Current line-width is 0'-⅛"
> Arc/Close/Halfwidth/Length/Undo/Width/<Endpoint of line>: **A** [enter]
> Angle/CEnter/CLose/Direction/Halfwidth/Line/Radius/Second pt/Undo/Width/<Endpoint of arc>: **D** [enter]
> Direction from start point: *(Select a point in the direction from the starting point you would like to proceed.)*
> <Endpoint>: *(Select desired endpoint of arc.)*
> <Endpoint of arc>: [enter]
> Command:

The process can be seen in Figure 13–8.

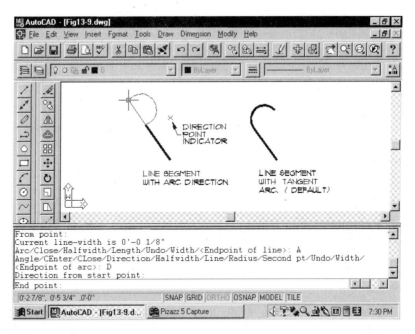

FIGURE 13–8. The Direction option of Arc will allow an arc to be draw that is not tangent to an existing pline.

Three-Point Arc

An arc may be drawn by entering a second point between each endpoint to allow for better placement of the arc. Once in the PLINE Arc option, the command sequence is:

> Command: **PL** [enter]
> From point: *(Select desired starting point.)*
> Current line-width is 0'-⅛" Arc/Close/Halfwidth/Length/Undo/Width/<Endpoint of line>: **A** [enter]
> Angle/CEnter/CLose/Direction/Halfwidth/Line/Radius/Second pt/Undo/Width/<Endpoint of arc>: **S** [enter]
> Second point: *(Select the desired point.)*
> <Endpoint>: *(Select desired endpoint of arc.)*
> Angle/CEnter/CLose/Direction/Halfwidth/Line/Radius/Second pt/Undo/Width/<Endpoint of arc>: [enter]
> Command:

The sequence can be seen in Figure 13–9.

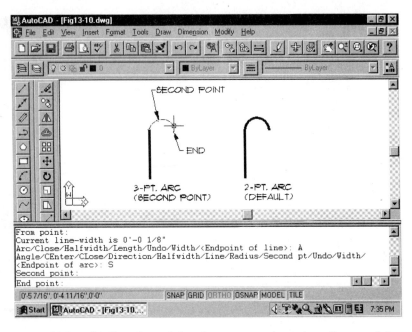

FIGURE 13–9. Using the S option of Arc draws an arc based on three points.

Closing a Polygon

As you draw with the LINE command, polygons are ended by typing **C** ⏎. This provides a line segment from the present location, back to the original starting point. A similar option is available with the PLINE Arc option. By entering **CL** ⏎ at the prompt, a series of polylines will be closed by an arc segment. Remember that CL must be entered when in the arc submenu rather than C to distinguish between CLose and CEnter. The final sequence to close a polyline is:

Angle/ CEnter/ CLose/ Direction/ Halfwidth/ Line/ Radius/ Second pt/ Undo /Width/<Endpoint of arc>: **CL** ⏎

The results can be seen in Figure 13–10.

UNDOING A PLINE

As with other drawing options, PLINE allows for removing the previous segment or segments in the reverse order from the way they were drawn. This can be done with the Undo option by entering **U** ⏎ at the prompt.

Arc/ Close/ Halfwidth/ Length/ Undo/ Width/<Endpoint of line>: **U** ⏎

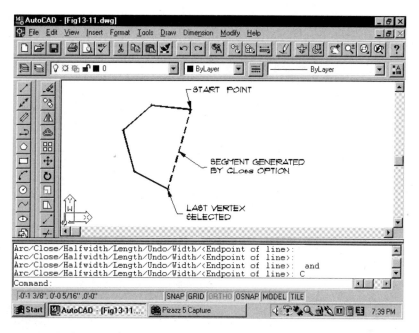

FIGURE 13–10. Polygons drawn using PLINES can be completed using the Close option.

This will remove the last polyline that was drawn, as seen in Figure 13–11. Undo may be used repeatedly until just one point is left in the polyline. Separate from the Undo option, the UNDO command can be used once the drawing command has been terminated. The UNDO command will remove an entire command sequence and will be covered in Chapter 14.

EDITING A POLYLINE

Polylines may be edited with the commands introduced in Chapters 11 and 12. Because polylines are segments drawn as a single entity, they are edited as a single entity. This can facilitate the selection process, since only one segment must be selected, rather than each of the segments that make up the entire object. This can be seen in Figure 13–12.

Most of the commands in Chapter 12 will modify a polyline in exactly the same way as another segment. FILLET and CHAMFER each have an added feature for a polyline.

FILLET

The FILLET command can't be used to fillet two different polylines in the same method used to FILLET two lines. However, the command can be used on closed

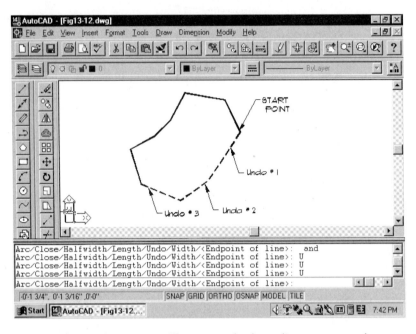

FIGURE 13–11. The Undo option will remove the last pline segment drawn.

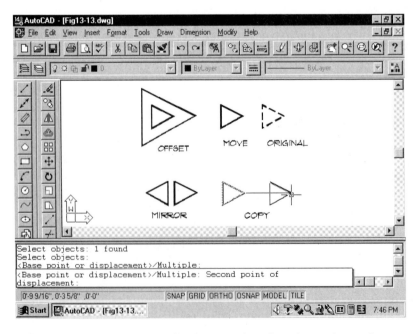

FIGURE 13–12. When editing a polyline, remember that the entire entity acts as one object.

polygons constructed of a polyline in two different methods. The command sequence to FILLET one corner of a closed polygon with a radius of ⅜" is:

Command: **F** `[enter]`
(TRIM mode) Current fillet radius = 0'-0"
Polyline/Radius/Trim/<Select first object>: **R** `[enter]`
Enter fillet radius <0'-0">: **.375** `[enter]`
Command: `[enter]`
(TRIM mode) Current fillet radius = 0'-0 3"
Polyline/Radius/Trim/<Select first object>: *(Select edge of object to receive fillet.)*
Select second object: *(Select second edge to receive fillet.)*
Command:

The fillet can also be applied to all edges of the polygon in one command sequence. Once the radius is set the command sequence is:

Command: **F** `[enter]`
(TRIM mode) Current fillet radius = 0'-3"
Polyline/Radius/Trim/<Select first object>: **P** `[enter]`
Select 2D polyline: *(Select object to receive fillets.)*
4 lines filleted *(Quantity will vary with each object.)*
Command:

The effects of the FILLET command on a polyline can be seen in Figure 13–13.

CHAMFER

Editing polylines with CHAMFER is similar to using the FILLET command. The editing options for CHAMFER are:

Command: **CHA** `[enter]`
(TRIM mode) Current chamfer dist1 = 0'-0 3", Dist2 = 0'-0 ⅜"
Polyline/Distance/Angle/Trim/Method/<Select first line>: *(Select object to receive chamfer.)*
Select second line: *(Select second edge to be chamfered.)*
Command:

Once the distances are set, the command will edit a polyline as two line segments if the default option is selected. If the Polyline option is selected, all corners of the selected polyline will receive chamfers. The command sequence is:

Command: **CHA** `[enter]`
(TRIM mode) Current chamfer dist1 = 0'-0 3",
Dist2 = 0'-0 ⅜"
Polyline/Distance/Angle/Trim/Method/<Select first line>: **P** `[enter]`
Select 2D polyline: *(Select object to receive chamfer.)*

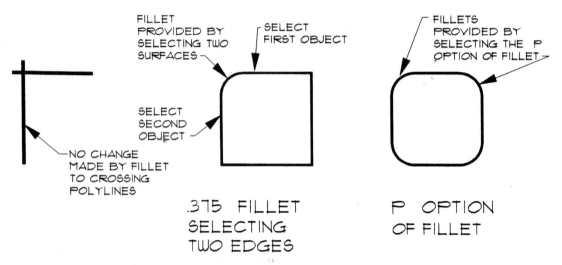

FILLET PROVIDED BY SELECTING TWO SURFACES

SELECT FIRST OBJECT

FILLETS PROVIDED BY SELECTING THE P OPTION OF FILLET

SELECT SECOND OBJECT

NO CHANGE MADE BY FILLET TO CROSSING POLYLINES

.375 FILLET SELECTING TWO EDGES

P OPTION OF FILLET

FIGURE 13–13. Two separate polylines are not affected by the FILLET command. The FILLET command can be used to modify polygons constructed of polylines or the Polyline option of PLINE can be used to fillet all edges of a polygon.

4 lines were chamfered
Command:

The effect of chamfering a polygon can be seen in Figure 13–14.

EXPLODE

One of the benefits of a polyline is also one of its drawbacks. Whether you've drawn two or 2,000 connected polyline segments, they all function as one entity. That's great if you want to move 2,000 lines at once by picking only one option, but a hindrance if you want to change just one of the segments. To overcome this obstacle, use the EXPLODE command. (This is not the long awaited command you long for when you're really frustrated with a computer. It only explodes groups of drawing entities, and doesn't even make a loud noise.) The command can be accessed by picking the Explode icon (the stick of dynamite) from the Modify toolbar by entering **X** ⟨enter⟩ or **EXPLODE** ⟨enter⟩ at the command prompt or by picking Explode from the Modify pull-down menu. The command sequence is:

Command: **X** ⟨enter⟩
Select objects: (*Pick objects to be edited.*)
Select object: ⟨enter⟩

Once the ⟨enter⟩ key has been pressed, the PLINE will be transformed to individual line or arc segments. If the polyline had a specified width, a message will be displayed at the command line:

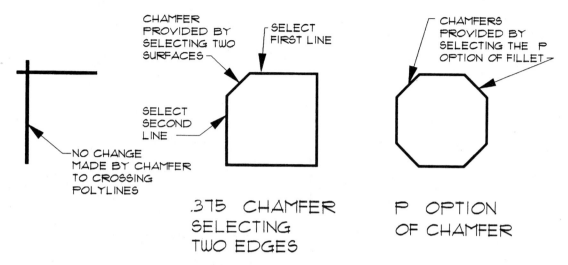

FIGURE 13–14. Two separate polylines are not affected by the CHAMFER command. The CHAMFER command can be used to modify polygons constructed of polylines or the Polyline option of PLINE can be used to chamfer all edges of a polygon.

> Exploding this polyline has lost width information.
> The UNDO command will restore it.
> Command:

If you can tolerate the loss of line width, proceed. If the line width is important to the drawing, typing UNDO at the command prompt will restore the polyline to its unedited form. The effects of EXPLODE on a polyline can be seen in Figure 13–15.

PEDIT

In addition to using the standard editing options, you can edit polylines by using the PEDIT command. The command can be accessed by picking the Edit Polyline icon from the Modify II toolbar, by entering **PE** [enter] or **PEDIT** [enter] at the command prompt, or by selecting Polyline from the Object flyout menu of the Modify pull-down menu. Options for the command will vary slightly depending on whether the polyline forms a closed shape or a line. Options for the command when a closed pline is selected are:

> Command: **PE** [enter]
> Select polyline: *(Select polyline to be edited.)*
> Open/Join/Width/Edit vertex/Fit/Spline/Decurve/Ltype gen/ Undo/eXit <X>:

If an open pline is selected, the Open option is replaced with the Close option.

Entering [enter] will exit back to the command line. Selecting objects may be done by any of the selection methods discussed in Chapter 11 or 12. Notice then when you are prompted to select objects, the cursor is exchanged for a selection box to allow

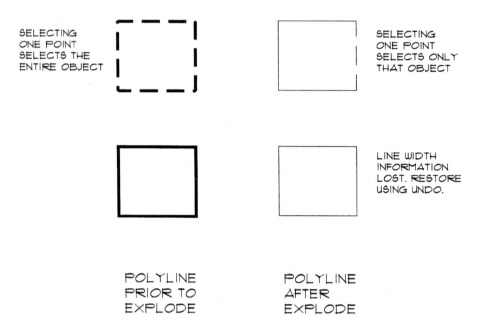

SELECTING
ONE POINT
SELECTS THE
ENTIRE OBJECT

SELECTING
ONE POINT
SELECTS ONLY
THAT OBJECT

LINE WIDTH
INFORMATION
LOST. RESTORE
USING UNDO.

POLYLINE
PRIOR TO
EXPLODE

POLYLINE
AFTER
EXPLODE

FIGURE 13–15. The EXPLODE option will allow segments of a polyline to be treated as individual entities.

you to pick polylines individually. Window or Crossing also are excellent for selecting polylines.

Occasionally, segments might be selected for PEDIT that are not polylines. When this happens, you will be given the prompt:

Object selected is not a polyline.
Do you want to turn it into one? <Y>

If the default is accepted, the segment will become a PLINE and the PEDIT prompt will be displayed allowing options of the new polyline to be altered. A CIRCLE cannot be changed to a polyline but it can be drawn using the 360° arc of PLINE Arc option. A DONUT also could be used to draw a circle that would have similar qualities to a polyline.

The editing options for a polyline include Open, Join, Width, Edit vertex, Fit curve, Spline curve, Decurve, Ltype gen, Undo, and eXit.

Open

The O (Open) option will remove the closing segment of a polyline. If a line is drawn back to the starting point without using Close, opening the polyline has no visible effect. The results can be seen in Figure 13–16. If O is entered for an open polyline, it will be closed.

FIGURE 13–16. A pline removed with the Open option.

Close

The CL (CLose) option creates a closing segment of the polyline while in the PLINE command routine. Once you've moved on to another command sequence, the PLINE must be edited with C (Close) option. If you select C on a closed polyline, it will be opened. Figure 13–16 shows the effects of the close command.

Join

This option adds Lines, Arcs, and other polylines that meet a selected polyline by typing **J** [enter] (Join) at the command prompt. Its option can be useful for joining two or more individual polylines so that they function as one polyline. This option may only be used with open polylines and cannot be used to join segments that do not touch the selected polyline. Objects that cross the polyline will not be joined.

Width

The W (Width) option will allow the current width of a polyline to be changed. Polylines that consist of segments with varying widths or tapers will be changed so that all segments are the new width. When [enter] is entered at the option prompt, a new prompt will be given requesting:

　　Enter new width for all segments:

The width may be entered by keyboard or by picking two points with the mouse. Once the width is altered, the PEDIT prompt will be displayed again allowing for other editing to take place. Press [enter] to return to the command prompt. Figure 13–17 shows the effects of the Width option.

Edit Vertex

Once a polyline has been formed, the shape may need to be altered. Altering the shape can be done by entering **E** [enter] at the option prompt. An X will be placed in the first vertex and a new prompt will be displayed with the following options:

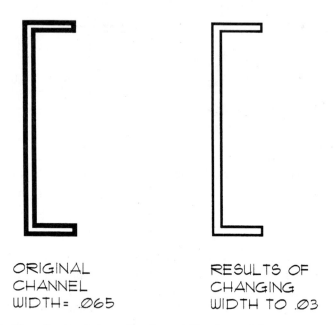

ORIGINAL
CHANNEL
WIDTH= .O65

RESULTS OF
CHANGING
WIDTH TO .O3

FIGURE 13–17 The effects of changing the width of a polyline.

Next/Previous/Break/Insert/Move/Regen/Straighten/Tangent Width/eXit <N>:

Unless the X is placed in the location you would like to edit, use the Next or Previous options to move the X around the Polyline to select the desired editing location.

NEXT (N) moves the X to the next vertex. By pressing 〖enter〗, you can move the X around the object, as seen in Figure 13–18.

PREVIOUS (P) moves the X to the previous vertex. By pressing 〖enter〗, you can move the X around the object in the opposite direction from the one used by Next. The option can be seen in Figure 13–18.

BREAK (B) will remove a portion of a polyline. The portion to be broken is selected by the N or P option. Once the X is in the desired position, the prompt reads:

Next/Previous/Go/eXit <N>:

As the X is now moved with N or P, any point between the original starting location and the present location will be removed. When the X is moved to the end of the segment to be used, enter **G** 〖enter〗. This will remove the desired segment. If Break is used on a closed polygon, the closing section of the polygon will be removed along with the selected portion. The process can be seen in Figure 13–19. The command sequence is:

Next/Previous/Break/Insert/Move/Regen/Straighten/Tangent/Width/eXit <N>:
    B 〖enter〗

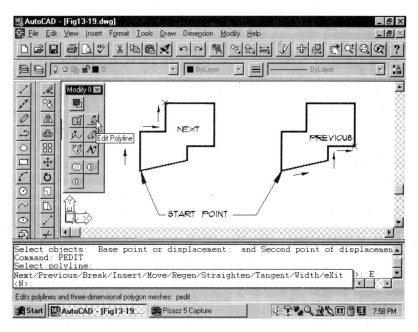

FIGURE 13–18 The vertex of a polyline can be edited by moving the "X" using the Next or Previous option.

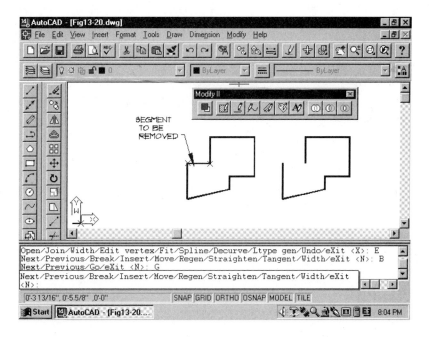

FIGURE 13–19 Removing a portion of a polyline using the Break option.

Next/Previous/Go/eXit <N>: [enter]
Next/Previous/Go/eXit <G>: [enter]

INSERT (I) will edit an existing polyline by adding a new vertex. The new vertex is added after the vertex marked by the X. The command sequence is:

Open/Join/Width/Edit vertex/Fit/Spline/Decurve/Ltype/Undo/eXit <X>: **E** [enter]
Next/Previous/Break/Insert/Move/Regen/Straighten/Tangent/Width/eXit
    <N>: [enter] (*Move X to the desired location.*)
Next/Previous/Break/Insert/Move/Regen/Straighten/Tangent/Width/eXit
    <N>: **I** [enter]
Enter location of new vertex: (*Pick desired location.*)

The process can be seen in Figure 13–20.

MOVE (M) will allow for moving an existing polyline vertex to a new location. The vertex to be moved must be the one currently marked by the X. The command sequence is:

Open/Join/Width/Edit vertex/Fit/Spline/Decurve/Ltype/Undo/eXit <X>: **E** [enter]
Next/Previous/Break/Insert/Move/Regen/Straighten/Tangent/Width/eXit
    <N>: [enter] (*Move X to the desired location.*)

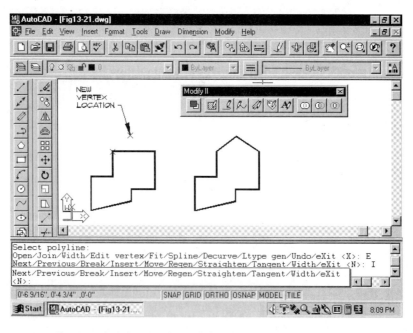

FIGURE 13–20 Altering the shape of a polyline by inserting a new vertex.

Next/Previous/Break/Insert/Move/Regen/Straighten/Tangent/Width/eXit
<N>: **M** [enter]
Enter new location: (*Pick desired location.*)

The process can be seen in Figure 13–21.

REGEN (R) will regenerate the edited polyline.

STRAIGHTEN (S) will straighten existing polyline segments that lie between two selected points. Move the X to the desired location to mark the start of the edit. The X can then be moved to mark the end of the edit. Any arcs or segments between the two marks will be deleted and replaced by a straight segment. The command sequence is:

Open/Join/Width/Edit vertex/Fit/Spline/Decurve/Ltype/Undo/eXit <X>: **E** [enter]
Next/Previous/Break/Insert/Move/Regen/Straighten/Tangent/Width/eXit
        <N>: [enter] (*Move X to the desired location.*)
Next/Previous/Break/Insert/Move/Regen/Straighten/Tangent/Width/eXit
        <N>: **S** [enter]
Next/Previous/Go/eXit <N>: [enter]
Next/Previous/Go/eXit <N>: [enter]
Next/Previous/Go/eXit <G>: [enter]

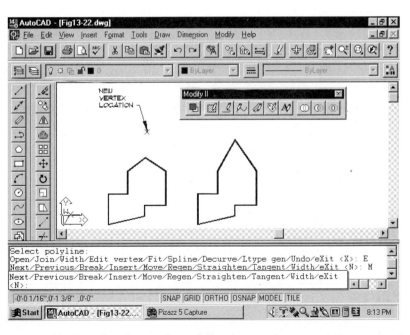

FIGURE 13–21 Altering the shape of a polyline by moving an existing vertex to a new location.

The process can be seen in Figure 13–22.

TANGENT (T) will allow you to attach a tangent direction to the current vertex for later use in Curve fitting. The command sequence is:

Open/Join/Width/Edit vertex/Fit/Spline/Decurve/Ltype/Undo/eXit <X>: **E** ⏎enter
Next/Previous/Break/Insert/Move/Regen/Straighten/Tangent/Width/eXit
    <N>: ⏎enter *(Move X to the desired location.)*
Next/Previous/Break/Insert/Move/Regen/Straighten/Tangent/Width/eXit
    <N>: **T** ⏎enter
Direction of Tangent:

Either a specific tangent angle may be specified from the keyboard or a point may be picked to mark the direction from the currently marked (X) vertex.

WIDTH (W) changes the starting and ending width of the segment following the marked vertex. You've explored the Width option of PEDIT, which changes the entire polyline. Width of the Edit Vertex options will allow only one segment to be edited. The command sequence is:

Open/Join/Width/Edit vertex/Fit/Spline/Decurve/Ltype/Undo/eXit <X>: **E** ⏎enter
Next/Previous/Break/Insert/Move/Regen/Straighten/Tangent/Width/eXit
    <N>: ⏎enter *(Move X to the desired location.)*

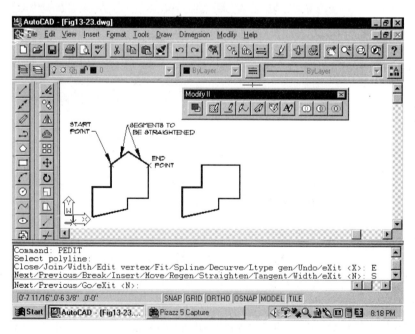

FIGURE 13–22 Removing an existing vertex using the Straighten option of PEDIT.

Next/Previous/Break/Insert/Move/Regen/Straighten/Tangent/Width/eXit
     <N>: **W** [enter]
Enter starting width <125>: **0** (*Enter desired width.*) [enter]
Enter ending width <0'-0">: **.375** [enter] (*Enter desired width.*)

The process can be seen in Figure 13–23.

Exit (X) exits from Vertex editing and returns you to the PEDIT prompt.

Fit Curve

The Fit curve option (F) of PEDIT allows straight line segments to be converted into curved lines. This can be especially helpful on a topography where lines typically are drawn from elevation to elevation. Fit curve also can be used to convert zigzag lines to smooth curves often used to represent insulation in section view, as seen in Figure 13–24. The command process is:

Command: **PE** [enter]
PEDIT Select polyline:
Close/Join/Width/Edit vertex/Fit/Spline/Decurve/Ltype gen/Undo/eXit
     <X>: **F** [enter]

As [enter] is entered, Fit curve transforms the points of the polyline into smoothed curves. The process can be seen in Figure 13–25.

Spline Curve

Unlike the Fit curve that passes through all the vertices, a Spline (S) curve only passes through the first and last points of the polyline. In between, the curve will be close to each vertex, but will not pass through them. The more control points you

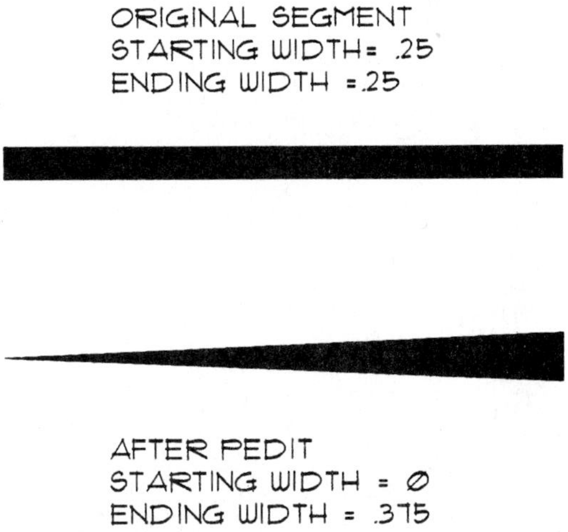

ORIGINAL SEGMENT
STARTING WIDTH= .25
ENDING WIDTH =.25

FIGURE 13–23. Altering the width of an existing pline using PEDIT W.

AFTER PEDIT
STARTING WIDTH = ∅
ENDING WIDTH = .375

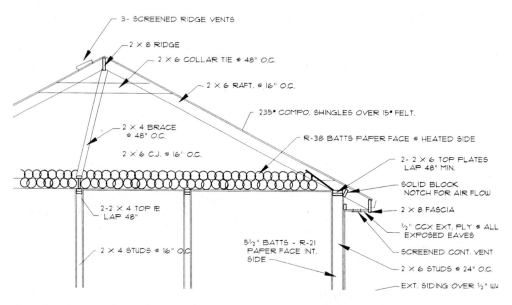

3- SCREENED RIDGE VENTS

2 X 8 RIDGE

2 X 6 COLLAR TIE @ 48" O.C.

2 X 6 RAFT. @ 16" O.C.

235# COMPO. SHINGLES OVER 15# FELT.

2 X 4 BRACE @ 48" O.C.

R-38 BATTS PAPER FACE @ HEATED SIDE

2 X 6 C.J. @ 16" O.C.

2- 2 X 6 TOP PLATES LAP 48" MIN.

SOLID BLOCK NOTCH FOR AIR FLOW

2-2 X 4 TOP PE LAP 48"

2 X 8 FASCIA

½" CCX EXT. PLY @ ALL EXPOSED EAVES

2 X 4 STUDS @ 16" O.C.

5½" BATTS - R-21 PAPER FACE INT. SIDE

SCREENED CONT. VENT

2 X 6 STUDS @ 24" O.C.

EXT. SIDING OVER ½" W

FIGURE 13–24. The ceiling insulation is drawn using the Fit Curve option of PEDIT. (Courtesy Residential Designs)

ORIGINAL PLINE Ø WIDTH

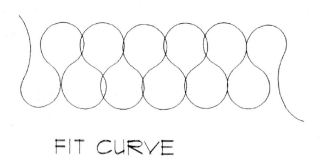

FIT CURVE

FIGURE 13–25. Changing a pline using the Fit Curve option.

specify, the smoother the Spline curve will be. Figure 13–26 shows a Spline curve. The command process is:

> Command: **PE** ⏎
> PEDIT Select polyline:
> Close/Join/Width/Edit vertex/Fit/Spline/Decurve/Ltype gen/Undo/eXit <X>: **S** ⏎

As ⏎ is entered, Spline curve transforms the points of polyline into smoothed curves. The Fit and Spline options are not often used on construction drawings. For further information remember to use the HELP menu in AutoCAD. You may also want to research the commands SPLINES, SPLINETYPE, SPLINESEGS, SPLINEDIT, and NURBS for related types of drawings entities.

Decurve

The D option can be used to remove any vertices that have been inserted by Fit curve or Spline curve and to straighten segments of a polyline. The command sequence is:

> Command: **PE** ⏎
> PEDIT Select polyline: *(Select polyline to be edited.)*
> Close/Join/Width/Edit vertex/Fit/Spline/Decurve/Ltype gen/Undo/eXit<X>: **D** ⏎

The effects can be seen in Figure 13–27.

FIGURE 13–26. Altering a pline using the Spline Curve option.

SPLINE CURVE

ORIGINAL PLINE Ø WIDTH

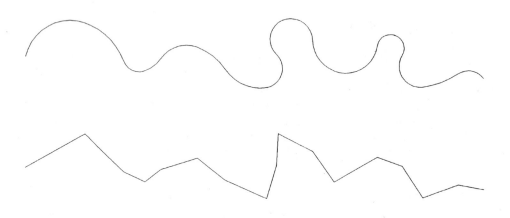

PLINE AFTER DECURVE

FIGURE 13–27. Decurve will remove any vertices that were inserted with Fit or Spline Curve.

LINETYPES

Figure 13–28 shows a polyline drawn with a center line with a long, short, long pattern. Notice that the pattern extends from vertex to vertex. Each segment is, in effect, a separate centerline. The polyline also can be drawn so that the pattern extends throughout the entire polyline, as seen in Figure 13–29. Ltype gen (L) of the PEDIT menu will adjust how the pattern is displayed. Toggled to the on position, Ltype gen will draw line patterns in a continuous pattern from beginning to end with no consideration to vertices. In the off position, line patterns will be based on each vertex. Ltype gen does not affect polylines with tapered segments.

The line pattern also can be adjusted by the PLINEGEN system variable. When set to 1, PLINEGEN draws a continuous pattern. With the variable set to 0, the line pattern is based on each vertex.

Undo

The Undo (U) option of PEDIT will undo the most recent PEDIT edit operation. Multiple entries of U will step back through the drawing, removing previous PEDIT entries.

Exit

The eXit (X) option is the default for PEDIT. Pressing [enter] will edit the PEDIT command and return to the command prompt.

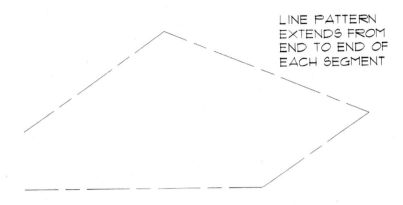

LINE PATTERN
EXTENDS FROM
END TO END OF
EACH SEGMENT

PEDIT LTYPE GEN OFF

FIGURE 13–28. Using the OFF setting of the L option of PEDIT will draw a linetype pattern from vertex to vertex.

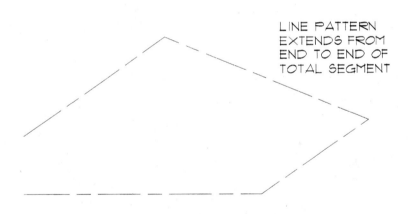

LINE PATTERN
EXTENDS FROM
END TO END OF
TOTAL SEGMENT

PEDIT LTYPE GEN ON

FIGURE 13–29. Using the ON setting of the L option of PEDIT will draw a linetype pattern from end to end of the entire line, disregarding the vertex.

CHAPTER 13 EXERCISES

E-13-1. Open a new drawing and draw a polyline polygon with five sides with a 0 width. Set the width of a PLINE as 1/16" wide and draw a polygon with a minimum of four sides. Use the close option to close the polygon. Change the width to .125 and draw three open polylines. Save the drawing as E-13-1.

E-13-2. Open a new drawing and draw a polyline with a beginning width of .125, a length of 1.5, and an ending width of .25. Continue from this segment with a line that is 1.5 long and .125 wide. Draw a third segment that will close a polygon with a starting width of .25 and ending with a 0 width. Save the drawing as E-13-2.

E-13-3. Open a new drawing and draw a circle with a 3" diameter and a thickness of 1/16". Use the three-point Arc option to draw a circle with a width of 1/8" and a radius of 1.5". Save the drawing as E-13-3.

E-13-4. Open a new drawing and use a width of 0 to draw a 3" long line. Draw a .125 wide arc with a radius of 1.75 and an angle of 60° on the right end of the original line segment. Use the right end of the straight line segment as the center of a 45° arc that ends at the left end of the arc segment. Save the drawing as E-13-4.

E-13-5. Open a new drawing and draw a four-sided polygon using polyline with a width of .125. Copy the polygon so that there are a total of five polygons. Edit one of the polygons so that the line width is .25. Edit another polygon so that the width is 0.065. Fillet two corners with a .5 radius. Fillet the remaining two corners with a .25 radius. Edit a third polygon so that the width of all lines is 0.0 wide. Provide a .4 × .25 chamfer at each corner. On the fourth polygon add a vertex at some point so that the polygon has five segments. Save the drawing as E-13-5.

E-13-6. Open Drawing E-12-9 and convert the lines that represent the footing to polylines. Set the width as .5". Set the line that represents the finish grade as .75. Copy the one-story footing and explode the drawing. Save the drawing as E-13-6.

E-13-7. Open Drawing E-12-8. Assume the light source to be in the upper left-hand corner to shade the drawing. Use .5 width polylines to create shade on all overhanging materials. On the left side of the horizontal trim, draw a shadow that tapers from 0 up to 1" wide. Save the drawing as E-13-7.

E-13-8. Open a new drawing and draw a series of seven zigzag polyline 0.0 width lines at approximately 15° from vertical. Make the line segments 10" long. Make two additional copies. Use Spline curve to edit one set of lines and Fit curve to edit the other set. Save the drawing as E-13-8.

E-13-9. Open a drawing E-13-6 and adjust the limits as required to draw a typical wall section that is similar to the attached drawing. Show:

a 2 × 6 sill with a ½" × 10" anchor bolt.
2 × 8 floor joist w/ 2 × 8 rim joist.
¾" plywood subfloor
8' high studs with a double top plate and a single base plate
2 × 6 ceiling joist
2 × 6 rafters @ a 27½° (5/12) pitch
2 × 8 fascia and solid eave blocking

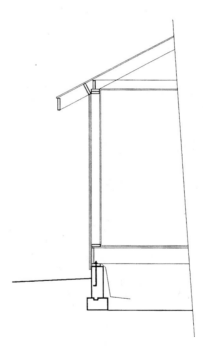

Use .25 wide polylines to represent any materials that would be cut by the cutting plane, such as the plate, sill, fascia, and blocking. Save the drawing as E-13-9.

E-13-10. Open a new drawing and set the units and limits to draw the attached grading plan. Lay out the grids and approximate the contour lines. Use a 6"-wide polyline to represent five foot intervals and a 2"-wide polyline to represent one foot intervals. Once all straight line segments have been drawn, edit the plines to provide the most accurate, smooth transitions as possible. Save the drawing as E-13-10.

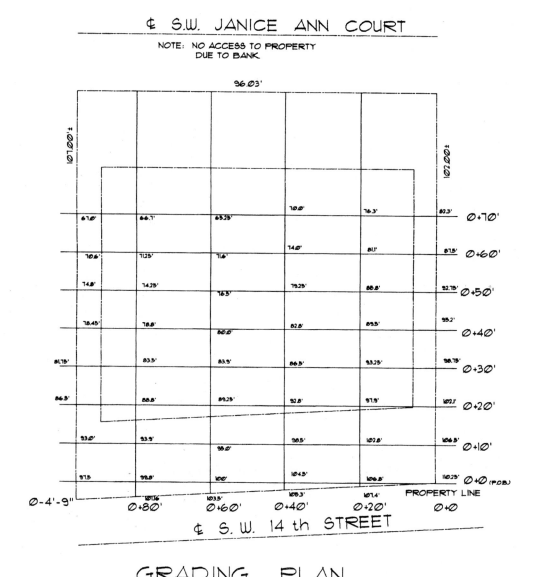

GRADING PLAN

1" ——— 1'-0"

CHAPTER 13 QUIZ

1. What does PLINEGEN control and what options are included?

   _____

   _____

2. Give the command sequence and show all options to draw a polyline with a width of .125.

   _____

   _____

   _____

   _____

3. List two options to control line width other than PLINE.

   _____

   _____

4. Major drawings for a residence include the site, floor, foundation plans and the elevations and sections. Find a set of professional drawings and list common ways Polylines can be used to enhance these drawings.

   _____

   _____

   _____

5. What option would allow a .25" wide line be drawn by entering .125?

   _____

6. Explain the difference between Spline and Fit curve.

   _____

   _____

   _____

7. Explain the difference between Open and Close pedit options.

   _____

   _____

   _____

8. How does EXPLODE affect a polyline?

_____

_____

_____

_____

9. To edit a vertex that is formed between the third and fourth lines drawn, what two options will be needed? _____

10. What option will allow an additional vertex to be added to a completed polyline?

_____

11. A PLINE has been drawn, but one vertex is ½" to the left of where it belongs. What option will fix the problem?

_____

_____

12. What PEDIT option will remove any vertices that have been inserted by Fit or Spline curves?

_____

_____

13. What PEDIT option will remove convert arcs to straight line segments? _____

_____

14. What option will display an X at the start of a PLINE segment, and what three options affect the X?

_____

_____

15. What is the default setting for PEDIT? _____

16. How does the Halfwidth option differ from the Width options.

_____

_____

17. A friend needs to draw a 2-inch-long pline parallel to an existing pline. How can this be done?

_____

_____

18. How can a circle be drawn using the PLINE command?

19. How can a pline arc be drawn in a clockwise direction?

20. How can the direction of an arc chord be controlled?

14

SUPPLEMENTAL DRAWING COMMANDS

Y ou've explored the methods for drawing and editing lines and polylines. This chapter will introduce methods of marking and dividing drawing space using the POINT, DIVIDE, and MEASURE commands. Five new methods of drawing lines will be introduced including the RAY, Construction Line, TRACE, and SKETCH. The final method of drawing lines to be presented is MLINES. The command can be used to draw groups of lines of varied quantities and type, Methods for editing Mlines will also be presented. Three commands will also be presented that can be used to edit any linetype. These commands are the U, UNDO, and REDO commands.

MARKING AND DIVIDING SPACE

Three commands are available in AutoCAD for marking and dividing space. Although these commands can't be used to draw lines, each can be used to mark space in a drawing and provide a location for placing lines.

POINTS

Often, a point may need to be marked on a drawing. The loads on a column on the upper level of a structure will need to be carried down through several floors into the foundation. AutoCAD will allow you to mark these load locations on a drawing by drawing a point. To place a point in a specific location, select the Point icon from the Draw menu or by selecting Point from the Draw pull-down menu, or type **PO** [enter] or **POINT** [enter] at the command prompt. This will produce a prompt:

Command: **PO** (enter)
Point:

The prompt is now waiting for coordinates to be typed. Points are usually selected by picking a point with the mouse. The shape that is used to identify a point can be altered using the Point Shape dialog box. The box can be seen in Figure 14–1. The box is displayed by typing **DDPTYPE** (enter) at the command prompt or by picking Point Style from the Format pull-down meu. The box allows the style and size of the point to be controlled. The active setting is shown in a black box, in this case a point is used for the marker. Notice the second box from the left in the top row contains no marker. If this option is selected, the point will be marked with an invisible marker. Although the point can't be seen, it will be selected when the Node option of OSNAP is active. The values for the shape and size can also be entered by keyboard using the PDMODE or PDSIZE commands. PDMODE and PDSIZE are the variables that control the appearance of Point entities.

PDMODE.   In addition to adjusting the shape of the POINT marker in the Point Shape dialog box, the shape can be adjusted by entering PDMODE (enter) at the command prompt. This will produce the prompt:

New value PDMODE<0>:

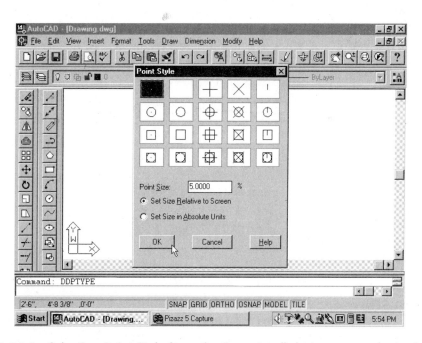

FIGURE 14–1 Selecting Point Style from the Format pull-down menu or by typing **DDPTYPE** (enter) at the command prompt will produce the Point Style dialog box. The box can be used to alter the size and shape of the point marker.

The five basic shapes are selected by entereing a number from 0–4.

There are five basic choices of how the point will be displayed.

Value	point drawn as:
0	(.) a dot is drawn for the point
1	( ) nothing is drawn
2	(+) a cross is drawn
3	(x) an X is drawn
4	(	) a vertical line is drawn

Each of these five basic shapes may be altered. This can be done by adding the numbers 32, 64, or 96 to the point value 0–5. The number 32 will place a circle around the point. A square will be placed around the point when 64 is added to the point number. Both a square and a circle will be added to the point display when the point value is added to 96. Typing 32 will draw a circle around a dot. Typing 67 will draw a box around an X. Each of the pont shapes and their corresponding numeric values are shown in Figure 14–2. Even though it is easier to set point styles from the dialog box, it may be faster to enter values by keyboard. Figure 14–2 shows the point to be entered and the numeric value to be entered.

PDSIZE. The size of the point marker can be adjusted from the Point Style dialog box or by keyboard. To adjust the size from the command prompt, type **PDSIZE** [enter] and provide a value for the marker size. No matter what method is used to access the variable, the same effect will be achieved based on the assigned value. A setting of 5 will produce a markers at 5% of the graphic area height.

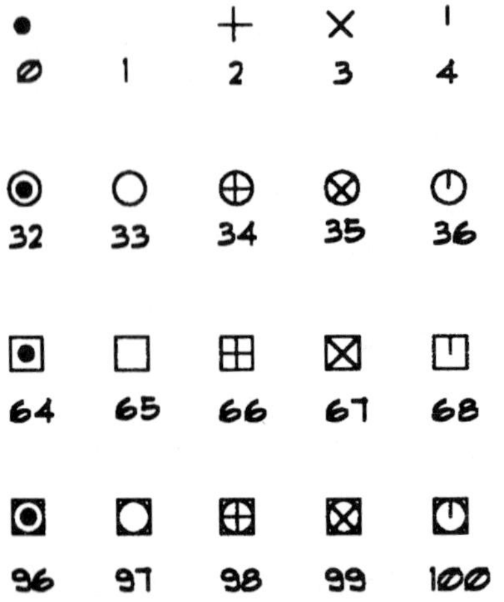

FIGURE 14–2 The shape of the marker can be entered by typing **DMODE** [enter] at the command prompt and then supplying the value of the desired marker.

If a positive size is entered for the variable, the number will define the size of the point. This can be seen in Figure 14–3. If a negative number is supplied, it is interpreted as a percentage of the viewport size. A negative number will adjust the PDSIZE size as you zoom in and out of the drawing.

DIVIDE

Although this command does not edit an object, DIVIDE will allow an entity to be divided into any number of segments of equal length for easy editing by placing markers along a line. This can be especially helpful when you are dividing space between two floors to locate where each stair tread and riser will be placed, or are locating stress points along a beam. DIVIDE will enable you to pick points on a Line, Arc, Circle, or Polyline.

Before using the DIVIDE command to mark an entity, you must decide how each mark will be indicated. The default value of <0> places an point controlled by the command PDMODE (see Figure 14–1).

Setting the value to 2 or 3 typically presents the most visible marker. Once the marker is set, the command sequence for DIVIDE can be started by picking Divide from the Point cascading menu in the Draw pull-down menu, or by entering **DIV** [enter], or **DIVIDE** [enter] at the command prompt. The command sequence is:

Command: **DIV** [enter]
Select object to divide: (*Pick the objects.*)

This is your opportunity to select a single entity to divide. If you pick an object that cannot be divided, an error message will be displayed and the command will be exited. Once an entity is selected, another prompt will be displayed.

<Number of segments>/Block: (*Enter number.*) [enter]

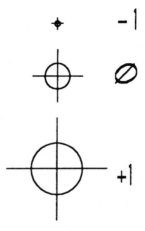

FIGURE 14–3 Typing **PDSIZE** [enter] at the command prompt allows the size of the point marker to be altered by providing a numeric value.

Entering a number between 2 and 32,767 and striking [enter] will cause the object to be divided into the desired number of segments. The entity is not physically divided into separate segments, but the markers are placed so that exact points can be selected. The process can be seen in Figure 14–4.

Another useful feature of the DIVIDE command is its ability to insert a group of entities called a block at a repetitive distance. Blocks will be covered further in Chapter 21, but the idea of inserting them is not difficult to remember. This can be done by dividing a line into the desired number of segments and then inserting the block at each marker. This can be done if the Block [enter] option is selected. This would produce the prompt:

<Number of segments>/Block: **B** [enter]
Block name to insert: (*Type name of desired block.*) [enter]
Align block with object? <Y> (*or N*) [enter]
Number of segments: (*Enter number.*) [enter]

Figure 14–5 shows the effect of aligning blocks with the object.

MEASURE

The MEASURE command is similar to the DIVIDE command. DIVIDE places a specified number of markers at equal distances along a segment. MEASURE places

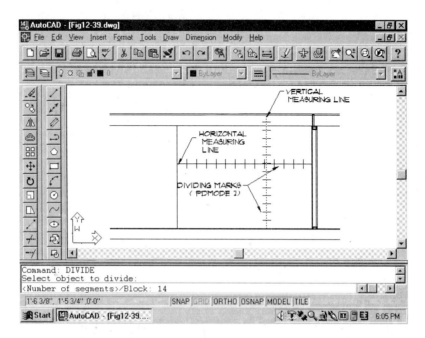

FIGURE 14–4 Dividing a horizontal line into fourteen segments.

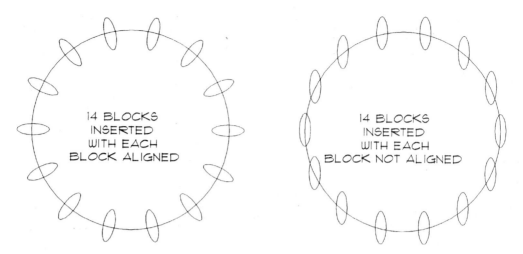

FIGURE 14–5 Predetermined drawing components can be inserted into a drawing at each specified DIVIDE marker using the BLOCK option of DIVIDE.

markers at specific distances along a Line, Arc, Circle, or Polyline. This command can be used whenever objects occur at a repetitive spacing, such as when you are placing studs at 16" o.c. along a plate or placing a manhole at every 527.25 feet along a sewer line. Figure 14–6 shows an example where MEASURE could be used as a drawing aid. The command can be started by picking Measure from the Point cascading menu in the Draw pull-down menu, or by entering **ME** [enter] or **MEASURE** [enter] at the command prompt. The command sequence to measure an object is:

Command: **ME** [enter]
Select object to measure: (*Select object.*)

This is your opportunity to select a single entity to measure. If you pick an object that cannot be measured, an error message will be displayed and the command will be exited. Once an entity is selected, another prompt will be displayed:

<Segment length>/Block: (*Enter distance.*) [enter]

Any specific distance may be entered through the keyboard or by specifying two points with the mouse. The markers will be located starting at the end of the line that was picked when the object was selected to be measured. This process can be seen in Figure 14–7.

To use a mouse to enter the distances, move the cursor to the desired location and press the pick button. This will produce the prompt:

Second point: (*Enter point or coordinates.*) [enter]

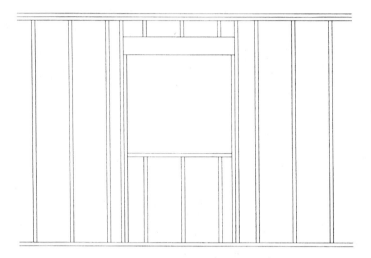

FIGURE 14–6 The MEASURE command can be used to place markers at a specific distance on a line, arc, circle, or polyline so that objects can be drawn.

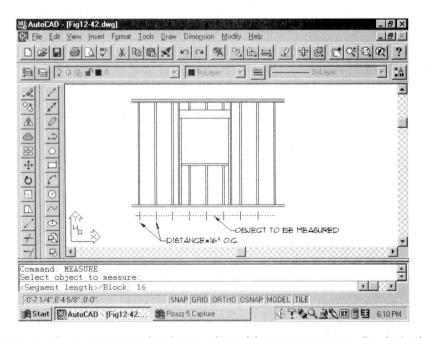

FIGURE 14–7 Once a segment has been selected for measurement, the desired distance to be measured can be entered by keyboard or by selecting the distance with the pointing device.

DRAWING LINES OF VARIED WIDTH AND LENGTHS

Each command introduced in this section can be used to provide an alternative to the LINE command. TRACE allows line segments to be drawn with a width similar to Plines, but they function individually. The SKETCH command allows free hand sketching to be inserted into a drawing. The RAY and XLINE commands allow semi-infinite and infinite lines to be added to a drawing.

TRACE

The TRACE command allows line segments to be drawn with a specified width. Line segments drawn using TRACE will appear as a polyline, but will be edited as a line segment. The command can only be started by typing **TRACE** [enter] at the command prompt.

Once the command sequence is entered, you will be prompted for a line width. The width can be provided by entering a distance (such as .125) or by picking two points on the screen and allowing AutoCAD to measure between them. The width can't be varied within a segment from starting to endpoint. Once the width has been determined, the command is similar to the LINE command. The command sequence is:

Command: **TRACE** [enter]
Trace width (current): **.125** [enter]
From point: (*Pick a point or enter coordinates.*)
To point (*Pick a point or enter coordinates.*)
To point: (*Pick a point or enter coordinates.*)
To point: (*Pick a point or enter coordinates.*) [enter]

Enter a new drawing, and follow the command sequence listed above. Notice that as the "To point" for the first segment was entered, nothing happened. The first line segment will not be displayed until the "To point" for the second segment is provided. The display sequence is always one sequence behind what you are entering to allow the corners to be mitered neatly. To end the sequence, press [enter] and the last segment will be completed. The process can be seen in Figure 14–8.

Line segments for POLYLINE and TRACE may be displayed as a solid black line or as pairs of lines by using the FILL option. Figure 14–9 shows the difference between a Trace segment displayed with Fill on and off. Fill can be adjusted by typing **FILL** [enter] at the command prompt and typing either **ON** or **OFF**. On large drawings, solid lines typically are drawn in the OFF position for faster REGEN speeds, and then toggled to ON for plotting.

SKETCH

Using the SKETCH command allows the equivalent of freehand drawings to be generated using a computer. Line segments are entered as the mouse is moved, rather than providing a "To point." The command is useful for entering signatures or other

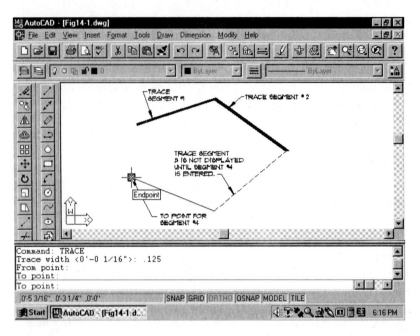

FIGURE 14–8 TRACE allows lines to be drawn that resemble polylines, but are edited as lines. As the lines are drawn, segments are not visible until the To point: for the next segment is specified.

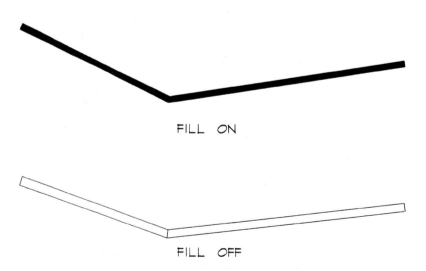

FIGURE 14–9 FILL allows PLINE and TRACE segments to be displaced as solid or hollow segments.

irregular material on a drawing. Because each tiny segment of the sketch is recorded as a line segment, drawing size can become extremely large in a very short time. SKETCH should be kept to a minimum to avoid filling up drawing space on a disk. The size of the drawing is limited only by your storage media.

The SKETCH command is accessed by typing **SKETCH** [enter] at the command prompt. Best results usually occur when ORTHO and SNAP are toggled to the Off positions allowing greatest flexibility in line placement. With ORTHO mode on, the SKETCH command will draw only horizontal and vertical lines. With the SNAP mode on, the record increment will equal the snap value.

Segment Length. As the command is entered, you will be prompted to provide information on the "record increment," which is the segment length to be used while you are sketching. The increment should be set small enough to provide the needed resolution; exceeding the resolution will only result in wasted disk space by storing lines that aren't necessary. The default line length is <0'-0 ⅛">. A new value can be entered by keyboard.

The increment that is chosen should be determined by the size of the reproduced drawing. If the drawing is to be reproduced at full scale on an 8½" × 11" sheet of paper, ⅛" will not produce high-quality resolution. As the mouse is moved, a line segment will be recorded for each movement greater than .1 of an inch. In a 10" wide space, 100 segments could be created. By changing the length to .5, a line could be generated with 20 segments in the same space, which would greatly decrease the resolution. However, if the drawing is going to be reproduced at a scale of ¼"=1'-0", a line segment of .5 would give extremely high resolution. A 1" line segment might be more appropriate. Figure 14–10 compares four record increments. The width can be entered by providing a value with the keyboard or by moving the mouse. To enter a value using the keyboard, enter the value followed by pressing the [enter] key. The command sequence is:

Command: **SKETCH** [enter]
Record increment <0'-0 1/8">: **.065** [enter]
Sketch. pen eXit Quit Record Erase Connect

AutoCAD is now ready to sketch. To set the line increment using the mouse, press the pick button when the prompt for the increment is provided. The sequence is:

Command: **SKETCH** [enter]
Record increment <0'-0 1/8">: *(Press the pick button.)*
Second point: *(Move the mouse to the disired location and press the pick button.)*
Sketch. pen eXit Quit Record Erase Connect

For a drawing requiring much accuracy, enter the line increment value using the keyboard. Once the width is entered, several options will be listed in the following prompt:

Sketch. Pen eXit Quit Record Erase Connect.

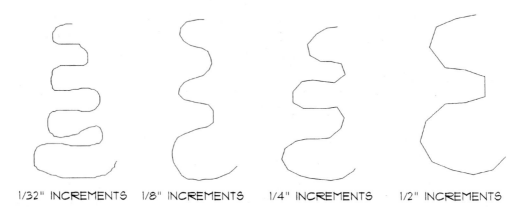

1/32" INCREMENTS   1/8" INCREMENTS   1/4" INCREMENTS   1/2" INCREMENTS

FIGURE 14–10. The smaller the value for the segment length, the smoother the resulting curves. A scale should be selected that provides the desired accuracy without requiring unneeded lines.

Once the line length is set, move the cursor into the drawing area with the mouse. You'll notice that nothing happens as the mouse is moved throughout the drawing area.

Pen (P). Think of the SKETCH command as sketching on paper with a pencil. The SKETCH command treats the cursor as a "pen" and allows for lifting the pen and moving it from one location to another without drawing a line. As SKETCH is started, the drawing pen is in the "up" position, allowing for selection of a starting location. Once the desired location is selected, the pen can be put down on the paper to allow for sketching. The pen is raised and lowered by pressing the mouse pick button or the P key on the keyboard. If you use the keyboard to control the pen, don't press [enter] after pressing the P key. For most users, skip the keyboard once you've entered the command and use the mouse to control the pen.

Each time the pick button is pressed, the pen is raised or lowered to allow for movement and drawing. The process can be seen in Figure 14–11.

"." Line to Point. The . (period) option can be used to draw a straight line from the endpoint of the last sketched line to the current location of the cursor. Sketch a few segments and raise the pen. Move the cursor to a new location and press the period key. This will produce a straight line from the end of the sketch segment to the new cursor location. The effect can be seen in Figure 14–12.

Record (R). As sketches are created, they are not actually part of the drawing until they are recorded. Enter a new drawing and sketch a few lines. You'll notice that the lines are green (color will vary depending on configuration of AutoCAD) even though you've done nothing to change the drawing color. The R option record all of the temporary sketch lines as a part of the drawing base and converts them to the present drawing layer and color. Once recorded, sketch lines cannot be edited

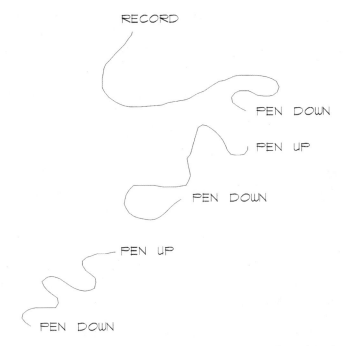

FIGURE 14–11 Controlling the pen position of UP/DOWN allows breaks to be created between segments.

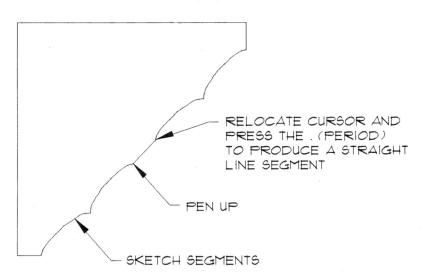

FIGURE 14–12 The . option of the SKETCH command allows a straight line to be drawn between the end of the last SKETCH segment and the present cursor location.

using the sketch subcommands. Sketched lines may be edited using any of the edit commands. "R" does not affect the position of the pen, and you can continue sketching after you've used it. Using [enter] will record the sketch segments and exit to the command prompt.

Record and Exit (X).   Using the X key will record the sketch segments to the drawing base, return the command prompt to the screen, and reestablish the command prompt. Using this option will display the number of lines recorded just above the command line.

Quit (Q).   Pressing the Q from the keyboard will function similarly to the ERASE command. Q will remove all temporary sketch segments entered since the last Record and return you to the command prompt. This option can be used if you sketch an object and then decide that you don't need it.

Erase (E).   This option allows for part of the sketch segments to be removed prior to recording. The option may be started by the E key and will remove a portion of the sketch from the selected point to the end of the sketch segments. Once the option is started a message is displayed:

    Erase: Select end of delete. < Pen up >

The pen will be placed in the up position to allow you to select the end of a segment. Move the cross-hairs to the portion of the line to be removed. The line will be displayed with the indicated portion of the line removed. If the removed section is acceptable, press the P key to erase the segment. If the selection is unacceptable, press the E key. The message "Erase aborted" will be displayed and the SKETCH prompt will be returned.

Connect (C).   Pressing the C key is similar to the END option of OSNAP when drawing lines, and will allow sketching to continue from the end of the last drawn sketch segment. When the option is selected, a prompt will be displayed:

    Connect: Move to endpoint of line.

and the cross-hairs will be displayed. When the cross-hairs are within one record increment of the end point, the pen will be placed in the down position at the end of the line. If the pen is already down when the Connect option is used, the prompt:

    Connect option meaningless when pen down.

will be displayed. To terminate the Connect command, type **C** a second time.

RAY

If you've done manual drafting, you've used light lines to plan your drawing prior to placing dark lines in graphite or ink. Those light lines are called construction or projection lines, and are only meant to be seen by the drafter. AutoCAD allows a similar line to be drawn using the RAY command. A ray is a semi-infinite line with a defined starting point that extends into infinity without changing the drawing area. Sounds technical, but they are really very helpful.

Rays can be especially helpful as construction lines in projecting features from one drawing to another. This often occurs when lines are projected from a floor, roof, or grading plan to develop exterior elevations. If you were to use a line to project from a floor plan to develop an elevation, it must have two endpoints. A zoom may be required to go from the "From point" to the "To point:" A transparent zoom ('Z) could be used, but this is time consuming when compared to a RAY. A ray will provide a guideline that will extend from the desired point indefinitely, no matter how many ZOOM P's are required.

The RAY command can be found in the Draw pull-down menu, or executed by typing **RAY** [enter] at the command prompt. The command sequence to draw a ray is:

Command: **RAY** [enter]
RAY from point: *(Select desired starting point of ray.)*
Through point: *(Select desired direction of ray.)*
Through point: [enter]
Command:

Figure 14–13 shows the layout of an elevation using the RAY command. A ray will show when the drawing is plotted. Rays should be kept on a separate layer from the

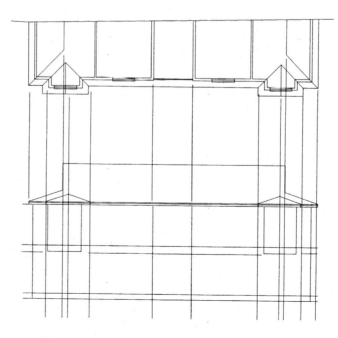

FIGURE 14–13 Rays and xlines make excellent projection lines. Because of their length, ZOOMs and STRETCHs can be minimized.

drawing being created, so they can be easily removed from the drawing base by freezing. They can also be erased easily using the Crossing Window or Fence option of ERASE.

XLINE(CONSTRUCTION LINE)

A line created using the XLINE command is similar to a ray except that it is infinite in length. With RAY, you picked a starting point and the line extended from that point. With XLINE, the line extends through the point an infinite distance in two directions and offers more options than with a ray. The command can be found by selecting Construction Line from the Draw pull-down menu, by selecting the Construction Line icon from the Draw toolbar, or by typing **XL** [enter] or **XLINE** [enter] at the command prompt. The command sequence is:

     Command: **XL** [enter]
     Hor/Ver/Ang/Bisect/Offset/<From point>: *(Select desired option.)*

Unlike RAY, six options are available to place the line. The default setting is to pick a From point:.

From Point.   By selecting a From point:, the line will be displayed through the point. Once the point is selected, the line will be rotated to the desired angle through the selected point. Once a line location is selected, the command sequence will allow for additional lines to be drawn through the From point:. The command sequence is:

     Command: **XL** [enter]
     Hor/Ver/Ang/Bisect/Offset/<From point>: *(Select desired point for line to pass through.)*
     Through point:
     Through point: [enter]
     Command:

Hor Option.   Choosing the Hor option will allow a horizontal xline to be drawn through a specified point. The line will remain horizontal no matter the setting of Ortho. The command sequence is:

     Command: **XL** [enter]
     Hor/Ver/Ang/Bisect/Offset/<From point>: **H** [enter]
     Through point: *(Select desired point for line to pass through.)*
     Through point: [enter]
     Command:

Ver Option.   The Ver option will pass a vertical line through the selected point. The command functions the same as the horizontal option once the Ver option is selected. The command sequence is:

Command: **XL** ⏎
Hor/Ver/Ang/Bisect/Offset/<From point>: **V** ⏎
Through point: *(Select desired point for line to pass through.)*
Through point: ⏎
Command:

Ang Option. The Ang option allows a specific angle for the xline to be specified. The selected angle is relative to horizontal. The angle option can be useful in laying out roofs or other inclined surfaces. The command sequence is:

Command: **XL** ⏎
Hor/Ver/Ang/Bisect/Offset/<From point>: **A** ⏎
Reference/Enter Angle>: **30** ⏎
Through point: *(Select desired point for line to pass through.)*
Through point: ⏎
Command:

If the Reference option of the command is selected, a line can be drawn at an angle to any line. This option could be used to draw a line 30° from a roof drawn at a 6/12 pitch if you can't remember what angle 6/12 represents and you don't feel like asking AutoCAD. Once the reference angle is selected, a location will need to pass the line through. The command sequence is:

Command: **XL** ⏎
Hor/Ver/Ang/Bisect/Offset/<From point>:**A** ⏎
Reference/Enter Angle>: **R** ⏎
Select a line object:*(Select line to serve as a reference.)*
Enter Angle>: **30** ⏎
Through point: *(Select desired point for line to pass through.)*
Through point: ⏎
Command:

Bisect Option. The Bisect option can be used to draw a construction line that bisects an existing angle. The xline is created by selecting three points. The first point to be selected is the Angle vertex point and the second and third are points located on the lines forming the angle to be bisected. The command sequence to bisect an angle is:

Command: **XL** ⏎
Hor/Ver/Ang/Bisect/Offset/<From point>: **B** ⏎
Angle vertex point: *(Select the point.)* **INT** ⏎
Angle start point: *(Select the end of one of the lines.)*
Angle end point: *(Select the end of the other.)*
Angle end point: ⏎
Command:

Figure 14–14 shows an xline used to bisect an angle.

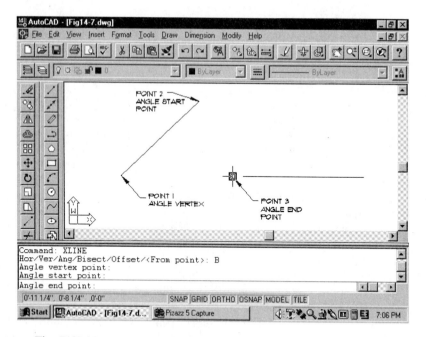

FIGURE 14-14 The BISECT option of XLINE can be used to divide an angle in half by selecting thee points.

Offset Option. The Offset option draws construction lines that are parallel to the selected line based on a specified offset distance or point. This option functions just as the OFFSET command. You'll be prompted for a line, a distance, and a side to offset. Select a through point, and you will be prompted for a line to be offset and for a through point. The command sequence to offset a line 12" is:

Command: **XL** ⏎
Hor/Ver/Ang/Bisect/Offset/<From point>: **O** ⏎
Offset distance or Through <Through>: **12** ⏎
Select a line object: *(Select the line to be offset.)*
Side to offset: *(Select the side of the offset.)*
Select a line object: ⏎
Command:

DRAWING WITH MULTIPLE LINES

In Chapter 7 you learned how to draw lines. In Chapter 12 you learned to use the OFFSET command. By combining these two commands and several editing commands, a floor plan was started. This system works, but it is rather cumbersome. The MLINE command, on the other hand, will allow you to draw up to 16 parallel lines at once. Each line is referred to as an *element* of the multiline. Drawing two to four lines will be a great aid when drawing construction-related drawings. Figure

14–15 shows an example of the use of MLINE. The command can be started by picking the Multiline icon from the Draw toolbar, by typing **ML** enter or **MLINE** enter at the command prompt or by selecting Multiline from the Draw pull-down menu. Each of these methods will display the following prompt:

Command: **ML** enter
Justification = Top, Scale = 1.00, Style = STANDARD
Justification/Scale/STyle/<From point>: *(Select a From point.)*

AutoCAD is waiting for you to select a From point: to start the command sequence using the same procedure used with the LINE command. Open a new drawing, enter the MLINE command and draw a few lines. Only two lines will be drawn as each segment is completed because the current STyle is standard (two lines). You'll notice that once a To point: is selected, that the next To point: prompt will change. The command prompts to select the four To point:'s of Figure 14–15 would be:

<To point>: *(Select a To point.)*
Undo/<To point>: *(Select another To point.)*
Close/Undo/<To point>: *(Select another To point.)*
Close/Undo/<To point>: *(Select C to close the shape.)* **C** enter

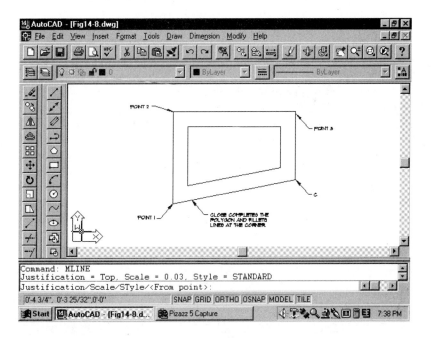

FIGURE 14–15 The MLINE command allows from two to sixteen lines to be drawn in a segment which functions as one line.

The Close option was used to close the object into a polygon. The Close and Undo options work with MLINE just as they do with LINE.

Setting the Justification

The term *justification* as it relates to MLINE describes where the lines will be placed relative to a base point as the From and To points are specified. The options include Top, Zero, Bottom with a default setting of Top. Figure 14–16 shows the effect of each point. With Top active, the top line of the multilines will be placed between the selected From and To points and other lines will be offset to go below the selected points. With Zero active, the selection points will serve as the midpoints for the multilines, with lines equally offset on each side of this theoretical centerpoint. Setting Bottom as the active selection will form the bottom line of the multilines between the selection points with other lines offset to the top.

The justification for multilines can only be selected once for each command sequence. The justification is altered by entering **J** [enter] at the prompt for MLINE. The sequence to change to Zero would be:

> Command: **ML** [enter]
> Justification = Top, Scale = 1.00, Style = STANDARD
> Justification/Scale/STyle/<From point>: **J** [enter]
> Top/Zero/Bottom <top>: **Z** [enter]
> Justification/Scale/STyle/<From point>:*(Select a From point.)*

Zero will now remain the default value until it is changed.

Setting the Scale

Scale controls the overall width of the multilines to be drawn. The default setting is 1.00. Keep in mind that 1.00 represents 1 unit, not 1 inch or 1 foot. With Top as the current justification value and 1 as the current scale value, a second line will be drawn 1 unit below the selection points. With the Zero option of justification current, one line will be ½ unit above and the other line will be ½ unit below for a total width of 1. Entering a value larger than one will enlarge the spacing between multilines.

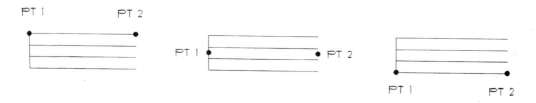

TOP JUSTIFICATION   ZERO JUSTIFICATION   BOTTOM JUSTIFICATION

FIGURE 14–16 The justification for MLINE determines where the line will be drawn relative to the selection point.

Entering a value of less the one will decrease the space between lines. Remember as you look at the lines in Figure 14–15, that the limits are fairly small and the setting of 1 creates what appears as wide lines. Changing the limits will allow a floor plan to be drawn, and the same lines will appear as single lines. The command sequence to adjust the scale to 6 so that 6" wide walls can be drawn would be:

Command: **ML** [enter]
Justification = Zero, Scale = 1.00, Style = STANDARD
Justification/Scale/STyle/<From point>: **S** [enter]
Set Mline scale <1.00>: **6** [enter]
Justification/Scale/STyle/<From point>: *(Select a From point.)*

The Multiline Style Dialog Box

The term *style* as it relates to MLINE refers to the number of lines to be drawn in the multiline pattern as well as how intersections and terminations will be displayed. The desired style must be created prior to using the MLINE command. The STyle option of MLINE only allows a created linestyle to be selected. Multiline styles can be created by selecting Multiline Style … from the Format pull-down menu, or by typing **MLSTYLE** [enter] at the command prompt. Each will produce the dialog box seen in Figure 14–17. Key components and their function of the Multiline Styles dialog box include:

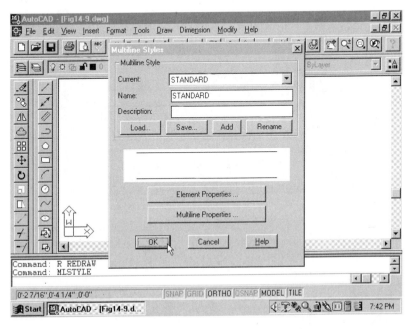

FIGURE 14–17 The Multiline Styles box can be used to create, save and load MLINE styles. Each style is comprised of the number, location, color and linetype of each element in a style. The box is accessed by typing **MLSTYLE** [enter] at the command prompt.

Current.   The Current edit box displays the name of the current multiline style. The current and only style at present is the default style of STANDARD. As other styles are created the arrow can be used to scan between style names and make other styles the current selection.

Name.   This box allows the name of the style being created to be assigned. The name is not specified until after the new style is created. The box can also be used to rename existing style names.

Description.   The description edit box allows a description of the style being created to be entered. Descriptions can contain up to 255 characters including spaces between words. As with other filenames and descriptions, keep them simple.

Load....   This button will display the Load Multiline Styles dialog box seen in Figure 14–18. Multiline styles will be saved as ACAD.MLN files within the current drawing. The File... button allows you to load a multiline style which has been saved in a separate .MLN file. Selecting the File... button displays a directory of available drives for selecting stored linetype styles.

Save....   This button allows styles to be saved or copied to a new location. Typing a name in the Name: edit box and selecting this button displays a directory of available drives for a destination for saving the new file as an .MLN file.

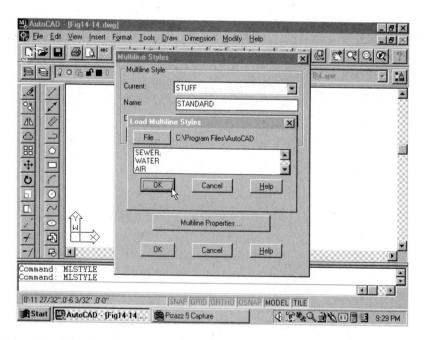

FIGURE 14–18   Selecting the Load... button in the Multiline Styles edit box allows different segment styles to be loaded into a drawing file.

Add.   This button allows a multiline style name to be added to the current multiline file.

Rename.   The Rename button allows the current style to be renamed. The Standard file can't be renamed.

Line Display Panel.   This panel displays the current linestyle.

Element Properties...   Selecting this dialog box will display the Element Properties dialog box shown in Figure 14–19. This box allows lines of various linetypes and colors to be used within a style.

Multiline Properties...   Selecting this button displays the Multiline Properties dialog box shown in Figure 14–20. This box allows properties such as joints, end caps and background fill to be set within a style. This box will be further discussed in the next section of this chapter.

Adjusting The Element Properties

Pressing the Element Properties... button allows the properties of a Multiline style to be adjusted as it is being created. Elements of the box include:

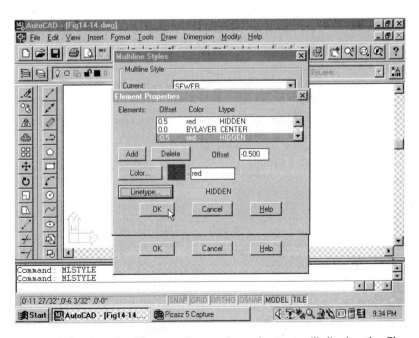

FIGURE 14–19   Selecting the Element Properties... button will display the Element Properties dialog box which allows the settings for a segment to be selected.

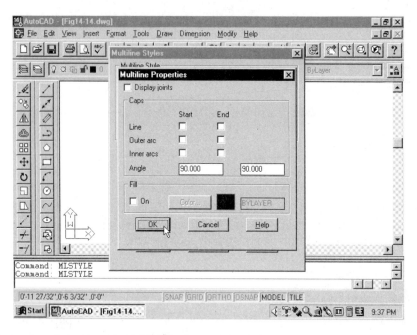

FIGURE 14–20 Selecting the Multiline Properties... button in the Multiline Styles edit box allows the properties for joints, end caps and fill to be selected for a style.

Elements. This box displays the offset size, color and linetype settings of each multiline style. Lines are listed in descending order of the offset distance.

Add. This button allows new lines to be added to the Multiline style.

Elements List box. Selecting Add will insert a new line with an offset distance of 0.0 to the Element Properties box. Once added, the properties of the current line can be altered.

Delete. This button allows the highlighted line style to be deleted from the Elements box.

Offset. this button allows the offset distance of the highlighted line to be altered. Offset distances are based on the location of the origin 0,0 and can be either positive or negative. To alter an offset value, select the desired line. The current offset will now be displayed in the Offset box. Use the backspace arrow to remove the existing value and type the desired value. The new value is added to the style by pressing the (enter) key.

Color... This button allows color to be assigned to a specified line. As the box is selected, a color palette will be displayed, allowing for color selection. Select the red tile in the upper left-hand corner and then select OK. The Element Properties box will be restored, and a red line will be displayed to the right of the Color... button.

Linetype...   This button allows a linetype to be assigned to the selected line of the multiline. This allows lines within the multiline to have varied line patterns. The default of BYLAYER, means the new line will be placed on the current layer (0) using a continuous linetype. Below the listing of the current linetype is a listing of each linetype that has been added to the drawing base (None). To draw with different linetypes you must load the lines to be drawn. Press the Load... button and the Load or Reload Linetype box will be displayed. Highlight the word CENTER by picking it with the mouse and then scroll down and select HIDDEN. Pick the OK box and the Linetype box will be redisplayed. The lines will be reflected in the Linetypes box.

Adjusting Multiline Properties

The Multiline Properties dialog box can be used to control how multilines will be displayed. Key components of this box include:

Display Joints.   In the current setting, as multilines are drawn, the joints are off. By selecting this box a check will be placed in the box and a line will be drawn between each line at each multiline intersection. Figure 14–21 shows the difference in line styles with joints off and on.

Caps.   This option has four selections for each end of each option. The standard setting is to have no caps. In the default all boxes are open. Highlighting a box will place a check in the selected box indicating that the box is on. The effects of each option can be seen in Figure 14–22.

Line.   With this option selected, a line will be drawn as a cap at the start or end of the multilines.

Outer Arc.   This option will draw a semicircle between the endpoints of the outer lines.

Inner Arc.   This option will draw a semicircle between the endpoints of the inner lines. If there are an odd number of inner lines, the centerline will not be capped.

Angle.   This option allows the angle of the cap line to be adjusted. Options include angles between 10 and 170 degrees.

JOINTS OFF   JOINTS ON

FIGURE 14–21   MLINE segments can be displayed so that their intersections are off or on.

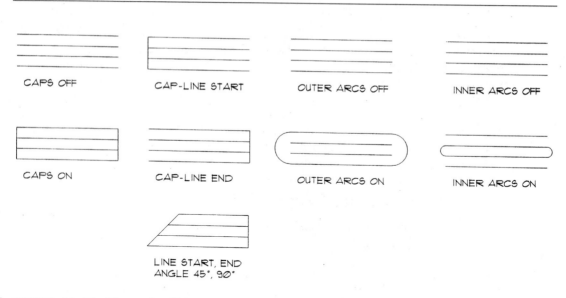

FIGURE 14–22 The ends of MLINE segments can be capped or uncapped. Caps can consist of a perpendicular or inclined line or with arcs from the outer or inner lines.

Fill. This option toggles Fill on and off. In the default setting of OFF, multilines will be drawn as two or more parallel lines. With Fill set to ON, multilines will resemble a polyline or trace line. Once toggled ON, the color button will be activated and a color can be selected for the fill. The effect of fill on a line can be seen in Figure 14–23.

Creating Multiline Styles

Now that the dialog box MLSTYLE has been explored, linestyles can be created that could be used on construction drawings. The following section will describe how to create an mline to represent a utility easement on a site plan. The exterior linetypes will be HIDDEN, the interior line will be a CENTERLINE. Each outer line will be red and the centerline will be black.

Open the Multiline Styles dialog box using one of the four methods presented earlier in the chapter and select the Element Properties... box. Select the desired offset val-

FIGURE 14–23 MLINE segments can be drawn as open or filled segments.

ues, color and linetypes for each element of the mline. Use offsets of .5, 0 and –.5. When the properties are acceptable, select the OK button.

Move the cursor to the Name: box and enter the desired name for the new style. For this example the name EASEMENT was used. Once typed, move the cursor to the Description: box and describe the line. A description of HIDDEN, CENTER, HIDDEN was used in the example. Now move the cursor to the Save... button and save the style using the Save Multiline Style box just as you would a drawing file. Pick the Load... button to display the Load Multiline Styles dialog box and select the new style and pick OK. Pick OK to exit the Multiline Styles dialog box. A style named easement has now been created and loaded into the drawing file. Now the style can be used with the MLINE command.

To draw multilines using the EASEMENT style, enter the MLINE command using one of the methods described earlier. The command sequence to draw lines using WALL would be:

Command: **ML** ⏎
Justification = Top, Scale = 1.00, Style = STANDARD
Justification/Scale/STyle/<From point>: **ST** ⏎
Mstyle name(or?): **EASEMENT** ⏎
Justification/Scale/STyle/<From point>: *(Select a From point.)*

If you have several linestyles and can't remember the names of them, choosing the **?** option will provide a display showing the name and description.

EDITING MULTILINES WITH GRIPS

Lines created using MLINE can't be edited using BREAK, CHAMFER, EXTEND, FILLET, or TRIM commands but are affected by the COPY, EXPLODE, MIRROR, MOVE and STRETCH commands. Multilines can also be edited using GRIPS. Placing the cursor over a multiline object will produce a grip at each endpoint of each segment. The grips are located based on the justification used when the lines were drawn. Examples of grip locations can be seen in Figure 14–24.

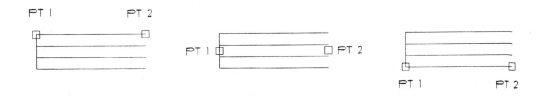

TOP JUSTIFICATION   ZERO JUSTIFICATION   BOTTOM JUSTIFICATION

FIGURE 14–24 Grips can be used to edit MLINE segments. The location of the grips will depend on which line justification has been selected.

EDITING MULTILINES USING MLEDIT

The MLEDIT command allows multilines to be edited by providing a dialog box which displays an icon of the edit to be preformed. The command can be accessed by selecting Edit Multiline... from the Modify pull-down menu, by picking the Edit Multiline icon from the Modify II toolbar or by typing **MLEDIT** [enter] at the command prompt. Each method will display the Multiline Edit Tools dialog box shown in Figure 14–25. This box allows editing of MLINE intersections, tees, corner joints and lines.

Editing Mline Crossing Intersections

Three types of intersections using MLEDIT include closed cross, open cross, and merged cross. Each are found in the left column of the Edit Tools dialog box.

Closed Cross. This option allows one group of multilines to remain as is, while the other pair is interrupted. Care must be taken in the order that objects are selected to be edited because the order determines which pair of lines will be edited. The first line selected will be trimmed but continue to function as one set of multilines. The second pair of lines selected will remain continuous. The command sequence to provide a closed cross is:

Command: **MLEDIT** [enter] *(Select the closed cross icon and OK.)*
Select first mline: *(Select the line to be trimmed.)*

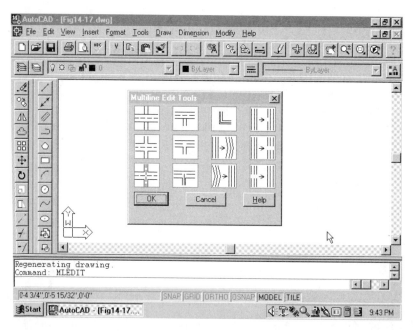

FIGURE 14-25 Type **MLEDIT** at the command prompt to display the Multiline Edit Tools box, which provides 12 options for editing.

Select second mline: *(Select the line to be continuous.)*
Select first mline (or Undo): [enter]

The command prompt will continue to be displayed allowing other intersections to be edited using the closed cross option. Selecting [enter] returns the command prompt. Selecting the U option allows the cross to be undone but restores the closed crossing prompt. Picking [enter] twice restores the Multiline Edit Tools box and allows a new option to be selected. The effect of selecting the crossing opting can be seen in Figure 14–26.

Open Cross.   This option will trim all lines of the multiline selected first, and only the outer lines of the second group. The command sequence is similar to the closed cross. The results can be seen in Figure 14–27. When only two lines are used, each pair of lines will be opened.

Merged Cross.   This option will trim the outer lines of both pairs of multilines and leave the inner lines to cross. The command sequence is similar to the closed cross. The results can be seen in Figure 14–28. When only two lines are used each pair of lines will be opened.

Editing Mline Tees

Three types of tees can be completed by using the icons in the second column from the left of the Edit Tools box. These include a closed, open and merged tee. Each com-

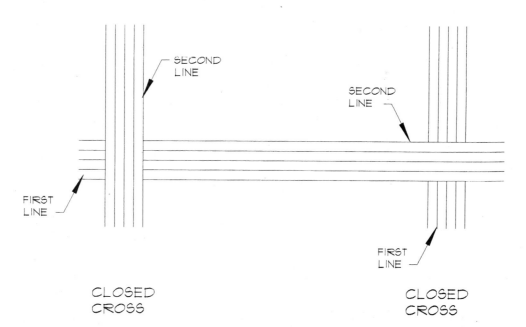

FIGURE 14–26 The results of a CLOSED CROSS edit depends on which line is selected first. The segment selected second will appear unaffected.

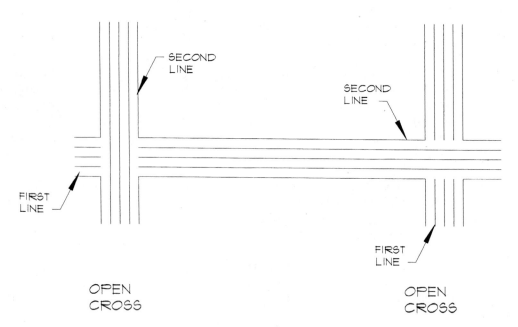

OPEN
CROSS

OPEN
CROSS

FIGURE 14–27 Mlines edited using OPEN CROSS will vary depending on the order of element selection.

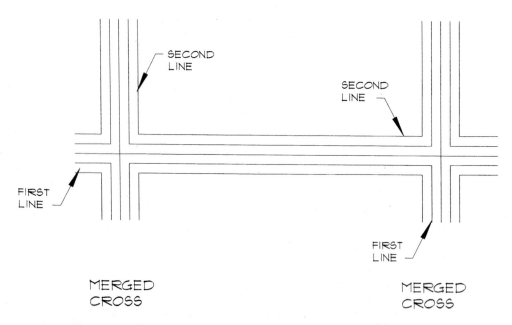

MERGED
CROSS

MERGED
CROSS

FIGURE 14–28 The MERGED CROSS option allows outer and inner lines to appear as if they had been filleted.

mand sequence is the same as its crossing counterpart. Examples of each tee can be seen in Figure 14–29.

Editing Mline Corners

Three types of corners can be formed using the icons in the second column from the right of the Edit Tools box. These include the top option of corner joint, the middle option of add vertex, and the delete option on the bottom.

Adding a Corner.   The Corner Joint forms a corner between two groups of multilines by trimming or extending each pair as required. An example can be seen in Figure 14–30.

Adding a Vertex.   The middle option can be used to added a vertex to a multiline. The vertex is inserted at the position where the multiline is picked but no apparent change is made. To see the effect of the change, move the cursor so that the grips of the desired line are displayed. Now it can be seen that a new vertex has been added along with a grip. The new grip can now be used to alter the shape of the multiline. The effect can be seen in Figure 14–31.

Delete a Vertex.   A similar process can be used to remove a vertex from the selected multiline. The vertex nearest to the selection point will be removed. The effects can be seen in Figure 14–32.

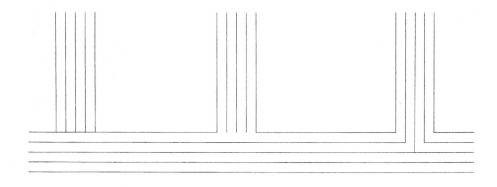

CLOSED   OPEN   MERGED
TEE   TEE   TEE

FIGURE 14–29   The CLOSED, OPEN and MERGED tee options.

MLEDIT CORNER

ORIGINAL INTERSECTION

FIGURE 14–30 The effects of CORNER on mlines.

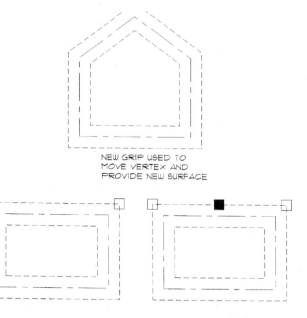

NEW GRIP USED TO
MOVE VERTEX AND
PROVIDE NEW SURFACE

ORIGINAL POLYGON

VERTEX ADDED
CENTER GRIP ACTIVE

FIGURE 14–31 A vertex can be added allowing the shape of mlines to be altered.

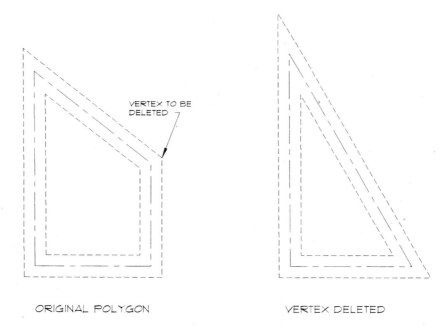

ORIGINAL POLYGON   VERTEX DELETED

FIGURE 14–32 The DELETE VERTEX can be used to remove a vertex and alter the shape of MLINE segments.

Editing Mline Lines

Each of the three types of edit groups for cutting and welding multilines are represented in the right column of the Edit Tools box. Examples of each can be seen in Figure 14–33.

Cut Single. This option preforms a function similar to BREAK. It can be used to remove a portion from one of the lines in an multiline group. A portion of the selected line will be removed but the balance of the elements remain unaffected.

Cut All. This option can be used to remove a portion from all elements of an mline group. A portion of the selected line will be removed between selected groups. Even though a portion of the line is removed, the two portions function as one mline.

Weld All. This option joins segments of an mline that have been cut. It only restores two segments of one mline and can't be used to join two separate mline groups.

Editing Mlines with EXPLODE

Each of the twelve options in the Edit Tool box allows Mlines to be edited but retain their properties. The EXPLODE command which was introduced in Chapter 13 can be used to turn MLINE groups into separate elements. When an Mline is exploded, all lines within the group will now function as if they had each been drawn using the

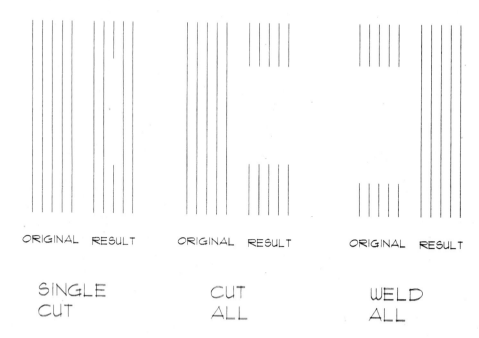

FIGURE 14–33 MLINE segments can be cut or joined using the line edit functions. The SINGLE CUT option can be used to remove a portion of one element of a segment. The option can be repeated so that more than one element of a segment can be altered. The CUT ALL option will remove a uniform length of all elements in a segment. The WELD ALL option can be used to restore lines in an MLINE segment.

line command. Color and linetype will remain unaffected but each of the standard edit commands can be used because each element now functions as an individual entity.

Editing Mlines from the Command Prompt

Although Mlines are easiest to edit using the Tool Box if you're a diehard typist, each of the edit commands can be entered at the command prompt. The command sequence is started by typing a minus sign followed by **MLEDIT**. Failure to enter the "-" will display the Edit tool box. The command sequence is:

Command: **-MLEDIT** ⏎
Mline editing option AV/DV/CC/OC/MC/CT/OT/MT/CJ/CS/CA/WA:

These letters represent each of the two options presented in the tool box and include:

CC- Closed Cross	CT- Closed Tee	CJ- Corner Joint	CS- Cut Single
OC- Open Cross	OT- Open Tee	AV- Add Vertex	CA- Cut All
MC- Merged Cross	MT- Merged Tee	DV- Delete Vertex	WA- Weld All

When a prompt is selected, you will be prompted to enter either select an mline or a point on an mline depending on the option.

THE U COMMAND

You've already used the Undo option to remove a drawing entity. The U option has been used in the middle of a command routine to remove an entity that may not be what you intended. The U command also can be used to backtrack through a command sequence once the sequence has been ended. Use the command carefully. Once used, objects are gone and will have to be redrawn if you want them back. Enter a new drawing, draw a series of several lines, and press [enter]. Draw a circle, a polygon, and an ellipse in the order listed. Now if you type **U** [enter] at the command prompt you will see:

    Command: **U** [enter]
    ELLIPSE

The last item drawn, the ellipse, will be removed from the screen. Notice that the most recent command (ellipse) will be listed and the command prompt will reappear. If the process is repeated,

    Command: **U** [enter]
    POLYGON

the polygon will be removed. By continuing to enter **U**, the entire drawing can be removed.

    Command: **U** [enter]
    CIRCLE
    Command: **U** [enter]
    LINE

UNDO

The UNDO command allows for several command sequences to be undone at one time and permits several operations to be carried out as the objects are undone. The UNDO command should give you the freedom to know you can try anything, and if the outcome is undesirable, enter UNDO at the command prompt to restore the drawing to its original state. The command sequence is:

    Command: **UNDO** [enter]
    Auto/Control/BEgin/Mark/Back/<Number>:

If a number is entered, such as **1** [enter], the last command will be undone. If you enter **3** [enter], the last three command sequences will be removed.

Auto

The Auto subcommand issues a prompt to toggle between on and off. When Auto is on, the default value, any group of commands that is used to insert an item or group of items, is treated as one item and removed by U. This will be helpful later when working with blocks, wblocks, and macros. If Auto is off, each command in a group of commands is treated as an individual one.

Control

The Control option is another method to limit the UNDO command; it also can be used to completely disable UNDO. The command sequence is:

Command: **UNDO** ⏎
Auto/Control/BEgin/End/Mark/<Number>: **C** ⏎
All/None/One<ALL>:

If All is selected, UNDO will not be limited, and it will erase back to the last MARK. The NONE will disable U and UNDO commands so that your machine will function as if these commands have never been placed in AutoCAD. Once disabled, U and UNDO can be reactivated by entering UNDO and typing **A** ⏎. Selecting the One option limits U and UNDO to a single operation before returning to the command line.

Begin/End

The Begin option of UNDO groups a sequence of drawing operations. The group of drawing command is defined by the use of the Begin option of UNDO and the end of the group is marked by the End option. With a specified group, AutoCAD will treat grouped commands as a single operation.

Back

The BACK option of UNDO will remove *everything* in the *entire* drawing. Not some, not most, we're talking *everything*. Do you get the feeling you need to use this option carefully? AutoCAD will even help you ponder the use of this option by displaying the prompt:

This will undo everything. OK? <Y>

If ⏎ is pressed, this will accept the default yes value, and really clean up your drawing. Entering **N** will ignore the Back option.

The BACK option also can be used with the MARK option so that the entire drawing is not erased.

Mark

This option will limit the distance that BACK will search through a drawing as it erases work. A MARK can be placed in a drawing prior to executing several commands. If the outcome of those commands is not what you expected, UNDO BACK will remove all of the commands until the MARK is reached. The command sequence is:

Command: **UNDO** `enter`
Auto/Control/BEgin/Mark/Back/<Number>: **M** `enter`
Command: **LINE** `enter` (*Now draw several features in an existing drawing.*)
Command: **UNDO** `enter`
Auto/Control/BEgin/Mark/Back/<Number>: **B** `enter`.
This will undo everything. OK? <Y> `enter`
Command:

Number

The Number option of UNDO will undo the specified number of drawing operations. The effect is the same as if the U command had be used the same number of times, but only a single regeneration is required.

REDO

Occasionally you will remove an entity with U or UNDO by mistake. The REDO command can be used to restore items that have been deleted. REDO will only restore entities, however, when used immediately after an UNDO or U. If U or UNDO is followed by a command such as LINE, and then REDO, REDO will have no effect on the material that was deleted prior to the UNDO. The command can be started by picking the Redo icon from the Standard toolbar or by entering **REDO** `enter` at the command prompt.

CHAPTER 14 EXERCISES

E-14-1. Start a new drawing and set the units and limits to be appropriate for this problem. Using the sketch as a guide, use the trace command to draw the outline of a 15-foot × 28-foot apartment unit. Make the 15-foot walls 6" thick, with the long walls 8" wide. Use 4" wide lines for interior walls and lay out a 12-foot × 15-foot bedroom, and a 5'-6" × 9-foot bathroom. No windows or doors are required at this time. Save the drawing as E-14-1.

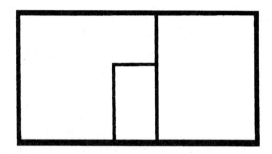

E-14-2. Use the guidelines presented in Exercise 14-1 and draw the apartment using MLINE. Save the drawing as E-14-2.

E-14-3. Open a new drawing and create mline segments with three, four and five elements. Use each of the cap options to terminate each segment. Save the drawing as E-14-3.

E-14-4. Use the MLSTYLE command and create an easement similar to the style that was created in this chapter. Save the drawings as E-14-4.

E-14-5. Draw a rectangle using the MLINE command and make two additional copies. Use one copy to add a vertex and change the object into a pentagon. Use the other copy to delete a vertex and change the object into a triangle. Save the drawing as E-14-5.

E-14-6. Draw at least three different groups of mlines using three different styles of lines with at least three lines in each style. Make enough copies of these lines and use each of the intersection, tee and corner options to edit each pair of lines. Save the drawing as E-14-6.

E-14-7. Enter a new drawing and draw a line 27'-6½" long. Divide the line into five equal segments. Copy the original line and place markers at 16" o.c. Save the drawing as E-14-7.

E-14-8. A stair is to be drawn between two floors of a townhouse with a distance of 9'-1½" from finish floor to finish floor. Plan a set of stairs with 7½" maximum rise and 10½" minimum run per step. Determine the required number of stairs and the total run. Draw horizontal lines to represent each floor. The upper floor level will be 13½" thick. Divide the space between floors into the required number of equal risers. Divide the total required run into the required individual steps. Using any editing process presented in the last two chapters, edit the grid that has been laid out to show the shape of the stairs. Draw a diagonal line that passes through the front edge of each step. Offset this line down 12" to determine the required depth of the stairs. Save the drawings as E-12-19.

Required risers. _____ Total run. _____

E-14-9. Use Figure 14–6 as a guide and draw an elevation showing an 8' high, 2 × 6 stud wall. Use (2)-2 × 6 top plates and a single base plate. Show an opening for a 6'-wide × 4'-high window. Use a 4 × 8 header over the window, and set the bottom of the header at 6'-8" above the floor level. Place the studs at 16" o.c. Save the drawing as E-14-9A. Make a copy of the elevation and show the wall at 9' high with the studs located at 24" o.c. Set the bottom of the header at 7-10" above the floor. Save the elevation as E-14-9B.

E-14-10. Use the SKETCH command and create a logo that can be used in the title block of a templet drawing. Save the drawing as E-14-10.

E-14-11. Open a new drawing and draw three lines 10'-0" long. Set the point marker to show a marker with a circle with an X in the center. Measure the line into 7 equal segments. Divide the second line into 16" long segments using a square around the cross-hair marker. Show the third line with perpendicular lines at 19.2" o.c. Save the drawing as E-14-11.

E-14-12. Use the SKETCH command and draw four curved lines using .065", .125", .25", and .5" long line segments. In at least two of the lines, include a straight line segment between the sketched segments. Save the drawing as E-14-12.

E-14-13. Open a new drawing and draw a rectangle. Place a point using the cross-hair surrounded by a circle and a square in the center of the rectangle. Draw lines of infinite length that pass through the point in a horizontal, vertical and at 30 degrees to the horizontal and vertical planes. Trim the lines that extend beyond the rectangle. Save the drawing as E-14-13.

E-14-14. Open a new drawing and draw a circle with a semi-infinite line extending outward from each quadrant. Draw lines that extend from the centerpoint to the edge of the circle and are placed at 15 degree intervals for a 360 pattern. Save the drawing as E-14-14.

E-14-15. Draw two lines that intersect. One of the lines is to be horizontal and the second line is to be 67 degrees above the first line. Draw a line that bisects the intersection of the two lines. Save the drawing as E-14-15.

CHAPTER 14 QUIZ

Enter a new drawing and draw the following entities in the order listed. Draw 4 pairs of line segments, a three-sided polygon, a circle, a polyline with three segments with a width of .125, and an ellipse. Copy the circle, move the ellipse, and enlarge the polygon to be twice its size.

1. If the U command is used after the polygon is enlarged, what will be affected?

2. If UNDO 2 is used after the polygon was drawn, what will be affected?

3. If a mark is placed after the circle is drawn, what will the effect be if the UNDO Back command is used after the circle is copied?

4. A whole drawing sequence is an experiment that you may want to delete. List the command and subcommands that would allow the entire sequence to be removed as one entity. Explain when these subcommands are used.

5. If 5 is used with the UNDO Number option, what will the effect be?

6. In using the UNDO Number 5 option, one too many objects was removed. How can this be corrected?

7. List the keys for controling the SKETCH subcommands.

8. Explain the difference between a TRACE and a POLYLINE line segment.

_____

_____

_____

9. What is the default setting for a sketch line if the units are set to architectural units with fractions of ⅟₁₆"? _____

For engineering units with four-place decimals? _____

10. What is the danger of having the sketch resolution too fine?

_____

_____

11. If the record increment is set to 1.25", what is the effect on the sketch if the mouse is moved 1"?

_____

_____

12. Several lines have been drawn using SKETCH and you would like to add to the drawing in a new location. How do you move the pen to the new location without drawing?

_____

_____

13. You've sketched a beautiful drawing and wish to save it as part of the drawing base. How can you do this?

_____

_____

14. Explain how the correct option of SKETCH will affect a drawing.

_____

_____

15. What is a RAY and how could it be used?

_____

_____

16. What is the difference between a ray and an xline?

_____

_____

17. What is the effect of HOR on a ray if Ortho is off?

18. Describe the difference between the two angle options of XLINE.

19. List and sketch an example of each of the Cap options of MLINE. Use segments with three and four elements for each option.

20. Describe the effect of each of the justifications on XLINE placement.

21. A scale of 4 has been selected to draw mline segments. Will the segment be 4", 4' or 4 meters wide?

22. Two mline segments need to be joined together. What command and options should be used?

23. What will the Element Properties box adjust?

24. Explain the process for selecting a color for an mline.

_____

_____

_____

25. What editing commands will affect Mlines?

_____

_____

_____

26. What is the difference between 5 and –5 as they relate to mlines?

_____

_____

27. Sketch the difference between a closed cross and an open cross in segments with three lines.

28. Sketch the difference between a merged tee and an open tee in segments with four lines each.

29. What will the letters "WA" do to the MLINE command?

_____

_____

30. What effect will EXPLODE have on MLINE segments?

_____

31. Describe the effects of PDSIZE and PDMODE on a drawing?

_____

_____

32. How do the values for PDSIZE affect the marker?

_____

_____

33. Explain the difference between DIVIDE and MEASURE on a line.

34. How can a line be created using TRACE and not be solid?

35. What commands and options are needed to complete E-14-15?

CHAPTER

15

LINES, LAYERS, AND COLORS

This chapter will examine the various types of lines available in AutoCAD as well as how to create your own linetypes to meet various project parameters by using the –LINETYPE command. The LTSCALE command allows you to alter the lengths of line patterns. Once you've explored the wide variety of linetypes, this chapter will provide a method of keeping drawing information organized by the use of layers, which can be activated or made invisible. This information will be altered using the LAYER command.

CONTROLLING LINETYPES

A linetype is a specific pattern of lines of varied length and spacing presented in a uniform pattern. Varied linetypes traditionally have been used throughout the architectural and engineering communities for years; examples of each can be seen in Figure 15–1. The most common types of lines used in construction drawings can be seen in Figure 15–2. Methods of adjusting linetypes were introduced in Chapter 7. Linetypes can also be loaded, altered, and created from the command prompt by typing **–LT** [enter] or **–LINETYPE** [enter] . The command sequence is:

    Command: **–LT** [enter]
    ?/Create/Load/Set:

Loading Linetypes

Just as with the dialog box presented in Chapter 7, linetypes contained in the .LIN library must be added to the current drawing base to be used in this drawing. Select-

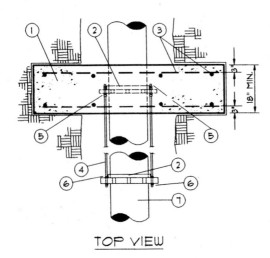

TOP VIEW

MATERIALS:

1. CONCRETE STRADDLE BLOCK.

2. UNI-FLANGE, SERIES 400C, CLASS 125

3. #4 REBAR EACH WAY, 12" O/C.

4. 3/4" ALL THREAD GALVANIZED STEEL
   TIE RODS, QUANTITY PER ENGINEER.

5. 3/4" GALVANIZED NUTS, 2-EACH SIDE

6. 3/4" GALVANIZED NUTS, 1-EACH SIDE

7. FLANGED FITTING

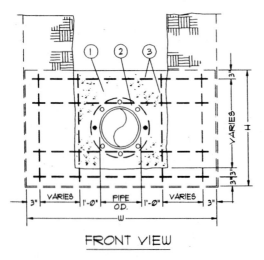

FRONT VIEW

NOTES:

1. STRADDLE BLOCKS SHALL BE DESIGNED
   INDIVIDUALLY BY THE ENGINEER AND
   SHALL BE BASED ON THE FOLLOWING:
   a.) 200 PSI WATER PRESSURE
   b.) SOIL BRG. CAPACITY, STEEL SIZE
       AND SPACING BY THE ENGINEER.

2. BEARING AREA OF BLOCK SHALL BE
   AGAINST UNDISTURBED SOIL.

3. STRADDLE BLOCK SHALL HAVE A
   MINIMUM OF 18" COVER

4. CONCRETE SHALL HAVE A MIN. 28-DAY
   STRENGTH OF 3000 PSI

5. ALL FITTINGS WITHIN THE CONC.
   SHALL BE WRAPPED IN PLASTIC OR
   BE COATED W/KOPPER'S 50.

6. STRADDLE BLOCK HEIGHT(H) &
   WIDTH(W) SHALL BE DETERMINED BY
   THE ENGINEER.

FIGURE 15–1 Varied linetypes and line widths are used to help distinguish construction components. (Courtesy Department of Environmental Services, Oregon City, OR)

ing the Load option allows the name of a linetype to be entered by keyboard. The command sequence is:

Command: **–LT** [enter]
?/Create/Load/Set: **L** [enter]
Linetype (s) to load:

The program is now waiting for you to enter the name of the desired linetype. To load a centerline, use the following sequence:

Command: **–LT** [enter]

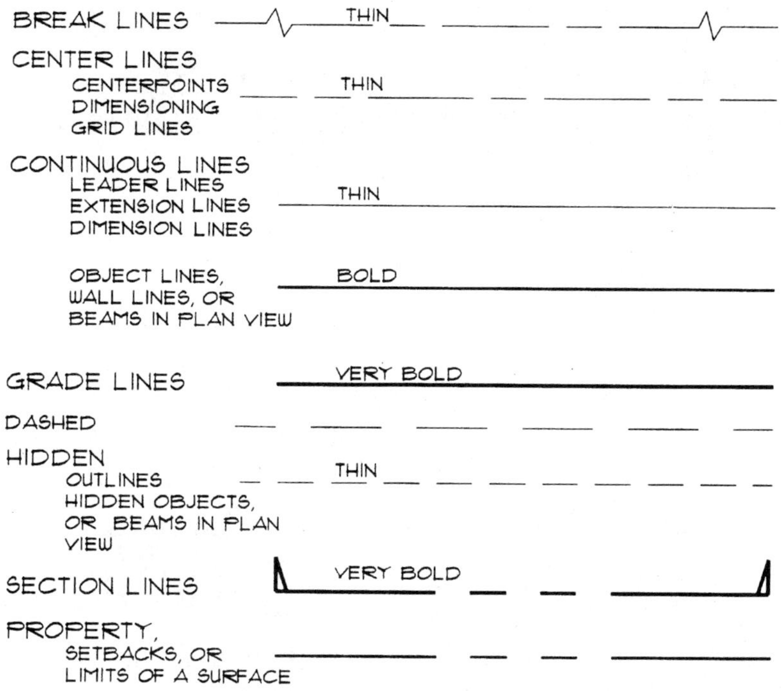
BREAK LINES
THIN
CENTER LINES
CENTERPOINTS
DIMENSIONING
GRID LINES
THIN
CONTINUOUS LINES
LEADER LINES
EXTENSION LINES
DIMENSION LINES
THIN
OBJECT LINES,
WALL LINES, OR
BEAMS IN PLAN VIEW
BOLD
GRADE LINES
VERY BOLD
DASHED
HIDDEN
OUTLINES
HIDDEN OBJECTS,
OR BEAMS IN PLAN
VIEW
THIN
SECTION LINES
VERY BOLD
PROPERTY,
SETBACKS, OR
LIMITS OF A SURFACE

FIGURE 15–2 Common linetypes used on constructoin drawings.

?/Create/Load/Set: **L** enter
Linetype (s) to load: **CENTER** enter
Linetype CENTER loaded.
?/Create/Load/Set:

If you're unsure of the name of a layer, picking the ? option will produce a display similar to the one shown in Figure 15–3 will be displayed. By responding again with the enter key or picking the OK button, the menu similar to Figure 15–4 will be displayed. As the enter is pressed again, the balance of the menu will be displayed. The command prompt will be redisplayed by pressing the enter key a final time. Pick the Close button to remove the display and restore the drawing area. Type the name or names of layers that you desire to load into the current drawing.

Setting Linetypes

The Set option allows standard AutoCAD linetype files to be added to a drawing file. Type will reflect the new line pattern. Objects drawn prior to the selection will not be affected. Objects drawn prior to the selection can be transformed to a different linetype using the CHANGE command, which will be discussed later in this chapter. The process to set a linetype is:

?/Create/Load/Set: **S** enter

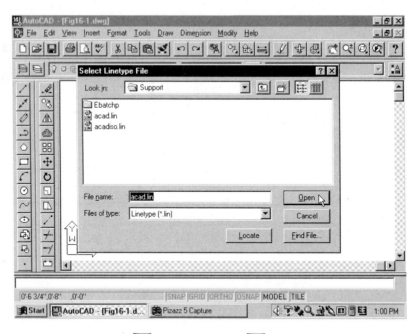

FIGURE 15–3 By typing **–LT** enter or **–LINETYPE** enter at the command prompt, and picking the ? option, the Select Linetype File is displayed.

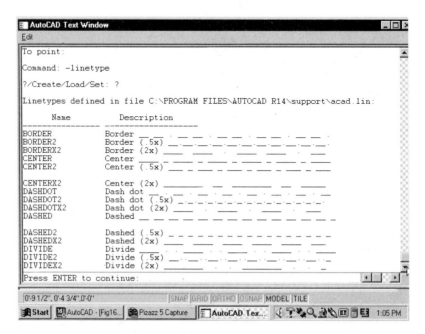

FIGURE 15–4 Standard linetypes available in AutoCAD. The balance of the display can be viewed by pressing the enter key.

New entity linetype (or ?) <BYLAYER>: **CENTER** [enter]
?/Create/Load/Set: [enter]
Command:

After the CENTER line pattern is set, the original prompt returns to allow for further alteration of LINETYPE. When you press the [enter] key, the command prompt is restored and drawing can continue.

By selecting the LINE command and drawing a few lines, you will produce lines created with a center line pattern. Although all lines were entered after the change in linetype are drawn using a centerline, not all lines appear as a centerline. Segments that are shorter than the pattern length in the library file will appear as continuous lines. The length of each segment can be altered by creating line lengths with the Create option or by using the LTSCALE command.

As the program is now set, this process must be repeated and CONTINUOUS entered at the prompt to return the drawing pattern to its original condition. The sequence to return to a continuous line is:

Command: **–LT** [enter]
?/Create/Load/Set: **S** [enter]
New entity linetype (or ?) <CONTINUOUS>: [enter]
?/Create/Load/Set: [enter]
Command:

Creating Linetypes

Between the multitude of linetypes contained in the ACAD.LIN file and contained in third party libraries, it's hard to imagine that someone would want to create their own line symbols. AutoCAD allows for the creation of your own linetypes using the Create option of –LINETYPE. Create will allow you to make variations of basic linetypes. A good example would be a centerline used on a site plan, floor plan, and foundation plan of a structure. Each drawing uses centerlines of the same line pattern but differing lengths. These drawings are typically combined by layers and stored in one file in many offices. Because they are in one file, one centerline length is not acceptable. LTSCALE can be used to adjust the length from one drawing to another, but it changes all line lengths. Create will overcome this limitation and will allow for three different centerlines to be created. The command sequence to create a new linetype is:

Command: **–LT** [enter]
?/Create/Load/Set: **C** [enter]
Name of linetype to create: **FNDBEAM** [enter]

Provide a name that will help you identify the linetype six months from now. In educational settings, projects usually are corrected within days if not hours of the original drawing session. In the professional setting, by the time consulting architects, engineers, contractors, building departments, financial backers, citizen advisory committees, etc., coordinate their contributions, the revision process can go on for months. Pick linetype names that will make sense after you've completely forgotten about the project. Names that describe the use or the pattern work best.

Once you've entered the name and pressed the [enter] key, the Create or Append Linetype File dialog box will be displayed. The dialog box can be used to create or append linetypes and will allow for selection of a filename and a storage location for the new file location. A typical location would be with other line files in the C:\Acadr 14\Support\Acad. lin file. A separate file should be created to store all custom linetypes. For now press the [enter] key. The box will disappear and the command prompt will change to:

Descriptive text:

The descriptive text is optional. To skip it, press the [enter] key. If used, it can be a series of dots, spaces, or dashes or a comment used to further describe the linetype being created. Up to 47 characters may be used as the descriptive text. An example of descriptive text is:

Enter pattern (*on next line*):
A,

AutoCAD is now waiting for you to enter the specifications for the line segments that will comprise the new linetype. Information can be placed in the specification in four forms, starting with the letter A. The A should be used at the beginning of all linetype input and will program the software to balance the line pattern between segment endpoints.

The three other symbols used in the linetype code are a negative number, a positive number, or a 0. A negative number is used to represent a space between line segments in the pattern. A positive number is used to represent the line segment in the pattern. The 0 is used to represent a dot. Figure 15–5 shows the effect of each entry on a line segment. The entry code for the linetype shown in Figure 15–5 is:

A,**48,–3, 6,–3, 6,–3,12,–3, 6,–3, 6,–3** [enter]
New definition written to File.
?/Create/Load/Set:

A linetype as shown in Figure 15–5 has now been created, but remember, it must be loaded into the drawing file before it can be used on a specific layer.

Loading Linetypes

Once a custom linetype has been created, it must be loaded into the drawing file. This can be done by using the Load option of LINETYPE. The command sequence is:

?/Create/Load/Set: **L** [enter]
Linetype(s) to load: **FNDBEAM** [enter]

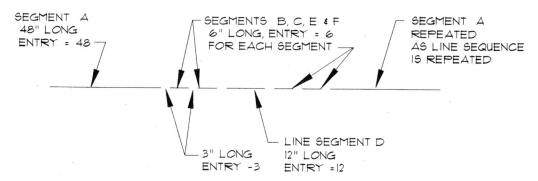

FIGURE 15–5 Positive numbers are used to represent line lengths and negative numbers are used to represent the space between line segments.

This sequence will search the ACAD.LIN file and load the FNDBEAM linetype into the current drawing file. The prompt will allow for several linetypes to be loaded from the same source at one time. To load multiple linetypes, type the linetype names separated by a comma between each name. A sample entry might include:

Command: **–LT** [enter]
?/Create/Load/Set: **L** [enter]
Linetype(s) to load: **CENTER,CENTER2,CENTERX2** [enter]

Complex Linetypes With Shapes

A complex linetype is a line that contains symbols in addition to the pattern formed by dots, dashes, and blank spaces. Users of older versions of the program will greatly appreciate that several complex linetypes are now part of the ACAD.LIN library. Open the Layer and Linetype properties dialog box, pick the Linetype tab, and then pick the Load... button to display the contents of the ACAD.LIN library. Scroll down to the end of the list. Notice lines such as Fenceline1, Batting, Hot_Water_supply, Gas_line and Zigzag are each complex linetypes. If the lines in the .LIN file do not met your needs, a complex linetype can be created using the guidelines contained in the Help file. The guidelines for creating a custom linetype can be accessed by opening the following areas of the Help menu in the order in which they are presented:

Customization Guide

Part I- Customization Reference

Chapter 2-Linetypes and Hatch Patterns

Linetype Definition Files

Complex linetypes

Although the entire process for creating a linetype will not be discussed, a few important things should be noted.

Before the AutoCAD shapes can be incorporated into a linetype, they must be loaded into the drawing file. This can be done by typing **LOAD** at the command prompt, which will produce the Select Shape File dialog box. Files can be loaded by selecting the LTYPESHP. SHX listing from the Support folder. The listing for shapes can now be displayed by typing **SHAPE** at the command prompt and choosing the ? option. Current shapes can be seen in Figure 15–6 or shapes can be drawn and added to the library.

Select the desired shape to be inserted into the linetype.

To insert a shape into a linetype, enter **SHAPE** (enter) at the command prompt. You will be prompted for several pieces of information which include: name, starting point, height, and rotation. The shape name is the name of the shape that you want to insert into the line. The shape should be one of the names loaded using the LOAD command. Next you'll be asked to provide the height and rotation angle for the symbol. The command prompt is:

Command: **SHAPE** (enter)
Shape name (or?): **TRACK1** (enter)
Starting point: (*Select desired point.*)
Height <0'-1">: (*Enter desired value or accept default.*) (enter)
Rotation angle <0>: (*Enter desired value or accept default.*) (enter)
Command:

Once the shapes have been loaded into the current .DWG file, a linetype can be created using a text editor. Don't skip over this little tidbit. A complex linetype can't be created at the command prompt of AutoCAD. Use a text editor such as Note Pad. To create a linetype called RAILWAY using a ¾" long line and ¹⁄₁₆" gap and the TRACK1 shape inserted in the line would require the following entry:

```
*RAILROAD, track1 line inserted into line
A,.75.-.065,[TRACK1,ltypeshp.shx,S=.1,X=-.065],-.065
```

You'll notice that the sequence is started by use of an astrerisk, followed by the name of the linestyle, and a description of the line. The second line of the sequence contains the information for the line segment length (the positive numbers) the size of

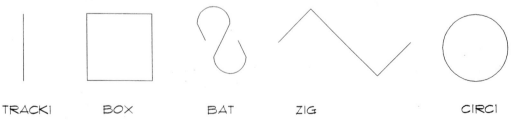

TRACK1   BOX   BAT   ZIG   CIRC1

FIGURE 15–6 Shapes contained in the LTYPESHP file.

the spaces (the negative numbers) and then information contained in [ ] brackets. The information in the brackets contains [the name of the desired symbol to be inserted, it's location, an S value which represents the scale factor of the symbol, and the X value which represents the X axis offset]. A Y= factor could also be provided to specify the offsetting of the symbol above or below the line.

Once the file has been saved, select the Exit option of File to return to your drawing in AutoCAD. The RAILROAD linetype has now been created and stored in the ACAD.LIN file. The style must now be loaded into your drawing file using the Load option of LINETYPE in order to be used.

The Create option of LINETYPE can be used to create the linetype. The command sequence to load and set the line shown in Figure 15–7 is:

Command: **–LT** [enter]
?/Create/Load/Set: **L** [enter]
Linetype(s) to load: **RAILROAD** [enter]
?/Create/Load/Set: **S** [enter]
New object linetype (or ?)<BYLAYER>: **RAILROAD** [enter]
?/Create/Load/Set: [enter]
Command:

Changing the Line Segment Size

The dashes that comprise the line pattern are currently defined as drawing units that may be unsuitable for your current drawing needs. The length of line segments throughout a drawing is controlled by the LTSCALE command. The command is started by typing **LTS** [enter] or **LTSCALE** [enter] at the command prompt. As a default, AutoCAD thinks of the selected length as one unit. This unit could be in inches, feet, millimeters, kilometers, or any other unit of measurement. The LTSCALE command will allow you to alter the selected unit length in a way similar to the way the SCALE command altered the size of objects. The distances specified in the linetype definition will be multiplied by the LTSCALE value to produce a new length for all patterns. The command sequence is:

Command: **LTS** [enter]
New scale factor<1>: **2** [enter]

By entering a value of two, all line segments and spaces will be doubled in size. The effect of LTSCALE on line segments can be seen in Figure 15–8. The line scale is determined by the size the drawing will be plotted at. Using an LTSCALE of 2 will make the line pattern twice as big as the line pattern created by a scale of 1. For a

FIGURE 15–7 A complex linetype using TRACK1.

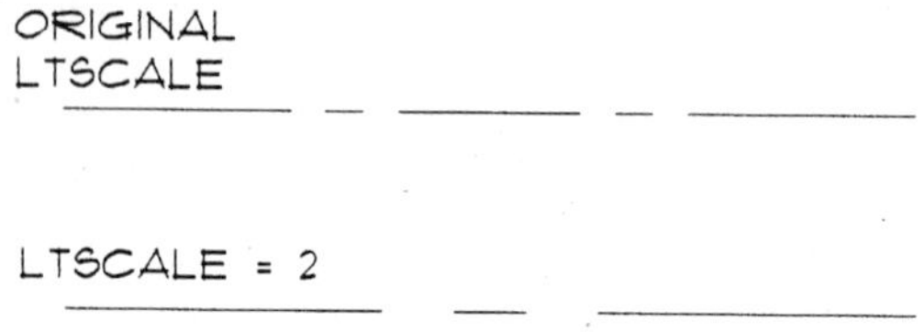

ORIGINAL
LTSCALE

LTSCALE = 2

FIGURE 15–8 The LTSCALE command allows the length of line segments to be altered.

drawing to be plotted at ¼"=1'-0", a dashed line that should be ⅛" when plotted, would need to be 3" long when drawn at full scale. This can be determined by multiplying .125" × 24 (the scale factor). The LTSCALE factor can be determined by using half of the scale factor. The drawing scale factor is always the reciprocal of the drawing scale. Using a drawing scale of ¼" = 1'-0" would equal .25" = 12". Dividing 12 by .25 produces a scale factor of 48 with an LTSCALE of 24. Keep in mind that when the LTSCALE is adjusted, it will affect ALL linetypes within the drawing. This is typically not a problem unless you are trying to create one unique line. Typically, the LTSCALE factor can be preset as prototype drawings are set up. In most offices, prototype drawings are established based on the size of the paper or the scale the finished drawing will be plotted at. Common plotting scales and their respective scale factors include:

ARCHITECTURAL VALUES		ENGINEERING VALUES	
Drawing scale	LTSCALE	Drawing scale	LTSCALE
1" = 1'-0"	6	1" = 1'-0"	6
¾" = 1'-0"	8	1" = 10'	60
½" = 1'-0"	12	1" = 100'	600
⅜" = 1'-0"	16	1" = 20'	120
¼" = 1'-0"	24	1" = 200'	1200
3/16" = 1'-0"	32	1" = 30'	180
⅛" = 1'-0"	48	1" = 40'	240
3/32" = 1'-0"	64	1" = 50'	300
1/16" = 1'-0"	96	1" = 60'	360

Changing Individual Line Segment Sizes

Using LTSCALE changes the values of all of the lines in a drawing. Using the CELTSCALE command will allow the scale of existing lines to remain unaltered, and change the scale of future lines. Typing **CELTSCALE** [enter] at the command prompt will produce the prompt:

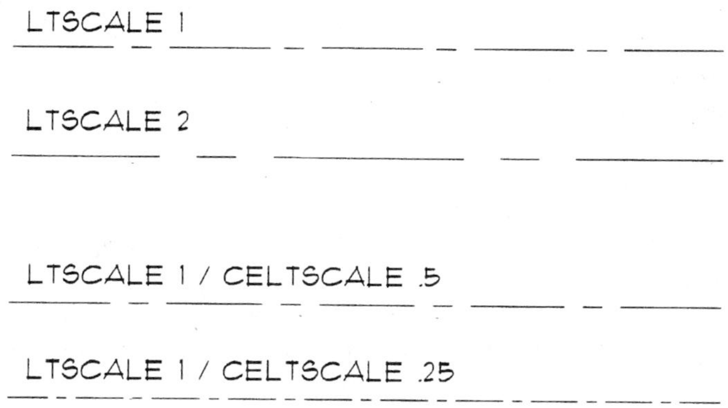

LTSCALE 1

LTSCALE 2

LTSCALE 1 / CELTSCALE .5

LTSCALE 1 / CELTSCALE .25

FIGURE 15–9 The effect of CELTSCALE and LTSCALE on line length.

New value for CELTSCALE <1.0000>:

The value that is assigned to CELTSCALE is used in combination with the value assigned to the LTSCALE. The CELTSCALE acts as a multiplier of the LTSCALE producing a net scale effect on lines being drawn. With an LTSCALE of 2, and a CELTSCALE of .25, the net scale factor would be 2 × .25 = .50. Figure 15–9 shows the effects of using CELTSCALE to control the current line width.

Paper Space Linetype Scale

The PSLTSCALE system variable allows the linetype scale to be standardized in paper space viewports. See Chapter 10 for a refresher of viewports if you haven't been using them to move through your drawings. With the PSLTSCALE variable set to 1, the line segments and spaces will have the same length in paper space viewports, regardless of the viewports' zoom scale factor. With the PSLTSCALE set to 1, the linetype scaling in paper space viewports is only based on the LTSCALE system variable.

With the PSLTSCALE variable set to 0, linetypes will be scaled according to the zoom factor of the viewport. The linetype scaling will be based on the LTSCALE value and multiplied by the zoom factor of the viewport.

NAMING LAYERS

Layers were introduced in Chapter 7 along with methods to create, alter, and control a layer. As your drawing ability has increased it's crucial to fully understand the importance of using layers. The use of layers will be increasingly important as the skills in the following chapters are introduced. Layers are used like separate sheets of paper to separate information within a drawing file. The base components of a floor plan can be placed on one layer; and text, dimensions, framing information, plumbing, etc can each be placed on separate layers to provide the information needed by various subcontractors. To help you control information, consideration must be given to layer names, controlling linetypes and colors by layer, as well as methods of displaying and hiding layer information.

Layer Names

Names for layers have not become standardized throughout the architectural or engineering communities. Most offices have developed a standard method of assigning names, so that all members of the staff have easy access to information on a drawing. Even within a firm, names will vary depending on the type of project to be drawn.

As you set up names for layers, use the same guidelines for naming layers that were used to name linetypes. Use a name that will still be meaningful in six months. This might include layer names such as UWALLS, MWALLS, LWALLS, UBEAMS, MBEAMS, LBEAMS. Up to 31 characters may be used in naming a layer. It also helps to keep layer names short so that alterations can be typed in quickly.

AIA Standardized Layer Names

As with other areas of architecture and engineering, there is no one standard method of naming and using layers on construction drawings. In 1989 the AIA established a guideline for architectural and engineering firms. The second edition of the AIA Cad Layer Guidelines was released in September of 1997. Copies of the new standard can be obtained by calling the A.I.A library at 1–800–365–2724 or from the AIA website at www.aia.org. The guidelines represent a new direction in layer and file labeling from the earlier AIA standard and is intended to coordinate with ISO (International Standards Organization) and CSI (Construction Specifications Institute) standards. The AIA has moved away from its original long and short format names and recommends a name comprised of a discipline code, and a major group name. A minor group name and a status name can also be added to the layer name depending on the complexity of the project. The new standard can be used to label both layers and drawing files with components of file names also used in the layer names. Because name components are common to both file and layer names, care must be taken to avoid creating project specific references within a layer name. Project designations should not be included in folder names. Use the guidelines presented in Chapter 5 for naming project folders, and the following guidelines for naming files and layers. Figure 15–10 shows example of the new guidelines and how they can be applied to layers and files names.

Naming Model/Sheet Files.   The guidelines that were introduced in chapter 5 work well for projects that do not require input from several different sub–contractors. When several firms are responsible for the project, one system that meets the needs of each discipline is required to name layers. The new AIA cad standard can be used to name layers and drawing files. Drawings files can be classed as either a model or sheet file. Model files are the components of a set of working drawings that are drawn full scale. These are the elements such as walls, doors and windows that

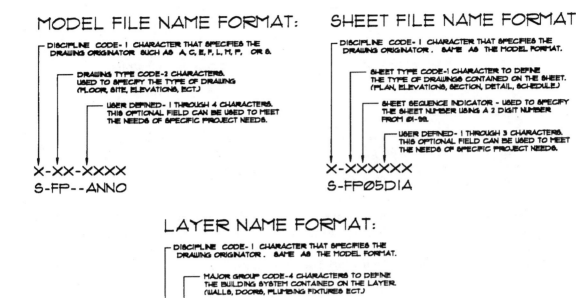

FIGURE 15–10 The AIA CAD Layer Guidelines require the use of codes to name layers and drawing files. By using a uniform system to name layers and files, architects, engineers and designers working for several different firms, can easily produce a coordinated set of drawings.

you've been drawing in model space. Chapter 5 introduced guidelines for naming model files. A sheet file is a completed drawing, such as a floor plan, that is prepared for plotting. A sheet file may contain one or more drawings that will be plotted on one sheet of paper. Chapter 23 will introduce merging a model file into a templet drawing using the XREF command. Sheet files allow for plotting in paper space at a scale of 1"=1". No matter the type of file, the new cad layer standards will provide a better means of tracking the contents of files and layers. The initial component of file and layer names is a discipline code.

Discipline Code. The discipline code is a character code that is used to identify the originator of the drawing. The discipline code will be the same for a model file, sheet file or layer name. The code allows instant recognition of the originator of the drawing or layer. The code may be either one or two characters. If only one character is used, it is followed by a hyphen. The first character represents one of the following 16 disciplines:

A	Architectural	M	Mechanical
C	Civil	P	Plumbing
E	Electrical	Q	Equipment

F	Fire–protection	R	Resource
G	General	S	Structural
H	Hazardous materials	T	Telecommunications
I	Interiors	X	Other disciplines
L	Landscape	Z	Contractor/ Shop drawings

The second character is either a hyphen or a user-defined modifier. Suggested modifiers might include **G** for graphics, **K** for kitchen or **V** for audiovisual equipment. A layer or file with a discipline code of **AG** would represent architectural graphics such as a rendering developed by the architectural firm.

Guidelines For Model File Names.   Figure 15–10 shows a sample of a model file name. The discipline code is followed by the drawing-type code. The AIA guideline lists eleven codes that are common to all disciplines. The codes are listed below and preceded by an asterisk that represents the space for the discipline code. Model file drawing type codes include:

*–FP	Floor Plan	*–SC	Section
*–SP	Site Plan	*–DT	Detail
*–DP	Demolition Plan	*–SH	Schedules
*–QP	Equipment Plan	*–3D	Isometric / 3D
*–XP	Existing Plan	*–DG	Diagrams
*–EL	Elevation		

In addition to drawing-type codes that apply to all disciplines, specific codes are listed for architectural, civil, electrical, fire protection, interiors, mechanical, plumbing, structural and telecommunications. Drawing types for the architectural field include A–CP for a ceiling plan and A–EP for enlarged plans. Structural drawing types are S–FP for a framing plan or S–NP for an foundation plan. See the AIA Cad Layer Guidelines for a complete listing of discipline specific codes.

User Definable Codes.   The final four digits of the model file name are defined by the user. Guidelines for naming files and folders that were presented in Chapter 5 can be used for the final four digits. The fifth-level floor plan for a structure could be saved with a name of A–FP–2.05.

Guidelines For Naming Sheet Files.   Figure 15–10 shows an example of a sheet file name. The discipline code is followed by a one-digit sheet-type code. The AIA guideline lists nine common codes that apply to all disciplines. Sheet type designators include:

0	General (symbols, legends, notes, etc.)	5	Details
1	Plans	6	Schedules / diagrams
2	Elevations	7	User defined

3	Sections	8	User defined
4	Large scale drawings that are not details	9	3D views

Sheet type designators are followed by a two-digit sheet sequence number with numbers assigned from **01** to **99**. The last component of a sheet title is a three-digit or letter user-defined code.

Sheet file **S–405CON** could be used to represent a sheet of concrete details that have been assembled for plotting by the structural engineer. The sheet would be page 5 in the drawing set.

Guidelines For Naming Layers. The method for naming layers is similar to naming files. A layer name can be composed of the discipline and major group name; a minor name and status code can be added to the sequence. Figure 15–11 shows some alternatives for using components with layer names. The discipline code is the same as the one- or two-character code that is used for model and sheet file names.

Major Group Name. The major group code is a four-character code that identifies a building component specific to the defined layer. Major group layer codes are divided into the major groups of architectural, civil, electrical, fire protection, general, hazardous, interior, landscape, mechanical, plumbing, equipment, resource, structural, and telecommunication. Codes such as ANNO (annotation), EQUIP (equipment), FLOR (floor), GLAZ (glazing), and WALL (walls) are examples of major

LAYER NAME FORMAT:

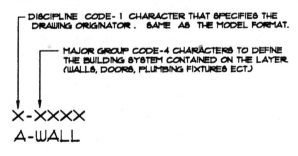

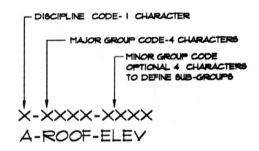

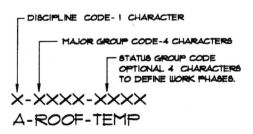

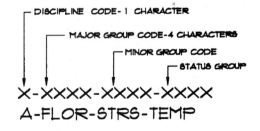

FIGURE 15–11 Based on the AIA Cad Layer Guidelines, layer names must fall into one of four categories.

group codes that are associated with the architectural layers. A complete listing of major codes for each group can be found in the AIA Layer Guidelines.

Minor Group Name. The minor group code is an optional four-letter code that can be used to define sub–groups to the major group. The code A–FLOR (architectural–floor) might include minor group codes for OTLN (outline), LEVL (level changes), STRS (stair treads or escalators), EVTR (elevators), or PFIX (plumbing fixtures). A layer name of A–FLOR–IDEN would contain room names, numbers and other related titles or tags. A complete listing of minor group codes specific to each discipline is listed in the AIA guideline.

Status. The status code is an optional four-letter code that can be used to define a sub–group of either a major or minor group. See Figures 15–11b and 15–11d. The code is used to specify the phase of construction. The layer names for the walls of a floor plan (**A–WALL–FULL**) could be further described using the status code of **NEWW** (new work) **EXST** (existing walls), or **DEMO** (demolition). A wall shown on the floor plan that is to be moved would be displayed on the **A–WALL–FULL–MOVE** layer.

CONTROLLING LAYERS

Layers can be controlled by using the Layer and Linetype Properties dialog box, by using the Layer Control menu display, or from the command line. The Layer and Linetype Properties dialog box shown in Figure 15–12 can be accessed by picking the Layer icon from the Object Properties toolbar, by entering **LA** [enter] **LAYER** [enter] or **DDLMODES** [enter] at the command prompt, or by picking Layer... from the Format pull-down menu. Chapter 7 introduced each element of the Layer and Linetype Properties. This chapter will introduce methods of setting and controlling each of the six layer elements.

Assigning Color by Layer

Once a layer has been created, colors and linetypes can be assigned by layers. See Chapter 7 for a review of how to create a layer.

Figure 15–13 shows an example of the dialog box with the addition of the MWALLS, UPWALLS, and BWALLS layers. Notice that the Current Layer is still 0, but new layers have been added. As new layers are created they are initially placed at the bottom of the layer list. As the screen is regenerated, the names will be displayed in alphabetical order. Once a layer has been created, the qualities for that layer color and linetype can be assigned.

The MWALLS layer is currently listed as black. Any lines that are created on this layer will be drawn as a black line. To alter the color of a layer, pick the layer so that the layer name is highlighted. The color of each layer can be altered by picking the color icon from the color list or by choosing Color from the Details box. The Details

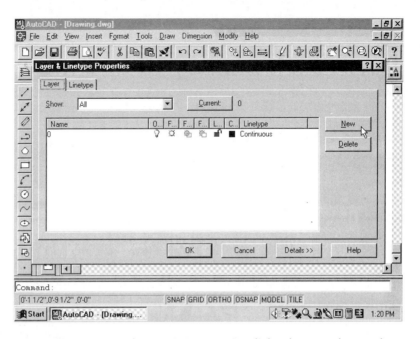

FIGURE 15–12 The Layer and Linetype Properties dialog box can be used to conrtol the characteristics of linetype and layers. It can be accessed by picking the Layer icon from the Object Properties toolbar or by entering **LA** [enter], **LAYER** [enter], or **DDLMODES** [enter] at the command prompt.

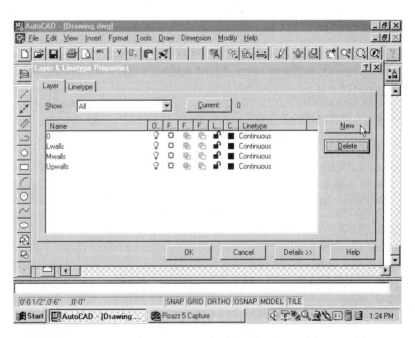

FIGURE 15–13 To create new layers press the New button. A layer with a name such as Layer 1 will be added. Methods of changing layer names will be discussed later in this chapter.

box is not active until a layer is highlighted. Picking the color icon for the selected layer will display the color palette seen in Figure 15–14. As the color for the desired color is selected from the palette, the name of the color will be displayed in the Details box, and the color will be displayed. The color for a layer can also be set by selecting the color icon in the Color: edit box or by selecting the arrow beside the edit box. Either method will produce the color menu shown in Figure 15–15. Color can be specified by picking the specific color icon or by entering the name or number of the color. Selecting the Other option of the menu will display the color palette shown in Figure 15–14. If you want to select another color, simply select another color chip and the display will be updated. When you're satisfied with the color, pick the OK button and the palette will be removed and the Layer Control dialog box will be redisplayed. The color of specific items on a layer can be altered by picking the Color icon from the Object Properties toolbar. Selecting this icon will allow the color of the layer to remain unchanged, but allow all items drawn after selecting the icon to have a color different than the specified layer color.

Objects and layers may be assigned any color you desire. It may be helpful to assign like objects of different layers a different color. For instance, if all plumbing fixtures are blue, it may not be readily apparent if the UPLUMBING fixtures are ON when only the lower-layer components are desired. Assigning a different color to the LPLUMB, MPLUMB, and UPLUMB layers may help you keep track visually of a drawing as different drawings are set to ON/OFF for viewing or plotting. Color as it relates to plotting will be introduced in Chapter 23.

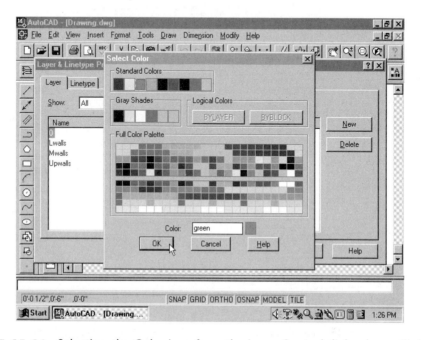

FIGURE 15–14 Selecting the Color icon from the Layer Control dialog box will display the Select Color palette, allowing the current layer color to be changed.

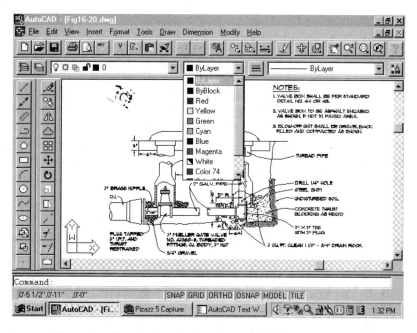

FIGURE 15–15   The color for a layer can be altered by selecting the Color icon from the Object Properties toolbar or the down arrow beside the box.

Assigning Linetypes to Layers

The current default for each layer or linetype is a continuous line. In Chapter 7, methods of drawing different linetypes were introduced. The LINETYPE command allows for setting and drawing different linetypes without regard to the layer. The Linetype option of LAYER will allow all lines on a specific layer to be drawn as one type of line. Linetypes can now be assigned to a specific layer, and all lines on that layer will be drawn with the selected linetype.

The Linetype can be altered using methods similar to altering layer color. Highlighting the name of the layer to be altered, and then selecting the linetype name will display the Select Linetype dialog box shown in Figure 15–16. The dialog box can be used to load and select linetypes. Selecting the Load...button will display the Load or Reload Linetypes shown in Figure 15–17 and display a list of linetypes contained in AutoCAD. Highlighting the names of the desired linetypes to load and selecting the OK button will remove the dialog box and restore the Select Linetype dialog box. As the box is restored, it will be updated to reflect the linetypes that were just loaded. One of these linetypes can now be selected from the Select Linetype dialog box and assigned to the layer that is highlighted in the Layer control Dialog box. Linetypes are assigned by picking the linetype and the OK button. As the button is picked, the Layer Control dialog box will be redisplayed reflecting the selected linetype.

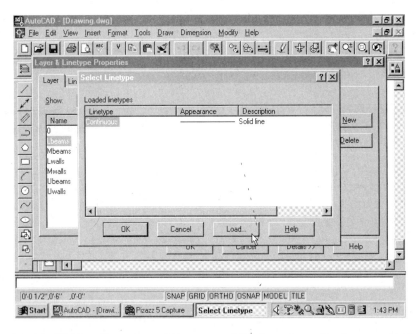

FIGURE 15–16 Picking the name of the current linetype from the Layer and Linetype Properties dialog box will display the Select Linetype dialog box, allowing for selection of one of the current linetypes.

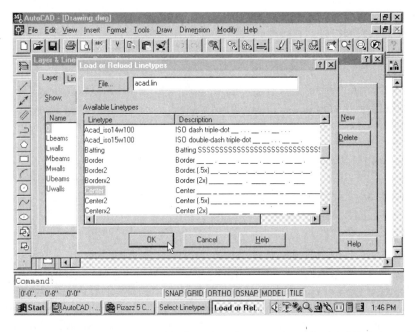

FIGURE 15–17 Linetypes can be added to the drawing base by highlighting the desired layer name with the cursor and picking the OK button.

Setting the Current Layer

Up to this point, you've created a layer with the name of MFLOOR and assinged the color red and a linetype of CENTER. If you exit the dialog box and start to draw, nothing will have changed. To draw red centerlines, the current layer must be altered. To make the MFLOOR layer current, highlight the name of the layer and pick the Current button. Selecting the OK button will keep this change and restore the drawing screen. As you enter the LINE command, segments will be drawn using the selected color and linetype that were just added.

Setting Layers On/Off

All of the new layers you have created up to this point are set in the ON option. If you were to switch to a layer, such as UPWALLS, and then draw, you would be able to see what is drawn. Although in a multilevel structure it is sometimes helpful to view the walls of several different levels to check alignment, typically you will want to have only one level of a structure current. By setting a layer to OFF, objects drawn on that layer will be invisible. The information is retained, but not displayed or plotted.

Figure 15–18 shows a sample listing for the layers contained in the drawing file for a simple residence. The drawing file contains the floor, framing, and electrical plans for each level as well as the foundation, roof, and site plans. Figure 15–19 shows the floor plan of a residence with all of the nonstructural material ON, and all of the structural material OFF. Figure 15–18 shows the current status for each layer for the residence shown in Figure 15–19. Figure 15–20 shows the same floor plan with the structural material layers toggled to ON, and the architectural information set to OFF. This particular drawing file contains information for the floor plan, framing plan, electrical plan, roof plan, foundation plan, and site plan. By use of the ON/OFF option, each drawing can be displayed and edited. Layer listings for the drawing file can be seen in Figure 15–18 and each plan can be seen in Figure 15–19 through 15–24.

Layer Off.   To turn layers off, enter the Layer and Linetype Properties dialog box and pick the ON/OFF icon for the desired layers to be turned off, followed by the OK button. The dialog box will be removed and any information that is stored on a layer that is off will be removed from the screen. If the current layer is selected to be turned OFF, you will be given the following prompt:

> WARNING: the current drawing layer is turned off.

If you pick the OK button, the dialog box is removed, and the items on the current layer are removed from the drawing screen. To restore the current layer, select OK so the drawing session can continue, and then reenter the Layer Control dialog box and restore the current layer. If you accept the Current layer as off, the current layer will be made invisible. Now if you draw objects, they will be drawn, but not displayed. This situation occurs more by accident than by design. If you find yourself

Layer name(s) to list <*>:

Layer name	State	Color	Linetype		Layer name	State	Color	Linetype
					-- Press RETURN for more --			
0	Off	4 (cyan)	CONTINUOUS		LWALLS	Off	1 (red)	CONTINUOUS
BASEDIM	Off	6 (magenta)	CONTINUOUS		LWINTEXT	Off	14	CONTINUOUS
EDGE	Off	3 (green)	CONTINUOUS		MSTAIRS	Off	3 (green)	CONTINUOUS
FIREPLACE	Off	7 (white)	CONTINUOUS		PLOTBORDER	On	7 (white)	CONTINUOUS
FNDBASEDIM	Off	3 (green)	CONTINUOUS		PLOTTEXT	On	12	CONTINUOUS
FNDBASEWALL	On	1 (red)	CONTINUOUS		PLOTTOPO	On	1 (red)	CONTINUOUS
FNDBEAMS	Off	5 (blue)	HIDDEN		PROPDIM	On	12	CONTINUOUS
FNDDIMEN	Off	5 (blue)	CONTINUOUS		PROPLINE	On	4 (cyan)	CENTER
FNDJOIST	Off	4 (cyan)	CONTINUOUS		PROPTXT	On	6 (magenta)	CONTINUOUS
FNDPIERS	Off	3 (green)	CONTINUOUS		ROOFPLAN	On	6 (magenta)	CONTINUOUS
FNDPILE	On	1 (red)	CONTINUOUS		SECTIONTAGS	Off	1 (red)	CENTER
FNDSHEAR	Off	7 (white)	CONTINUOUS		TEXT	Off	2 (yellow)	CONTINUOUS
FNDTEXT	Off	5 (blue)	CONTINUOUS		UDOORWIN	Off	5 (blue)	CONTINUOUS
FNDUPSTEEL	Off	7 (white)	HIDDEN		UPBEAM	Off	3 (green)	HIDDEN
FNDWALLS	On	1 (red)	CONTINUOUS		UPCABS	Off	5 (blue)	CONTINUOUS
-- Press RETURN for more --					-- Press RETURN for more --			
FOUTLINE	On	5 (blue)	OUTLINE		UPDECK	Off	4 (cyan)	CONTINUOUS
GRADES	Off	6 (magenta)	CONTINUOUS		UPDIMEN	Off	3 (green)	CONTINUOUS
GRID	Off	7 (white)	CONTINUOUS		UPFURN	Off	3 (green)	HIDDEN
LBEAMS	Off	5 (blue)	HIDDEN		UPJOIST	Off	5 (blue)	CONTINUOUS
LCABS	Off	5 (blue)	CONTINUOUS		UPPERPTLOADS	Off	5 (blue)	CONTINUOUS
LDOORWIN	Off	4 (cyan)	CONTINUOUS		UPPLUMB	Off	2 (yellow)	CONTINUOUS
LFURN1	Off	6 (magenta)	HIDDEN		UPSHEAR	Off	7 (white)	CONTINUOUS
LJOIST	Off	6 (magenta)	CONTINUOUS		UPTEXT	Off	11	CONTINUOUS
LOUTLINE	On	11	OUTLINE		UPVAULT	Off	14	HIDDEN
LOWCOLUMNS	Off	1 (red)	CONTINUOUS		UPWALLS	Off	1 (red)	CONTINUOUS
LOWDECK	Off	6 (magenta)	CONTINUOUS		USTAIR	Off	4 (cyan)	CONTINUOUS
LOWDIMEN	Off	2 (yellow)	CONTINUOUS		UWINTEXT	Off	3 (green)	CONTINUOUS
LPLUMB	Off	4 (cyan)	CONTINUOUS					
LSHEAR	Off	7 (white)	CONTINUOUS		Current layer: PLOTBORDER			
LTEXT	Off	4 (cyan)	CONTINUOUS					
					?/Make/Set/New/ON/OFF/Color/Ltype/Freeze/Thaw:			
-- Press RETURN for more --								

FIGURE 15–18 The ? option will produce a listing of current layer options. (Courtesy Residental Designs)

drawing a line, but nothing is displayed, check the listing to see if the current layer has been switched OFF accidently.

Layer On. Layers that have been made invisible by the OFF option can be restored by the ON option. Enter the Layer and Linetype Properties dialog box and highlight the desired layers to be set to ON. Picking the ON/OFF icon followed by the OK button will restore the selected layers, remove the dialog box, and display the information on the selected layers.

Thawing and Freezing Layers

If you have set layers to OFF, you removed information from the display screen but not from the drawing base. The information merely became invisible. If you had requested a REGEN, it would have been performed with all of the information on the ON layers redisplayed. Information on layers that were set to OFF also would have to be processed. On a project the size of one of the chapter exercises, this might take an extra .0578532 seconds. On a project the size of the structure used throughout this chapter, the REGEN could take several seconds. It might not seem like a long time, but if the time is cutting into your coffee break, you can see why it would

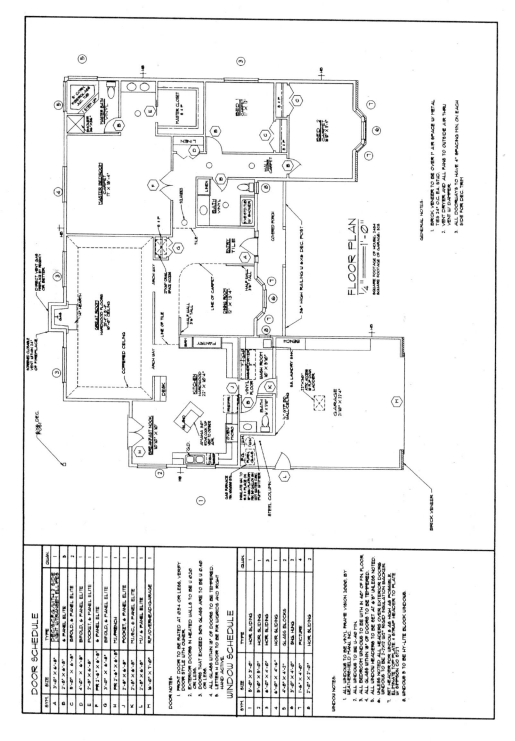

FIGURE 15–19 Each of the drawings seen in Figure 15–19 through 15–24 are contained in the same drawing file. Each drawing is stacked over each other. By use of a layer matrix, layers can be easily selected to display the desired drawing. (Courtesy Tereasa Jefferis)

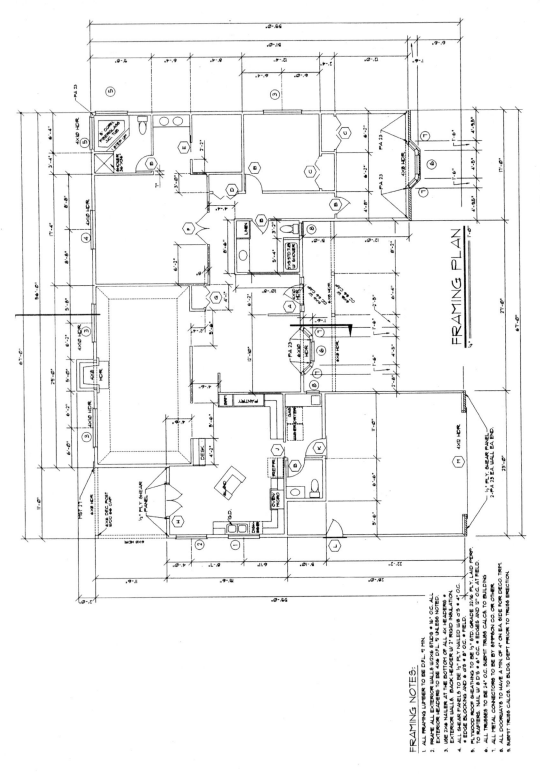

FIGURE 15-20 The framing plan for the residence seen in Figure 15-19. (Courtesy Tereasa Jefferis)

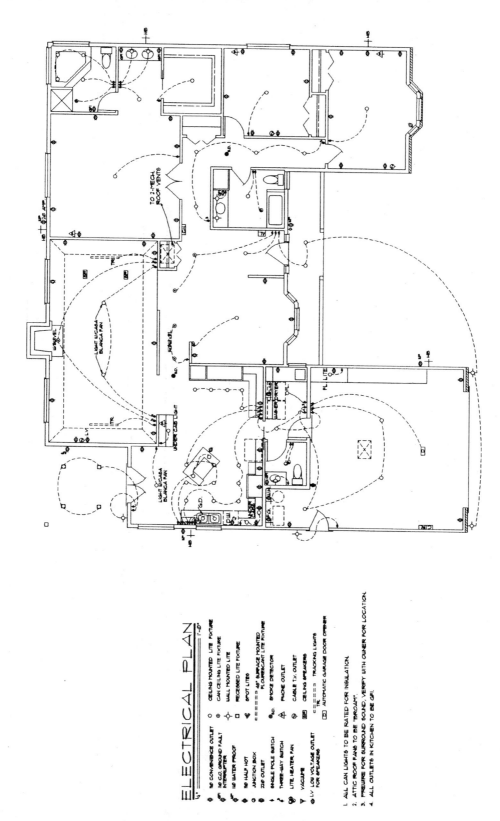

ELECTRICAL PLAN
1'-0"

$\ominus$	110 CONVENIENCE OUTLET
$\ominus^{GF}$	110 G.D. GROUND FAULT INTERRUPTER
$\ominus^{WP}$	110 WATER PROOF
$\ominus^{HH}$	110 HALF HOT
$\ominus$	JUNCTION BOX
$\ominus$	220 OUTLET
S	SINGLE POLE SWITCH
S_3	THREE-WAY SWITCH
S_{HF}	LITE HEATER FAN
$\blacktriangledown$	VACUUME
$\ominus^{LV}$	LOW VOLTAGE OUTLET FOR SPEAKERS
$\bigcirc$	CEILING MOUNTED LITE FIXTURE
$\oplus$	CAN CEILING LITE FIXTURE
$\square$	WALL MOUNTED LITE
$\heartsuit$	RECESSED LITE FIXTURE
Ψ	SPOT LITES
$\sqsubset\!\!=\!=\!=\!\!\sqsupset$	48" SURFACE MOUNTED FLOURESCANT LITE FIXTURE
$\bigcirc_{SD}$	SMOKE DETECTOR
$\triangle$	PHONE OUTLET
$\triangleleft$	CABLE T.V. OUTLET
$\boxed{CS}$	CEILING SPEAKERS
$\sqsubset\!=\!\!=\!\!=\!\!\sqsupset$ TR	TRACKING LIGHTS
$\boxed{}$	AUTOMATIC GARAGE DOOR OPENER

1. ALL CAN LIGHTS TO BE RATED FOR INSULATION.
2. ATTIC ROOF FANS TO BE THROUGH.
3. PREWIRE FOR SURROUND SOUND, VERIFY WITH OWNER FOR LOCATION.
4. ALL OUTLETS IN KITCHEN TO BE GFI.

FIGURE 15–21 An electrical plan uses the walls, windows, doors, and cabinets of the floor plan for a drawing base. (Courtesy Tereasa Jefferis)

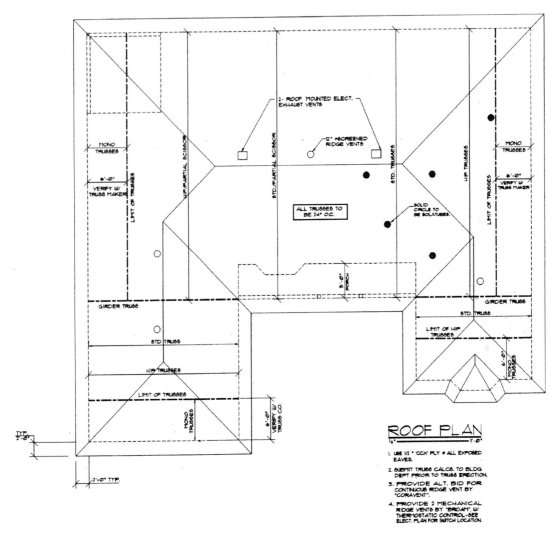

2- ROOF MOUNTED ELECT. EXHAUST VENTS

12" #SCREENED RIDGE VENTS

MONO TRUSSES

MONO TRUSSES

HIP/PARTIAL SCISSOR

STD/PARTIAL SCISSOR

STD. TRUSSES

HIP TRUSSES

6'-0" VERIFY W/ TRUSS MAKER

6'-0" VERIFY W/ TRUSS MAKER

LIMIT OF TRUSSES

LIMIT OF TRUSSES

ALL TRUSSES TO BE 24" O.C.

SOLID CIRCLE TO BE SOLATUBES.

GIRDER TRUSS

GIRDER TRUSS

STD. TRUSS

STD TRUSS

LIMIT OF HIP TRUSSES

HIP TRUSSES

MONO TRUSSES

LIMIT OF TRUSSES

MONO TRUSSES

6'-0" VERIFY W/ TRUSS CO.

3'-0" PORCH

TYP. 2'-0"

2'-0" TYP.

ROOF PLAN
¼" = 1'-0"

1. USE 1/2 " 'CCX' PLY @ ALL EXPOSED EAVES.
2. SUBMIT TRUSS CALCS. TO BLDG DEPT PRIOR TO TRUSS ERECTION.
3. PROVIDE ALT. BID FOR CONTINUOUS RIDGE VENT BY "CORAVENT".
4. PROVIDE 2 MECHANICAL RIDGE VENTS BY "BROAN". W/ THERMOSTATIC CONTROL-SEE ELECT. PLAN FOR SWITCH LOCATION.

FIGURE 15–22 None of the layers on the floor, framing or electrical plans are used on the roof plan, but the walls serve as an outline for the roof. Using OSNAP, and the LINE command, with the walls thawed, the outline can be quickly drawn, before freezing all other drawings. (Courtesy Tereasa Jefferis)

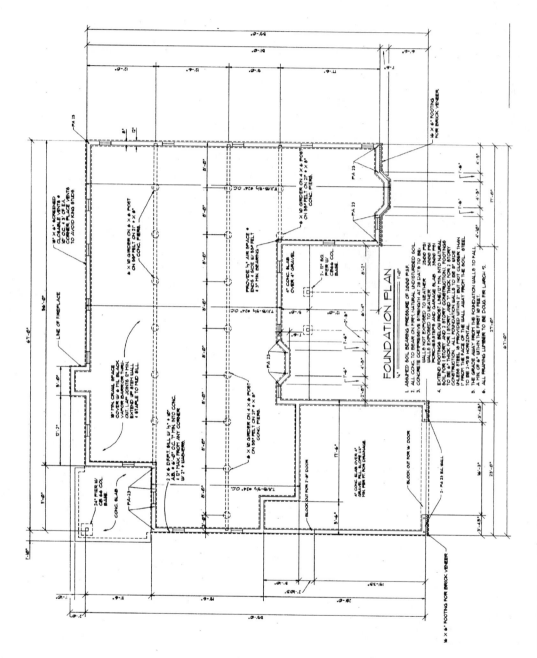

FIGURE 15–23 The foundation can be drawn using the walls of the floor plan as a guide. With careful planning and good layer control, many of the dimensions on the floor plan can be used on the foundation. (Courtesy Tereasa Jefferis)

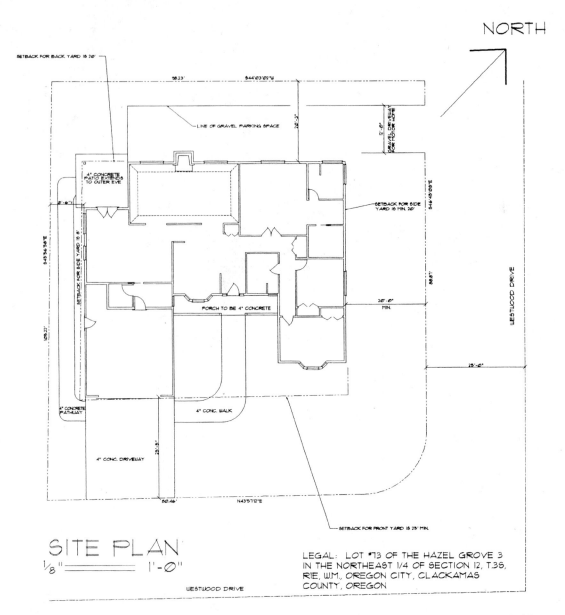

NORTH

SITE PLAN
1/8" ──────── 1'-0"

LEGAL: LOT #73 OF THE HAZEL GROVE 3
IN THE NORTHEAST 1/4 OF SECTION 12, T.3S,
RIE, W.M., OREGON CITY, CLACKAMAS
COUNTY, OREGON

FIGURE 15-24 A site plan often requires careful use of varied linetypes and CELTSCALE. (Courtesy Tereasa Jefferis)

be nice to be able to do things the fastest way possible. Using the Freeze/Thaw options, you will be able to control the extent of information to be scanned during the REGEN process. Layers that are in the Thaw mode are processed for a REGEN, information that is FROZEN is not processed during a REGEN, greatly reducing the time required to REGEN.

Freeze. Open the Layer and Linetype Properties dialog box and select the layer that you would like to Freeze, and pick the Freeze button. The sun icon (Thaw) will be changed to an icicle (Freeze). Picking the OK button will activate the selection, restore the drawing screen, and remove the objects on the selected layer from the display.

Thaw. The Thaw option of the LAYER command will allow for specified frozen layers to be thawed. To thaw a layer that is frozen, open the Layer and Linetype Properties dialog box and select the icicle icon of the layer to be thawed. As the icicle is picked, it will change to the sun icon, indicating that the layer has been thawed. Picking the OK button will activate the selection, restore the drawing screen and display the material on the layer that had been frozen.

Controlling Layer Display in the Current Viewports

Viewports were introduced in Chapter 10. Layer visibility can be controlled in the current viewport without effecting the inactive viewports. Layers can be used to hide information in one viewport and accent other information in a second viewport. This can be done by freezing or thawing layers in specific viewports, controlling the display of individual viewports, or limiting the number of viewports that are active. To freeze a layer in the current viewport, open the Layer and Linetype Properties dialog box, select the layer to be frozen, and then select the Freeze in Current Viewport icon. Selecting the OK button will carry the results of picking the ON/OFF option of the Current Viewport icon.

Controlling Layer Display in New Viewports

Layer values can be set for future viewports to be created. To restrict the QUESTIONS layer in all viewports, enter the Layer and Linetype Properties dialog box and pick the layer name. Click the Freeze in New Viewports icon and the pick the OK button. This will now restrict the display of the QUESTIONS layer in all new viewports that are created.

Safeguarding Files

The final pair of options of the LAYER command are Lock and Unlock. As the name implies, a lock is used to protect something. AutoCAD allows you to lock layers so that they cannot be edited accidentally. Information on a locked layer is still visible; it's just protected. To lock a layer, highlight the desired layer and pick the Lock icon, followed by the OK button.

To Unlock a layer, highlight the desired layer and pick the Unlock button, followed by the OK button. This process will reverse the consequences of the Lock example just completed. Remember that locking a layer is not a security device, but a method of protecting information on a layer from careless editing. Entities on a locked layer will not be included in a selection set of an edit command. Locking a layer will allow you to view the information on the layer, but will not effect it as you edit a drawing. Locked entities can still be selected for OSNAP locations, thawed or frozen, or altered using the Linetype or Color icons.

Filters

Filters provide the ability to display certain layer names in the listing box, based on a specific attribute. Layers can be filtered based on color, frozen/thaw, linetype, lock/unlock, and layer name. The sequence is controlled by opening the Layer and Linetype Properties Dialog box and selecting the arrow beside the Show: edit box. This will produce the menu shown in Figure 15–25. Picking the Set Filter dialog option will produce the menu shown in Figure 15–26 and allow the parameters of the filter to be set. In the example, the filter is based on the HIDDEN linetype. Selecting the OK button will produce a layer listing of layers created using the specified criteria. (See Figure 15–27.) All of the existing features are still displayed on the screen, they just will not be listed in the Layer display until the filter is altered. Opening the Show: menu and selecting the All option will restore the total layer listing.

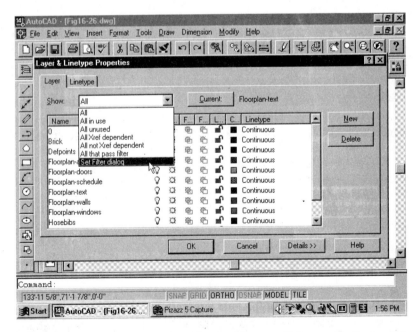

FIGURE 15–25 Selecting the arrow beside the Show: edit box will display this menu which can be used to established filters to search for layers.

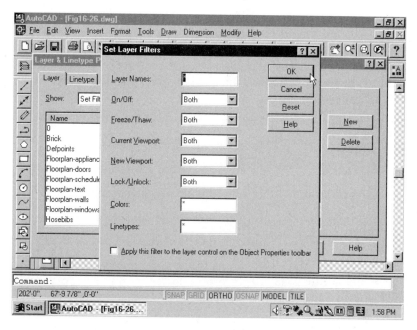

FIGURE 15–26 Selecting the Set Filter dialog option will produce this menu allowing the scope of the filter to be defined. Names for layer, colors, and linetypes can each be defined using wild cards.

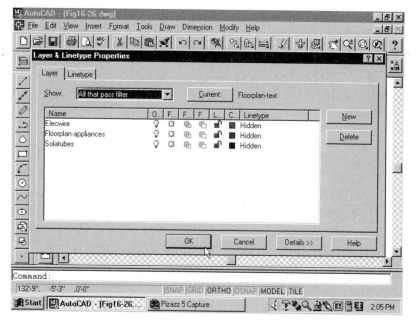

FIGURE 15–27 The layers defined, based on the linetype HIDDEN.

The Layer name:, Color:, and Linetype: boxes each contain an asterisk in their default settings. Wild card characters can be used in each of these three boxes to determine patterns within one of the three filter methods. A specific name or a combination of wild card symbols can be used to select the filter set. The * symbol is used to match any string, and can be used anywhere in the search string. Other common wild cards include:

# (pound)	Matches any numeric digit
@ (at)	Matches any alpha character
. (period)	Matches any nonalphanumeric character
* (asterisk)	Matches any string, anywhere in the string search
? (question)	Matches any single character
~ (tilde)	Matches anything but the pattern

See Chapter 8 of the Users Guide in the Help menu for a complete list of wild cards. The symbols are used in place of a letter or number in establishing the search parameters. Entering ### would provide a filter based on any three digit layer names.

Renaming Layers

The current name of a layer can be altered to better define the use of the layer at any time throughout the life of a drawing. Only the 0 layer and XREF layers can't be renamed. XREF layers are layers that are created on a drawing that is attached to the current drawing, and will be explained in Chapter 23. Layers are renamed inside of the Layer and Linetype properties dialog box by picking the name of the layer to be renamed, and then picking it again. As it is picked the second time, the name will be highlighted and a box will be placed around the original name. Pressing the backspace arrow while the name is highlighted will remove the exiting name and allow a new name to be entered. Pressing the [enter] key once will highlight the name again and allow the process to be repeated. Pressing the [enter] key a second time accepts the name, removes the dialog box, and restores the drawing area.

Deleting Layers

An unused layer can be deleted from the drawing base at any time throughout the life of the drawing with the exception of the 0 layer or an XREF layer. Layers that are referenced to a BLOCK and a layer named DEFPOINTS can not be deleted even if the layer is empty. Blocks will be introduced in Chapter 21. DEFPOINTS are created as dimensions are added to a drawing and will be discussed in Chapter 20. Unused layers can be deleted from inside of the Layer and Linetype properties dialog box by picking the name of the layer to be removed, and then picking the Delete button. As the button is selected, the layer will be removed from the list.

CONTROLLING LAYERS FROM THE COMMAND PROMPT

Layers can be controlled at the command prompt by typing **–LAYER** [enter]. This will produce the following prompt:

Command: **–LA** [enter]
?/Make/Set/New/ON/OFF/Color/Ltype/Freeze/Thaw/LOck/Unlock:

The Question Symbol (?)

Typing ? at the prompt line will produce the prompt:

Layer name(s) for listing <*> [enter]

Pressing [enter] will produce a list of currently defined layers showing their status. In the default setting, the ? Listing will be:

Layer name	State	Color	Linetype
0	On	7(white)	Continuous

Current Layer: 0

?/Make/Set/New/ON/OFF/Color/Ltype/Freeze/Thaw/LOck/Unlock:

The initial prompt is returned to allow the command to continue.

Entering the name of a specific layer at the question mark symbol will cause information for that layer to be displayed. Entering a null reply to the Layer names list prompt will produce a list of information about all current layers.

Figure 15–18 shows the layer listing for the floor plan shown in Figure 15–19.

Creating New Layers

Layers can be created by using the Make or the New LAYER options. The Make option will create a new layer and replace the current layer with the new layer. The New option will create a new layer, but will leave the current layer as the current drawing layer.

New Layers.   The New option can be used to create one or more new layers without affecting the current layer. The command sequence to create the New layers named MWALLS, BWALLS, and UPWALLS is:

Command: **–LA**
?/Make/Set/New/ON/OFF/Color/Ltype/Freeze/Thaw/LOck/Unlock: **N** [enter]
New layer Name(s) <0>: **MWALLS, UPWALLS, BWALLS,** [enter]
?/Make/Set/New/ON/OFF/Color/Ltype/Freeze/Thaw/LOck/Unlock:

Several new layers may be created at the prompt as long as each name is separated by a comma.

Make Layers.   Using the Make option will create a specified new layer and change the current layer from 0 to the new layer. The command sequence to Make a new MDOORWIN is:

Command: **–LA** (enter)
?/Make/Set/New/ON/OFF/Color/Ltype/Freeze/Thaw/LOck/Unlock: **M** (enter)
New Current Layer <0>: **MDOORWIN** (enter)

The Make option also can be used to change the current drawing layer quickly. By entering MWALLS at the Make prompt, the current layer of MDOORWIN will be changed to MWALLS.

Setting a Layer

The Set option is similar to the Make option and can be used to change the current drawing layer. The Set option will not create new options however, so names that are entered for the Set option prompt must already exist. The prompt is:

?/Make/Set/New/ON/OFF/Color/Ltype/Freeze/Thaw/LOck/Unlock: **S** (enter)
New Current Layer <MWALLS>: **BWALLS** (enter)

Setting Layers On/Off

To turn selected layers off, enter **OFF** at the layer option prompt. The sequence to turn the BWALLS and UPWALLS to the OFF setting is:

?/Make/Set/New/ON/OFF/Color/Ltype/Freeze/Thaw/LOck/Unlock: **OFF** (enter)
Layer name(s) to turn off:

Layers that have been made invisible by the OFF option can be restored by the ON option. The prompt is:

?/Make/Set/New/ON/OFF/Color/Ltype/Freeze/Thaw/LOck/Unlock: **ON** (enter)
Layer name(s) to turn On: **UWALLS** (enter)

Assigning Layer Color

The Color option of Layer will allow you to assign color to each layer. The sequence to change a current color is:

?/Make/Set/New/ON/OFF/Color/Ltype/Freeze/Thaw/LOck/Unlock: **C** (enter)
Color: **RED** (enter)

Color Prompt. The first prompt of Color: is your opportunity to pick any of the 256 options based on your machine's limitations. You may enter the color name, such as red, the first letter of the color name (R), or the color number (1). If a positive color number is used as the color entry, the color will be assigned to the specified layers, with the layers left ON. If a negative sign precedes the number, such as –1, the color red will be assigned to the specified layers and these layers will be set to the OFF designation. The negative sign also can be used preceding the letter or name options.

Assigning Color to Layers. The second prompt of:

Layer name(s) for color 1 (red) <MWALLS>:

allows you to specify which layers are to be displayed as the selected color. Pressing the [enter] key will set the current layer at the specified color. Existing layers can be assigned the color also by listing their names separated by a comma. To set the wall layers as red, and OFF, the command sequence is:

?/Make/Set/New/ON/OFF/Color/Ltype/Freeze/Thaw/LOck/Unlock: **C** [enter]
Color: **–1** [enter]
Layer name(s) for color 1 (red) <MWALLS>: **MWALLS,BWALLS,UPWALLS**
[enter]

Assigning Linetypes to Layers

The current default for each layer or linetype is a continuous line. Linetypes can now be assigned to a specific layer, and all lines on that layer will be drawn with the selected linetype. The command sequence to set the linetype of MBEAMS layer to a dashed line is:

?/Make/Set/New/ON/OFF/Color/Ltype/Freeze/Thaw/LOck/Unlock: **L** [enter]
Linetype (or ?)<Continuous>: **DASHED** [enter]
Layer name for linetype DASHED <MWALLS>: **MBEAMS** [enter]

If the linetype has not been used in the current drawing, it will be loaded automatically from the ACAD.LIN file. If you're unsure of the name of the linetype, enter ? at the prompt:

Linetype (or ?)<Continuous>: **?** [enter]

Freeze/Thaw

As the Freeze option is entered, you will be asked to supply the names of layers to be frozen. The prompt is:

?/Make/Set/New/ON/OFF/Color/Ltype/Freeze/Thaw/LOck/Unlock: **F** [enter]
Layer name(s) to Freeze: **BWALLS, MWALLS** [enter]

Enter the name of any desired layers to be frozen. Any layer may be frozen except the current layer.

The Thaw option of the LAYER command will allow for specified frozen layers to be thawed. The command sequence is:

?/Make/Set/New/ON/OFF/Color/Ltype/Freeze/Thaw/LOck/Unlock: **T** [enter]
Layer name(s) to Thaw: **BWALLS, MWALLS** [enter]

This command sequence will reverse the consequences of the Freeze example just completed.

ALTERING LINETYPE, COLOR AND LAYERS

AutoCAD allows individual properties within a layer to be altered as well as changing the properties of the entire layer. The qualities of an entire layer can be altered by using the Layer Control Icon from the Object Properties toolbar. Individual entities of a layer can be changed by using the Make Object's Layer Current icon, by using the MATCHPROP, the DDMODIFY, or CHANGE commands.

Layer Control Menu

The Layer control menu shown in Figure 15–28 can be used to alter the visibility of an existing layer. The menu is displayed by picking the Layer Control arrow on the right side of the box in the Object Properties toolbar. This menu can be used to toggle ON/OFF, THAW/FREEZE, and LOCK/UNLOCK and to set the current layer. The visibility of layers is adjusted by picking the ON/OFF or THAW/FREEZE icon. As the icon for each layer is selected, the layer will be displayed or removed depending on the setting of the icon. Layers are locked and unlocked by picking the icon of the desired layer. The current layer is altered by picking the name of the layer that you want to make current. As soon as the layer name is selected, the menu will be closed, the layer will be changed and the drawing area will be restored.

Making an Object's Layer Current

The Make Object's Layer Current icon is located on the far left end of the Object Properties toolbar. The icon can be used to alter the current layer. Picking the icon displays the following prompt at the Command prompt:

Command:ai_molc
Select object whose layer will become current:

The cursor will be changed to a pick box as the program waits for you to select a drawing entity.

Picking a drawing entity will make the layer that contains the selected entity the current entity.

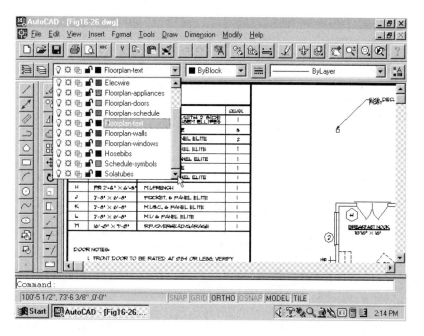

FIGURE 15–28 The Layer Control menu can be accessed by selecting the Layer Control arrow in the Object Properties toolbar. The menu can be used to toggle On/Off, Thaw/Freeze, and Lock/Unlock.

The command prompt will then display the name of the current layer, and provide the command prompt so that drawing may continue.

MATCHPROP

The properties of one layer entity can be transferred to another object by the use of Match Properties. The Match Properties icon is found in the Standard toolbar, or the command can be started by entering **MA** [enter] or **MATCHPROP** [enter] at the command prompt. Either method will produce the following prompt:

> Command: **MA** [enter]
> Select Source Object:*(select the object with the properties that you would like to copy)*
> Current active settings = color layer ltype ltscale thickness text dim hatch
> Settings/<Select Destination Object (s)>: *(select the object that you would like to change)*
> Settings/<Select Destination Object (s)>: [enter]
> Command:

The sequence will copy the properties of the first object to the second object including the layer, the linetype and the color. Instead of picking the object that you would like to change, entering **SETTINGS** [enter] at the second prompt will produce the Property Settings dialog box shown in Figure 15–29. The box can be used to specify the qualities that will be copied to selected objects. One or more of the properties can be assigned including the Color, Layer, Linetype, Ltscale, Thickness, as well as several other qualities that will be discussed in later chapters. Once the qualities have been selected, pick the OK button and resume the command sequence by picking the object to receive the new properties.

DDMODIFY

The DDMODIFY command can be used to control the properties of existing objects. The command can be started by picking the Properties icon from the far right side of the Object Properties toolbar, by selecting Properties... from the Modify pull-down menu, or by typing **MO** [enter] or **DDMODIFY** [enter] at the command prompt. Either method will produce the following prompt:

Command: **MO** [enter]
Select object: *(Select the object to be modified.)*

Once an entity is selected, the Modify Line box is displayed. The exact name of the box will vary depending on the shape of the entity that is selected. Select a circle,

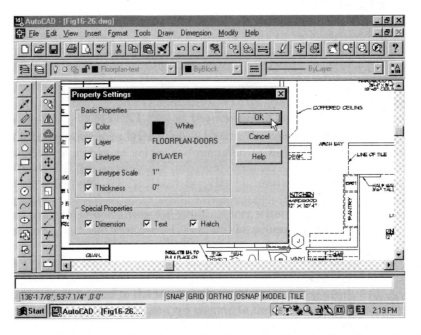

FIGURE 15–29   Selecting the Matchprop icon from the Standard toolbar or by typing **MA** [enter] or **MATCHPROP** [enter] at the command prompt will display the Property Settings dialog box that will allow specific properties of a layer to be altered without effecting other layer qualities.

and the Modify Circle dialog box is displayed. No matter the name, the contents remain the same. (See Figure 15–30.)

The box allows for altering the Color, Layer, Linetype, Handle, Thickness, and Ltscale. To alter the color of a line, pick the line to be changed, pick the Color: button, and then pick the desired color from the color pallette. Once the selection is made, pick the OK button to remove the palette, and return to the Modify box. Alter other properties as desired and pick the OK button to close the dialog box and return to the drawing.

CHANGE

The final command to be considered when setting linetype, color, or layers is CHANGE. Each of the attributes related to layers can be altered within the LAYER command. When a layer color is changed from red to blue, all lines on the specified layer become blue. The CHANGE command will allow for only selected entities to be altered. The command is entered by typing **–CH** [enter] or **CHANGE** [enter] at the command prompt. The command sequence is:

Command: **–CH** [enter]
Select object: (*Pick object to be altered.*)
Properties/<Change point>: **P** [enter]

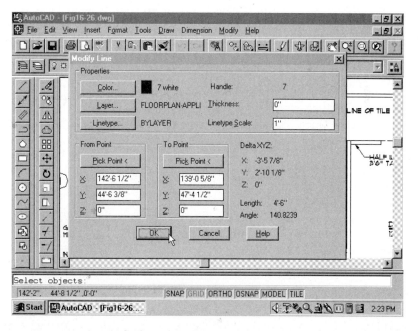

FIGURE 15–30 Typing **MD** [enter] or **DDMODIFY** [enter] at the command prompt will display the Modify dialog box that will allow specific properties of a layer to be altered.

Most of the changes required will be to properties. Enter **P** [enter] at the prompt. This will produce the prompt:

Change what property (Color/Elev/LAyer/LType/LtScale/Thickness)?:

Enter the capital letters to represent the option to be altered.

Color. Enter **C** [enter] at the prompt to allow the color of selected objects to be altered. The command sequence is:

Change what property (Color/Elev/LAyer/LType/LtScale/Thickness)?: **C** [enter]
New color (by layer): **G** (*green*) [enter]
Change what property (Color/Elev/LAyer/LType/LtScale/Thickness)?: [enter]
Command:

Notice that the original prompt was returned to allow for other options to be altered. Pressing [enter] twice will return the command prompt automatically. The selected objects will now be redisplayed as green.

LAyer. Enter **LA** [enter] at the prompt to change the layer location of selected objects to be altered. Some CAD users find drawing several objects with no regard to the proper layer location faster than constantly switching to the right layer. Once drawn, several objects can be transferred to the desired layer. The command sequence is:

Change what property (Color/Elev/LAyer/LType/LtScale/Thickness)?: **LA** [enter]
New layer (0): **MWALLS** [enter]
Change what property (Color/Elev/LAyer/LType/LtScale/Thickness)?: [enter]
Command:

At the New layer prompt, the default is the current layer location of the object to be changed. In this case, MWALLS will be the layer to which the selected objects will be relocated. Once transferred, entities will take on the characteristics of the new layer.

Linetype. Enter **LT** [enter] at the prompt to change the linetype of selected objects to be altered. The command sequence is:

Change what property (Color/Elev/LAyer/LType/LtScale/Thickness)?: **LT** [enter]
New linetype (BYLAYER): **CENTER** [enter]
Change what property (Color/Elev/LAyer/LType/LtScale/Thickness)?: [enter]
Command:

The specified line will be redisplayed as a center line.

LtScale . LTSCALE was introduced earlier in this chapter as a method of changing the size of the pattern in a linetype. LTSCALE changes the scale of all linetypes within a drawing.

The LtScale option of CHANGE can be used to alter the scaling of individual lines in a drawing.

The command sequence is:

Change what property (Color/Elev/LAyer/LType/LtScale/Thickness)?: **S** [enter]
New linetype scale <0'-1"> **4** [enter]
Change what property (Color/Elev/LAyer/LType/LtScale/Thickness)?: [enter]
Command:

The pattern of the selected line will now be four times larger than it had been.

Elev, Thickness. These options are only used in 3D applications and will not be covered in this text.

Change Point. The initial default for CHANGE is change point. If the default is accepted, the point selected is used to modify a specific attribute of the selected object. The operation performed depends on the type of entity selected.

Line. If the selected entity to be changed is a line, the endpoint of the line nearest to the CP (change point) will be moved to the changed point. Figure 15–31 shows the effect of CP on a line.

Circle. If the selected entity to be changed is a circle, the radius of the selected circle is altered to pass through the CP. Figure 15–32 shows the effect of CP on a circle.

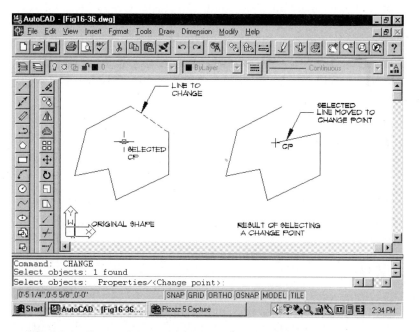

FIGURE 15–31 The effects of CP on a line.

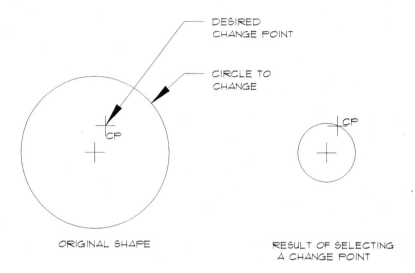

FIGURE 15–32 If the selected entity is a circle, the radius of the circle will pass through the CP.

CHAPTER 15 EXERCISES

E-15-1. Enter a new drawing and load CENTER, CENTER2, CENTERX2, DASHED, DASHED2, AND HIDDEN. Draw a line with each linetype long enough so that the line pattern can be seen. Save the drawing as E-15-1.

E-15-2. Open E-12-4. Create layers for WALL with red lines and DOORWIN with green lines. Move the existing entities to the appropriate layer and save the drawing as E-15-2.

E-15-3. Open drawing E-9-4. Create layers for the specified layers and options:

PROPLINE	red	Centerx2	property line
TREES	Green	Dashed	tree shape
T-LAYOUT.	7	Dot2	point/baseline

Set the LTSCALE appropriate for future plotting at a scale of 1" = 10'-0"

E-15-4. Open drawing E-9-1. Assign six different layers each with a different linetype and color. Change each of the existing shapes to a different layer. Draw a different geometric shape on the remaining layers. Use PLINE and OSNAP. Save the drawing as E-15-4.

E-15-5. Open drawing E-15-1, set the LTSCALE, and set the scale values appropriate to ¼" = 1'-0". Set the Units to Architectural, with ¹⁄₁₆" fractional value. Set the limits to 96 feet × 144 feet. Draw a 2" wide border, 24" inside the limits of the paper. Draw the border 48" inside the left side of the limits. Establish layers that would be suitable for a two-level residential floor plan, including walls, dimensions, architectural notes, structural text, joist, beams, furniture, appliances, planting, plumbing, cabinets, section tags, doors and windows, decks at each level, electrical fixtures, and electrical wiring. No provisions need to be made for other drawings that typically are grouped with a floor plan. Assign colors to each layer. Assign linetypes appropriate to the layer contents. Save the drawing as FLOOR14. This will become the basis for future architectural plan drawings that are to be plotted at ¼" = 1'-0".

E-15-6. Open drawing E-12-11 and create separate layers using an AIA long format so that walls, cabinets, fireplace, door and windows, and plumbing fixtures can be separated from each other. Assign each layer a different color. Add center, and hidden lines to the drawing base. Add a dishwasher to the kitchen using hidden lines. Save the drawing as FLRPLN.

E-15-7. Open a new drawing and create three different linetypes using TRACKS, BATTING, and FENCELINE1. Use a different line length for segments and different colors and layers for each linetype. Save this drawing as E-15-7.

CHAPTER 15 QUIZ

1. How many linetypes are in the ACAD.LIN library?

2. List four commands that can be used to alter the appearance of a line.

3. List four entities that are affected by a change in linetype.

4. List the command sequence to draw with a centerline and a hidden line.

5. Explain the following symbols used to create a linetype.

    A,24,–2,4,–2,4,–2

    A –

    24 –

    –2 –

    4 –

6. What would be the line scale value for a line that is to be plotted at a scale of 1" = 20'?

_____

_____

7. What command is used to set the line length value for plotting?

_____

_____

8. Give the command and the setting required to have the line pattern scale scaled according to the zoom factor of the viewport.

_____

_____

9. List the four common monitor options for displaying color and circle the capabilities of the monitor you are using.

_____

_____

10. List the corresponding name to the following color numbers for a 16-color display.

    1–

    4–

    6–

    13–

    15–

11. What is descriptive text as it relates to linetype?

_____

_____

12. What is the length limit of a layer name display on the status line and the command prompt?

_____

_____

13. When describing the STATE of a layer, what three qualities are considered?

_____

_____

14. List the four groups of information to be used in layer names recommended by the AIA.

    _____

    _____

15. What change properties apply to 2D drawing?

    _____

    _____

16. Describe guidelines for naming a model file using the AIA guidelines.

    _____

    _____

17. You are working on a drawing containing several different linetypes and would like to find out their names. What command and option should be used?

    _____

    _____

18. List the preferred LTSCALE for the following scales.

    Full size

3" = 1'-0"	1" = 1'-0"
¾" = 1'-0"	⅜" = 1'-0"
¼" = 1'-0"	⅛" = 1'-0"
¹⁄₁₆" = 1'-0"	1" = 10'-0"
1" = 20'-0"	1" = 60'-0"

19. Sketch examples of five types of lines typically found on construction drawings.

20. Write the information needed to describe a line being created to represent sewer lines on a plot plan comprised of long, short, short, short lines. The long line is to be ¾" long, the short lines are to be ¼", ³⁄₁₆", ⅛" long. Each space between lines is to be ¹⁄₁₆".

    _____

    _____

21. Where are shape files stored in AutoCAD?

22. List the five types of shapes typically used in construction drawings.

23. Describe guidelines for naming a sheet file using the AIA guidelines.

24. What will you be able to do after typing DDltype at the command prompt?

25. Describe the difference between the effects of LTSCALE and CELTSCALE.

26. A layer has a title of S-ANNO-SYMB. Who created the drawing and what do you think it would contain?

27. What information is in the minor group portion of a layer name?

28. You need to erase some information that is on a layer that is currently frozen. What's your plan?

29. Describe what you see as advantages of working with layers at the command prompt and the dialog box.

30. You're about to start the plans for a one-level residence that will require a site plan, floor plan, framing plan, electrical plan, foundation plan, exterior elevations and sections. How many drawing files do you expect to use, and how will you create layers for these files. (It's not in the chapter, you've got to search through your cranium.)

CHAPTER

16

PLACING PATTERNS
IN DRAWING OBJECTS

Various materials are represented by different line or shape patterns. This is seen most often when objects are seen in section view. Concrete material is typically filled with a pattern of dots and triangles; soil as a series of groups of perpendicular lines; and masonry as diagonal or crossing diagonal lines. These patterns are referred to as hatch patterns and can be created using either the BHATCH or the HATCH commands. A solid pattern can be placed in an object using either the BHATCH or SOLID command. The pattern should be placed on a layer with a minor code of PATT. Figure 16–1 shows an example of a drawing with hatch and solid patterns added to define each material. The BHATCH command adds patterns to a drawing by use of the Boundary Hatch dialog box. The HATCH command adds patterns to a drawing from the command prompt.

HATCHING METHODS

Hatch patterns are repetitive patterns of lines, dots, and other symbols used to represent a surface or specific material. Hatch patterns are used on sections and details, although plan views and elevations also use them. Hatch patterns typically are used to distinguish between masonry and wood walls on architectural plan views, as seen in Figure 16–2. Hatch patterns can be used on elevations to represent shades and shadows, as seen in Figure 16–3. The majority of uses for hatch patterns, however, will be found in details and sections.

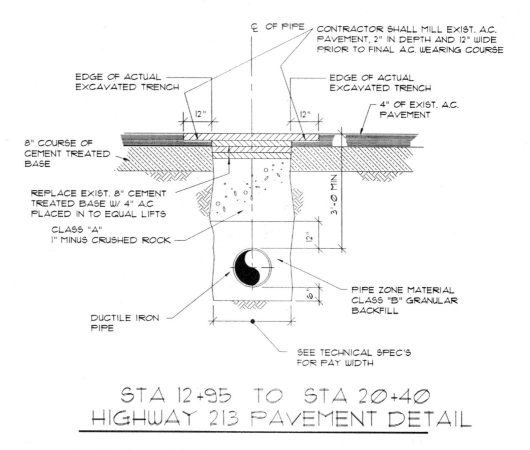

FIGURE 16–1 Shading and drawing patterns to represent materials can be produced by using the BHATCH, HATCH, and SOLID commands. (Courtesy Department of Environmental Services, City of Gresham, OR)

USING BOUNDARY HATCH

The **BHATCH** command can be started by selecting Hatch from the Draw toolbar, by typing **BH** [enter] or **BHATCH** [enter] at the command prompt, or by selecting Hatch... from the Draw pull-down menu. Each method will produce the display shown in Figure 16–4.

As you prepare to hatch an area using BHATCH, you can take several steps to speed up the process. Zooming into an area rather than keeping the entire drawing in view will eliminate the number of lines that must be examined to determine the boundary set that will define the hatch area. Speed also can be gained if extra layers are turned OFF or Frozen so that objects on these layers are not examined as part of the selection set.

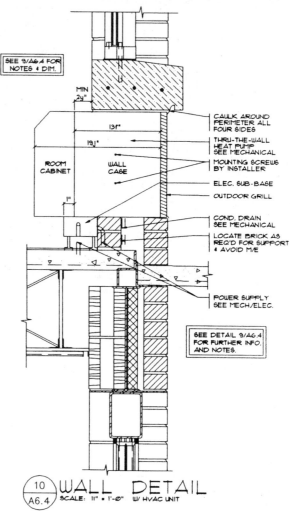

FIGURE 16–2 Hatch patterns are used to distinguish various materials from each other. (Courtesy G. Williamson Archer, AIA of Archer & Archer, P.A.)

SEE 9/A6.4 FOR NOTES & DIM.

MIN 24"

CAULK AROUND PERIMETER ALL FOUR SIDES

THRU-THE-WALL HEAT PUMP SEE MECHANICAL

MOUNTING SCREWS BY INSTALLER

ELEC. SUB-BASE

OUTDOOR GRILL

COND. DRAIN SEE MECHANICAL

LOCATE BRICK AS REQ'D FOR SUPPORT & AVOID M/E

POWER SUPPLY SEE MECH/ELEC.

ROOM CABINET

WALL CASE

13f"

19f"

1"

SEE DETAIL 9/A6.4 FOR FURTHER INFO. AND NOTES.

10
A6.4

WALL DETAIL

SCALE: 11" = 1'-0"   W/ HVAC UNIT

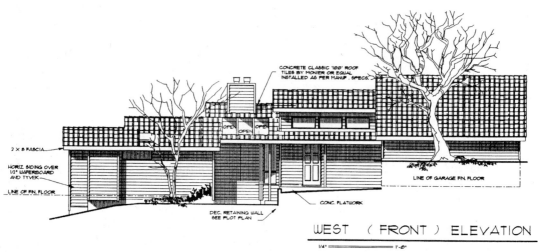

CONCRETE CLASSIC '100' ROOF TILES BY MONIER OR EQUAL INSTALLED AS PER MANUF. SPECS.

2 X 8 FASCIA

HORIZ. SIDING OVER 1/2" WAFERBOARD AND TYVEK

LINE OF FIN. FLOOR

LINE OF GARAGE FIN. FLOOR

OPEN OPEN OPEN

CONC. FLATWORK

DEC. RETAINING WALL SEE PLOT PLAN

WEST (FRONT) ELEVATION

1/4" = 1'-0"

FIGURE 16–3 Hatch patterns have been used to represent the tile roof, wood siding, and the shadows. (Courtesy Residential Designs)

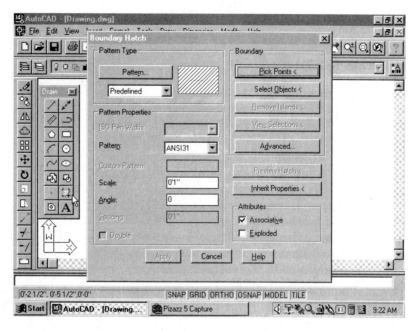

FIGURE 16-4 Hatch patterns can be added to a drawing using the Boundary Hatch dialog box. The box can be accessed by picking Hatch from the Draw toolbar, by typing **BH** [enter] or **BHATCH** [enter] at the command prompt, or by selecting Hatch... from the Draw pull-down menu.

Selecting a Pattern

Notice that the ANSI31 hatch pattern is the default setting and is shown in the Pattern: box of the Pattern Properties portion of the dialog box. A different pattern can be selected for a BHATCH and it will remain the default until a new pattern is selected. A new pattern can be selected by picking the Pattern edit box. Selecting this box presents three options. The default of Predefined allows patterns stored in the ACAD.PAT file to be selected. With Predefined active, moving the cursor to the pattern image box and clicking allows for scrolling through the AutoCAD hatch patterns. With the cursor in the image box, each click of the mouse will display a new pattern. The name of that pattern will be updated in the Pattern: box as each new image is displayed. The pattern can also be changed by selecting the Pattern: button. Pick the button to display a listing of the hatch pattern names and a display of each pattern similar to the list shown in Figure 16–5. The slide bar can be used to move through the list to select the name of the desired hatch pattern. If you're unsure of the pattern name, pick the desired hatch pattern from the icon display. To move through the icons, press the Next button at the bottom of the display. Once an icon is selected for use in a drawing, press the OK button to return to the dialog box.

Selecting the down arrow displays the listings shown in Figure 16–6. For most uses, the predefined patterns will be all that are ever needed. The program does allow

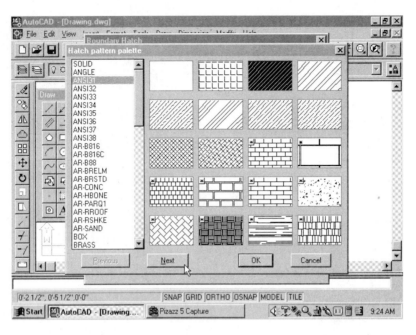

FIGURE 16–5 The Hatch pattern palette can be used to select the pattern to be used for the BHATCH command. The pattern can be selected by picking either the name or the icon and then picking the OK button.

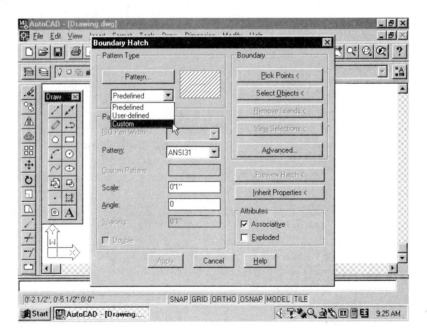

FIGURE 16–6 Picking the arrow beside the pattern icon will display options for selecting the pattern type.

user-defined and custom patterns to be created. A user defined patten is created using the current line style. Custom patterns allow drawing patterns to be saved in the acad.pat or in its own PAT file. Methods of creating patterns will be discussed as the HATCH command is explored.

Scale Pattern

This option sets the size of the pattern elements with the default of one drawing unit. The relationship of the pattern can be controlled by the drawing scale factor. Scale factor was first introduced in Chapter 15 with the LTSCALE command and will be discussed in greater detail in Chapter 23 when PLOTTING is introduced.

To alter the pattern scale, move the cursor to the Scale: edit box and press the mouse pick button. Use the backspace arrow to remove the current scale, and then provide the desired scale by keyboard. Once the scale is entered, move the cursor as needed to adjust other pattern values. Enter **XP** after the value to allow the pattern to be reproduced based on the plotted scale factor. Once a scale of 1XP has been entered, AutoCAD will calculate the actual scale factor for future hatch patterns. Figure 16–7 shows an example of three different hatch scale patterns. Picking the largest scale pattern that is still visually appealing will save REGEN time. Drawing a line the length of the desired pattern to be inserted can help to ease the scale selection. If you need a hatch pattern to represent a concrete block, draw a 16-inch-long line below the area to receive the hatch. Later in the command process, you'll have a chance to preview how the pattern will appear. The 16" line will serve as a reference to verify that the scale selection is accurate.

Pattern Angle

The pattern angle default is 0. By accepting the default, the pattern will be reproduced as shown in Figure 16–8. Entering a value other than zero allows the hatch

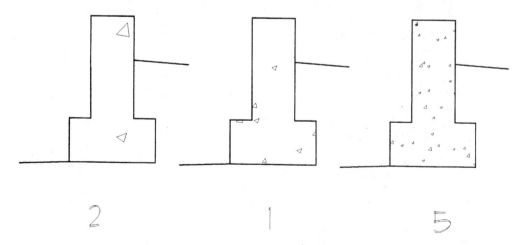

FIGURE 16–7 The effect of altering the scale factor. Always use the largest size possible to save regen time.

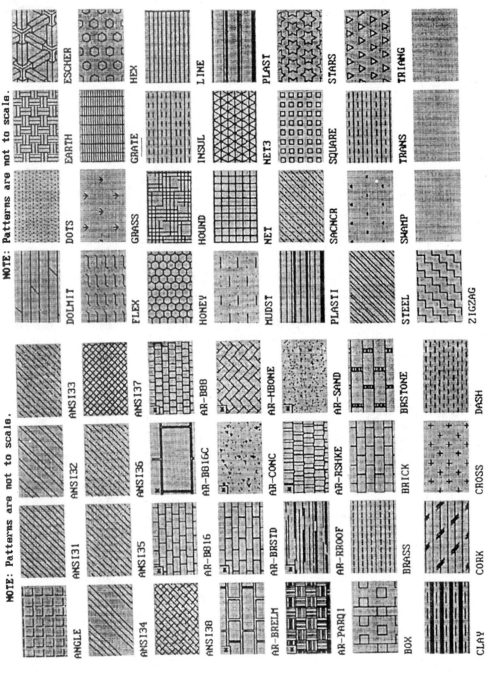

NOTE: Patterns are not to scale.

NOTE: Patterns are not to scale.

ANGLE · ANSI31 · ANSI33 · ESCHER

ANSI34 · ANSI32 · ANSI35 · DOLMIT · EARTH · HEX

ANSI38 · ANSI35 · ANSI36 · FLEX · DOTS · GRATE · LINE

AR-BRELM · AR-BB16 · AR-BBB · ANSI37 · HONEY · GRASS · INSUL · PLAST

AR-PARQ1 · AR-BRSTD · AR-BB16C · AR-HBONE · MUDST · HOUND · NET3 · STARS

AR-CONC · AR-SAND · PLAST · NET · SQUARE

AR-RROOF · AR-RSHKE · BRSTONE · STEEL · SACNCR · SWAMP · TRANS · TRIANG

BOX · BRASS · BRICK · CLAY · CORK · CROSS · DASH · ZIGZAG

FIGURE 16-8 Hatch icons are displayed in the Pattern Type icon box. Each time the icon box is picked, the next icon is displayed.

568

pattern to be rotated. By entering an angle of 45°, pattern ANSI31 can be used to represent a shadow, as seen in Figure 16–9. The angle value is adjusted using the same method used to adjust the scale value.

Selecting Objects to be Hatched

Once the pattern, scale, and angle for hatch have been selected, use one of the methods from the Boundary portion of the dialog box to choose the area to be hatched. For now, the options are limited to Pick points and Select objects. If the Pick Points< button is selected, only one point will need to be selected. If the Select Objects< button is selected, any of the selection methods introduced in Chapter 11 can be used to define the area. Other methods of selecting the boundary can be selected using the BOUNDARY command that will be discussed later in this chapter.

Using the Select Objects Option.   Figure 16–10 shows the process for selecting an object to be hatched using the Select Objects< option. In Figure 16–10a, the fire-

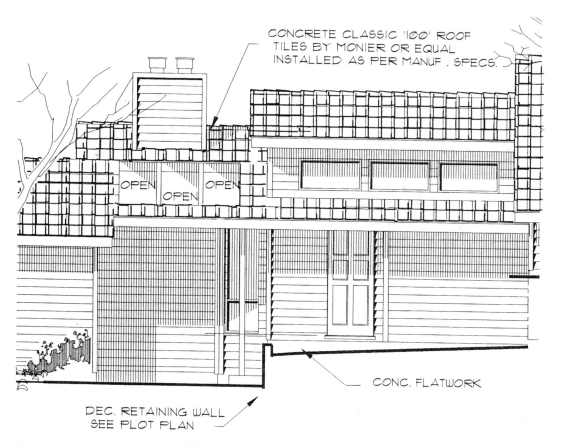

FIGURE 16–9 Hatch pattern ANSI31 is drawn at a 45° angle by default. Providing a rotation angle of 45° will reproduce the pattern in a vertical position. (Courtesy Residential Designs)

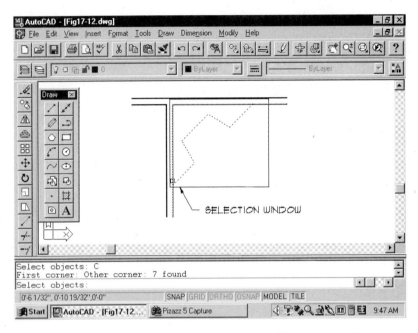

FIGURE 16–10a Using the Select Object option, an object can be selected using any of the normal selection methods such as Window or Crossing.

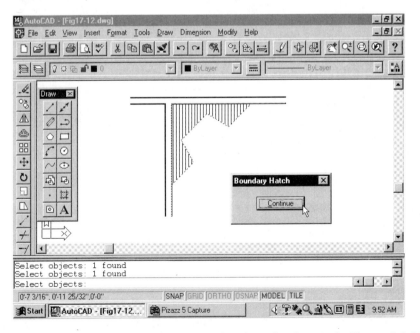

FIGURE 16–10b The resulting hatch pattern for the selection set in Figure 16–10a.

place was selected using the Window selection option. Figure 16–10b shows the hatch pattern that would result. The object to be hatched does not need to form a closed boundary, or may contain an internal area referred to as an island that will not be hatched. Figure 16–11 shows the results of selecting an object with an island using the Window selection method.

If the object to be hatched is surrounded by a polyline, the object can be selected by picking the object at any point as seen in Figure 16–12. The pattern will be the same as the pattern in Figure 16–10b. Once the object to be hatched is selected, press the ⏎, button to restore the Boundary Hatch dialog box. Pressing the ⏎, button a second time will apply the pattern. Figure 16–13 shows the results of selecting an object with an island to be hatched by selecting an edge.

Using the Pick Points Option. The Pick Points< option can be used to hatch drawing objects that form an enclosed boundary.

Once you've selected the Pick Points < button, the dialog box will be removed and the drawing will be restored. A prompt is now displayed requesting "Select internal point." Once selected, the command line will display the prompt:

Selecting everything visible...
Analyzing the selected data...
Analyzing internal islands...
Select internal point:

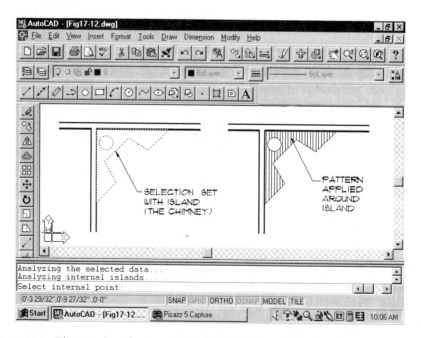

FIGURE 16–11 The results of using the BHATCH Select Object option on an object with islands. The placement of the pattern can be altered by using the Style option.

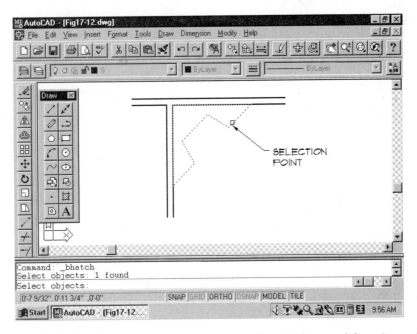

FIGURE 16–12 When the boundary of an object is formed by a polyline, it can be selected by picking a single point. The resulting pattern will be the same as the pattern in Figure 16–10b.

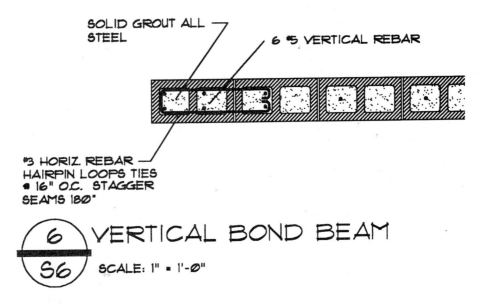

FIGURE 16–13 The resulting hatch pattern for a selection set with islands.

Because the Pick points < box was selected, the object to be hatched can be selected by selecting one point that lies within the object. As the point is selected, the boundary of the area to be hatched will be displayed as dashed lines. AutoCAD is waiting for you to select additional objects to be hatched. Select [enter] to redisplay the Boundary Hatch dialog box. Press [enter] a second time to apply the hatch pattern to the selected object. Figure 16–14 shows the resulting hatch pattern.

Using the Preview Option

If you're unsure of any of the pattern parameters, select the Preview Hatch < button prior to applying the hatch pattern. This removes the dialog box and displays the

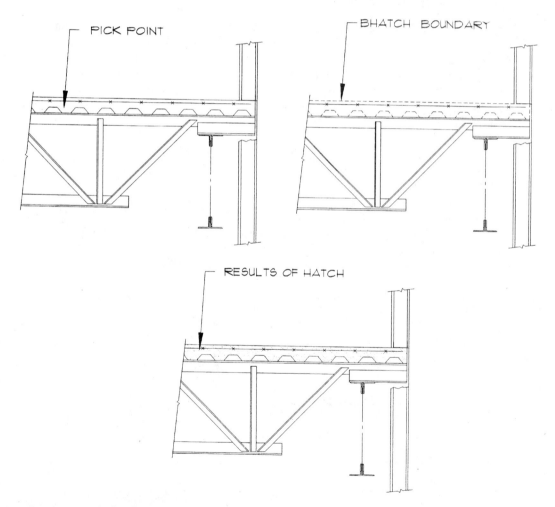

FIGURE 16–14 Using BHATCH to add a pattern to a concrete floor slab. Once the area to receive the pattern is selected the boundary will be highlighted. The AR-SAND pattern at a scale of .5 and an angle of 0 was selected. With this pattern, the angle is unimportant.

object with the hatch pattern as seen in Figure 16–15. Notice that the boundary is still dotted, which is a reminder that the hatch pattern has not yet been added to the drawing file. Select the Continue option to restore the Boundary Hatch dialog box. If the pattern is acceptable pick the Apply button or the [enter] key. If the preview did not display the intended results, any of the pattern options can be altered. In Figure 16–16, a new hatch pattern was selected by returning to the Hatch options dialog box. The Preview Hatch option was selected. Using BHATCH Preview to test alternatives can go on indefinitely. The final change is to adjust the Angle. The results can be seen in Figure 16–17. Since this view was acceptable, [enter] was used to return to the Boundary Hatch dialog box, and Apply button was selected.

Removing Islands

Figure 16–18 shows a concrete floor slab with four islands. If the boundary is selected using the Select Objects < option, each of the islands will be hatched. If the object is selected using the Pick Points < option, the Remove Islands < option box will be activated. Islands are automatically excluded from the hatch boundary. Selecting this box allows islands to be selected to be included in the hatch boundary. The example on the right side of Figure 16–18 shows the results of removing two of the islands from the boundary set.

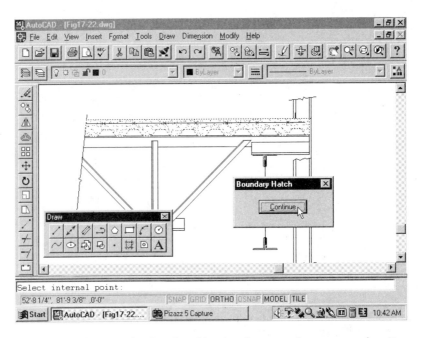

FIGURE 16–15 By selecting the Preview Hatch < button, the proposed pattern will be displayed.

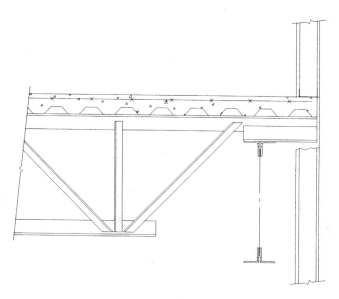

FIGURE 16–16 If the Preview is acceptable, it can be added to the drawing base by picking the Apply box. A new pattern can be installed by returning to the Hatch Options dialog box.

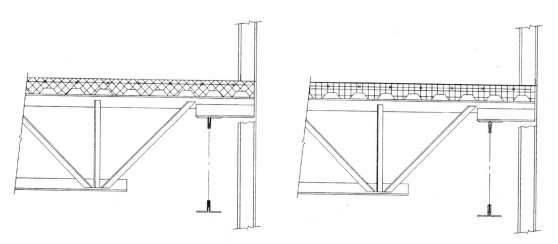

FIGURE 16–17 The preview option can be used to change the pattern, the scale and the angle options.

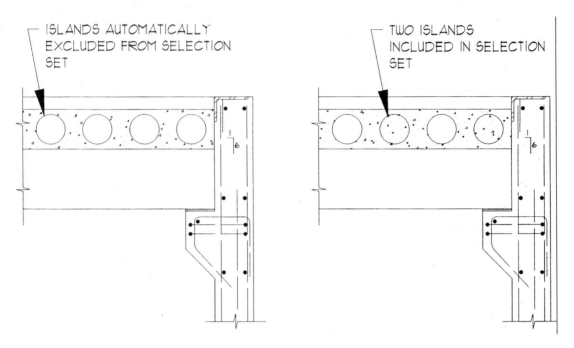

ISLANDS AUTOMATICALLY
EXCLUDED FROM SELECTION
SET

TWO ISLANDS
INCLUDED IN SELECTION
SET

FIGURE 16–18 The Remove Islands < option can be used to control the placement of the pattern within islands within the boundary.

View Selection <

With all of the options available for altering the boundary or patterns used in a BHATCH you might lose track of what options you've selected. The View Selections < option allows the display of the hatch boundary to be displayed.

Boundary Errors

Occasionally BHATCH will produce an error message when the Pick Points < option is used. The message shown in Figure 16–19 occurs when there is a break in the boundary. Select the OK button to resume the command. Selecting the Select Object < option may allow the object to be hatched if the break is small. If the error is too great, the error will need to be fixed before the pattern can be applied. In this example, the hole is obvious. For smaller errors, you may need to exit the BHATCH and use a ZOOM to find the error.

Advanced Options

Selecting the Advanced... button displays the Advanced Options dialog box which can be used to increase the efficiency of selecting bhatch boundaries. Unless you're working on a very large drawing, you will probably not need this option. The BHATCH command evaluates all objects displayed on the screen when you select

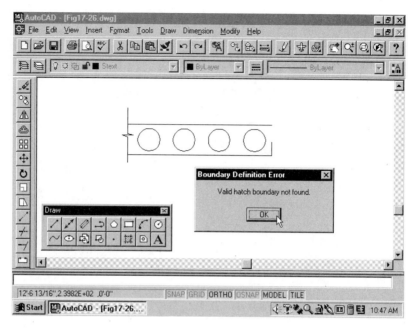

FIGURE 16-19 Proceeding with a BHATCH once this message is displayed may produce an erratic pattern.

objects to be hatched using the Pick Point < method. If you are working on a very large structure, the Advanced... options can speed the boundary selection process. The dialog box can be seen in Figure 16–20. The Advanced Options dialog box is divided into the areas of Define Boundary Set, Boundary Options, and Boundary Style.

Define the Boundary Set

This portion of the dialog box defines how the selection set for the hatch pattern will be selected. By fine tuning the selection set, the BHATCH command functions quicker because AutoCAD has to examine fewer objects to define the selection set. The default selection to define boundaries is From Everything on Screen, and Island Detection. Using these two options, the outermost boundary of the object will be used to define the selection set, and internal areas will be removed and not receive the hatch pattern.

Selecting Make New Boundary Set < button will remove the dialog box from the screen and allows a new boundary to be selected. The existing boundary set will be discarded and you will be allowed to construct a new boundary set to define the area to be hatched. Once selected, the dialog box will be returned and the default setting is now From Existing Boundary Set. An unlimited number of boundary sets can be selected. The last set selected will remain the default until a new boundary set is selected. When the From Everything on Screen button is selected, the selection set

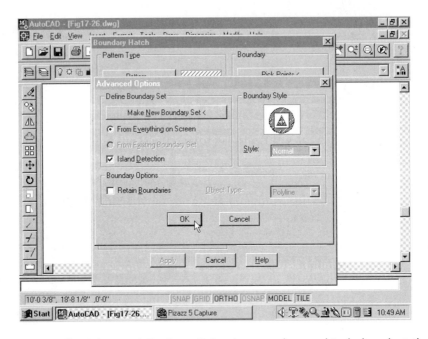

FIGURE 16–20 The Advanced Options dialog box can be used to help select the boundary.

will be created from everything that is visible in the current viewport. Selecting the option when there is a current boundary set will discard the current selection and place the pattern in everything in the current viewport. The normal setting is to place the pattern based on the From Existing Boundary Set. The final option of this portion of the box is the Island Detection box. In its current state, internal features (islands) are selected to be part of the boundary set. When the option is not current, AutoCAD sends an invisible object called a *ray* to the nearest object it can find and then traces the boundary in a counterclockwise direction. You might be tempted to store this with other boring facts about AutoCAD that you wish you never knew, but it really is important when using BHATCH. If the ray hits text or an internal boundary of an object, the boundary definition error is displayed.

Controlling Style. The Style option of the BHATCH command allows three options to be used for displaying the hatch pattern. Each of these options can be controlled using the Advanced Options of the Boundary Hatch dialog box. The Normal setting is the default and the image tile reflects how the pattern will be displayed. Picking the arrow beside the box displays the options shown in Figure 16–21. Options may be changed by selecting one of the options from the list, or by clicking on the icon. Each time the icon is selected, the display is changed to reflect the current style.

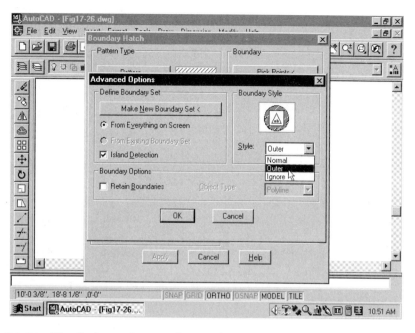

FIGURE 16–21 The Style: option can be used to control which boundaries will be observed as the pattern is displayed.

Normal. Figure 16–22(a) shows how hatching sets made up of multiple objects will be hatched when the default of Normal style is selected and each of the objects is selected as a boundry. Normal hatching style works inward, starting at the area boundary, and proceeds until another boundary is found. The pattern is turned off until another boundary is discovered.

Outer. Figure 16–22(b) shows how hatching sets made up of multiple objects will be hatched when the Outer option is selected and each of the objects is selected as a boundary. The Outer hatching style works inward, starting at the area boundary, and proceeds until another boundary is found. The pattern is turned off at the second boundary.

Ignore. Figure 16–22(c) shows how multiple objects will be hatched when the Ignore option is selected. The Outer hatching style works inward, starting at the area boundary, and passes through all objects within that boundary.

Boundary Options. The third area of the Advanced Options dialog box controls boundary options for applying the hatch pattern. The options for the boundary include Retain Boundaries and Object Type. The Retain Boundaries option specifies whether or not a temporary boundary will be added to the drawing. Boundaries for hatch patterns can be specified by Object Type. These options are not active unless

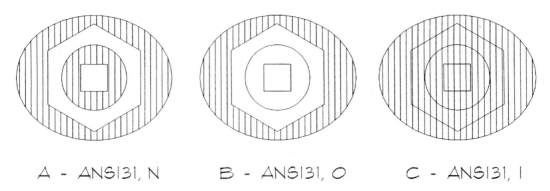

A - ANSI3I, N   B - ANSI3I, O   C - ANSI3I, I

FIGURE 16–22 The style option controls how the pattern will be displayed around multiple object sets. The N (normal) option works inward until another boundary is found. The pattern is not displayed until another boundary is found. The O (outer) option will only display the pattern in the outer boundary area. The I (ignore) option ignores all boundaries and hatches all areas within the outer boundary.

the Retain Boundaries option is active. Object Type controls the type of new boundary to be created. Options are Region and Polyline.

Inherit Properties

This option is a great option to use to hatch an object using an existing pattern when you can't remember the qualities of the existing hatch pattern. Picking the Inherit Properties < button will close the Boundary Hatch dialog box and return the drawing area with the prompt:

Select hatch object: (*select the hatched pattern that you would like to duplicate*)

Once the pattern is selected, the Boundary Hatch dialog box will be restored and the Pattern Properties portion of the box will be revised to reflect the properties of the selected pattern.

Attributes

The final portion of the Boundary Hatch dialog box to be examined is the Attributes box. The Attributes box contains toggles for Associative and Explode.

Associative Patterns. With the Associative box active, as an object containing a hatch pattern is edited, so is the hatch. With Associative deactivated, if an object with a hatch pattern is altered, the object is altered, but the hatch pattern remains unaffected. Figure 16–23 shows the affect of stretching two steel beams and the affect of an associative pattern.

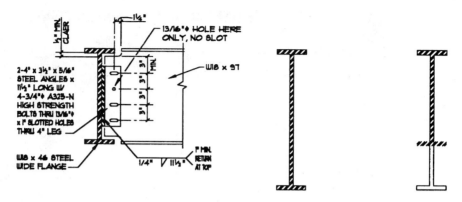

FIGURE 16–23 With Associative active, stretching a steel beam will alter the hatch pattern. With the box inactive, the beam is stretched, but the hatch pattern is unaltered.

Exploded. Pick a hatch pattern to be edited, and the entire pattern is selected. This option allows the hatch pattern to be created using individual line segments. With the Exploded option active, each segment of the pattern will respond as an individual entity. To edit the pattern, selection options such as Window or Crossing would need to be used. The EXPLODE command can also be used to alter the qualities of a hatch pattern block. The command will be introduced in Chapter 21.

USING THE HATCH COMMAND

The HATCH command allows patterns to be applied to an object from the command prompt. HATCH works well once a pattern has been created, and you want to reuse the pattern. Although the process should not be the first choice of applying hatch patterns for new users of AutoCAD, the command allows patterns to be placed in a drawing without the use of the preview option. When you're unsure of the results of creating a pattern, stay with the BHATCH command.

The HATCH command can be started entering **–H** [enter] or **HATCH** [enter] at the command prompt. Three prompts will be given to complete the command sequence. The sequence is:

> Command: **–H** [enter]
> Enter pattern name or [?/Solid/User defined]<ANSI31>: **AR-CONC** [enter]
> Scale for pattern <1.000>: [enter]
> Angle of pattern <0>: [enter]
> Select object: *(Select object to be hatched.)*
> Command:

The command sequence and results can be seen in Figure 16–24.

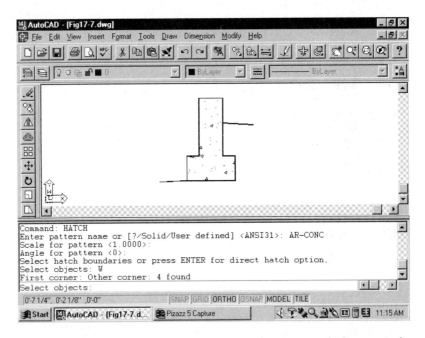

FIGURE 16–24 The HATCH command requires a pattern, a scale factor, and an angle of rotation to be selected.

Selecting Hatch Value

By responding to the pattern prompt with a question mark, a listing of hatch patterns shown in Figure 16–25 will be displayed. This list will be helpful once you've gained experience with hatching and recognize the pattern names. Figure 16–8 shows the standard hatch patterns available through AutoCAD. These patterns can be accessed through the Help command by selecting Standard Libraries from Command Reference. For now, type in the pattern name **AR-CONC**. This will produce the prompt:

Scale for pattern <1.000>: ⏎

The Scale prompt sets the size of the pattern elements with the default of one drawing unit just as the option for BHATCH. The pattern angle default is 0. Entering a value other than zero allows the hatchet pattern to be rotated. By entering an angle of 45°, pattern ANSI31 can be used to represent repetitive vertical lines.

Selecting Objects to be Hatched

Once the three options controlling the hatch pattern have been set, you will be prompted to provide the object to be hatched. The prompt is:

Select objects:

SOLID	-	SOLID FILL	FLEX	-	FLEXIBLE MATERIAL
ANGLE	-	ANGLE STEEL	GRASS	-	GRASS AREA
ANSI 31	-	ANSI IRON, BRICK, STONE, MASONRY	GRATE	-	GRATED AREA
ANSI 32	-	ANSI STEEL	HEX	-	HEXAGONS
ANSI 33	-	ANSI BRONZE, BRASS, COPPER	HONEY	-	HONEYCOMB PATTERN
ANSI 34	-	ANSI PLASTIC, RUBBER	HOUND	-	HOUNDSTOOTH CHECK
ANSI 35	-	ANSI FIRE BRICK, REFRACTORY MATERIAL	INSUL	-	INSULATION MATERIAL
ANSI 36	-	ANSI MARBLE, SLATE, GLASS	LINE	-	PARALLEL HORIZONTAL LINES
ANSI 37	-	ANSI LEAD, ZINC, MAGNESIUM, SOUND/HEAT/	MUDST	-	MUD AND SAND
		ELEC INSULATION	NET	-	HORIZONTAL / VERTICAL GRID
ANSI 38	-	ANSI ALUMINUM	NET3	-	NETWORK PATTERN 0-60-120
AR-B816	-	8X16 BLOCK ELEVATION STRETCHER BOND	PLAST	-	PLASTIC MATERIAL
AR-B816C	-	8X16 BLOCK ELEVATION STRETCHER BOND	PLASTI	-	PLASTIC MATERIAL
		WITH MORTAR JOINTS	SACNCR	-	CONCRETE
AR-B88	-	8X8 BLOCK ELEVATION STRETCHER BOND	SQUARE	-	SMALL ALIGNED SQUARES
AR-BRELM	-	STANDARD BRICK ELEVATION ENGLISH BOND	STARS	-	STAR OF DAVID
		WITH MORTAR JOINTS	STEEL	-	STEEL MATERIAL
AR-BRSTD	-	STANDARD BRICK ELEVATION STRETCHER BOND	SWAMP	-	SWAMPY AREA
AR-CONC	-	RANDOM DOT AND STONE PATTERN	TRANS	-	HEAT TRANSFER MATERIAL
AR-HBONE	-	STANDARD BRICK HERRINGBONE PATTERN @ 45°	TRIANG	-	EQUILATERAL TRIANGLES
AR-PARQ1	-	2X12 PARQUET FLOORING PATTERN OF 12X12	ZIGZAG	-	STAIRCASE EFFECT
AR-RROOF	-	ROOF SHINGLE TEXTURE	ACAD_ISO02W100	-	DASHED LINE
AR-RSHKE	-	ROOF WOOD SHAKE TEXTURE	ACAD_ISO03W100	-	DASHES SPACE LINE
AR-SAND	-	RANDOM DOT PATTERN	ACAD_ISO04W100	-	LONG DASHED DOTTED LINE
BOX	-	BOX STEEL	ACAD_ISO05W100	-	LONG DASHED DOUBLE LINE
BRASS	-	BRASS MATERIAL	ACAD_ISO06W100	-	LONG DASH TRIPLICATE DOTTED LINE
BRICK	-	BRICK OR MASONRY-TYPE SURFACE	ACAD_ISO07W100	-	DOTTED LINE
BRSTONE	-	BRICK AND STONE	ACAD_ISO08W100	-	LONG DASHED SHORT DASHED LINE
CLAY	-	CLAY MATERIAL	ACAD_ISO09W100	-	LONG DASHED DOUBLE-SHORT-DASHED LINE
CORK	-	CORK MATERIAL	ACAD_ISO10W100	-	DASHED DOTTED LINE
CROSS	-	A SERIES OF CROSSES	ACAD_ISO11W100	-	DOUBLE-DASHED DOTTED LINE
DASH	-	DASHED LINES	ACAD_ISO12W100	-	DASHED DOUBLE-DOTTED LINE
DOLMIT	-	GEOLOGICAL ROCK LAYERING	ACAD_ISO13W100	-	DOUBLE-DASHED DOUBLE-DOTTED LINE
DOTS	-	A SERIES OF DOTS	ACAD_ISO14W100	-	DASHED TRIPLICATE-DOTTED LINE
EARTH	-	EARTH OR GROUND (SUBTERRANEAN)	ACAD_ISO15W100	-	DOUBLE-DASHED TRIPLICATE-DOTTED LINE
ESCHER	-	ESCHER PATTERN			

FIGURE 16–25 By responding to the Pattern prompt with **?** (enter), a written list of hatch patterns will be displayed. (Courtesy AutoDESK, Inc.)

and any of the selection methods introduced in Chapter 11 may be used to define the objects to be hatched. Figure 16–26 shows the completed process for hatching the masonry walls of a fireplace shown in plan view.

One of the difficulties of the HATCH command is deciding how to select objects to be hatched when the area is an irregular shape. Problems arise depending on the order that lines are drawn in the area to be hatched and whether OSNAP has been used. Boundaries enclosing a hatch pattern must intersect or the HATCH pattern will exceed the boundaries. Figure 16–27 shows an example of the problems that can be encountered if the hatch boundary is not closed.

Users of older versions of AutoCAD have typically required the area to be hatched to be retraced with a polyline with a width of 0, thereby solving both problems. Using a polyline for the boundary also simplifies the boundary selection process, since only one point on the boundary must be picked. If you did not create an object to be hatched using a polyline, AutoCAD will allow you to create a pline border in the middle of the hatch sequence using a process called *direct hatching*. The polyline created in direct hatching will have a 0 width and is not permanent. This option is useful if you want to hatch a portion of an object that is not outlined by a line, such as the zone map shown in Figure 16–28. The command sequence to hatch the floor plan using the direct hatching option is:

Command: **–H** (enter)
Enter pattern name or [?/Solid/User defined]<ANSI31>: (enter)

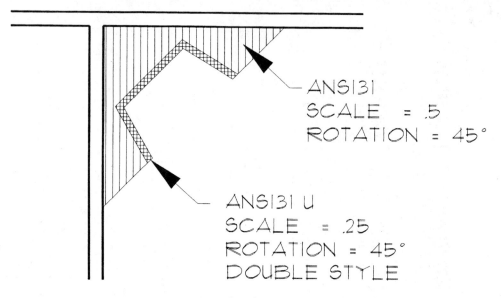

ANS131
SCALE = .5
ROTATION = 45°

ANS131 U
SCALE = .25
ROTATION = 45°
DOUBLE STYLE

FIGURE 16–26 Hatch parameters for a fireplace.

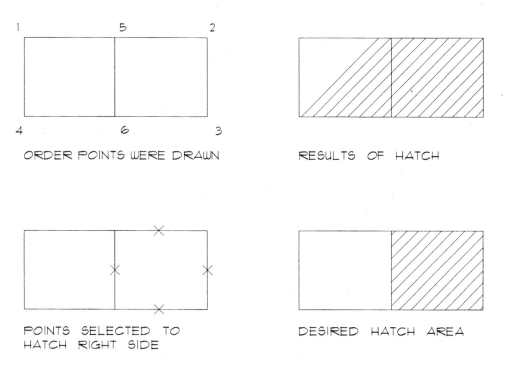

ORDER POINTS WERE DRAWN

RESULTS OF HATCH

POINTS SELECTED TO
HATCH RIGHT SIDE

DESIRED HATCH AREA

FIGURE 16–27 The relationship of boundary lines to each other can often affect the HACTH pattern that is created.

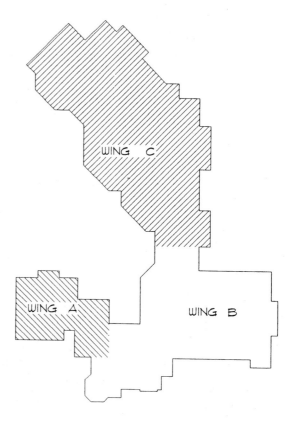

FIGURE 16–28 Creating a polyline while in the HACTH command can often be helpful in defining the limits of the HATCH. Accepting the default option of N for Retain polyline will remove the pline once the pattern is displayed.

Scale for pattern <1.000>: [enter]
Angle for pattern <0>: [enter]
Select hatch boundaries or RETURN for direct hatch option,
Select objects: [enter]
Retain polyline? <N> [enter]
From point: (*Select first point of hatch boundary.*)
Arc/Close/Length/Undo/<Next point> (*Select second point of hatch boundary.*)
Arc/Close/Length/Undo/<Next point> (*Select third point of hatch boundary.*)
Arc/Close/Length/Undo/<Next point> (*Select forth point of hatch boundary.*)
Arc/Close/Length/Undo/<Next point> (*Select fifth point of hatch boundary.*)
Arc/Close/Length/Undo/<Next point> **C** [enter]
From point or RETURN to apply hatch: [enter]

As you apply the hatch pattern, the polyline is removed. The polyline will be left if you enter **Y** at the Retain polyline? <N> prompt. The BHATCH command, can also be used to solve most of the boundary selection problems associated with the HATCH command.

Hatching Around Text

Although TEXT will not be introduced until Chapter 18, you will need to know how to avoid having text swallowed by the hatch pattern. By picking the boundary to be hatched and including the text as part of the boundary, you will automatically place an invisible box around the text that will act as part of the boundary. Figure 16–29 shows an example of this process. If you fail to select the text while selecting the hatch boundary, the hatch pattern will cover the text.

Creating Simple Hatch Patterns

One of the options for the HATCH process is "U." The U option of Hatch allows you to create simple patterns by answering prompts for angle, spacing and single or double patterns. The command sequence is:

Command: **–H** ⌨
Enter pattern name or [?/Solid/User defined]<ANSI31>: **U** ⌨

When you respond to the prompt with a **U**⌨, the balance of the hatch prompts will be altered to allow for the pattern to mirror itself, providing a doubled image or a simple pattern to be created. The command sequence is:

Command: **–H** ⌨
Enter pattern name or [?/Solid/User defined]<ANSI31>: **U** ⌨
Angle for crosshatch lines <0>: ⌨
Spacing between lines <1.0000>: ⌨
Double hatch area <N>: **Y** ⌨
Select objects: (*Pick desired object.*) ⌨

SELECTION SET

RESULTS OF HATCH SELECTION

FIGURE 16–29 Text will automatically be excluded from the area to be hatched if the text is selected as part of the selection set.

Select objects: [enter]
Command:

The results of using the U option with pattern ANSI31 can be seen in Figure 16–30.

Hatching Style

The Style option of the first Hatch prompt determines how surrounding areas of the hatch boundary will be affected. The three hatching style options are the same as the BHATCH command.

Normal. Normal hatching style works inward, starting at the area boundary, and proceeds until another boundary is found. The pattern is turned off until another boundary is discovered. The start of the command sequence to HATCH the objects is:

Command: **HATCH** [enter]
Enter pattern name or [?/Solid/User defined]<ANSI31>:N[enter]

Outer. The Outer hatching style works inward, starting at the area boundary, and proceeds until another boundary is found. The pattern is turned off at the second boundary. The start of the command sequence to HATCH the objects is:

Command: **HATCH** [enter]
Enter pattern name or [?/Solid/User defined]<ANSI31>:O[enter]

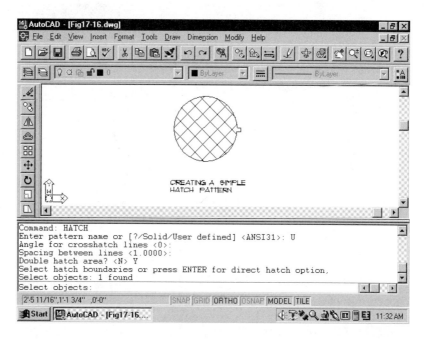

FIGURE 16–30 Selecting the **U** option when prompted for a pattern will allow simple patterns to be created.

Ignore. The Outer hatching style works inward, starting at the area boundary, and passes through all objects within that boundary. The start of the command sequence to HATCH the objects is:

Command: **HATCH** ⏎
Enter pattern name or [?/Solid/User defined]<ANSI31>: **I** ⏎

Editing Hatch Patterns

Features of hatch patterns created with BHATCH and HATCH can be edited using the HATCHEDIT command. The command can be started by selecting Edit Hatch from the MODIFY II toolbar, by selecting Hatch... from the Object flyout menu of the Modify pull-down menu, or by typing **HE** ⏎ or **HATCHEDIT** ⏎ at the command prompt. Each method will produce the prompt:

Command: **HE** ⏎
Select hatch object: (*Select hatch pattern to be edited.*)

As the pattern is selected, a Hatchedit dialog box similar to the one seen in Figure 16–31 will be displayed. The exact display will vary depending on which pattern is selected. The box is very similar to the Boundary Hatch box except the Boundary options are not active. Each of the features of the pattern can be altered and function as they did in the Boundary Hatch box.

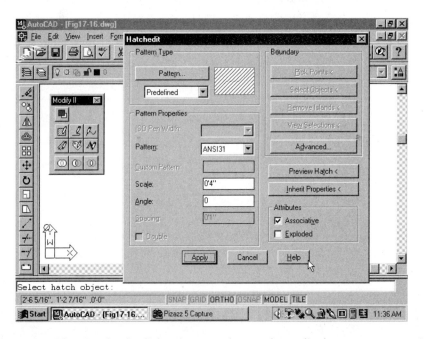

FIGURE 16–31 The Hatchedit dialog box can be used to edit the pattern once it has been added to the drawing base.

SOLIDS

The SOLID command allows objects that are outlined to be filled in. The fill pattern is provided when you specify three or four points to define the limits of the fill boundary. The results are an area defined with straight lines producing either a triangular or quadrilateral shape. Although the edge is a straight line, a curved surface can be filled by producing a series of small triangles. The command can be started by selecting 2D Solid from the Surfaces toolbar, by entering **SOLID** [enter] at the command prompt or by selecting 2D Solids from Surfaces of the Draw pull-down menu. The command sequence is:

Command: **SO** [enter]
First point: (*Pick point 1.*)
Second point: (*Pick point 2.*)
Third point: (*Pick point 3.*)
Fourth point: (*Pick point 4 if desired or* [enter] *to form a triangle.*)
Third point: (*Pick* [enter] *to end sequence.*) [enter]
Command:

The command sequence is repeated until the key [enter] is pressed.

The order in which the points are picked is of great importance. Figure 16–32 shows the results of a three-point sequence for selecting fill points. Figure 16–33 shows the results of three different sequences for entering points to define the fill. To define an

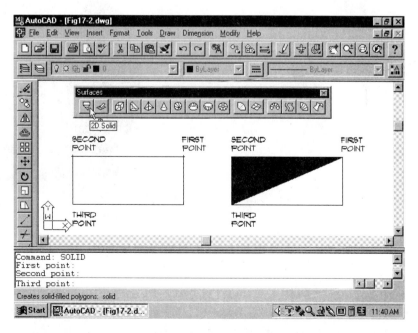

FIGURE 16–32 The order that points are entered affects the SOLID command.

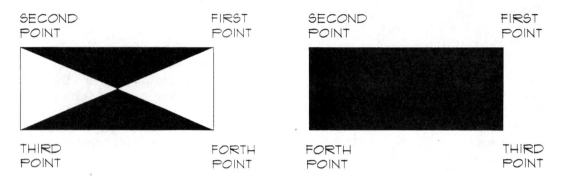

SECOND POINT   FIRST POINT   SECOND POINT   FIRST POINT

THIRD POINT   FORTH POINT   FORTH POINT   THIRD POINT

FIGURE 16–33 Changing the order of selection alters the FILL pattern that will be displayed.

irregular-shaped object, a combination of three- and four-point fills can be used. Figure 16–34 shows the process for filling an irregular-shaped object. This same area can be easily filled using the Solid option of the BHATCH command.

Once the object has been filled, faster operating times can be achieved by using the FILL OFF setting. In the OFF setting, only the outline of the fill pattern will be displayed.

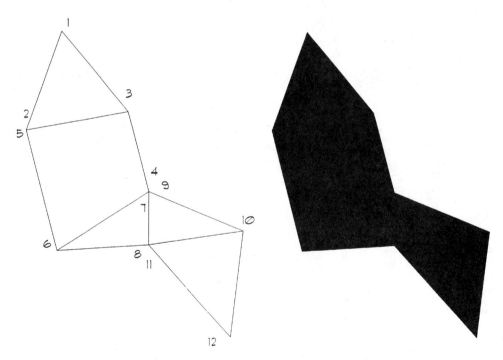

FIGURE 16–34 An irregular shape can be filled by using a series of triangles, or by using the Solid pattern of the BHATCH or HATCH commands.

CHAPTER 16 EXERCISES

E-16-1. Enter a new drawing and use the following coordinates to draw the required object.

Start: 1,1	B. = @ 3,0	C. = @ –5,.25	D. = @ –.75,1	E.= @ –1,1
F. = @ .75,1	G. = @ .5,1	H. = @ 0,1	J. = @ .25,.25	
K. = –2.25,0	L. = @ .25,–.25	M. = @ 0,–1	N. = @ .75,–1	
P. = @ –1,–1	Q.= @ 0,–1	R. = @ 1,–.75	S. = C	

Make a copy of the completed shape. Use the SOLID command to fill one of the objects. Turn Fill to OFF. Use the ANSI31 hatch pattern and adjust the scale factor to a suitable scale to HATCH the object so that it appears solid. Save the drawing as E-16-1.

E-16-2. Enter a new drawing and use the drawing as a guide to draw a 15" diameter circle. Draw a 12" diameter circle with a centerpoint 10" above the first centerpoint. Array the 12" circle so that there are a total of six circles. Copy the pattern so that you have a total of three circular patterns. Create layers to separate the object from the hatching. Use the BHATCH command to form three completely different hatching styles. Use different hatch patterns for each area to be hatched. Save the drawing as E-16-2.

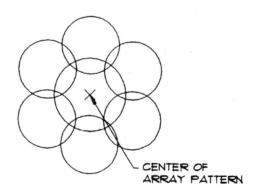

CENTER OF
ARRAY PATTERN

E-16-3. Enter a new drawing and draw a red square, surrounded by a blue circle, surrounded by a green octagon, surrounded by a cyan triangle. Copy this pattern so that there are a total of three copies. Create a separate layer for hatching. Use pattern ANSI34 and create a double hatching pattern. Provide examples of Normal, Outer, and Ignore styles of hatching. Save the drawing as E-16-3.

E-16-4. Open a new drawing and draw a one-story footing. Use a 6" wide stem wall extending from the natural grade 8". Support the wall on a 12" × 6" foundation. Show a 2 × 4 key, a 2 × 6 plate with a ½" diameter × 10" anchor bolt. Place wood, steel, and concrete on separate layers with different colors. Use varying widths of polylines to distinguish each material. Use EARTH to represent the soil and AR-CONC. to represent the footing and stem wall. Copy the drawing and show a 2 × 10 floor joist with ¾" plywood floor sheathing. Show the bottom of a 2 × 6 stud wall with single-wall construction. On the second copy, show a post and beam floor with 2" decking and a 6" deep beam. Save the drawing as E-16-4.

E-16-5. Draw the plan view of a 5' wide masonry fireplace. Use hatch patterns to distinguish between the fire brick of the firebox, and normal brick of the chimney. Save the drawing as Figure 16–5.

E-16-6. Open drawing 15-3 and create a layer for oak (green), maple (red), and Dutch elm (yellow). Change the trees from their existing layer to these new layers. Tree E is an oak, A is a maple, and the other three are Dutch elm. Hatch each type of tree with a different pattern. Save the drawing as E-16-6.

E-16-7. Open drawing E-9-5 and create a white SHADE layer. Hatch the glass area with DOTS, rotated at 45°. Adjust the scale so that the glass appears gray rather than black or white. Save the drawing as E-16-7.

E-16-8. Open drawing PB-FND (E-11-7). Create layers for FNDWALLS, FNDFOOT (dashedx2), FNDPIER (hidden), FNDCONC, FNDTXT, and FNDBEAM (center). Assign each layer a different color. Assume the wall to be made from concrete masonry units and hatch the wall with an appropriate pattern. Offset the wall lines to form a 16" wide footing centered on the wall. Trim the footings as required and change the footing lines to the appropriate layer. Save the drawing as E-16-8.

E-16-9. Open drawing E-14-9 and complete the drawing by providing the section. Use approximate sizes if dimensions are missing. Provide suitable hatch material and save as E-16-9.

E-16-10. A residence is 24 foot × 36 foot wide. The ridge runs 36 foot long and the roof is a 6/12 pitch with 24" overhangs, with a 2 × 6 fascia. The roof has a 12" overhang at the gable end walls. The finish floor is 18" above the finish grade and the plate height for the walls is 8'-0". You are to design an elevation of the 36-foot front side, which must include:

a. 3' × 6'-8" front door (show some type of appropriate decorative pattern).
b. 8' × 5' sliding/picture window with grids.
c. 18" × 60" shutters each side of window.
d. 12"-wide × 6"-deep stucco columns each side of window shutters with 6" horizontal siding above and below the window and shutters.
e. a 24"-wide × full-height area of brick on each side and above the front door (remember, brick should be approximately 8" long).
f. Hatch the roof with AR-ROOF.
g. The area (36") between the doorway brick and the stucco columns is to be AR-RSHKE. (Rows should be approximately 10" wide.)
h. All other areas to have vertical siding @ 8" o.c.
i. Show a shadow that extends 6" below the top of the door and window and show shade that would result from the sun being above and to the left of this gorgeous structure. (Hey! Be glad it's not in your neighborhood!).

Create a separate layer for each material and save this work of art as E-16-10.

CHAPTER 16 QUIZ

1. A circle surrounds a square, which encloses a hexagon. What style is used to hatch the circle and hexagon?

2. In the object described in question 1, what effect will Ignore have on each shape?

3. What shapes are used by the SOLID command to fill an object?

4. List the dialog boxes that are associated with BHATCH.

5. How can the Hatch Pattern Palette box be accessed?

6. List the options and the default values to control a hatch pattern that is created with the keyboard.

7. How can the scale factor of the BHATCH pattern be controlled by the plotted scale factor?

8. List three advantages of BHATCH over HATCH.

9. What process is used to keep the hatch pattern from crossing text?

_____

_____

10. Explain the differences in using the Pick option and the Select Objections options.

_____

_____

11. What is direct hatching?

_____

_____

12. Describe the process to create a hatch pattern of vertical lines at $\frac{1}{16}$" spacing.

_____

_____

13. What is the major difference between the BHATCH and HATCH commands.

_____

_____

14. What is the effect of using the *ANSI32 pattern?

_____

_____

15. What effect will the EXPLODE command have on a hatch pattern?

_____

_____

16. What does associative have to do with BHATCH

_____

_____

17. Where are hatch patterns stored?

_____

_____

18. You've defined a hatch pattern, but you're not sure how it will appear. List two methods to view the pattern.

_____

_____

19. What is an island?

_____

_____

20. What is the major area that the advanced options control?

_____

_____

CHAPTER

17

INQUIRY COMMANDS

As the drawings you're working with get more complex, you may find it hard to keep track of some of the drawing parameters you have established. The Inquiry toolbar and the Inquiry menu in the Tools pull-down menu contain five commands used to obtain information about a drawing. These commands are the DISTANCE, AREA, Mass Properties, LIST and LOCATE. Because Mass Properties is primarily used with 3D objects, it will not be covered in this chapter. If you're really curious and have an abundance of time, the command can be researched in the Help menu by searching for MASSPROP. The Inquiry menu in the Tools pull-down menu also contains the TIME, and STATUS commands. In addition to these commands for determining information about an objects geometry, the program's calculator will be examined as the CAL command is explored.

DISTANCE

The DIST or DISTANCE command measures the distance and angle between two selected points. This command can be especially helpful when used with SNAP or OSNAP locations to measure exact locations. DIST can be useful in determining the size of a room during the design stage, or checking an angle of an object from a baseline. The command can be accessed by selecting the Distance icon from the Inquiry toolbar, by picking Distance from the Inquiry option of the Tools pull-down menu, or by typing **DI** [enter] or **DIST** [enter] at the command prompt. The command sequence is:

Command: **DI** [enter]
Dist First point: (*Pick a point.*)
Dist Second point: (*Pick second point.*)

The resulting display for the points in Figure 17–1 would be:

Distance = 13'-11"
Angle in XY plane = 225.0
Angle from XY plane = 0.0
Delta X = –9'-10½"
Delta Y = –9'-10½"
Delta Z = 0'-0"
Command:

AREA

The AREA command can be used to determine the area and perimeter of an object. By using OSNAPS, the AREA command can be used to determine exact square footage sizes of structures and, at the same time, to determine wall lengths. A running total can be kept, and areas can be added or deleted from the total. The command can be accessed by selecting the Area icon from the Inquiry toolbar, by picking Area

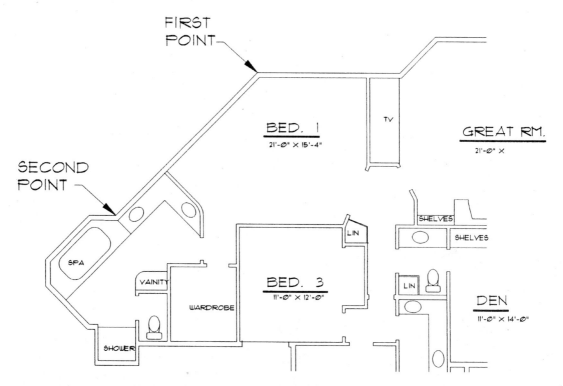

FIGURE 17–1 The DIST command measures the distance and angle between two selected points. (Courtesy Residential Designs)

from the Inquiry option of the Tools pull-down menu, or by typing **AA** [enter] or **AREA** [enter] at the command prompt. As the command is entered, four options are given:

<First point>/Object/Add/Subtract:

First point

When you accept the AREA First point option, you will be asked to supply a series of "Next points" to specify the area to be defined. Points may be picked with the pointing device or selected by any of the other selection methods. Using OSNAP will greatly increase the accuracy in computing the area. Figure 17–2 shows an example of computing the area of a retaining wall by using the point option. The command sequence is:

Command: **AA** [enter]
<First point>/Object/Add/Subtract: *(Pick first point.)*
Next point:

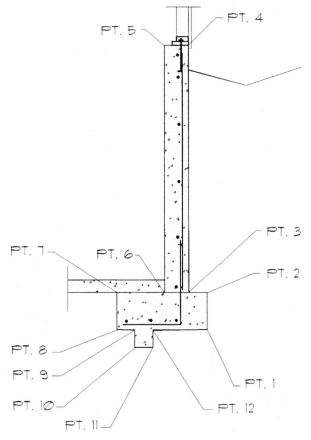

FIGURE 17–2 The AREA command can be used to compute the area of a specified shape. Notice that the area and the perimeter are given at the Command Line.

Continue to select points as required. In the example in Figure 17-2, press the ⏎ key at Point 12. This will produce a display listing the area (in square inches and square feet) and the perimeter. The same results will occur if Point 13 (Point 1) is entered as the ending point.

Object

By selecting a point, the Object option will compute the area of a circle, ellipse, spline, or an object composed of closed polylines. Picking a circle will cause the area and the circumference to be displayed. When a closed polyline is selected, the area and perimeter will be displayed. If an open polyline is selected, the area is computed based on an imaginary line extending from each end of the polyline. The command sequence is:

Command: **AA** ⏎
<First point>/Object/Add/Subtract: **O** ⏎
Select object: (*Select object.*)

Once the object is selected, a display of the area and circumference will be displayed. If an open polyline is selected, the program will calculate the area as if a line were drawn between the endpoints. (See Figure 17–3.) The line that is projected by AutoCAD is not included in the calculation of the area or perimeter. The command will list the area confined by the polyline and the length of the line.

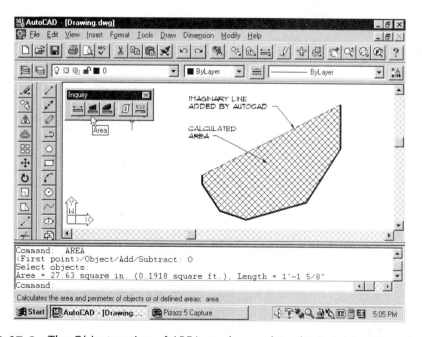

FIGURE 17–3 The Object option of AREA can be used to obtain information about an open or closed polyline. AutoCAD will provide the line required to close the polygon and calculate the area without displaying the line.

Add/Subtract

Using the Add option will allow each area measured in this mode to be displayed and added to a running total. The running total will be displayed after each addition. To begin a running total, select the Add option before picking the objects to be added. Using the Subtract option provides the opposite effect of Add. Each area measured in the Subtract mode will be displayed and then subtracted from the running total. With both the Add and Subtract options, the prompt will be altered to reflect the active mode. The selected option will stay in effect until the command is canceled. Figure 17–4 provides an example of computing the usable area of a parcel of land with the Add and Subtract options. The area of the lot was determined using running OSNAPS to pick each corner of the lot, and then subtracting the area of the residence and a garage, which are both outlined by a polyline. The command sequence is:

Command: **AA** enter
<First point>/Object/Add/Subtract: **A** enter

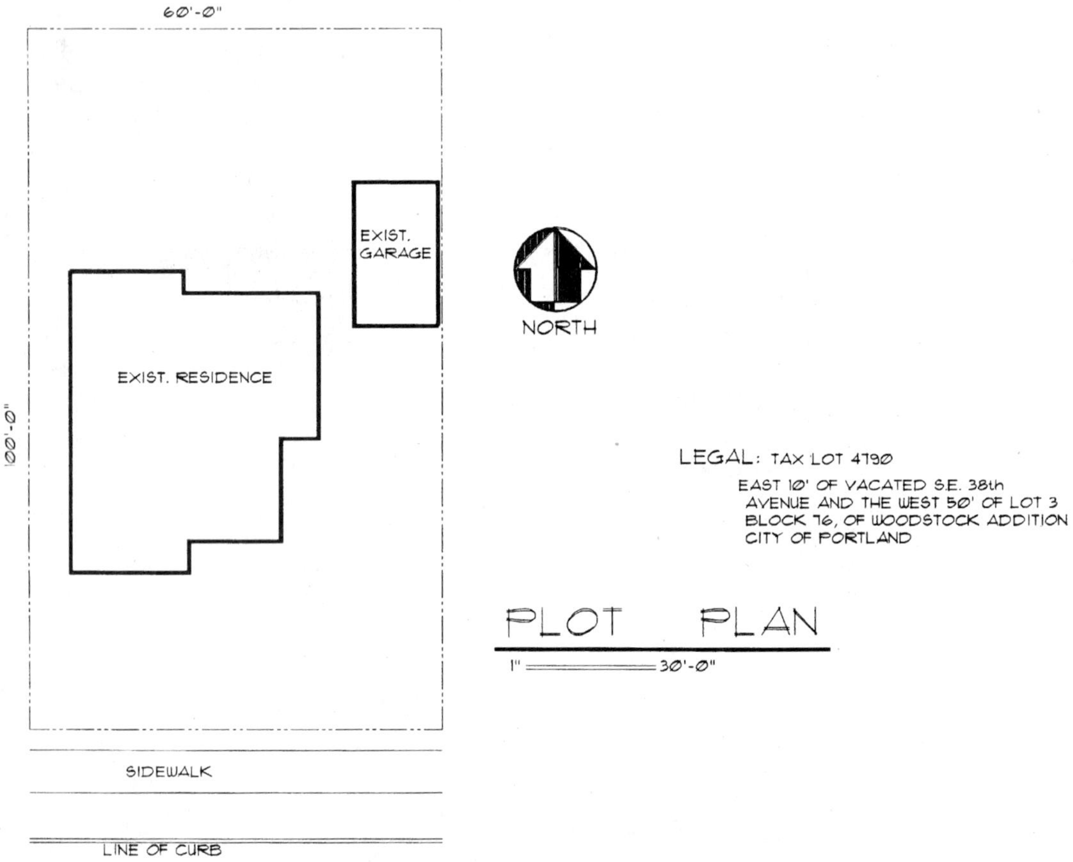

FIGURE 17–4. The ADD/SUBTRACT options can be used with the AREA command to add or subtract areas from a running total. (Courtesy Residential Designs)

<First point>/Object/Subtract:
(ADD mode) Next point: (*Pick point 1.*)
(ADD mode) Next point: (*Pick point 2.*)
(ADD mode) Next point: (*Pick point 3.*)
(ADD mode) Next point: (*Pick point 4.*) ⏎
Area = 864000.0 square in. (6000.000 square ft.), Perimeter = 320'-0"
Total area = 864000.0 square in. (6000.000 square ft.)
<First point>/Entity/Subtract: **S** ⏎
<First point>/Entity/Add: **E** ⏎
(Subtract mode) Select circle or polyline: (*Pick first object.*)
Area = 189154.2 square in. (1313.571 square ft.), Length = 118'-3½"
Total area = 674845.8 square in. (4686.429 square ft.)
(Subtract mode) Select circle or polyline: (*Pick second object.*)
<First point>/Entity/Subtract: **E** ⏎
(SUBTRACT mode) Select circle or polyline:
Area = 36365.9 square in. (252.541 square ft.), Length = 65'-7"
Total area = 638479.9 square in. (4433.888 square ft.)
(Subtract mode) Select circle or polyline: ⏎
<First point>/Entity/Add: ⏎
Command:

LIST

The LIST command displays a listing of information about an entity in the drawing. The command can be accessed by selecting the List icon from the Inquiry toolbar, by picking List from the Inquiry option in the Tools pull-down menu, or by typing **LI** ⏎, **LS** ⏎, or **LIST** ⏎ at the command prompt. The command sequence is:

Command: **LI** ⏎
Select objects: (*Pick an object.*)

One or several entities may be selected for listing, although the information may not fit on the screen all at once. To view additional pages, press the ⏎ key to resume the output display. Pressing **F2** will terminate the listing and return the command prompt. The window can also be closed by picking the Close button in the title bar.

Figure 17–5 shows a partial floor plan with a line selected for LIST and the resulting LIST for this line.

The information that is displayed depends on the type of entity that is selected, but the entity's type, position relative to the current UCS, the layer location, and either model or paper space are always listed. The color and linetype are listed if either is not set by layer. Four values are always given to describe a line. The X and Y values specify the horizontal and vertical distances of the "from point:" and the "to point" from the 0,0 point when the limits were established. The line in Figure 18–1 is 32'-1" to the right of the X axis. Notice that distance of the from point and the to

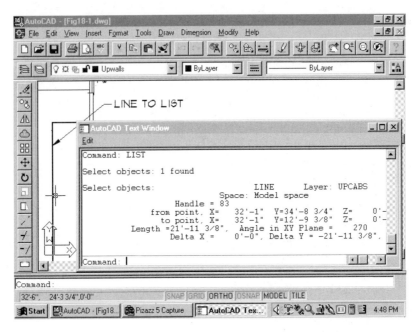

FIGURE 17–5 The LIST command can be used to supply information about a drawing entity.

point of the Y axis is 34'-8¾" and 12'-9⅜" respectively. The difference between these two values is 21'-11⅜" which is the specified length. The angle is 270°, indicating a vertical line. Figure 17–6 shows an example of each of the values on a vertical and an inclined line.

Listing circles will provide information based on the centerpoint and radius, the area, and the circumference. Listing a circle is an easy way to determine the area of a pier for doing concrete estimates. An example of a listing for the circle in Figure 17–7 is:

CIRCLE	Layer: FNDFOOT		
	Space: Model space		
	Handle = 2D		
centerpoint,	X = 119'-8″	Y = 103'-6″	Z = 0'-0″
radius	0'-9″		
circumference	4'-8⁹⁄₁₆″		
area	254.47 sq in.	1.7671 sq ft	

The LIST command can also be used to provide information about text in the drawing base. Text will be introduced in the next chapter. The listing to describe the text for the sliding glass door in Figure 17–5 is:

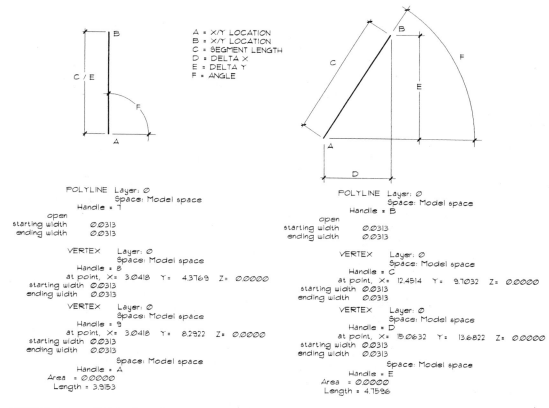

A = X/Y LOCATION
B = X/Y LOCATION
C = SEGMENT LENGTH
D = DELTA X
E = DELTA Y
F = ANGLE

POLYLINE Layer: Ø
        Space: Model space
    Handle = 7
  open
starting width   Ø.Ø313
ending width   Ø.Ø313

    VERTEX   Layer: Ø
             Space: Model space
    Handle = 8
    at point, X= 3.Ø418 Y= 4.3769 Z= Ø.ØØØØ
starting width Ø.Ø313
ending width   Ø.Ø313
    VERTEX   Layer: Ø
            Space: Model space
    Handle = 9
    at point, X= 3.Ø418 Y= 8.2922 Z= Ø.ØØØØ
starting width Ø.Ø313
ending width   Ø.Ø313
             Space: Model space
      Handle = A
Area = Ø.ØØØØ
Length = 3.9153

POLYLINE Layer: Ø
        Space: Model space
    Handle = B
  open
starting width   Ø.Ø313
ending width   Ø.Ø313

    VERTEX   Layer: Ø
             Space: Model space
    Handle = C
    at point, X= 12.4514 Y= 9.7Ø32 Z= Ø.ØØØØ
starting width Ø.Ø313
ending width   Ø.Ø313
    VERTEX   Layer: Ø
            Space: Model space
    Handle = D
    at point, X= 15.Ø632 Y= 13.6822 Z= Ø.ØØØØ
starting width Ø.Ø313
ending width   Ø.Ø313
             Space: Model space
      Handle = E
Area = Ø.ØØØØ
Length = 4.7596

FIGURE 17–6 LIST values for two lines.

TEXT	Layer: UPWINTXT
	Space: Model space
	Handle = 10
Style =	ARCHSTYLFont file = archstyl
start	point, X=27'-1 ⅞" Y=17'-3 ¾" Z= 0'-0"
height	0'-6"
text	12'-0" SLDG. TEMP.
rotation	angle 90
width	scale factor 1.000
obliquing	angle 0
generation	normal

DBLIST

The LIST command is used to obtain information about selected entities. The DBLIST command will produce information about the entire drawing. This com-

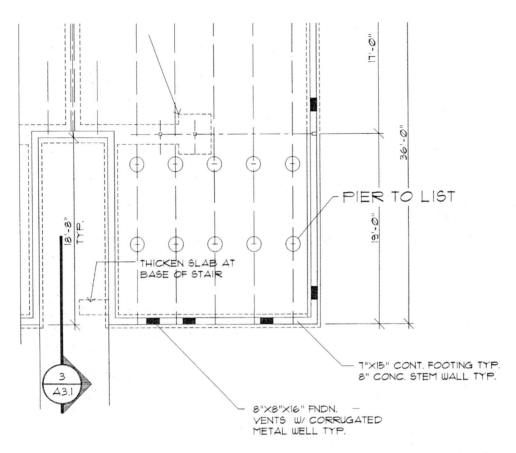

FIGURE 17–7 The LIST command will provide information based on the centerpoint, the radius, and the diameter. (Courtesy Scott R. Beck, Architect)

mand can be useful when comparing drawing sequences or information contained in two similar drawings. Information is presented in the same order that is used for LIST, one page at a time. The command can only be accessed from the command line. The command sequence is:

Command: **DBLIST** (enter)

Pressing the (enter) key will advance through the listing. Pressing the ESC key will terminate the scrolling of the listing, and picking F2 key will remove the text window and restore the drawing screen. The window can also be removed by picking Hide from file bar of the window control. Listings of both the DBLIST and LIST commands can be printed, if your terminal is connected to a printer, by pressing the PRINT SCREEN key.

IDENTIFYING A POINT

Information about a specific point in a drawing can be obtained by using the IDentify point command. This command will allow you to determine the exact location of a specific point. The command is named Locate Point in the Inquiry toolbar, and is called ID Point in the Inquiry menu of the Tools pull-down menu. The command can also be accessed or by typing **ID** [enter] at the command prompt. The sequence is:

> Command: **ID** [enter]
> Point: (*Pick a point.*)

Entering a point will display the corresponding X, Y, and Z coordinates. This can be helpful when determining the dimension of an object.

The ID command also can be used to identify a specific location in the drawing. By entering ID at the command prompt, coordinates can be supplied, and that point will be identified in the drawing. This can be useful with relative coordinates entry methods to start an entity at a specific location.

The ID listing for the location of intersection of the rail in Figure 17–5 would be:

> Command: **ID** [enter]
> Point: (*Select desired point. For this example the intersection*
> *if the handrail in Figure 18–1 was used.*)
> of X = 31'-11 ½"   Y = 34'-10 ¼"   Z = 0'-0"

TIME

The current time is displayed at the bottom of the drawing screen in the status bar. The TIME command uses the date and time maintained by the computer to keep track of the time related to the drawing session. The command can be accessed by selecting Time from Inquiry of the Tools pull-down menu, or by typing **TIME** [enter] at the command prompt. Before the command can function properly, the date and time must be set as the computer is configured. If the date or time are inaccurate, they can be set by accessing Date/Time of the Control Panel from the My Computer. If you reenter a drawing, AutoCAD can now provide the current time, drawing creation time, last update time, and time elapsed while editing the drawing. Each may seem unimportant in an educational setting, but the information is highly useful in keeping track of hours spent on a project for billing purposes and minor bookkeeping functions, such as determining profit margins.

Time Display Values

The time display is accessed by typing **TIME** [enter] on the command line. The display will resemble:

> Current time: 16Aug 1997 @ 11:15:25.510
> Times for this display:

Created:	20 Jul 1996 at 30:3106.970
Last updated:	11 Sep 1996 at 2:2506.970
Total editing time:	0 days 3:27:38.175
Elapsed timer (on)	0 days 1:59:51.510
Next automatic save in:	0 days 01:59.20.70
Display/On/Off/Reset:	
Command:	

Times are listed as day, month, year, hour, minute, seconds, and milliseconds.

The Current Time is the current date and time. Created time is marked when the drawing was initially started. The Last updated time lists the last editing session when either the END or SAVE command was used with the default filename. The total Editing time is the amount of time used editing the current drawing since it was created until SAVE is used. If QUIT is used to end a drawing session, the time in that session is not added to the accumulated time. Time spent printing and plotting is excluded from the Editing time. This time is updated continuously and cannot be reset or stopped. This timer is most helpful to supervisors to track efficiency and log total hours for billing.

The User Elapsed time can be reset and turned off or on. This timer is best suited for individual use each day to keep track of your time if you will be working on several different jobs. The Automatic Save time indicates when the next save is scheduled to occur (see Chapter 4).

Prompt Options

After the time display, four options are listed on the prompt line. The Display option will redisplay the TIME display and update the current values. The On option will activate the User Elapsed timer. The default value is set to ON when a drawing is started. The OFF option will stop the User Elapsed timer. The Reset option clears the User Elapsed timer to zero. The command line can be returned by a null response or the F2 key.

STATUS

The STATUS command will list the current value settings of a drawing. The command can be accessed by selecting Status from the Data pull-down menu, or by typing **STATUS** [enter] at the command prompt. The listings of the drawing shown in Figure 17–7 can be seen below.

3931 entities in FOUND		
Model Space Limits are:	X: 0'-0"	Y: 0'-0" (off)
	X: 275'-0"	Y: 250'-0"
Model space uses	X: 45'-3⅛"	Y: 48'-7½"
	X: 194'-11⅜"	Y: 152'-7⅜"
Display shows	X: 89'-5"	Y: 83'-6¾"

	X: 145'-0⅝"	Y: 123'-9 358"	
Insertion base is	X: 0'-0"	Y: 0'-0"	Z: 0'-0"
Snap resolution is	X: 0'-1"	Y: 0'-1"	
Grid spacing is	X: 2'-0"	Y: 2'-0"	

Current space: Model space
Current layer: FNDFOOT
Current color: BYLAYER—   1
Current linetype: BYLAYER—   DASHED
Current elevation: 0'-0"   thickness: 0'-0"
Fill on   Grid off   Ortho on   Qtext off   Snap off   Tablet off
Object snap modes: None
Free dwg disk (C:) space: 2047.7 Mbytes
Free temp disk (C:) space: 2047.7 Mbytes
Free physical memory: 2.6 Mbytes (out of 31.5M)
Press ENTER to continue: [enter]
Free swap file space: 1996.6 Mbytes (out of 2016.5M).
Command:

Several of the options, such as color, linetype, snap, grid, and ortho, should be very familiar by now. Other terms, may seem a little foreign. An important value to keep track of is Free Disk. The Free Disk value refers to the space that is available on the drive that contains the drawing. When this space is used up, AutoCAD will terminate the drawing session and save what has been done up to the point of termination. If you're working on the hard drive, this will rarely be a problem unless you never remove damaged files.

USING THE CALCULATOR IN AUTOCAD

The calculator of AutoCAD can be used just as a hand held calculator and also as a means of accurately locating points in a drawing. Typing **CAL** [enter] at the command prompt starts the geometry calculator that will compute standard mathematical formulas. The prompt at the command line is:

Command: **CAL** [enter]
>> Expression:

Peforming Mathematical Calculations

Mathematical calculations can now be completed by AutoCAD but specific symbols must be included as the values are entered. Symbols recognized by AutoCAD include:

+	Add	–	Subtract
*	Multiply	/	Divide
^	Exponents	( )	Grouped expressions

To add 4 and 4, enter **4+4** [enter] at the prompt. To subtract 50 from 100, enter **100–50** [enter] at the command prompt. To multiply 50 by 50 would require **50*50** [enter] to be entered. Dividing 100 by 50 would require an entry of **100/50** [enter] . Of course, none of these problems require a calculator, but the command is very useful for preforming advanced calculations without having to clean up your work area to find a pocket calculator. By entering **CAL**[enter] at the command prompt, the calculator can be used transparently. If the command line currently displays only one line of text, the answer will not be displayed.

In addition to preforming simple calculations, advanced formulas can be executed using the calculator. Parentheses are used to group symbols and values of complex mathematical equations into sets. CAL evaluates expressions based on standard mathematical procedures with expressions in parentheses first, starting with the innermost set. Operations are preformed by exponents first; followed by multiplication, division, addition, and subtraction. Operations of equal precedence are done from left to right.

Entering Feet and Inches in the Calculator. Numbers are entered into the calculator just as they are at the command prompt to enter coordinates. Ten feet three inches would be entered as 10'-3" or 10'3". If you are curious about how many inches that would be equal to, press the [enter] key. Feet/inch values will be converted to inch values. The process is:

```
Command: CAL [enter]
>> Expression: 10'-3" [enter]
123.0
Command:
```

Entering Angles. The default units of entering angle values are decimal degrees. Angles are entered using a format of 45d30'30". The minute (') and second (") symbols can be omitted if the values are zero. If the value is less than one degree, enter the value as 0d40'10". Angles can be entered in radians if the value is followed by the letter r (5.5r) or in grads when followed by the letter g (15g). No matter how the values are entered, they will be converted to decimal degrees.

Converting Units of Measurement. As metric dimensioning becomes more important in the construction industry, the need to convert values will increase. The calculator can be used to convert units of measurement from one system to another. Enter the value in the following format: **CVUNIT(value, from unit, to unit).** The process to convert 120 inches to millimeters is:

```
command: CAL [enter]
>> Expression: CVUNIT(120,in,mm) [enter]
3048
Command:
```

Additional values accepted by the calculator can be found in the CAL Help menu by researching Standard Numeric Functions.

Using the Calculator to Create or Edit Points

In addition to the normal functions of a calculator, the CAL command can be used to create or edit a drawing. Common functions include:

Function	Calculation to be performed
CUR	Allows a point to be selected using a cursor.
DIST	*(P1,P2)* Distance between point P1 and P2.
DPL	*(P,P1,P2)* Distance between point P and line P1-P2.
ILL	*(P1,P2,P3,P4)* Gives the intersection between lines P1-P2 and P3-P4.
PLD	*(P1,P2,Dist)* Point on a line P1-P2, that is the specified distance from P1.
PLT	*(P1,P2,T)* Point on line P1-P2 defined by the parameter T.
RAD	Gives the radius of a selected circle.
ROT	*(P,Origin, Angle)* Rotates point P through an angle about the origin.
ROT	*(P,P1ax, P2ax)* Rotates point P through angle using line P1-P2 as the rotation axis.

In addition to the functions listed above, some commonly used functions have shortcuts that combine some of the previous functions with the Endpoint snap mode. The shortcuts include:

Function	Shortcut for	Calculation to be performed
DEE	Dist(end,end)	Distance between two end points.
ILLE	ill(end,end,end,end)	The intersection of two lines defined by four endpoints.
MEE	(End +end)/2	Midpoint between two endpoints.
NEE	Nor(end,end)	Unit vector in the XY plane and normal to two endpoints.
VEE	Vec(end,end)	Vector from two endpoints.
VEE1	Vec1(end,end)	Unit vector from two endpoints.

To determine the length of a line would require the following sequence:

```
Command: CAL [enter]
>> Expression: DEE [enter]
>> Select one endpoint for DEE: (Select desired endpoint.)
>> Select another endpoint for DEE: (Select the other endpoint.)
10.2500 (Length given for specified line.)
Command:
```

To determine the intersection of two lines would require the following sequence:

Command: **CAL** [enter]
>> Expression: **ILLE** [enter]
>> Select one endpoint for ILLE:First line: *(Select first endpoint.)*
>> Select another endpoint for ILLE: First line: *(Select the other endpoint.)*
>> Select one endpoint for ILLE:Second line: *(Select first endpoint.)*
>>Select another endpoint for ILLE: Second line: *(Select the other endpoint.)*
(12.25 5.25,0.0) *(Intersection given for specified lines in XYZ coordinates.)*
Command:

To determine the radius of a circle would require the following sequence:

Command: **CAL** [enter]
>> Expression: **RAD** [enter]
Select circle, arc, or polyline segment for RAD function: *(Select circle.)*
1.13702 *(Radius for specified circle.)*
Command:

Using Object Snap Modes. Object snap modes can be used as part of arithmetic expressions in place of point coordinates. Prompts are automatically given for the OSNAP mode that is needed to pick the specified point. Only the first three letters are required for the prompts. Prompts accepted by AutoCAD include ENDpoint, INSert, INTersection, MIDpoint, CENter, NEArest, NODe, QUAdrant, PERpendicular, and TANgent.

Using the Calculator Effectively. So far the program calculator has been used as a tool to solve math problems. Its real value is in locating features within other features. Figure 17–8 shows an example of how the calculator can be used to place a circle at the centerpoint of a rectangle using the MEE function. MEE will locate the center between two endpoints. The opposite intersections will be used as endpoints. The command sequence is:

Command: **C** [enter]
CIRCLE 3P/2P/TTR/<Center point>: '**CAL** [enter]
>> Expression: **MEE** [enter]
Select one endpoint for MEE: *(Select one corner for endpoint.)*
Select another endpoint for MEE: *(select opposite corner for endpoint.)*
Diameter/<Radius> <0'-2">: *(Select point or provide coordinates for radius.)*
Command:

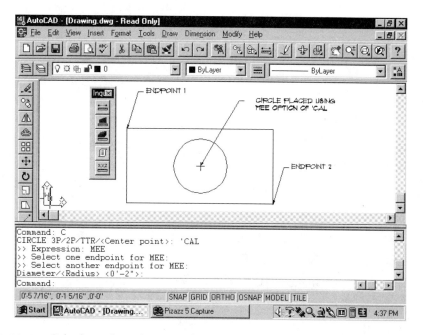

FIGURE 17–8 Calculator functions can be used to locate objects in a drawing. The MEE function is used to locate the centerpoint of a circle in the center of a rectangle.

CHAPTER 17 EXERCISES

E-17-1. Open exercise E-16-9 and determine the area of one side of the cross-sectional view of the manhole cover. Assume all unmarked radii to be .25". Maintain a thickness of ⅞" at the cover support. Compute the:

Surface area of one side the cover _____

Sectional area _____

Perimeter of section _____

Manhole surface area _____

Circumference _____

E-17-2. A steel fabricating company needs to order enough primer to cover 5,000 21½" × 8½" steel plates. Each plate has four ¾" diameter holes with the centerpoints located 2" down, 2" from plate edge and 3" o.c. The plates will be made of ½" steel. Draw the top, front and side views of the plate and determine the following information:

Surface area 1 side _____

Surface area 2 sides _____

Perimeter _____

Area of edges _____

Area of each hole _____

Total area of holes _____

Total area of 1 plate _____

Total area of 5,000 plates _____

E-17-3. Enter drawing E-17-2. Copy the listed times required to complete the project.

Created _____

Last updated _____

Total editing time _____

Terminate the drawing session using END or save.

E-17-4. Open E-15-3 and provide the following information.

Drawing name _____

Total entities _____

Limits _____   _____

_____   _____

Current layer _____

Current linetype _____

Free disk space _____

Status of: Grid _____   Snap _____   Ortho _____

Qtext _____   Tablet _____

Terminate the drawing with SAVE.

E-17-5. Open drawing E-9-5 and determine the surface area of the window.

AREA _____   PERIMETER _____

What is the distance between the end points of the inner side of the interior arch? _____

List the values that describe the outer arch at the top of the window.

_____

_____

_____

List the total editing time _____

Current drawing session time _____

Terminate the drawing using Save or End.

CHAPTER 17 QUIZ

1. Give four types of information that will be given if a circle is selected for a list.

   _____

   _____

   _____

   _____

2. Explain the difference between LIST and DBLIST.

   _____

   _____

   _____

   _____

3. Write the name of the command and describe two different uses for "Identifying a Point."

   _____

   _____

4. What are the two methods of selecting objects to determine their area?

   _____

   _____

   _____

5. What command is best suited to determine the length of a line between two points?

   _____

   _____

6. What is the process to set the internal clock of your computer?

   _____

   _____

7. What value of the status listing keeps track of available drawing space?

_____

_____

8. What is the meaning of DELTA X and DELTA Y?

_____

_____

9. You've opened a drawing, made a few changes, and then Discard the drawing. Will the time spent in this drawing be added to the accumulated time?

_____

10. Explain each of the categories used to measure time.

_____

_____

11. Describe how the AREA command can be used on an open polyline segment.

_____

_____

12. You need to determine the area of a rectangle. Explain the difference in selecting 3 and 4 points.

_____

_____

13. List the symbols used by calculator for add, subtract, multiply, divide, and set.

_____

_____

14. Determine the following values for your work station:

Free drawing space _____

Free temp disk space_____

Free physical memory _____

15. Determine how many millimeters are in 500 inches and how many feet are in 3000 mm. Explain the entries needed to determine the answer.

_____

_____

16. What is the circumference of a 4"-diameter circle to the nearest $\frac{1}{16}$"?

_____

17. What function of the calculator should be used to determine the radius of an existing circle?

    _____

    _____

18. Research the functions required for the calculator to determine the following functions:

    The cosine of an angle _____

    The square root of a positive number _____

19. What will ILLE calculator function do?

    _____

    _____

20. Research and describe the calculator function to rotate a point about an axis.

    _____

    _____

    _____

SECTION 4

PLACING TEXT AND DIMENSIONING

Placing Text on a Drawing

Introduction to Dimensioning

Placing Dimensions on Drawings

CHAPTER

18

PLACING TEXT ON A DRAWING

• •

If you have previous drafting or design experience you will still remember the long hours of practice required to develop your lettering skills. Throw away all the templates, lettering guides, and rub-on letters of manual drafting, because AutoCAD has a better way. Figure 18–1 shows an example of stock notes drawn using AutoCAD. Instead of lettering, you'll now be placing text using either the TEXT, DTEXT, QTEXT, or MTEXT commands. This chapter will introduce you to the basic TEXT, DTEXT, and MTEXT options, as well as methods to change the STYLE variables, load different FONTS, change existing text styles, edit existing text, and adjust existing text for quicker regenerations.

TEXT

Before text is placed on a drawing, several important factors should be considered. Each time a drawing is started, consider who will use the drawing, the information the text is to define, and the scale factor of the text. Once these factors have been considered, the text values can often be placed in a templet drawing and saved for future use.

General Text Considerations

Architectural and engineering offices often use two different styles of lettering. Many engineering offices that do civil, municipal, or government projects tend to use a simple block lettering shape. Architectural and engineering offices dealing with construction projects typically use a style of lettering that features thin vertical strokes, with thicker horizontal strokes to give a more artistic flair to the drawing. Other common variations are compressed and elongated letter shapes. In addition

WOOD:

1. MANUFACTURE OF THE STRUCTURAL GLU-LAMINATED TIMBER SHALL BE BY AN AITC APPROVED FACILITY AND SHALL BE IN CONFORMANCE WITH AMERICAN NATIONAL STANDARDS ANSI.AITC A190.1-1983. APPEARANCE GRADE SHALL BE INDUSTRIAL UNLESS NOTED OTHERWISE ON FRAMING PLANS.

2. ALL GLU-LAMINATED ADHESIVES SHALL BE FOR WET-USE CONFORMING TO ANSI. AITC A190.11983 SEC. 4.5.1.2 AND ASTM D2559-82..

3. ALL GLU-LAM BEAMS, JOIST, POST AND BLOCKING AT EDGES OF STRUCTURE SHALL BE PRESSURE TREATED W/ 0.05 PCF K-520 COPPER NAPHTHENATE IN LIGHT SOLVENT, IN ACCORDANCE WITH AWPA STANDARDS C-14, UNLESS NOTED. ALL CUTS AND BOLT HOLES SHALL BE MADE PRIOR TO TREATMENT, UNLESS NOTED. INADVERTENT FIELD DRILL HOLES AND CUTS SHALL BE REPEATEDLY FLOODED WITH 2% COPPER NAPHTHENATE, PRIOR TO FASTENER INSTALLATION IN CONFORMANCE WITH AWPA STANDARDS.

4. MINIMUM LUMBER GRADES SHALL BE AS FOLLOWS UNLESS NOTED ON PLANS:
   GLU-LAM BEAMS-SIMPLE SPANS. . . . . COASTAL D-FIR 24F-V4
   GLU-LAM CANTILEVERS . . . COASTAL D-FIR 24F-V8
   STUDS, JOIST, AND BLOCKING . . . COASTAL D-FIR #2 AND BETTER
   4 × 6 × POST AND BEAMS . . . COASTAL D-FIR #1 AND BETTER
   DECKING . . . CEDAR #1 OR BETTER

5. PRE-MANUFACTURED FRAMING CONNECTORS SHALL BE BY SIMPSON CO. OR APPROVED EQUAL. JOIST HANGER NAILS SHALL BE COMMON. CONNECTORS SHALL BE ATTACHED TO FRAMING MEMBERS WITH THE MAXIMUM SIZE AND NUMBER OF NAILS OR BOLTS AS SPECIFIED BY SIMPSON.

FIGURE 18–1 A partial listing of standard construction notes. (Courtesy Residential Designs)

to the basic shape of the letters, some offices use a forward or backward slant to make their office lettering more distinctive. You are unlikely to find all of these variations on one professional drawing. Figure 18–2 shows some of the variations that can be found on drawings.

Text Similarities

With all of the variation in styling, two major areas are typically uniform throughout architectural and engineering offices. Text height and capital letters are as close to an industry standard as architects and engineers get. Text placed in the drawing area is approximately ⅛" high. Titles are usually between ¼" and 1" in height, depending on office practice. The other common feature is that text almost always comprises capital letters, although there are exceptions based on office practice. The letter "d" is always printed in lowercase when representing the penny weight of a nail. A typical use would resemble:

USE (3)-16d @ EA. RAFT./PL CONNECTION.

Occasionally manufacturers use lowercase letters to represent the size of special nails or fasteners specific to their products. With the CAPS key activated, all lettering will be capitalized unless the SHIFT key is pressed. Pressing the SHIFT key will make any letters typed while the key is depressed lowercase letters. If you have typ-

ARCHITECTURAL OFFICES USE MANY
STYLES OF LETTERING.
SOME TEXT IS ELONGATED.
SOME OFFICES USE A COMPRESSED STYLE.
*NO MATTER THE STYLE THAT IS USED,
IT MUST BE EASY TO READ.*

FIGURE 18–2 Some common types of lettering found on construction drawings.

ing experience, it will take some time getting used to not having to press the SHIFT key at the start of each sentence.

Numbers are generally the same size as the text they are placed with. When a number is used to represent a quantity, it is generally separated from the balance of the note by a dash. This can be seen in the nailing specification above. Another common method of representing the quantity is to place the number in parentheses () to clarify the quantity and the size. This would resemble:

USE (3)-3/4" DIA. M.B. @ 3" O.C.

Fractions are generally placed side by side (3/4") rather than one above the other (¾) so that larger text size can be used. When a distance such as *one foot two and one half inches* is to be specified it may be done as 1'-2 1/2". The ' (foot) and " (inch) symbols are always used with the numbers separated with a dash. If metric sizes are to be represented on drawings, the use of a comma as a number separator is somewhat different than with English units. Metric sizes of four digits are often written with no comma. *Three thousand one hundred and fifty millimeters* is written as 3150 mm. When a number five digits or larger must be specified, a space is generally placed where the comma would normally be placed. *Fifty six thousand, three hundred and forty five millimeters* is written as 56 345mm. Other common methods of placing numbers will be presented in Chapter 19 as dimension methods are introduced.

Text Placement

Text placement refers to the location of text relative to the drawing and within the drawing file. Care should be taken with the text orientation so that it is generally placed parallel to the bottom or parallel to the right edge of the page. When describing structural material shown in plan view, text is ideally placed parallel to the member being described. Figure 18–3 shows methods of placing text in plan views. On plan views such as Figure 18–4 text is typically placed within the drawing, but arranged so that it does not interfere with any part of the drawing. Text is generally placed within 2" of the object being described. A leader line is used to connect

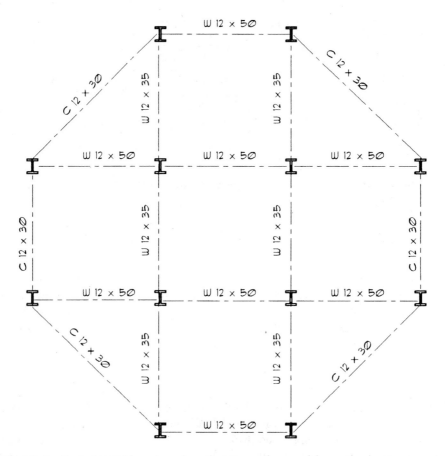

FIGURE 18–3 Text should be placed so that it can be read from the bottom or right side of the drawing.

the text to the drawing. The leader line may be either a straight line or an arc depending on office practice.

On details, text may be placed within the detail if large open spaces are part of the drawing. It is preferable to keep text out of the drawing. Text should be aligned to enhance clarity and may be either aligned left or right. Figure 18–5 shows an example of a detail using good text placement.

The final consideration of text placement is to decide on layer placement. Text should be placed on a layer with a major group code of ANNO. Minor group codes include KEYN (key notes), LEGN (legends and schedules), NOTE, NPLT (non-plotted text), REDL (redline), REVS (revisions), SYMB (symbols) TEXT, and TTLB (border and title block).

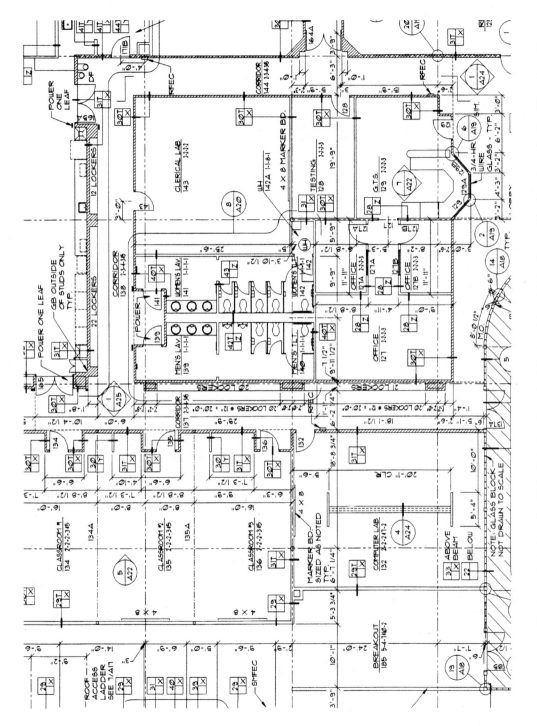

FIGURE 18-4 Text can be placed within the drawing, but it should not interfere with drawing clarity. (Courtesy Tom Kuhns, Michael & Kuhns, Architects, P.C.)

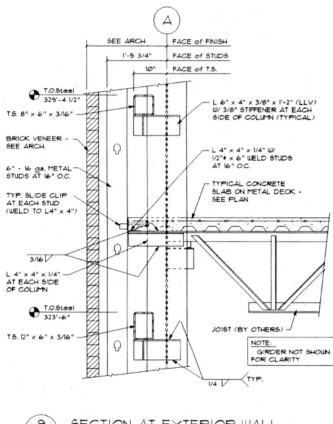

SECTION AT EXTERIOR WALL

FIGURE 18–5 Text should be aligned so that it can be quickly read. Courtesy Van Domelen / Looijenga / McGarrigle / Knauf, Consulting Engineers.

Types Of Text

Text on a drawing is considered to be either a general or local notes. *General notes* may refer to an entire project or to a specific drawing within a project. The notes shown in Figure 18–1 are general notes that specify the materials of a framing plan. *Local notes* refer to specific areas of a project such as the notes in Figure 18–4. Practical uses for including each into templet drawings will be discussed later in this chapter.

Lettering Height and Scale Factor

One of the first prompts you'll be given as you place text will be for the height. Manually, the height might have been ⅛" high text, but in AutoCAD you're drawing in real (model) space. In Chapter 15, the scale factor required to make line segments the desired size when plotted was considered. Similar factors must be applied to text height to produce the desired ⅛" high text when plotting is finished. Although text

height can be adjusted before plotting, it is best to multiply the desired height (⅛")
by the scale factor.

The text scale factor is the reciprocal of the drawing scale. For drawings at
¼" = 1'-0", this would be

$$\frac{4}{1} \times \frac{12}{1} = \frac{48}{1} \quad \text{or 48.}$$

By multiplying the desired height of ⅛" (.125) × 48 (the scale factor), you see that the
text should be 6" tall. Quarter-inch-high lettering should be 12" tall. You'll notice
that the text scale factor is always twice the line scale factor. Other common text
scale heights include:

ARCHITECTURAL VALUES		ENGINEERING VALUES	
Drawing scale	Text Scale	Drawing scale	Text Scale
1" = 1'-0"	12	1" = 1'-0"	12
¾" = 1'-0"	16	1' = 10'	120
½" = 1'-0"	24	1' = 100'	1200
⅜" = 1'-0"	32	1' = 20'	240
¼" = 1'-0"	48	1' = 200'	2400
¹⁄₁₆" = 1'-0"	64	1' = 30'	360
⅛" = 1'-0"	96	1' = 40'	480
³⁄₃₂" = 1'-0"	128	1' = 50'	600
¹⁄₁₆" = 1'-0"	192	1' = 60'	720

Working at a scale of 1" = 20', ⅛" high lettering would be required to be drawn as 30"
high text (.125" × 240).

USING THE TEXT COMMAND

The text command can be selected from the Draw pull-down menu by selecting the
Single-Line Text option, or by typing **TEXT** [enter] at the command prompt. The
results of the TEXT command will be slightly different when the command is
selected from the Draw pull–down menu or the command prompt. The TEXT command from the command prompt will be discussed first. The results of selecting the
command from the pull–down menu is similar to the DTEXT command and will be
discussed later in this chapter. Use the following sequence to start the command
from the command prompt:

Command: **TEXT** [enter]
Justify/Style/<Start point>:

Once you know the other options, the Text command may be started with any one of the options. Since the "Start point" is the default, it is a logical place to start.

Start Point. The name says it all. This is your opportunity to dictate where the start of the text will be placed. Indicating a point with the mouse will make that point the lower left corner of the text location. The relation of the text to the starting point can be seen in Figure 18–6. Once the start point has been picked, a prompt for the text height will be displayed.

Height. The second option to be selected specifies the height of the text, with the default of 0'-0 ³⁄₁₆" in height. If you'll be plotting this drawing at ¼" = 1'-0", enter **6** 🄴ⁿᵗᵉʳ for the height. If you have not adjusted the initial drawing limits, any text you create is going to appear to be of gigantic proportions.

Rotation Angle. The third option to be adjusted prior to entering text is the rotation angle of the text. This can be visualized by thinking of a line extending from the start point at the specified angle, with the text above this line. Figure 18–7 shows an example of text rotated at various angles. Although text typically is read looking from the bottom or right side of the drawing, this option will allow text to be aligned with specific objects of a drawing, as seen in Figure 18–8. The default of 0 rotation can be accepted by pressing the 🄴ⁿᵗᵉʳ key, or a numeric value can be entered.

Placing Text

The entire command sequence for placing text is:

    Command: **TEXT** 🄴ⁿᵗᵉʳ
    Justify/Style? (Start point.): (*Pick point.*)
    Height: <0'-0 3/16">: **6** 🄴ⁿᵗᵉʳ
    Rotation angle: <0>: 🄴ⁿᵗᵉʳ
    Text:

The desired text can now be entered by keyboard. Be sure to have the CAP key activated. Notice that as you type, nothing is displayed in the screen. The text can be seen at the TEXT display line, which has replaced the command prompt. When the text entry is completed, press the 🄴ⁿᵗᵉʳ key to display the text on the screen. To add additional lines of text, press the 🄴ⁿᵗᵉʳ key a second time and the new start position

FIGURE 18–6 The default placement for TEXT is to place text on the right side of the "Start point." This is specified as "Left Justified."

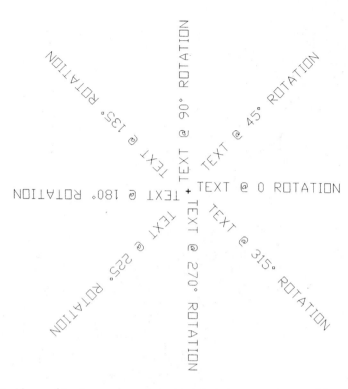

FIGURE 18–7 Lines of text may be rotated so that they can be read from any angle.

will be adjusted to be directly below the first line of text. The prompts for start point, height, and rotation angle will be skipped and the values for the first line of text will be reused. The spacing between lines of text will be half the height of the text. The command sequence is:

Command: **TEXT** [enter]
Justify/Style? (Start point): (*Pick point.*)
Height: <0'-0 1/4">: **6** [enter]
Rotation angle: <0>: [enter]
Text: **FINISH FLOOR** [enter]
Command: [enter]
TEXT Justify/Style? (Start point): [enter]
Text: **ELEV. 100.00'** [enter]
Command:

Figure 18–9 shows the results of this command sequence. Notice that when the [enter] key is pressed the first time, the last line of text entered will be highlighted. Don't worry, you aren't about to destroy it. The original text will be restored when the [enter] key is pressed a second time. As you work with multiple lines of text, each succeeding line will have all of the attributes of the original text. The DTEXT and MTEXT

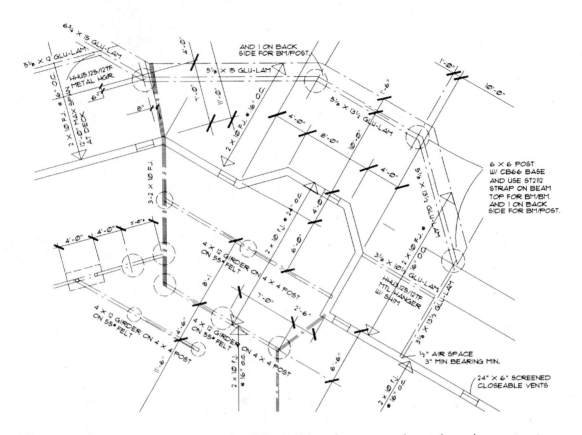

FIGURE 18–8 Text should be rotated to be parallel to the structural member when a structure is an irregular shape. (Courtesy Residential Designs)

FIGURE 18–9 The results of the TEXT command using a specified start point, a height of 6, and a rotation angle of 0.

commands, which will be discussed later in this chapter, can also be used to make multiple lines of text.

Justifying Text

To justify text is to determine where and how the text will be located within a defined space. The default for placing text is for left-justified text, which places the left edge of each new line of text directly in line with the line of text above. The lower left starting point is defined, and text will be placed to the right of the starting point. This method is used on construction drawings, such as sections, and details where large amounts of written information can be neatly aligned. Figure 18–10 shows an example of left-justified text. Thirteen other options are available for aligning text. Each option can be accessed by the following sequence:

Command: **TEXT** (enter)
Justify/Style/<Start point>: **J** (enter)
Align/Fit/Center/Middle/Right/TL/TC/TR/ML/MC/MR/BL/BC/BR: **R** (enter)

Once you feel comfortable with each option, the option can be accessed without entering the Justify menu by typing the letter of the option at the command prompt. The process to align text from the right can be started by the following sequence:

Command: **TEXT** (enter)
Justify/Style/<Start point>: **R** (enter)
End point:

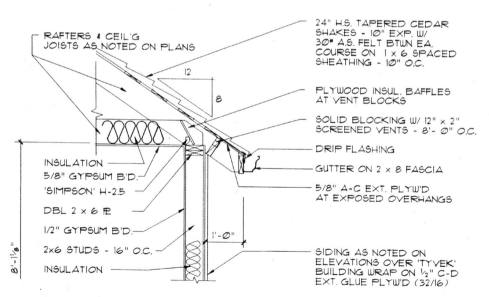

FIGURE 18–10 Left justified text helps add to the clarity of a drawing. (Courtesy Piercy & Barclay Designers, Inc., A.I.B.D.)

Even though Right is not one of the listed options, R will be accepted and will change the prompt to the End point prompt. This prompt will then be followed by the height and rotation prompts.

Align.   The aligned option will allow two useful options to be accessed. The starting and ending point can be selected so that text can be aligned with an inclined object, similar to how the Rotation option is used. This method of selection is superior to using Rotation when the angle is not known. The second benefit of Align is the ability to pick the start and endpoints of the line of text. The drawback to this option is that text height is assigned based on the amount of text to be placed between the start and end points. Figure 18–11 shows an example of aligned text. The command sequence is:

Command: **TEXT** ⏎
Justify/Style/<Start point>: **J** ⏎
Align/Fit/Center/Middle/Right/TL/TC/TR/ML/MC/MR/BL/BC/BR: **A** ⏎
First text line point: (*Pick starting point.*)
Second text line point: (*Pick ending point.*)
Text:

Fit.   The Fit option is very similar to the align option. Using Fit, the text height will remain constant throughout several different lines of text, but the width will be varied. This option uses a command sequence similar to Align, but allows for a height to be specified prior to placing the text. Figure 18–12 shows an example of text placed using the Fit option. The command sequence for Fit is:

Command: **TEXT** ⏎
Justify/Style/<Start point>: **J** ⏎
Align/Fit/Center/Middle/Right/TL/TC/TR/ML/MC/MR/BL/BC/BR: **F** ⏎

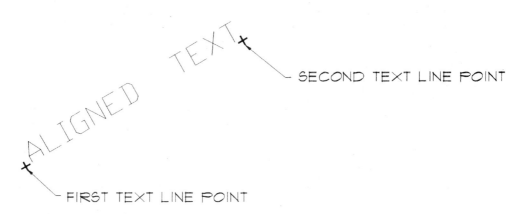

FIGURE 18–11 Text can be aligned by specifying two points. The desired text will now be placed between these points. If more than one line of text is required, the height of each line may vary.

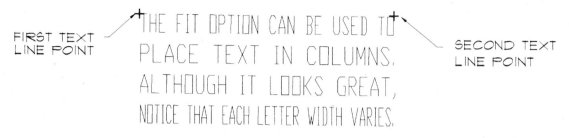

FIRST TEXT LINE POINT

SECOND TEXT LINE POINT

FIGURE 18–12 The FIT option will place text between two specified points, but the height will remain constant if more than two lines of text are required.

> First text line point: (*Pick starting point.*)
> Second text line point: (*Pick ending point.*)
> Height<0'-0 3/16">: **6** [enter]
> Text:

Center. The Center text option will place a line of text centered on a specified point, as seen in Figure 18–13. The command sequence is:

> Command: **TEXT** [enter]
> Justify/Style/<Start point>: **J** [enter]
> Align/Fit/Center/Middle/Right/TL/TC/TR/ML/MC/MR/BL/BC/BR: **C** [enter]
> Center point: (*Pick a point.*)
> Height: <0'-0 3/16">: **6** [enter]
> Rotation angle: <0>: [enter]
> Text:

To place several lines of text using the Center option, press the [enter] key twice after entering the first line of text. Move the cursor to the desired spacing below the first

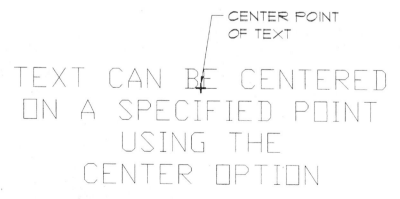

CENTER POINT OF TEXT

FIGURE 18–13 The CENTER option allows text to be spaced equally around a specified center point.

FIGURE 18–14 The MIDDLE justified option will stagger text along an imaginary line.

line and enter the location using the pick button. Type the second line of type and press ⏎ when the line is complete. This process can be continued indefinitely.

Middle. This option is similar to the effects of Center, but text is centered vertically and horizontally around the start point, as seen in Figure 18–14. The command sequence is similar to the sequence used for the Center option, once the Middle point is selected.

Right. This option is the reverse of the default left-justified option. Figure 18–15 shows an example of right-justified text.

Letter Options. The remaining nine options can be entered by typing the designated letters at the Justify prompt. The letters represent:

TL-top left	TC-top center	TR-top right
ML-middle left	MC-middle center	MR-middle right
BL-bottom left	BC-bottom center	BR-bottom right

The effects of each can be seen in Figure 18–16. The command sequence is similar to the sequence used for left-justified text, although the start point is referred to by the new location. For instance, the command sequence for BR is:

Command: **TEXT** ⏎
Justify/Style/<Start point>: **J** ⏎
Align/Fit/Center/Middle/Right/TL/TC/TR/ML/MC/MR/BL/BC/BR: **BR** ⏎

FIGURE 18–15 The RIGHT justified option will align the right ends of lines of text.

FIGURE 18–16 The start points for each of the nine letter options for placing text.

Bottom right point: (*Pick desired point.*)
Height: <0,-0 1/4">: **6** [enter]
Rotation angle: <0>: [enter]
Text:

The results of this sequence can be seen in Figure 18–17.

Style Option

The last option of TEXT to be considered is the Style option, which will determine the appearance of the text characters. The option is selected by entering **S** at the first text prompt. The sequence and options include:

Command: **TEXT** [enter]
Justify/Style/<Start point>: **S** [enter]
Style name (or ?) <STANDARD>:

Standard will remain the default style until the STYLE command is used to create other styles. The STYLE command will be introduced later in this chapter.

FIGURE 18–17 BR (bottom right) justification is often used when notes must be placed in the left side of an object.

By selecting the question (?) option, the following prompt will be displayed:

Text style(s) to list <*>: [enter]

This will produce a display showing the current text style qualities.

Style name: STANDARD Font files: TXT
Height: 0'-6" Width factor: 1.0000 Obliquing angle: 0
Generation: NORMAL
Current text style: STANDARD
Command:

LETTERING FONTS

The word *"font"* is written in an italic font. The term *font* is used to describe a particular shape of text which includes all lower and upper case text and numerals. The shape of the letters is one of the defining differences between architectural and mechanical lettering. Construction drawings usually contain text fonts that are more artistic than the gothic block lettering used on mechanical drawings. Most offices use different fonts throughout a drawing to distinguish between titles and text, or to highlight specific information.

The fonts contained in AutoCAD can be seen in Figure 18–18. The illustrations in this chapter have been created using the default font of TXT. TXT is simple font design, with letters consisting only of straight line segments. Because of its simplicity, it is quick to regenerate and plot. Many engineering offices use fonts such as ROMANS, ROMANC, ROMANT, or ITALICC to letter drawings. In addition to these fonts, some architectural offices use fonts that more closely resemble freehand lettering, such as City Blueprint and Country Blueprint for notes on a drawing and ROM, ROMB, or SASBO for titles. Examples of these fonts can be seen in Figure 18–19. It is important to select fonts that are easy to read and quickly reproduced by a plotter. The complexity of a font also affects the amount of storage space required, the speed of regens, and the plotter speed.

Lettering Programs

One of the benefits of AutoCAD is that many third-party suppliers have developed material to supplement the standard program. Many offices add text libraries to the basic program. You'll notice that much of the text used throughout the illustrations in this text is not found in the AutoCAD font files. This lettering is produced using material created by third-party vendors. Consult advertisements in local trade magazines for other available text libraries. In Release 14, the Text Style dialog box can be used to select the font to be used with each lettering style. The box will be discussed in detail shortly, but it's important to remember that in addition to the fonts contained in AutoCAD, TrueType fonts contained in additional programs in your

txt — The quick brown fox jumped over the lazy dog.   ABC123

monotxt — The quick brown fox jumped over the lazy dog.   ABC123

Simplex fonts

romans — The quick brown fox jumped over the lazy dog.   ABC123

scripts — *The quick brown fox jumped over the lazy dog.*   *ABC123*

greeks — Τηε ϑυιχκ βροων φοξ δυμπεδ οϵερ τηε λαζψ δογ.   ABX123

Duplex font

romand — The quick brown fox jumped over the lazy dog.   ABC123

Triplex fonts

romant — The quick brown fox jumped over the lazy dog.   ABC123

italict — *The quick brown fox jumped over the lazy dog.*   *ABC123*

Complex fonts

romanc — The quick brown fox jumped over the lazy dog.   ABC123

italicc — *The quick brown fox jumped over the lazy dog.*   *ABC123*

scriptc — *The quick brown fox jumped over the lazy dog.*   *ABC123*

greekc — Τηε ϑυιχκ βροων φοξ δυμπεδ οϵερ τηε λαζψ δογ.   ABX123

cyrillic — Узд рфивк бсоцн еоч йфмпдг охдс узд лащш гож.   АББ123

cyriltlc — Тхе цуичк брошн фож щумпед овер тхе лазй дог.   АБЧ123

Complex fonts

romanc — The quick brown fox jumped over the lazy dog.   ABC123

italicc — *The quick brown fox jumped over the lazy dog.*   *ABC123*

scriptc — *The quick brown fox jumped over the lazy dog.*   *ABC123*

greekc — Τηε ϑυιχκ βροων φοξ δυμπεδ οϵερ τηε λαζψ δογ.   ABX123

cyrillic — Узд рфивк бсоцн еоч йфмпдг охдс узд лащш гож.   АББ123

cyriltlc — Тхе цуичк брошн фож щумпед овер тхе лазй дог.   АБЧ123

Gothic fonts

gothice — The quick brown fox jumped over the lazy dog.   ABC123

gothicg — The quick brown fox jumped over the lazy dog.   ABC123

gothici — The quick brown fox jumped over the lazy dog.   ABC123

FIGURE 18–18 Standard fonts contained in Release 14. (Courtesy Autodesk, Inc.)

cibt   The quick brown fox jumped over the lazy dog.   ABC123

cobt   The quick brown fox jumped over the lazy dog.   ABC123

rom   The quick brown fox jumped over the lazy dog.   ABC123

romb   The quick brown fox jumped over the lazy dog.   ABC123

sas   The quick brown fox jumped over the lazy dog.   ABC123

sasb   The quick brown fox jumped over the lazy dog.   ABC123

saso   The quick brown fox jumped over the lazy dog.   ABC123

sasbo   The quick brown fox jumped over the lazy dog.   ABC123

te   THE QUICK BROWN FOX JUMPED OVER THE LAZY DOG.   ABC123

tel   THE QUICK BROWN FOX JUMPED OVER THE LAZY DOG.   ABC123

teb   THE QUICK BROWN FOX JUMPED OVER THE LAZY DOG.   ABC123

eur   The quick brown fox jumped over the lazy dog.   ABC123
à á â ã ä å æ ç è é ê ë ì í î ï ð ñ ò ó Û Ý ß Ø µ ¶
© ™ ® ¢ £ ¤ ¥ ¦ § ± † ‡ ¶ ¿ ¡

euro   The quick brown fox jumped over the lazy dog.   ABC123
à á â ã ä å æ ç è é ê ë ì í î ï ð ñ ò ó Û Ý ß Ø µ ¶
© ™ ® ¢ £ ¤ ¥ ¦ § ± † ‡ ¿ ¡

pan   The quick brown fox jumped over the lazy dog.   ABC123
ə √ Æ æ Ŋ ŋ ¿ ʰ ə ɗ θ ß Ð Ð Ș ¶ á á â ã ä ā ā ' a a'
Á Ä Ã Ä

suf   The quick brown fox jumped over the lazy dog.   ABC123
£ ɗ ø ± ß ‹ › ¶ ( ) ∈ Ł + ɦ Ɩ ¿ ª á á â ã ' a a' a Á Ä

FIGURE 18–18 Standard fonts contained in Release 14. (Courtesy Autodesk, Inc.)

THE STYLE 'STANDARD' IS COMPOSED
OF THE TXT FONT.

THE STYLE 'NOTES 'IS COMPOSED
OF THE ROMANS FONT.

THE 'TITLERESPONSE' STYLE IS COMPOSED
OF THE ROMANC FONT.

**THE TBLOCK STYLE IS COMPOSED
OF THE FONT ROMANT.**

*THE STYLE 'GENERALNOTES' IS
COMPOSED OF THE FONT ITALICC.*

**THE 'TITLEBLOCK' STYLE IS COMPOSED OF
THE FONT ROMB.**

**THE STYLE 'ENGINEERSTAMP'
IS COMPOSED OF THE SASBO FONT.**

THE STYLE 'FNDNOTES' IS
COMPOSED OF THE FONT COUNTRYBLUEPRINT

THE 'TEXT' STYLE IS COMPOSED OF THE CITYBLUEPRINT FONT.

THIS TEXT IS A SAMPLE OF TEXT DEVELOPED
BY A THIRD PARTY DEVELOPER . THE FONT IS
'ARCHSTYL'.

FIGURE 18–19 Common fonts used in architectural and engineering offices.

computer can be used in AutoCAD drawings. Methods of accessing these files will be discussed as ways of creating text styles are introduced.

WORKING WITH STYLE

The term *style* in CAD drafting describes the characteristics of a group of text such as the font, height, width, and the angle used to display the text. One style can be used for all drawing text, a second style used for the drawing titles, and a third style used for all information in the title block. The STYLE command can be used to add different fonts to your drawing as well as change the properties of the letter shapes. As styles are created, they should be stored in drawing templets for future use with all similar projects. Assigning a style name, selecting a font, and defining STYLE properties are the three areas to be considered when setting a new style of text. The command can be accessed by picking Text Style... from the Format pull-down menu, or by typing **ST** [enter] **or STYLE** [enter] at the command prompt.

Naming a Style

Both options will produce the dialog box shown in Figure 18–20 that can be used to name a style, assign a font, control the style options, and preview the results.

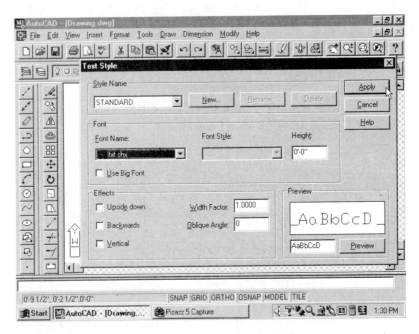

FIGURE 18–20 Picking Text Style... from the Format pull-down menu or by typing **ST** enter at the command prompt produces the Text Style dialog box that can be used to name and control text styles.

New styles can be created using the Text Style dialog box by picking the New.... button. This will produce the New Text dialog box shown in Figure 18–21. The default style is named Style1. Accept the name by picking the OK button. The New Text Style box will be closed and STYLE1 is now displayed as the current name in the Style Name edit box. Picking the Cancel button will remove the New Text Style box and allow a different aspect of the STYLE command to be addressed, or allow the command to be terminated. The name STYLE1 can be renamed to something more meaningful by picking the Rename... button and using the Rename Text Style dialog box that can be seen in Figure 18–22. STYLE1 can also be altered before it is accepted as a style name by using the same procedure to rename a style name. Use the cursor to delete the name, and then type in the desired name. Selecting the OK button will remove the dialog box and change the name STYLE1 to the name provided. Picking the Cancel button will remove the Rename Text Style box and allow a different aspect of the STYLE command to be addressed, or allow the command to be terminated.

The third option in the Style Name box is the Delete...button. Selecting this button allows a style to be deleted from the current list of text styles.

The name of the text style to be created should reflect the use of the text. The name may contain up to 31 letters, numbers, and special characters. The same guidelines that were used to name linetypes and layers should be used to name text styles.

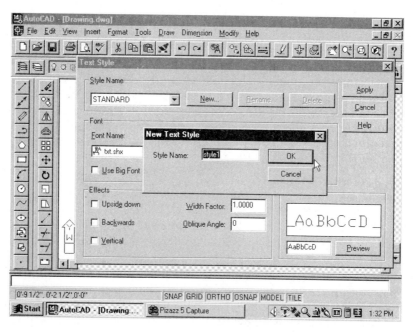

FIGURE 18–21 Picking the New... button produces the New Text dialog box that allows a new text style to be named.

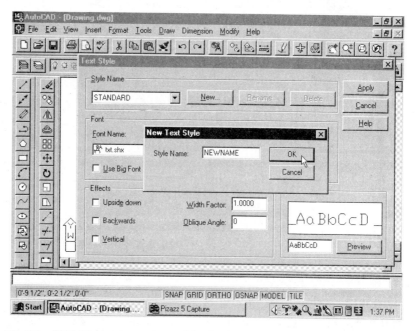

FIGURE 18–22 STYLE1 can be renamed in the New text dialog box before the name is assigned, or can be alterd by picking the Rename... button.

Some offices use the name of the text font for the style name. Names such as title-block, gennotes (general notes), fndnotes (foundation notes), or sitenotes are examples of names that could be used to describe different styles of text. These names could be abbreviated further as TB, GEN, FND, and ST to speed typing text selections. Names that include the text font and usage are helpful. Much of the lettering in this text was created using a font called *archstyl,* which is created by a third-party vendor for use with AutoCAD. You can create sytles in the templet drawing such as, *archtitl, archtxt,* and *italitxt.* These would be used for creating titles and text using the font archstyl, and a text style using the font italic for drawing text.

Choosing a Font

The second aspect of defining a text sytle is choosing a font. Once the style name has been entered, selecting the down arrow beside the Font Name: edit box will produce a menu of all of the text fonts available to you. A sample of the menu can be seen in Figure 18–23. The display shown in the menu will be altered depending on the status of the Use Big Fonts check box. With the box inactive, all fonts in all programs on your computer will be displayed. With the box active, only the fonts contained in AutoCAD14 will be displayed in the menu. The AutoCAD fonts are preceded by Divider icon and followed by .SHX. Other fonts available to AutoCAD are preceded by the TrueType icon. As the name of the desired font to be used is selected from the

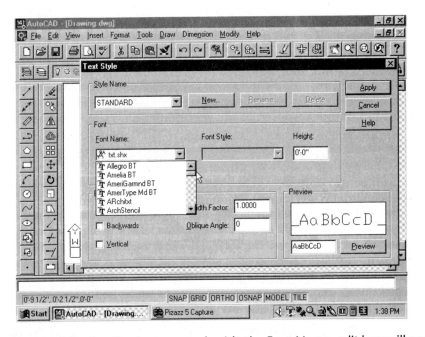

FIGURE 18–23 Selecting the down arrow beside the Font Name: edit box will produce a listing of the AutoCAD and the True Type fonts files. Scroll through the desired list and select the name of the font to be used with the style being created.

menu, the name will be displayed in the edit box, the menu will be closed, and a sample of the font will be displayed in the Preview edit box.

Defining Style Properties

Once the Style has been named and the font has been assigned five options are available to further define the Style being created. These options included Upside down, Backwards, Width Factor, Oblique Angle, and Vertical. The value for Height in the Font box is also used to control the appearance of the style.

Height. If a height is entered at this prompt, all text created using this Style of text will be the same height. If all text throughout a drawing is to be the same size, the height variable of the STYLE command can be set so that the variable will not have to be set each time the font is used. The height of text is best controlled with the Height option of the command used to create the actual text. Entering a height now will delete the Height prompt in the TEXT, DTEXT, and MTEXT command sequences when this Style is used. Responding with a 0 will allow the height to be determined each time this Style is used.

Upside-Down. The default for this option will place text in the normal position. Activating this box will place text upside-down. Although this option is not typically used in construction drawings, an example of upside-down text can be seen in Figure 18–24.

Backwards. The default value for this option will produce text that is read from left to right. Activating this box will produce text that is read from right to left or backwards. Although this option is not used often, it could be useful if a drawing is to be plotted on one side of a sheet of vellum or Mylar®, and text on the other side. Figure 18–25 shows an example of text printed backwards.

FIGURE 18–24 Text may be printed in both normal and upside-down orientations.

NORMAL ORIENTATION

BACKWARDS ORIENTATION

FIGURE 18–25 Text can be printed in both the normal and backward position.

WIDTH FACTOR = .5

NORMAL WIDTH FACTOR = 1.00

WIDTH FACTOR =2.0

FIGURE 18–26 Adjusting the Width factor will affect the appearance of text.

Width Factor. Font characters are displayed and plotted based on a ratio of their height and width. No matter what the actual height is, the height is assigned a value of one. The default value for the width factor is also 1, making the width equal to the height of the letter. If a value of .5 is entered, text will be half as wide as it is tall. A value of 2 will produce an elongated style of lettering. Figure 18–26 shows a comparison of various width factors.

Obliquing Angle. This option will allow text to be tilted from true vertical. The default angle of 0 will display text in the normal position, with the vertical leg of each letter in a vertical position. Entering a positive value will slant text to the right. Entering a negative number will produce a backslant. Figure 18–27 show examples of altering the obliquing angle.

Vertical. This option is not active if a TrueType font has been selected. The option will display .SHX fonts in the vertical position as shown in Figure 18–28.

Once all of the parameters have been selected for the style, the options of APPLY and CLOSE remain to set the style. Choosing the Apply button will apply any style changes made in the dialog box to the text in that current style within this current drawing. Figure 18–29 shows an example of text created using the TXT.SHX font and the same text after the font was changed to ITALIC.SHX with a width factor of 1.5. The text was altered by changing the appropriate values, selecting the apply button, and then picking the Close button. Choosing the Close option will accept the current style values and remove the dialog box. Existing styles will remain unchanged, and new text will be added using the current settings.

NORMAL OBLIQUING ANGLE = 0°

POSITIVE OBLIQUING ANGLE = 15°

NEGATIVE OBLIQUING ANGLE = −15°

FIGURE 18–27 The Obliquing angle of the STYLE command affects the angle of the vertical stroke of each letter.

V E R T I C A L T E X T O R I E N T A T I O N

FIGURE 18–28 Placing text in a vertical position may be required on some construction drawings.

TEXT STYLES CAN BE CHANGED ONCE THEY HAVE
BEEN CREATED. THIS IS TXT.SHX FONT.

*TEXT STYLES CAN BE CHANGED ONCE THEY
HAVE BEEN CREATED. THIS IS ITALIC WITH
A WIDTH FACTOR OF 1.5.*

FIGURE 18–29 Text created using the THX.SHX font was changed to ITALIC.SHX with a width factor of 1.5 by changing the appropriate values and then selecting the Apply button.

Special Characters

In addition to the symbols on a keyboard, it is often necessary to print special symbols, such as a degree or diameter symbol. This can be done by typing a code letter preceded by two % (percent) symbols. Common symbols of AutoCAD include:

%%C = diameter symbol

%%D = degree symbol

%%O = overscores text

%%P = plus/minus symbol

%%U = underscore text

%%% = percent symbol

The symbols for overscoring and underscoring text are toggles. Type **%%U** preceding the text to be underscored to activate underscoring. The symbol must also be typed at the end of the phrase to turn off underscoring. An example of the sequence would be:

%%U<u>PROVIDE AN ALTERNATE BID FOR ALL ROOFING MATERIALS</u>%%U

As the text line is displayed at the command line, the %%U will be displayed in addition to the desired text. When the ⏎ key is pressed, the %%U will be removed from the screen and the text will be underscored. Although a single percent symbol can be created by using the percent key (Shift + 5), when the percent symbol must proceed another special character, %%% must be used. To write 50°+/–5% you must enter the following sequence:

50%%D%%P5%%%

Special symbols can also be created using a three-digit code. Common symbols found on construction drawings include:

%%035 = #   %%047 = /

%%037 = %   %%060 = <

%%038 = &   %%061 = =

%%040 = (%%062 = >

%%041 =)   %%064 = @

Remember that these symbols apply to the fonts in AutoCAD and do not apply to text created by third-party vendors.

EDITING EXISTING TEXT

Text can be edited by using the backspace key and entering corrections before making the text part of the drawing base. Text can be edited once it is part of the drawing base using the DDEDIT command. The command can be started by selecting the Edit Text icon from the Modify II toolbar, by selecting Text... from Object of the Modify pull-down menu, or by entering **ED** ⏎ or **DDEDIT** ⏎ at the command prompt. Typing **EDIT** ⏎ will produce the following prompt.

Command: **ED** ⏎
<Select an annotation object> /Undo:

You may select either a line of text or an attribute definition. Editing attributes will be introduced in Chapter 22. For now, select the text to edit by moving the cursor to the line of text and select it with the pick button. The error does not have to be selected, only the line of text containing the error. As the text is selected, a dialog box is presented containing the line of text. Figure 18–30a shows a line of text in a drawing and the line contained in the Edit Text dialog box. Because the line of text is too long to fit in the edit display, the left and right arrow keys can be used to alter the line of text that is displayed.

Once the proper portion of the text line is displayed, editing can be performed by moving the arrow to the desired location and using the pick button. In the example, the following note is used:

aLL FRAMINGG LUMBER IS TOBE D.F.L. # 2 OR BETTER UNLESSNOTED.

The "A" of "all" should be in capital letters, FRAMING is misspelled, and a space should be placed between "TO BE" and "UNLESS NOTED." To fix the "a" move the arrow to the right side of the "a" and press the mouse pick button. Use the backspace arrow to eliminate the "a". Without moving the cursor, the error can be fixed by typing an **A**.

Move the arrow to move the cursor to the right side of "FRAMINGG" and press the pick button. Press the backspace key, to eliminate the last "G". Now if the cursor is placed between the "O" and "B" of "TOBE", a space can be inserted by pressing the spacebar. The text should now match Figure 18–30b. "UNLESSNOTED" can be cor-

FIGURE 18–30a The DDEDIT command will allow a string of text to be edited. Once selected, the desired text string will be placed in the Edit Text box.

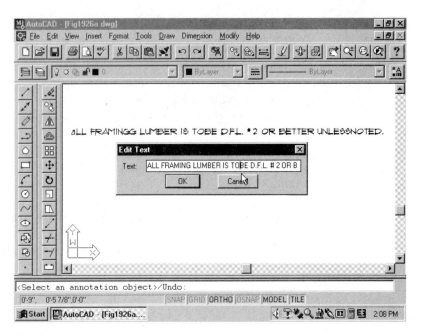

FIGURE 18–30b The "a" was edited by moving the cursor to the right of the letter and pressing the pick button on the mouse. Once selected, the error can be removed with the backspace arrow, and the desired correction made at the key board. Even though the "A" and the word FRAMING have been corrected, these changes are not reflected in the drawing until the OK is selected.

rected in the same method by pressing the END key to scroll that area of text into the edit box. The corrected text will now read as it was intended. See Figure 18–30c. If the edit is not what you desired, the Undo option can be used to restore the line of text to its previous status. Once you are satisfied with the text, pick the OK box to return to the drawing. The DDEDIT command is still active. If no other editing needs to be done, press the [enter] key to return the command line.

Text and the Mirror Command

The MIRROR command was introduced in Chapter 11. If you mirror an object that includes text, the text will also be reversed. To avoid the problems this will create use the MIRRTEXT command. The command is started by entering **MIRRTEXT** [enter] at the command prompt. This will produce the prompt:

> Command: **MIRRTEXT** [enter]
> New value for MIRRTEXT <1>:

ALL FRAMING LUMBER IS TO BE D.F.L. # 2 OR BETTER UNLESS NOTED.

FIGURE 18–30c The note shown in Figure 18–30a is now as it was intended.

Entering [enter] at the prompt will accept the default value and the text will be reversed as the object is mirrored. Entering **0** [enter] at the prompt will mirror the object and the text, but keep the text readable.

DTEXT

The DTEXT command creates dynamic text, which can be seen on the screen as it is entered at the keyboard. Seeing the text placed in the drawing field as it is typed offers advantages over the TEXT command for most situations. DTEXT also allows some basic editing to be performed and multiple lines of text to be entered while in the same command sequence. The command sequence for DTEXT is identical to the Text command sequence and can be accessed by selecting the Single Line Text from Text of the Draw pull-down menu or by typing **DT** [enter] or **DTEXT** [enter] at the command prompt.

The default of *Start Point:* is used to pick the starting location, and new text will be placed left-justified. As the prompt for Text: is displayed the cross-hairs will display the starting point at the lower left corner of the text. This is where the start of a new line of text will be displayed. If you don't like the location, move the cursor to a new location and select that location with the pick button.

Once the desired location is selected, text can be entered by keyboard. Although the text will be displayed in the drawing area, depending on the ZOOM setting, you may be unable to read it. Either cancel the command and ZOOM into the area where you will be adding the text or read the text at the TEXT prompt line that replaces the Command line. If any special characters are used, they will show up on the screen, but not at the text prompt. To enter multiple lines of text, simply press the [enter] key to move the text box to the start of a new line.

Once the desired text has been entered, press the [enter] key twice to exit DTEXT. If just a few words of text have been entered, the screen will blink and the command prompt will return. If several lines of type have been added to the drawing, your heart might stop when you end the command sequence. All of the text will start scrolling off the screen until it has disappeared, but it will be regenerated almost instantly. Pressing the ESC key will destroy any text entered during the active command sequence.

QUICK TEXT

Quick text is not an additional way to create text, but a very useful way to display it. As you're editing drawings containing large amounts of text, you'll find that REGEN times start to slow. The QTEXT command will display lines of text as rectangles, which are much quicker to regenerate. The height of the rectangle is the same as the text height, but the length of the rectangle is typically longer than the line of text. The command sequence is:

Command: **QTEXT** (enter)
ON/OFF <OFF>:

With QTEXT set to ON, text will retain its current setting until a REGEN is performed. Figure 18–31 shows an example of QTEXT in both the ON and OFF settings.

CREATING TEXT WITH MTEXT

You may be feeling overwhelmed with text by now. You've worked with TEXT, STYLE, DTEXT, and QTEXT and you may be wondering why you need another way to enter text into a drawing. The answer is because each method offers more options than the previous method. TEXT places text on a drawing, but it is not seen on the screen as it is created, and it is cumbersome to edit. DTEXT can be seen as it is created, giving you a sense of control. DTEXT is ideal for placing local notes in and around the drawing. Paragraphs created in TEXT and DTEXT can be very frustrating to edit if they need to be moved and resized. QTEXT is a tool for displaying text and offers no benefits for creating or editing text.

MTEXT is ideal for placing large areas of text near a drawing. MTEXT can be created in the Edit MText dialog box which offers greater flexibility in creation and editing of the text. Using MTEXT is similar to using a word processing program within AutoCAD. MTEXT also offers advantages when editing the drawing. Lines of text placed with TEXT or DTEXT act as individual lines when they are moved. Text placed with MTEXT functions as one entity. Multiline text also allows text to be easily underlined and varied fonts, colors, and heights to be assigned throughout the text. MTEXT is ideal for entering and editing large bodies of text such as those seen in Figure 18–32. A paragraph can be moved by picking one character rather than the entire paragraph. The command can be accessed by selecting the Multiline Text icon from the Draw toolbar, by selecting Multiline Text... from the Draw pull-down menu, or by typing **T** (enter) or **MTEXT** (enter) at the command prompt. Each will produce the following prompt:

Command: **T** (enter)
Specify first corner:

By default, AutoCAD is waiting for you to select where you would like to place the text. Once you select the first corner, you will be prompted for:

Specify opposite corner or [Height/Justify/Rotation/ Style /Width]:

Although you can respond to the second prompt by providing a value for one or more of the five options, picking the second corner to define the text window width will produce the Multiline Text Editor box and allow each of these values and many more to be easily set. Notice in Figure 18–33, as the box is being dragged to indicate the second corner, an arrow is shown indicating the direction that excess text will flow out of the box. Methods of altering the direction will be introduced shortly. By select-

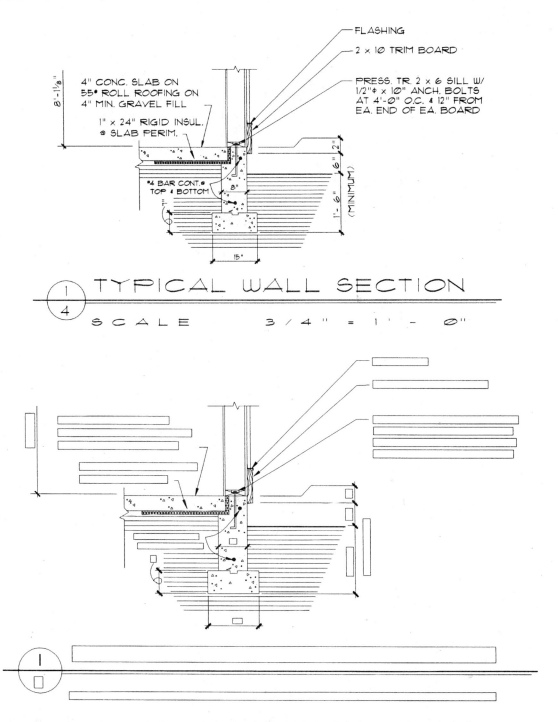

FLASHING

2 x 10 TRIM BOARD

PRESS. TR. 2 x 6 SILL W/ 1/2"⌀ x 10" ANCH. BOLTS AT 4'-0" O.C. & 12" FROM EA. END OF EA. BOARD

4" CONC. SLAB ON
55# ROLL ROOFING ON
4" MIN. GRAVEL FILL

1" x 24" RIGID INSUL.
@ SLAB PERIM.

8'-1¹⁄₈"

#4 BAR CONT. @
TOP & BOTTOM

8"

2"

1'-6" 6"

(MINIMUM)

15"

1/4

T Y P I C A L   W A L L   S E C T I O N

S C A L E   3 / 4 "   =   1 ' - 0 "

1

FIGURE 18–31 QTEXT can be used to speed the REGEN of drawings. The upper drawing shows an example of normal text; the lower drawing shows the effect of QTEXT. (Courtesy Piercy & Barclay Designers, Inc., A.I.B.D.)

STEEL JOIST
1. STEEL JOISTS SHALL BE DESIGNED, FABRICATED AND ERECTED PER SJI STANDARD SPECIFICATION.
2. CAMBER ALL JOISTS WITH STANDARD CAMBERS PER ABOVE STANDARD SPECIFICATION UNLESS NOTED ON THE DRAWINGS.
3. PROVIDE BRIDGING AS CALLED FOR BY ABOVE SPECIFICATIONS. DO NOT USE SLOTTED HOLES IN BOLTED CONNECTIONS UNLESS APPROVED BY ENGINEER.
4. PROVIDE SHOP DRAWINGS AND CALCULATIONS THAT INDICATE JOIST SIZES, LOADING, MEMBER SIZES, PANEL POINT LOCATIONS, CAMBERS, BRIDGING AND ANY OTHER INFORMATION THAT MAY BE PERTINENT TO THE JOB. SHOP DRAWINGS SHALL BE STAMPED BY AN ENGINEER (THE PRODUCT ENGINEER) REGISTERED IN THE STATE OF OREGON.
5. PROVIDE INSPECTION REPORT PREPARED BY THE PRODUCT ENGINEER CERTIFYING THAT THE JOISTS HAVE BEEN FABRICATED ACCORDING TO THE DESIGN ASSUMPTIONS AND REQUIREMENTS. REPORT IS TO ACCOMPANY THE JOISTS ON DELIVERY TO THE PROJECT.
6. JOIST ERECTOR SHALL EXERCISE EXTREME CARE DURING ERECTION OF JOISTS TO PREVENT THE JOISTS FROM BUCKLING LATERALLY. USE SPREADER BARS FOR LIFTING JOISTS AND PROVIDE LATERAL BRACING AS NECESSARY. REMOVE ANY DAMAGED JOISTS FROM THE JOB SITE. DO NOT ATTEMPT TO REINFORCE DAMAGED JOISTS.

STEEL FLOOR DECK
1. STEEL FLOOR DECK TO BE VERCO 22 GAGE 'TYPE B' FORMLOK OR EQUIVALENT, FOR UNSHORED CONSTRUCTION.
2. WELD DECK TO SUPPORTS AT ALL VALLEYS AND TO MARGINAL MEMBERS AT 24" O.C. WITH 1/2" DIA. FUSION WELDS UNLESS OTHERWISE NOTED ON DRAWINGS.
3. BUTTON PUNCH SIDE SEAMS AT 24" ON CENTER MAXIMUM.

STEEL ROOF DECK
1. STEEL ROOF DECK TO BE VERCO 20 GAGE 'TYPE HSB-36' OR EQUIVALENT.
2. WELD DECK TO SUPPORTS AT ALL VALLEYS AND TO MARGINAL MEMBERS AT 24" O.C. WITH 1/2" DIA. FUSION WELDS UNLESS OTHERWISE NOTED ON DRAWINGS.
3. BUTTON-PUNCH SIDE LAPS AT 24" ON CENTER MAXIMUM.

FIGURE 18–32 MTEXT is ideal for creating large quantities of text.

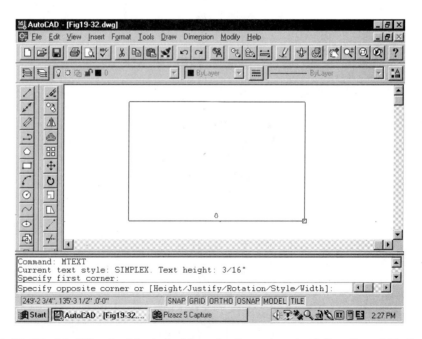

FIGURE 18–33 Multiline text is started by selecting a box to define the width that the text will occupy. If the text does not fit into the box, it will overflow the bottom of the box in the current setting (the down arrow). The direction of overflow can be adjusted.

ing a second point with the cursor, AutoCAD is asking you to provide a window to place your text in. The width of the window you specify is very important. The width of the window will control the width of your paragraph. Extra text will spill out of the top or bottom of the window depending on the justification, so that the depth of the window is not critical.Once the other corner is selected, the Multiline Text Editor shown in Figure 18–34 will be displayed. Text can now be entered into the edit window using similar procedures used with other text editors. AutoCAD allows a third-party text editor to be used by selecting the desired editor in the Preferences dialog box of the Tools pull-down menu. See Chapter 9 of the User's Guide in the Help menu to select a third-party editor. The balance of this chapter will introduce the many capabilities of the program editor.

As text is entered at the keyboard, it will be displayed in the edit box of the text editor. Text will wrap to the next line as the desired width is reached, just as with any word processing program. Notice in Figure 18–34 that the editor contains three tabs for controlling text features. The default tab controls Characters, with tabs for Properties and Find/Replace.

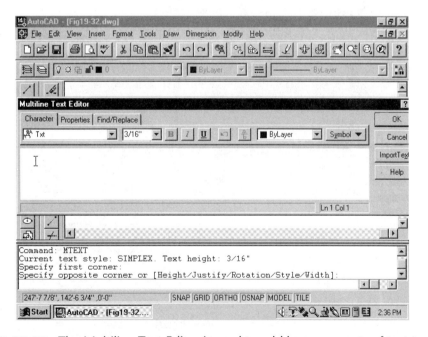

FIGURE 18–34 The Multiline Text Editor is used to add large amounts of text to a drawing. By using the Character, Properties and Find/Replace tabs, text can be assigned various colors, heights, underlined, and altered. In addition to using the AutoCAD text editor, a third-part editor can be used by selecting the desired editor using Preferences... in the Tools pull-down menu.

Controlling Text Characters

As with TEXT and DTEXT, the qualities of MTEXT can be easily altered. Qualities controlled by the Character tab include the font, height, bold, italic, underline, undo, stack, color, and special symbols. This dialog box also includes the question mark button in the title bar. Pressing the **?** key will produce a question mark icon beside the cursor. Now if any of the buttons are selected, an on screen explanation will be displayed. Figure 18–35 shows an example of selecting the Stack button.

Font. Pressing the down arrow beside the Font edit box displays a menu similar to the font menu displayed with the STYLE command. The menu allows a font for new text to be selected or changes the font of selected text. The current font for MTEXT is the same as the current font for the STYLE command. With MTEXT, the style can be altered at anytime throughout the use of the command including mid sentence. Figure 18–36 shows an example of text that was altered to provide different fonts within the same command sequence.

Height. Just as with the TEXT and DTEXT commands, this option controls the height of the text. The default value is the same as the current style. The value for text height can be changed at any point of the command allowing words to be printed at varied heights. Picking the down arrow will display the heights of all current styles. To provide a new height, pick the height edit box and click the cursor to highlight the current height. Remove the current height with the back space arrow and

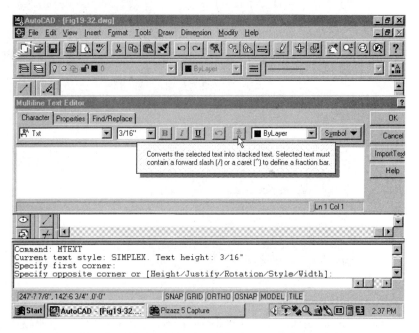

FIGURE 18–35 Pressing the **?** button in the Title bar and then picking a box or button in the Multiline Text Editor will provide a brief explanation of the function.

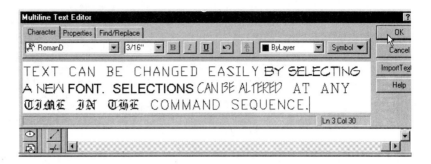

FIGURE 18–36 The text font of one word or a string of words can be altered at any point of the drawing life.

then enter the desired height value. Text height can be altered before the text is added to the paragraph by highlighting the text to be altered, and then providing the required height in the edit box. Figure 18–37 shows an example of how text height can be altered.

Bold. Picking the Bold button toggles between bold and normal text. The options is only active for TrueType fonts. Bold text can be created by picking the Bold button before text is entered. Existing normal text can be changed to bold by highlighting the desired text to be changed, and then pressing the Bold button. Figure 18–38 shows an example of text that was altered to provide bold text.

Italic. Selecting the Italic key toggles italic formatting for text. As with other options, Italic can be applied to text being created as well as used to modify existing text. The option is only available for TrueType fonts.

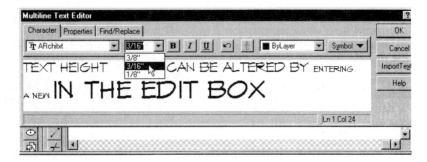

FIGURE 18–37 MTEXT height can be altered at any time throughout the life of the drawing. Picking the down arrow will display the heights of existing styles. A new height can be entered by picking the Height box and replacing the current value with the desired value.

FIGURE 18–38 Text can be highlighted by picking the Bold, Italic, or Underline buttons. One or more of the features can be used in combination to effect the font.

Underline. Picking this button allows the underline function to be toggled on and off. This is so much easier than **%%something** that is required with the text command.

Undo. This option can be used to undo the last sequence of the Text Editor. Changes can be made to the text or to one of the formatting options.

Stack. The Stack option can be used to place a portion of text over other text. This is useful to write fractions such as ¼ rather than 1/4 or to write equations. With the text written as 1/4, highlight the fraction by pressing and holding the pick button and moving the cursor over the text. Once the desired text is selected, picking the Stack button will stack the indicated text. See Figure 18–39.

Color. Picking the Color button allows the color of the text to be altered. By default, MTEXT will be assigned the color of the current layer. To select a new text color, pick the down arrow beside the color edit box and pick the desired color. Select-

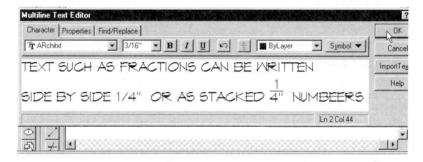

FIGURE 18–39 Text such as fractions can be written side by side or stacked depending on office preference.

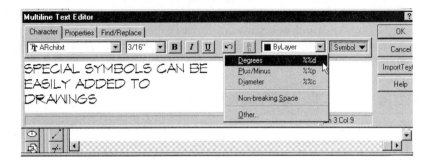

FIGURE 18–40 Using MTEXT, special characters such as the degree, plus/minus, and diameter symbols can be added without using any %% special character.

ing the Other color option will produce the Color Palette and allow for a full range of colors to be selected.

Symbol. Picking the down arrow beside the Symbol button displays the menu shown in Figure 18–40. The degree, plus/minus, and diameter are often used on construction documents. Either can be inserted into MTEXT by picking the down arrow to display the menu, and then picking the desired symbol. If this seems too easy, the symbols can also be inserted using the special characters that were discussed as TEXT was introduced. Picking the Other... option will display the menu shown in Figure 18–41 and allow one of the symbols from the Character Map to be inserted. Symbols will vary depending on the current font.

Control Buttons. As with other dialog boxes, control buttons are provided to implement the desired settings. This box features the OK, Cancel, Import Text, and Help buttons. The results of the OK, Cancel, and Help buttons should be familiar by now. The Import Text... button allows ASCII (American National Standard Code for

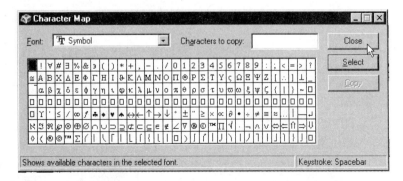

FIGURE 18–41 Selecting the Other option from the Symbols menu displays the character map to be displayed.

Information Interchange) and RTF (Rich Text Format) text files to be inserted into a drawing file. This may seem unimportant as you're learning AutoCAD now, but it means you can create schedules or large bodies of text up to 16KB using your second favorite software, and then import the file into AutoCAD once you feel more comfortable with MTEXT.

Adjusting Text Properties

Selecting the Properties tab of the Multiline Text Editor will provide options for controlling text properties by changing the style, justification, width, and rotation. Figure 18–42 shows the Properties tab. The tab can be selected at any point of the MTEXT command.

Style.   The Style option allows an existing TEXT style to be applied to a new or selected text body. Character formats for font, height, bold, and italic can be overridden by applying a new style to existing multiline objects. Information can be typed in one style, and altered before the text is placed in the drawing base. To alter the current text style, pick the Properties tab and then pick the Style down arrow to display a listing of current text styles. Figure 18–43 shows an example of the Style menu.

Justification.   Justification with MTEXT is similar to that of the TEXT command. Picking the Justification option allows the alignment of text to be altered.

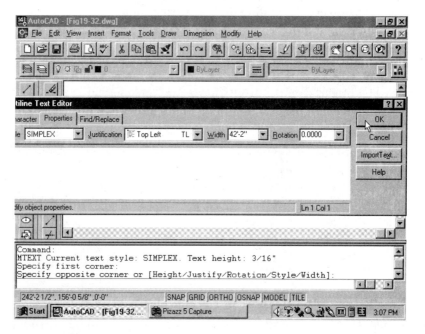

FIGURE 18–42 Picking the Properties tab displays this menu for altering MTEXT.

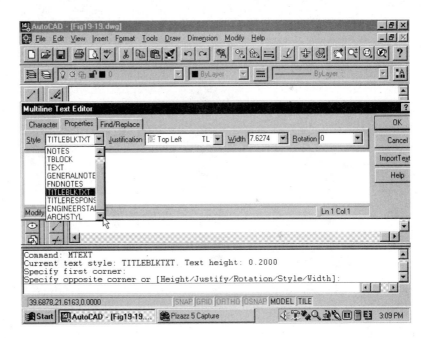

FIGURE 18–43 Picking the down arrow for the Style displays a menu of current text styles. Styles for MTEXT are created using the same procedures used to create a style for TEXT and DTEXT.

The default setting places the cursor in the top left corner of the text box. Selecting the Justification down arrow will produce the menu shown in Figure 18–44. Figure 18–45 shows examples of how text will be placed based on each setting.

Width. This option allows the width of the text to be redefined. Selecting the down arrow beside the Width edit box will display a menu of current MTEXT widths. One of the current widths can be selected or a new value can be added to the edit box.

Rotation. This option functions just as the option did with the TEXT command, to set the text rotation angle. Selecting the Rotation option will display a listing of angles in 15 degree increments. One of these options can be selected from the menu, or any numerical value between 0 and 360 can be assigned. The option can be used with the current text before it is added to the drawing or existing text can be selected and altered.

Using Find/Replace

Selecting the Find/Replace tab of the Multiline Text Editor displays the options shown in Figure 18–46. The options help create search patterns for words or specified strings of text. Options include the Find box and button, the Replace box and button, Match Case, and Match Word.

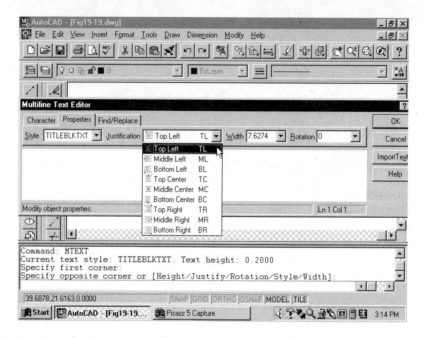

FIGURE 18–44 Selecting the Justification down arrow will produce a list of options for arranging MTEXT that are similar to the TEXT justification options.

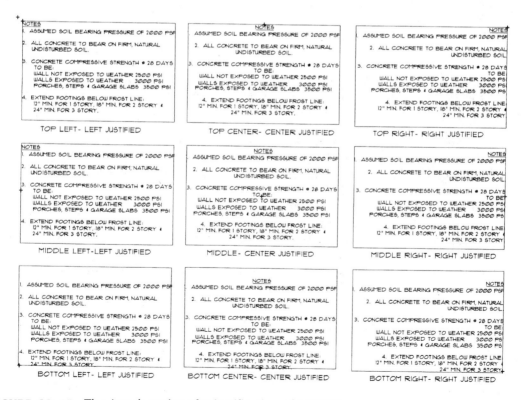

FIGURE 18–45 The Attach options for justification of MTEXT.

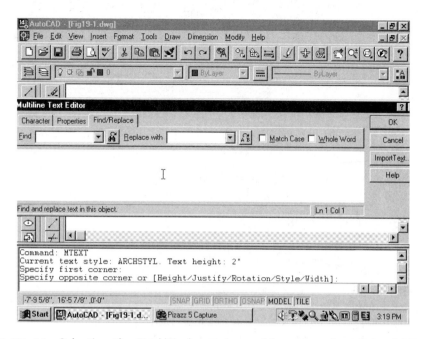

FIGURE 18–46 Selecting the Find/Replace tab provide options for editing MTEXT.

Find. The Find box can be used to enter a word or string of text that you wish to find. This can be especially helpful in finding a key word in a long list of specifications. To find a word, pick the Find box to place the cursor in the box and then enter the desired word. Figure 18–47 shows an example of using Find to locate the word **JOISTS**. Pressing the Find button on the right end of the Find edit box starts the search for the text in the Find box. Press the ⟨enter⟩ key to find the next occurrence of the selected word.

Replace. The Replace option can be used in conjunction with Find to alter a word or word string. Once a word has been found, enter a word in the Replace box and press the Replace button. Figure 18–48 show the results of selecting **Joist** and replacing it with the word **trusses.**

Match Case. With this box checked, only the exact word with the same case will be identified. With the box inactive, placing **Truss** in the find box would identify **truss** in the body of text. With the Match Case box active, **truss** would not be identified. Only uses of the word or string that are identical will be located.

Whole Word. With this box inactive, any use of the selected word will be identified with the Find option. With **truss** indicated in the Find box, **Truss–joist** would be identified in the document. Selecting this option, will identify the word only if it is a single word. When the word is part of another text string it will be ignored.

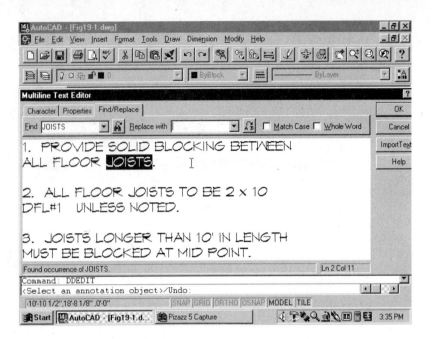

FIGURE 18–47 Picking the Find box allows a word or word string to be specified for a word search. The word JOIST has been selected for the search. Pressing the Find button will highlight the first use of the specified word. Pressing the [enter] key will highlight the next occurence of the selected word or string.

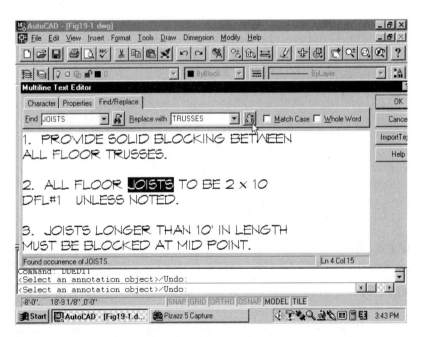

FIGURE 18–48 With the word JOISTS as the active word for the search, the first ocurence of the word was altered to read TRUSSES by entering TRUSSES in the Replace with box and pressing the Replace with button.

EDITING MTEXT

The contents of an MTEXT object can be edited using the COPY, ERASE, GRIPS, MIRROR, MOVE, and ROTATE commands. The text in MTEXT can be edited using either DDEDIT or DDMODIFY commands.

Using DDEDIT

To use the DDEDIT command with MTEXT, you start the same way as when editing TEXT or DTEXT. The quickest method of accessing the command is from the command prompt, but the command can also be accessed by selecting Edit Text from the ModifyII toolbar, or by selecting Text... from Object of the Modify pull-down menu. The command sequence is:

Command: **ED** [enter]
<Select an annotation object>/Undo:

When the paragraph to be edited is selected, the drawing will be removed from the screen display and the Multiline Text Editor will be displayed. The mouse can also be used to move the cursor to the desired location. When the desired changes are made, the drawing screen can be returned by picking the OK button. As the screen is restored, the prompt for selecting additional annotation is still displayed. The program is waiting for you to select additional text to be edited. Press [enter] a second time to exit DDEDIT and restore the command prompt.

Editing with DDMODIFY

The DDMODIFY command is started by entering **MO** [enter] or **DDMODIFY** [enter] at the command prompt. The sequence is:

Command: **MO** [enter]
Initializing... DDMODIFY loaded.
Select object to modify:

Selecting an MTEXT paragraph will display the Modify MText dialog box shown in Figure 18–49.

The Properties portion of the box displays the current settings for Color and Layer. Each option is used as described in Chapter 15. The Insertion Point portion of the box allows you to provide new coordinates for the justification point by picking the desired coordinate box and entering the new coordinate. The box also allows a new point to be picked by selecting the Pick Point < button. As this option is selected, the dialog box is removed from the screen and the selected text is displayed. A drag line is displayed from the justification point and you are prompted for a New Insertion Point: at the command line. When the new point is selected, the dialog box is returned to the screen.

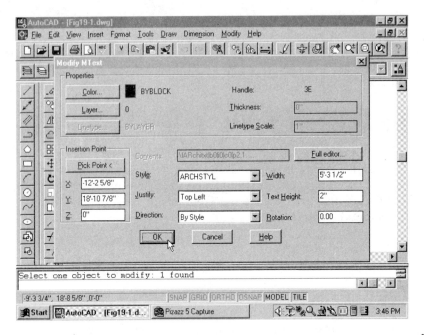

FIGURE 18–49 The DDMODIFY command can be used edit MTEXT. Enter **MO** ⌐enter⌐ at the command prompt to produce the Modify Text dialog box.

Selecting the Full editor... button displays the Multiline Text Editor dialog box. Selecting the OK button once the desired properties are edited, will restore the Modify MText dialog box. When all modifications have been made, selecting the OK button will restore the drawing screen and make the selected changes.

SPELL CHECKER

In addition to using DDEDIT and DDMODIFY to edit text, the SPELL command can be used to correct spelling errors in TEXT, DTEXT, and MTEXT. SPELL can be accessed by selecting Spelling from the Tools pull-down menu, selecting the Spell icon from the Standard toolbar, or by entering **SP** ⌐enter⌐ or **SPELL** ⌐enter⌐ at the command prompt. Each option will prompt you to select objects to be corrected for spelling. Once a word or body of text is selected, a Check Spelling dialog box similar to Figure 18–50 will be displayed.

If no spelling errors are contained in the selected text, the display shown in Figure 18–51 will be displayed. Selecting ENTER will return you to the drawing. Selecting ⌐enter⌐ ⌐enter⌐ will remove the AutoCAD alert and reenter the SPELL command so that additional text can be checked.

If errors are found in a selection, the line of text will be displayed in the Context box of the Check Spelling dialog box. The misspelled word will be displayed in the

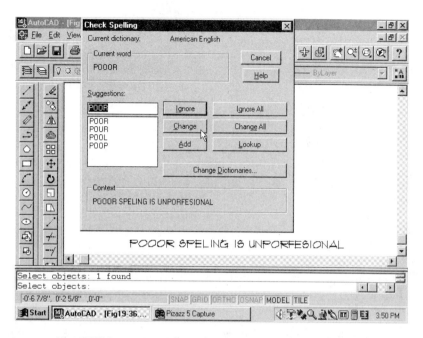

FIGURE 18–50 The SPELL command can be used to check TEXT, DTEXT and MTEXT. The current word "pooor," four suggestions, and the current context of the word are all displayed.

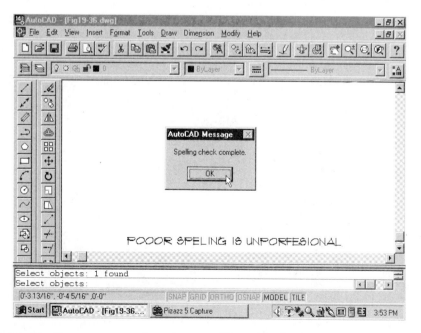

FIGURE 18–51 If no errors are found, you will see this display.

Current word box. A possible listing of alternatives can be found in the Suggestions: box. If the current word is as you would like it, pick the Ignore option and the spell checker will move to the next word identified by the checker. If you would like to use one of the suggested words, move the cursor to highlight the word and then select Change. The editor will proceed through your text until the alert seen in Figure 18–51 is displayed.

Changing the Dictionary

Selecting the Change Dictionaries... box will produce the dialog box shown in Figure 18–52. The Main dictionary box will allow you to change the language that is stored in the dictionary. The Custom dictionary and Custom dictionary Words boxes can be very useful to add words, acronyms, client names or vendor names specific to your area of expertise to the dictionary. When creating a custom dictionary, names may contain up to 8 characters using the same guidelines that were presented in Chapter 2 for naming drawing files. An extension of .CUS must be added after the name. To enter a custom name for an architectural dictionary, type **ARCHWORD.CUS** in the Custom dictionary box. If several custom dictionaries have been created, the Browse... box can be used to select the desired dictionary.

To add words to a specific dictionary, pick the Custom dictionary Words box. Names of up to 32 characters can be entered into the box. Multiple names can be entered at one time when they are separated by a comma. Selecting the Add box will add the names to the existing list of names contained in the dictionary. Figure 18–53 shows

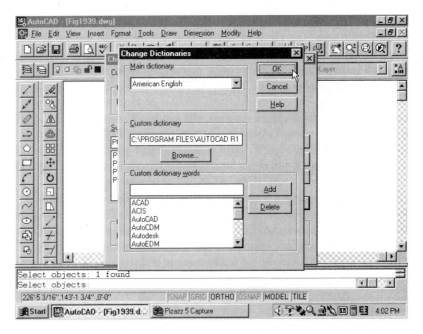

FIGURE 18–52 A separate dictionary can be created to contain words specific to your field.

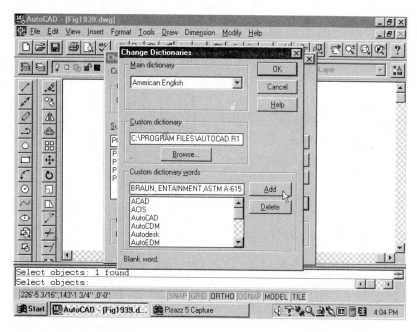

FIGURE 18–53 Three words are to be added to the dictonary. A partial listing of the contents is shown in the lower left corner of the dialog box.

a custom dialog box for the ARCHWORD.CUS dictionary with three words to be added to the existing list.

INCORPORATING TEXT INTO TEMPLET DRAWINGS

Much of the text used to describe a drawing can be standardized and placed in a templet drawing. Text can also be saved as a WBLOCK and inserted into drawings as needed. WBLOCKS will be discussed in Chapter 21. Figure 18–54 shows an example of notes that an office includes as standard notes for all site plans. These notes can be typed once and saved as a separate drawing so that they do not have to be retyped with each use. The wblock can then be either inserted into every site plan or nested in the templet drawing on the SITETXT layer. The layer can be frozen, thawed when needed, and then either the DDEDIT or DDMODIFY command can be used to make minor changes based on specific requirements of the job. Notice that several notes, and each of the schedules include the text `xxx'. These are notes that have been set up as blocks and stored with attributes that can be altered for each usage. Attributes will be discussed in Chapter 22.

Many drawings, such as sections, contain basically the same notes that are displayed as local notes. Although sections for a tilt-up structure will be radically different from a section for a steel rigid frame, a templet sheet can be developed for each type of construction. The actual drawing may look different, but the notes to specify standard materials are similar for most buildings using the same type of con-

GENERAL NOTES:

1. ENGINEER NOT RESPONSIBLE FOR LAND SURVEY OR TOPOGRAPHY.
2. ALL EXTERIOR SIGNS TO COMPLY W/ COUNTY SIGN ORDINANCE.
3. LOWER LEVEL FRAMED IN STEEL.
4. ALL STEEL COLUMNS TO BE 3" DIA. X 60" STEEL COLUMN FILLED W/ CONC. EMBEDDED 24" INTO GRADE.
5. SQUARE FOOTAGE OF LOT--00,000 SQ. FT.
6. SQUARE FOOTAGE OF STRUCTURE--00,000 SQ. FT. APPROXIMATELY.
7. 00.00% OF THE LOT IS COVERED.

UTILITY NOTES:

1. BALANCE OF WATER LINES TO BE SHOWN ON FLOOR PLANS.
2. CLEAN OUT @ ••••

PLANTER NOTES:

1. ALL PLANTERS TO BE CONSTRUCTED W/ A SEPARATE PERMIT.
2. PLANTER ON WEST SIDE ALONG FLOYD MILLER BLVD. TO BE 44' X 7.5' X 3' CONCRETE BLOCK W/ SIMILAR PLANTER IN THE SOUTHWEST CORNER BY THE STAIRWAY.

PARKING NOTES:

1. PARKING TO BE 4" MIN. CONC. W/ 6" X 6"--4 X 4 WWM OVER 4" COMPACTED GRAVEL.
2. EVERY TWO PARKING SPACES TO HAVE A 6" X 6" WHEEL STOP W/ (2) ½" X 12" STEEL DOWELS @ EA. WHEEL STOP.
3. PARKING SPACES TO BE MARKED WITH 4" WIDE WHITE STRIPES.

PARKING SPACES		
FULL-SIZE	9' X 20'	xx
COMPACT	7'-6" X 15'	xx
HANDICAPPED	12' X 20'	xx

	OCCUPANCY OF THE STRUCTURE	MAXIMUM ALLOWABLE FLOOR AREA (SQ. FT.)	TOTAL OCCUPANTS (FOR BLDG.)
FIRST FLOOR	xx	xx,xxx	xx
SECOND FLOOR	xx	xx,xxx	xx
THIRD FLOOR	xx	xx,xxx	xx

FIGURE 18–54 Common sets of notes can be stored in a templet drawing. Text that is subject to change with each usage is marked with "000" or "xxx".

½" CD APA 32/16 PLY. ROOF SHTG. INT. GR. W/ EXT. GLUE. LAY FACE GRAIN PERP. TO JOISTS & STAGGER JOINTS. USE: 8d COM. NAILS @ 6" O.C. BOUNDARY & EDGES, 8d COM. NAILS @ 12" O.C. FIELD. SEE ROOF-FRAMING PLANS FOR SPECIAL NAILING.

'TRUS-JOIST' 16" TJI 35" @ 24" O.C.

MINERAL CAP SHEET OVER 2 LAYERS OF ASB. FELT AS PER JOHN MANSVILLE SPEC. #406

PROVIDE ½" GYP. BD. DRAFTSTOPS FROM CEILING TO PLY ROOF SHTG. FOR EVERY 3,000 SQ. FT. MAX.

1½" 100#/FT LT. WT. CONC. OVER 15# ASPH. SATURATED FELT W/ 6" X 6" W-4" x 4" WWM IN SLAB OVER PLY. SHTG.

3/4" CD APA 42/20 T&G PLY FLR. SHTG., INT. GRD. W/ EXT. GLUE. LAY FACE GRAIN PERP. TO JOISTS & STAGGER JOINTS. USE: 10d COM. NAILS @ 6" O.C. BOUNDARY & EDGES, 10d COM. NAILS O.C. FIELD. SEE FLOOR FRAMING PLANS FOR SPECIAL NAILING.

LINE OF SUSP. CEILING (CLASS 'A')

1½" 100#/FT LT. WT. CONC.

3/4" CD APA 42/20 T&G PLY FLR. SHTG.

R-25 RATED INSULATION

PROVIDE ½" GYP. BD. DRAFT STOPS FROM CEILING TO PLY FLOOR SHTG. FOR EVERY 1,000 SQ. FT. MAX.

2-2 x 4 DFL TOP PLATES LAP 48" MIN. W/ 10-16d COM.

⅝" TYPE "X" GYP. BD.

2 x 4 STUDS @ 16" O.C. UNLESS NOTED

2-2 x 6 DFL TOP PLATES LAP 48" MIN. W/ 12-16d COM.

R-21 RATED INSULATION

2 x 6 SOLE PLATES.

R-38 RATED INSULATION

FIGURE 18–55 Local notes can also be stored in a templet drawing and moved as needed.

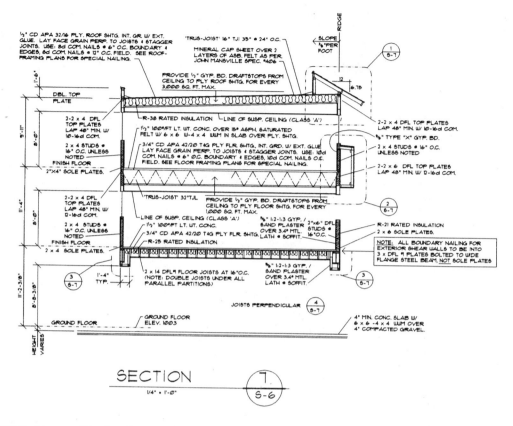

FIGURE 18–56 The local notes from Figure 18–54 can be moved and placed as needed.

struction. Local notes can also be placed in the templet drawing as shown in Figure 18–55. These notes can also be stored as a WBLOCK, moved to the desired drawing, thawed, moved into the needed position, and edited as seen in Figure 18–56. This can greatly increase drafting efficiency.

CHAPTER 18 EXERCISES

E-18-1. Open a new drawing and type the following text using left-justified text with a height of .125″and a TrueType font.

All sheathing should be ½″ standard grade 32/16 ply interior type with exterior glue. Lay perpendicular to rafters and stagger all joints. Nail with 8d's @ 6″ o.c. at edge, blocking and beams, and 8d's @ 12″ o.c. @ field. Use common wire nails.

Save the drawing as E-18-1.

E-18-2. Open a new drawing and make ROMANS the active font, with an obliquing angle of 10° and a height of .125″.

Use 8 × 8 × 16 grade "A" concrete block units with a triple score. Use # 5 diameter rebars @ 24″ o.c. each way solid-grout all steel cells.

Save the drawing as E-18-2.

E-18-3. Open a new drawing and make CIBT the active font and a height of .125″. Use a rotation angle of 35°. Write the following note:
5⅛ × 13½″
exposed glu-lam
ridge beam f:2200

Save the drawing as E-18-3.

E-18-4. Open a new drawing and make ROMANT the active font with a height of .125″. Use a width factor of 1.25. Use the center option to type the following text:

ATTENTION ALL CADD
OPERATORS.
DUE TO
BUDGETARY PROBLEMS
YOUR MOM DOES NOT WORK HERE.
PLEASE CLEAN UP
YOUR OWN MESS.

Save the drawing as E-18-4.

E-18-5. Open a new drawing and create a style for TEXT using ROMANS as the font with a height of .125″. Use a width factor of 1.00. Create a second style named TITLES using ROMANT as the font, with a height of .25 and a width factor of 1.5. Assign each style a layer named for the style and provide a separate color for each. Use MTEXT to create the following text:

INSULATION NOTES

1. Insulate all exterior walls with 5½″ high-density fiberglass batt insulation, R-21 min.
2. Insulate all flat ceiling joist with 12″ R-38 batts (no paper facing required).
3. Insulate all vaulted ceiling with 10¼″ high-density, paper-faced fiberglass R-38 min. with 2″ air space above.
4. Insulate all wood floors with 8″ fiberglass batts, R-25 with paper face. Install plumbing on heated side of insulation.

CAULKING NOTES

Caulk the following openings with expanded foam or backer rods. Elastomatic, copolymer, siliconized acrylic latex caulks may also be used where appropriate.

Any space between window and door frames.
Between all exterior wall sole plates and plywood sheathing.
On top of rim joist prior to plywood floor application.

Wall sheathing to top plate.
Joints between wall and foundation.
Joints between wall and roof.
Joints between wall panels.

Save the drawing as E-18-5.

E-18-6. Open drawing E-18-5 and make the following changes to the insulation notes.

1. Change wall insulation to reflect 6″ fiberglass batt insulation, R-19 min.
2. Change all flat ceiling to 12″ R-38 batts with paper facing.
3. Change all vaulted ceiling to 10″ paper-faced fiberglass R-30 min. with 2″ air space above.
4. Change wood floors to 6″ fiberglass batts, R-19 with paper face.

Save the drawing as E-18-6.

E-18-7. Open drawing E-14-8. Use a text font other than txt and label the drawing using drawing E-14-8 as a guide. Complete the drawing and add all hatch patterns. Place the required text box on a separate layer from other text. Use the LINE command to add all leader lines. Save the drawing as E-18-7.

E-18-8. Open E-16-10 and label the materials that have been drawn using a suitable font. Use QTEXT to disable the text. Create a separate layer for all text and save the drawing as E-18-8.

E-18-9. Open drawing E-16-9. Create separate layers, styles, and fonts for titles and text and label the drawing. Use straight lines to connect text to the object being described. Save the drawing as E-18-9.

E-18-10. Open drawing FLOOR14 and design a 2″ wide title block along the right side of the page. Design space for page number, date, revision date, client information, and designer information. Since you are the designer, provide a company name and logo and your name, address, and phone number. Create new layers for the title block lines and text that will remain constant, and a separate layer for text that will vary for each job. Use a minimum of three different text fonts. Save the drawing as FLOOR14.

E-18-11. Type the following note: <u>PROVIDE A 5° +/−.5° BEND AT ALL</u> #6 DIA. STEEL REINFORCING. STEEL TO BE WITHIN 1% OF SPECIFIED LENGTH. Save it as a E-18-11.

E-18-12. Create a line of text and make a copy of the text. Mirror one of the lines using a variable of 1 and one of the lines with a variable of 0. Save the drawing as E-18-12.

E-18-13. Use the MTEXT command to make a list of a minimum of three guidelines that describe the creation of MTEXT. Use a second text style and justification point and type a second list of guidelines for describing DTEXT. Save the drawing as E-18-13.

E-18-14. Open drawing E-18-13 and make a copy of each set of text. Edit the new text to add additional words in the existing sentences and change the color, style and width of text. Save the drawing as E-18-14.

CHAPTER 18 QUIZ

1. What command should be used as text is added to a drawing so that it can be seen as it is typed?

2. What text option will allow text to be placed between two selected points, at any angle?

3. What text option can be used to place multiple lines of text in a column?

4. If text is to be .125 high on a drawing that is to be plotted at a scale of 1″ = 20′-0″, what should the text height be?

5. Explain the difference between rotation angle and obliquing angle.

6. What is the difference between a style and a font?

7. What symbol is needed to produce a degree symbol?

8. What commands are used to correct errors in text?

9. What effect will setting the style height option have on future text?

10. Explain how to correct the following problem. VERIFY EXATC hEIGHT AT JOBSITE.

11. When are lowercase letters appropriate for text on construction drawings?

12. Write the required entry to correctly specify three bolts that are three quarters of an inch in diameter at six inches on center. Include all special characters required if the style is set to be Romans.

_____

13. A beam is located in an opening in a wall that is at 45° to horizontal. How should the text be orientated to describe the beam?

_____

14. List and describe the two types of notes generally found on construction drawings.

_____

15. What should be considered when choosing a font?

_____

_____

16. What are the effects of setting the text height in the STYLE command as opposed to setting the height in the DTEXT?

_____

_____

17. What advantages does DTEXT offer over TEXT?

_____

_____

18. When would QTEXT be used?

_____

_____

19. What is MTEXT best suited for?

_____

_____

20. What should be the value for MIRRTEXT if you want to mirror the text with the object?

_____

21. What is the effect of using the **?** option from the title bar of the Multiline Text Editor?

_____

_____

22. Your instructor wants you to create a body of text that starts in the upper right hand corner of a templet drawing. The notes will be 4" wide and be ½" down and 4½" from the border. What are the commands and options, and what is the proper sequence to place these notes?

_____

_____

23. You've decided to dig deeper into the MTEXT command. Where can you find help from AutoCAD?

_____

_____

24. Use the HELP command and research how the text boundary of MTEXT will plot.

_____

_____

25. What is the process required to save a paragraph created using MTEXT.

_____

_____

26. A paragraph of MTEXT has been created using ROMANS but it should be in italics. How can this be fixed?

_____

27. Explain the major differences between DDEDIT and DDMODIFY on MTEXT.

_____

_____

28. What does the .CUS extension designate?

_____

29. What is the advantage of placing stock notes on a templet drawing?

_____

_____

30. A client has decided to change from floor joist to open-web trusses after all of the written specifications have been added to the drawing. How can this change be easily included?

_____

_____

CHAPTER

19

INTRODUCTION TO DIMENSIONING

In addition to the visual representation, and the text used to describe a feature, dimensions are needed to describe the size and location of each member of a structure. Figure 19–1 shows a floor plan for a dormitory and the dimensions used to describe the location of structural members.

In this chapter you will be introduced to the basic principles of dimensioning and to the many commands that can be used to place dimensions. Several methods will be used to enter the dimension command. Because so many options are available to place dimensions, this chapter will examine only the default values for dimensioning. Chapter 20 will introduce methods of creating different styles of dimensions and describe how to adjust the variables that control the size and placement of the dimensions.

Because so many options are included in the software, some options that do not apply to the construction industry will be skipped entirely. Feel free to explore these options on your own, or consult the Help menu.

BASIC PRINCIPLES OF DIMENSIONING

To make better use of the software's potential, you need to understand several basic dimensioning concepts and terms before you explore the dimension commands. These principles include linetypes, placement of dimension features, and locating exterior and interior drawing features.

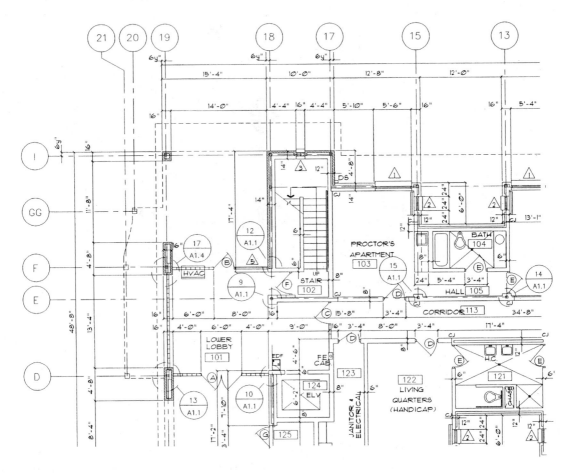

FIGURE 19–1 Dimensions are used throughout construction drawings to show the location of each system to be installed. (Courtesy G. William Archer, AIA, Archer and Archer, P.A.)

Dimensioning Components

Dimensioning features include extension and dimension lines, text, and line terminators. Each component can be seen in Figure 19–2.

Extension Lines. Extension lines are thin lines that extend out from the object being described, and set the limits for the dimensions. Extension lines usually are offset about ⅛″ from the object, and extend past the last dimension line between ¹⁄₁₆″ to ⅛″. Each can be seen in Figure 19–3.

Two different types of linetypes typically are used for extension lines. Solid lines are used when dimensioning to the exterior face of an object, such as a wall or footing. A centerline is used when dimensioning to the center of a wood wall or the center of other objects. Figure 19–4 shows examples of each.

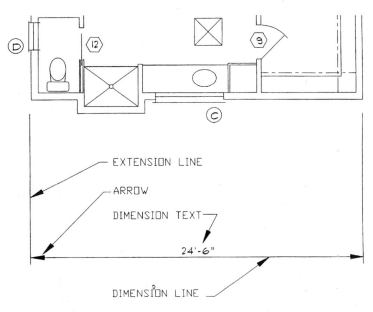

FIGURE 19–2 Dimensions are composed of the dimension text, a dimension line, extension lines, and arrows.

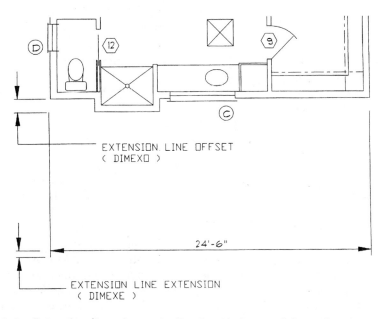

FIGURE 19–3 Extension lines are typically placed about ⅛" from the object, and extend about ⅛" past the dimension line.

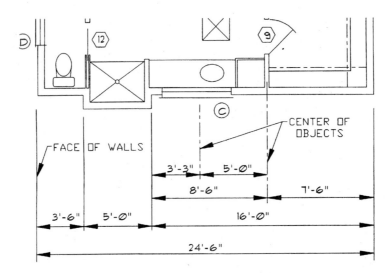

FIGURE 19–4 Solid extension lines are used to dimension to the edge of a surface; centerlines are used to dimension to the center of objects.

Dimension Lines. Dimension lines are thin lines used to show the extent of the object being described. Exact location will vary with each office, but dimension lines should be placed in such a way that there is room for notes, but still close enough to the features being described that clarity will not be hindered. Guidelines for placement will be discussed later in this chapter.

Text. The text for dimensioning typically is placed above the dimension line, and centered between the two extension lines. On the left and right sides of the structure, text is placed above the dimension line and rotated so that the text can be read from the right side of the drawing page. Examples of each placement can be seen in Figure 19–5. On objects placed at an angle other than horizontal or vertical, dimension lines and text are placed parallel to the oblique object as seen in Figure 19–6.

Often not enough space is available for the text to be placed between the extension lines when small areas are dimensioned. Although options vary with each office, several alternatives for placing dimensions in small spaces can be seen in Figure 19–7.

Terminators. The default method for terminating dimension lines at an extension line is with an arrowhead. Many offices don't use arrowheads when doing manual drafting because of the time it takes to produce them. Instead of arrows, a tick mark or dot has become the most commonly used substitute for manually produced drawings. With CAD-produced drawings, any of the three options can be produced easily, although the tick mark is quickest to plot. A fourth option is a thickened tick mark, which offers good contrast between lines, closely resembles its manual counterpart,

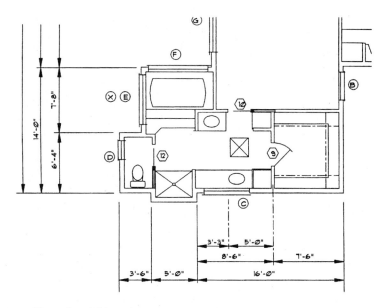

FIGURE 19-5 Text should be placed above the dimension line, and read from the bottom or right side of the drawing.

and can be plotted quickly. All four options can be seen in Figure 19–8. Release 14 also offers several new options that will be introduced in the next chapter.

DIMENSION PLACEMENT

Construction drawings requiring dimensions typically consist of plan views, such as the floor, foundation, and framing plans; and drawings showing vertical relationships, such as exterior and interior elevations, sections, details, and cabinet drawings. Each type of drawing has its own set of problems to be overcome in placing drawings. No matter the drawing type, dimensions should be placed on a layer titled ***-ANNO-DIMS** with the letter of the proper originator used to replace the asterisk.

Plan Views

Placing dimensions in plan view can be divided into the areas of interior and exterior dimension. Whenever possible, dimensions should be placed outside the drawing area.

Exterior Dimensions. Most offices start by placing an overall dimension on each side of the structure that is approximately 2″ from the exterior wall. Moving inward with approximately ½″ between lines, would be dimension lines that are used to describe major jogs in exterior walls, the distance from wall to wall, and the distance

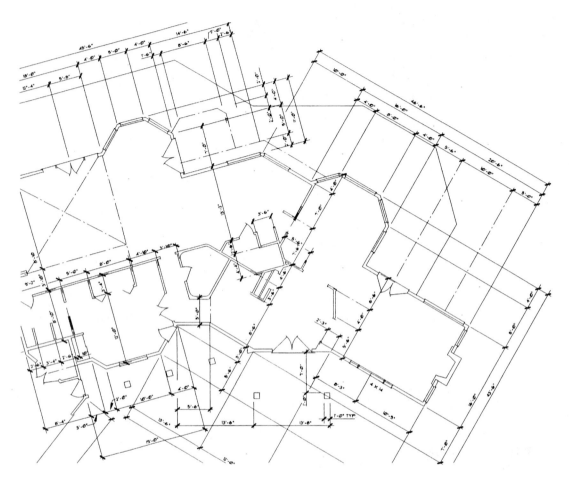

FIGURE 19–6 When dimensioning objects on an angle, dimension lines and text are placed parallel to the inclined surface. (Courtesy Residential Designs)

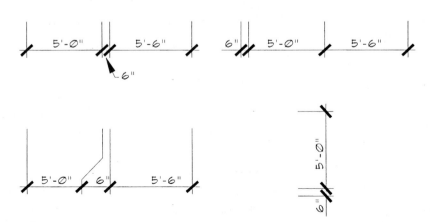

FIGURE 19–7 Common professional alternatives for placing dimensions in small spaces.

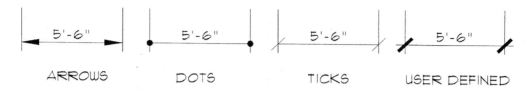

FIGURE 19–8 Alternatives for terminating dimension lines.

from wall to window or door to wall. Examples of placing these four dimension lines can be seen in Figure 19–9.

Two different systems are used to represent the dimensions between exterior and interior walls. Architectural firms tend to represent this distance from edge to edge of walls as seen in Figure 19–10. Engineering firms tend to represent the distance from exterior edge to center of interior wood walls using methods shown in Figure 19–11. Concrete walls are dimensioned to the edge by both disciplines, as seen in Figure 19–12. Concrete footings normally are dimensioned to the center. Examples of dimensioning concrete footings and walls can be seen in Figure 19–13.

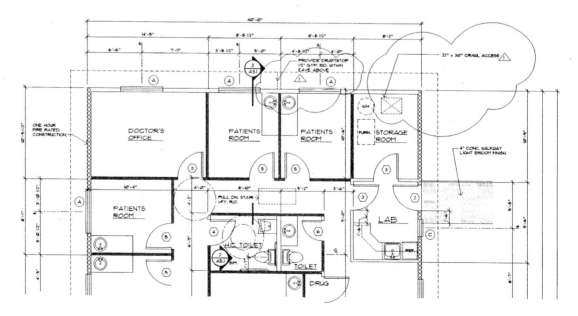

FIGURE 19–9. Floor plans typically are dimensioned by placing an overall dimension, dimensions to locate major jogs, wall-to-wall dimensions, and wall-to-opening dimensions. (Courtesy Scott R. Beck, Architect)

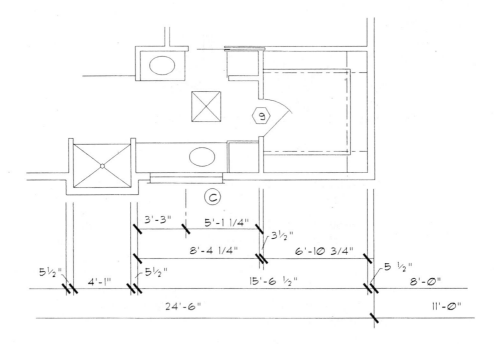

FIGURE 19–10 Some architectural firms dimension from edge to edge of interior walls.

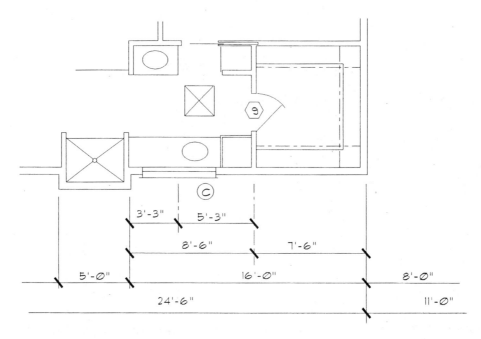

FIGURE 19–11 Engineering firms often dimension to the center of interior walls.

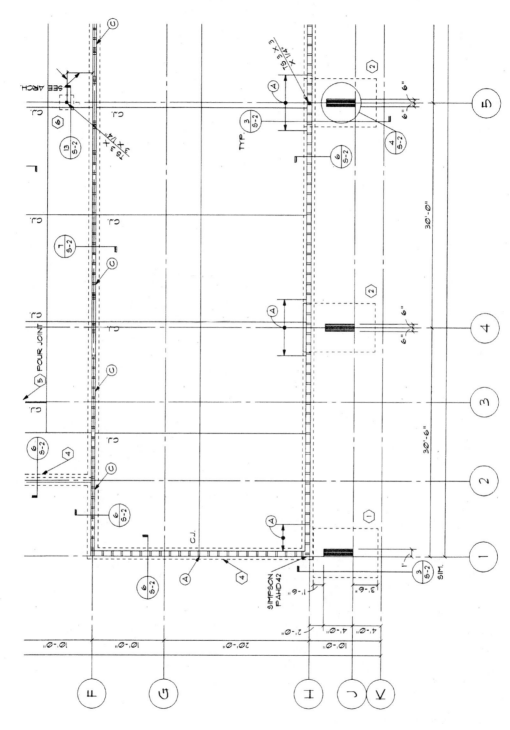

FIGURE 19–12 Concrete walls are dimensioned from edge to edge. (Courtesy of Van Domelen/Looijenga/McGarrigle/Knauf Consulting Engineers)

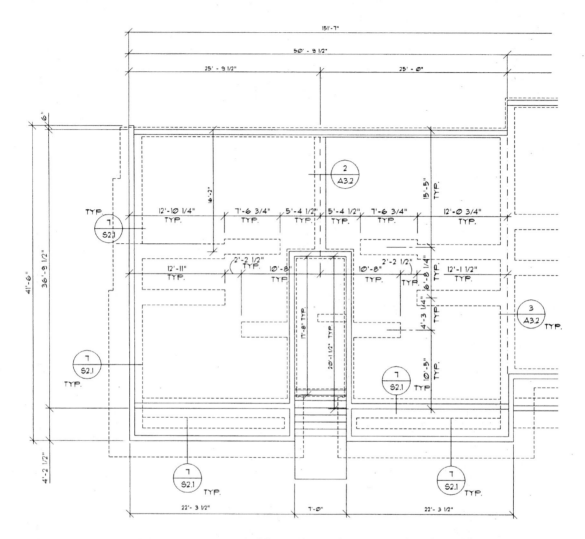

FIGURE 19–13 Interior concrete piers and footings are dimensioned to their centers. (Courtesy Scott R. Beck, Architect)

Interior Dimensions. The two main considerations in placing interior dimensions are clarity and grouping. Dimension lines and text must be placed so that they can be read easily and so that neither interferes with other information that must be placed on a drawing. Information also should be grouped together as seen in Figure 19–14, so that construction workers can find dimensions easily.

Vertical Dimensions

Unlike the plan views, which show horizontal relationships, the elevations, sections, and details each require dimensions that show vertical relationships, as seen

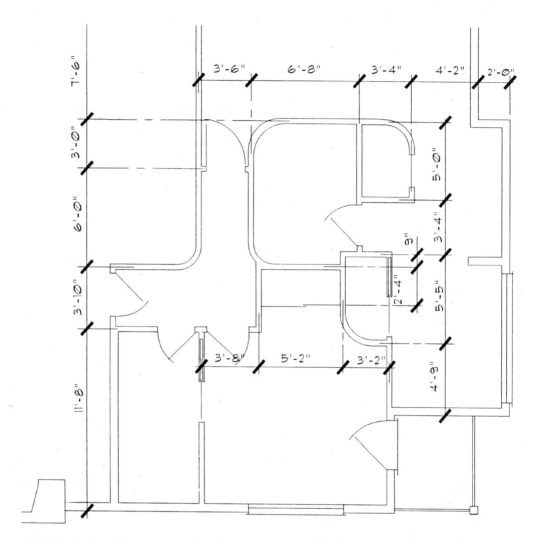

FIGURE 19–14 Dimensions placed on the inside of a drawing should be grouped together. (Courtesy Residential Designs)

in Figure 19–15. Typically, these dimensions originate at a line that represents a specific point, such as the finish grade, a finish floor elevation, or a plate height.

ACCESSING DIMENSIONING COMMANDS

Several methods are available for accessing the commands that control dimensions. Commands can be accessed using the Dimensioning toolbar, or by selecting an option from the Dimension pull-down menu. The pull-down menu and toolbar can

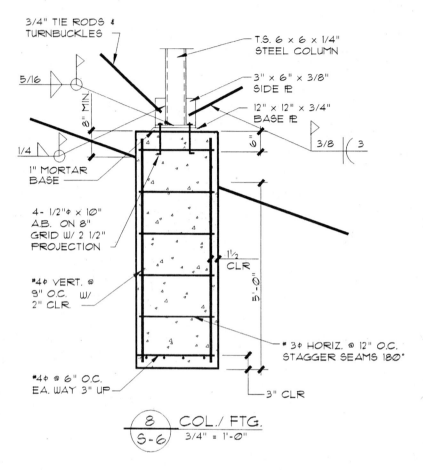

FIGURE 19-15 Elevations, sections, and details each require dimensions to show vertical relationships. (Courtesy Residential Designs)

be seen in Figure 19–16. Each of the options presented in these menus will be presented in this chapter. Dimensions can also be entered at the command prompt using two different options. Menus are also available in dialog boxes, which will be presented in Chapter 20.

DIMENSIONING OPTIONS

Eleven methods are available for placing dimensions on a drawing. The Linear and Aligned options are used for providing linear dimensioning. Diameter, Radius, Angular, and Center are used to describe circular features. Three other options, Baseline, Continue, and Ordinate, are used to place dimensions and can be combined with other commands. Ordinate dimensioning usually isn't used on construc-

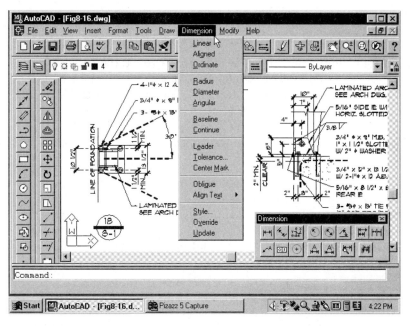

FIGURE 19–16 The commands for placing dimensions can be selected from the Dimension toolbar, pull–down menus, or entered by keyboard.

tion drawings and will not be covered. The LEADER command can be used for referencing text and dimensions to a drawing. The Tolerance... option can be used to access geometric dimensioning and tolerancing symbols. These symbols are rarely used and will not be discussed in this text. Style: can be used to access dimensioning dialog boxes that are used to save dimensioning styles. Styles will be discussed in Chapter 20.

LINEAR DIMENSIONS

Linear dimensions are used to dimension straight surfaces common to construction features. Selecting Dimensioning from the Dimension toolbar or pull-down menu will produce the command prompt, which will be discussed shortly. Linear dimensions can also be started by typing **DLI** [enter], **DIMLIN** [enter], or **DIMLINEAR** [enter] at the command prompt. This will produce the prompt:

> Command: **DLI** [enter]
> DIMLINEAR
> First extension line origin or RETURN to select.

AutoCAD now offers two options to place a dimension.

Select Extension Line Origin. The start and end of a line or distance to be dimensioned can be specified. The command sequence for specifying each endpoint is:

First extension line origin or RETURN to select: (*Pick point.*)
Second extension line origin: (*Pick point.*)
Dimension line location (Mtext/Text/Angle/Horizontal/Vertical/Rotated):
  (*Select/dimension line location.*)

The default requires you to move the cursor and select the desired location to place the dimension line.

The OSNAP command can be very helpful in selecting exact endpoints. As the second extension line location is picked, the dimension and extension lines are dragged into position to allow for visual inspection, but the coordinate display can be useful in spacing dimension lines. The dimension line location should be selected to allow room for all future notes. Place the dimension line in the desired location and press the pick button. The process can be seen in Figure 19–17.

Text. One of the strengths of AutoCAD is also one of its weaknesses. AutoCAD is extremely accurate. Sometimes the dimension text that will be assigned is too accurate for the drawing purpose. Chapter 20 will introduce methods of controlling dimension accurately. If you haven't been accurate as you created the drawing, the

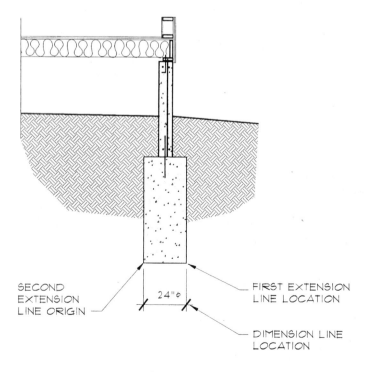

SECOND
EXTENSION
LINE ORIGIN

24"ϕ

FIRST EXTENSION
LINE LOCATION

DIMENSION LINE
LOCATION

FIGURE 19–17 Horizontal and vertical dimensions can be placed using the DIMLIN command. The dimension is placed by picking the starting location for each extension line and then selecting the location for the dimension line. As the location for the second extension line is selected, the dimension is displayed and can be dragged into position.

text assigned to the dimension will not produce the desired results. Choosing the Text option before the dimension line location is selected allows the dimension text to be altered by entering the desired option by keyboard. To change the prompt from 3½" to 4" you would use the following sequence:

Dimension line location (Mtext/Text/Angle/Horizontal/Vertical/Rotated): **T** [enter]
Dimension text <3.5>: **4** [enter]
Dimension line location (Mtext/Text/Angle/Horizontal/Vertical/Rotated):
   (Select the desired location.)
Command:

Pressing the Enter key returns the dimension and allows it to be dragged into position. Pressing the space bar and the Enter key rather than assigning a value at the prompt for Dimension text displays the extension lines, dimension lines, and terminators but no text is assigned. This option can be useful for specifying the limits of a specific product, as seen in Figure 19–18.

Mtext.   Selecting the Mtext option allows the dimension value to be replaced by a body of text in a similar manner as the Text option. Entering **M** [enter] at the dimension prompt will produce the display shown in Figure 19–19 with the cursor in front of the brackets. The brackets represent the default dimension value. With the cursor

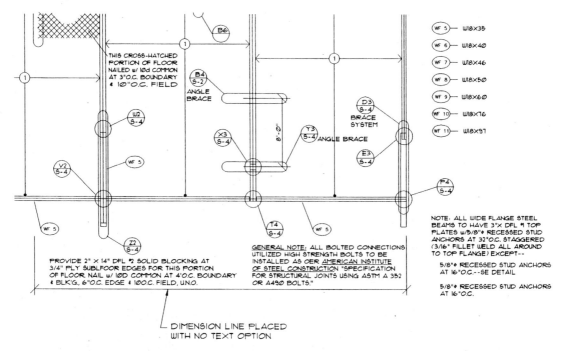

FIGURE 19–18 Rather than entering a text value, pressing the Space key followed by the [enter] key allows the dimension line to be placed with no dimension.

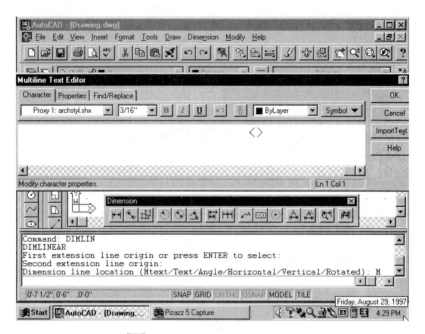

FIGURE 19–19 Entering **M** [enter] at the command prompt will produce the Multiline Text Editor.

preceding the brackets, text will be written followed by the dimension. An entry of **F.O.S. <>** would be displayed as **F.O.S. 40'–0"**. An entry of **<>F.O.S.** will be displayed as **40'–0" F.O.S.** Using the cursor to remove the brackets will remove the default dimension value and allow only text to be created.

Angle. Enter **A** [enter] at the line location prompt to provide an opportunity to alter the angle that the text will be displayed. Although this is not often done, it could be especially helpful to make a dimension for a small space standout. Figure 19–20 shows an example of text rotated at 45 degrees to the dimension line. The command sequence, once the two endpoints are selected, would be:

> Dimension line location (Mtext/Text/Angle/Horizontal/Vertical/Rotated): **A** [enter]
> Enter text angle: **45** [enter]
> Dimension line location (Mtext/Text/Angle/Horizontal/Vertical/Rotated): *(Select the desired location.)*
> Command:

Horizontal/Vertical. Entering the **H** [enter] or the **V** [enter] option at the prompt allows only the specified dimension direction to be entered. AutoCAD automatically places a vertical dimension if two vertical endpoints are selected. This option is outdated in AutoCAD Release 14, but is available to make users of older versions feel at home.

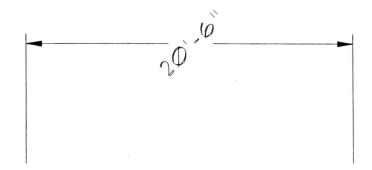

FIGURE 19–20 The text angle can be rotated from the dimension line using the **A** option. Although not a common method of dimensioning, rotated text can be used to help draw attention to the dimension.

Rotated. Entering **R** [enter] at the prompt rotates the extension and dimension lines as a specified angle from the desired surface. This option can be used to draw attention to a dimension describing a small space. Use this option with the space key or [enter] to place the extension and dimension lines for referencing a note to a specific area. The sequence to place the lines at 20 degrees to a vertical surface would be:

> Dimension line location (Mtext/Text/Angle/Horizontal/Vertical/Rotated): **R** [enter]
> Dimension line angle: **20** [enter]
> Dimension line location (Mtext/Text/Angle/Horizontal/Vertical/Rotated): *(Select
>   the desired location.)*
> Command:

Figure 19–21 shows an example of using the rotated option.

The Return Option. Up to this point, each of the options discussed were made available once the two origins for the extension line endpoints were selected. By selecting the return option, a specific line can be selected, and the endpoints will be determined automatically. The command sequence is:

> Command: **DLI** [enter]
> DIMLINEAR
> First extension line origin or Return to select: [enter]

Once [enter] is pressed, the cross-hairs are changed to a pick box and a specific entity may be selected:

> Select object to dimension:
> Dimension line location location (Mtext/Text/Angle/Horizontal/Vertical/
>   Rotated): *(Select the desired location.)*
> Dimension text <24>:
> Command:

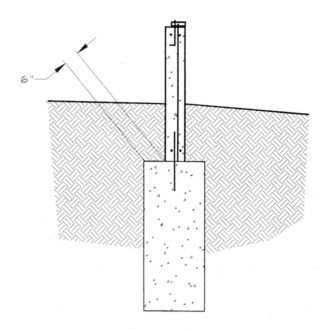

FIGURE 19–21 The extension lines, dimension lines, and text can be rotated from the surface to be dimensioned using the **R** option.

Rather than selecting two line origin points only one object is selected, and the extension lines, dimension line and dimension text are dragged into position.

> Select object to dimension: (*Pick object to be dimensioned.*)
> Dimension line location

This process can be seen in Figure 19–22. The effects of the DIMLINEAR command on vertical dimensions are shown in Figure 19–23. If the area to be dimensioned is smaller than the length required to place the arrows and text, each will be placed outside the extension lines. Methods of altering placement will be discussed in Chapter 20.

Aligned Dimensions

This method of placing dimensions is useful when the object being described is not parallel to the drawing borders. An example of aligned dimensions can be seen on the foundation plan shown in Figure 19–24. The command sequence is similar to the DIMLINEAR command. The ALLIGNED command can be selected from the Dimensioning toolbar, the pull-down menu or started at the command prompt by entering **DAL** [enter], **DIMALI** [enter] or **DIMALIGNED** [enter]. The command sequence is:

> Command: **DAL** [enter]
> First extension line origin or press ENTER to select: (*Select first point.*)

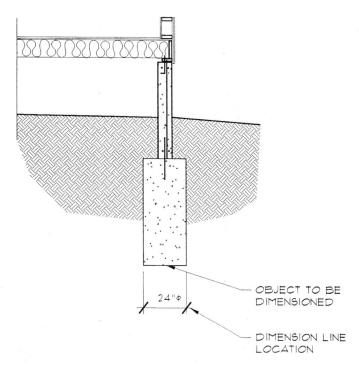

FIGURE 19–22 Choosing the Return option rather than selecting an extension line location allows an object to be selected. Extension lines will automatically be placed at each end of the selected object.

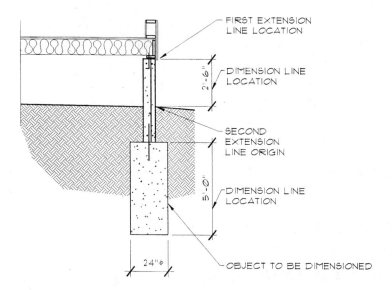

FIGURE 19–23 Placement of vertical dimension lines using DIMLINEAR using the return option to place the lower dimension and two point selection method on the upper dimension.

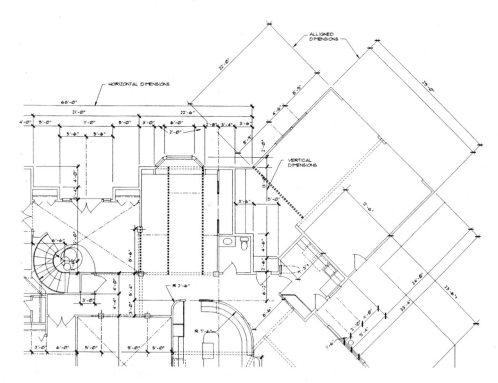

FIGURE 19–24 When a surface is not horizontal or vertical, the DIMALIGNED command can be used to place dimensions. (Courtesy Residential Designs)

> Second extension line origin: *(Select second point.)*
> Dimension line location (Mtext/Text/Angle):
> Dimension text <45'-0">:
> Command:

If a location is selected for the dimension line, the line will automatically be placed parallel to the surface defined by the endpoints. Selecting the Mtext or Text option allows the default text value to be altered. Selecting the Angle option allows the angle of the text to be altered. Each of the options is the same as the corresponding options for DIMLINEAR. Selecting the Return option automatically places the extension lines at the ends of the selected line and allows the dimension to be dragged into position.

The effects of DIMALIGNED can be seen in Figure 19–25.

ORDINATE DIMENSIONS

Ordinate dimensioning consists of placing dimensions without the use of dimension lines and arrows. An example of ordinate dimensions can be seen in Figure 19–26.

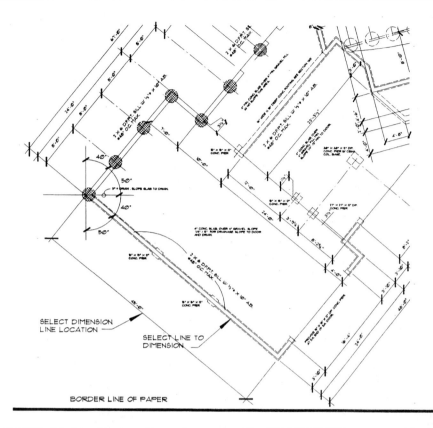

FIGURE 19–25 Placing dimensions using the DIMALIGNED command is similar to using the DIMLINEAR command.

They are not typically used in construction, but are occasionally used in placing vertical dimensions on details and sections. The command can be accessed from the Dimensioning toolbar and pull-down menu or from the command prompt by typing **DOR** enter , **DIMORD** enter or **DIMORDINATE** enter.

> Command: **DOR** enter
> DIMORDINATE
> Select feature: *(Select the surface to be dimensioned.)*

AutoCAD is prompting for the location that will serve as a base. On a section, this is often the finished floor elevation. It will be helpful to move the 0,0 location of the world coordinate system from the lower corner of the drawing to a corner of the object being dimensioned. The Ortho mode should also be set to ON. Once the initial edge has been selected, the prompt will be altered to read:

> Leader endpoint (Xdatum/Ydatum/Mtext/Text): *(Pick the endpoint of the extension line.)*

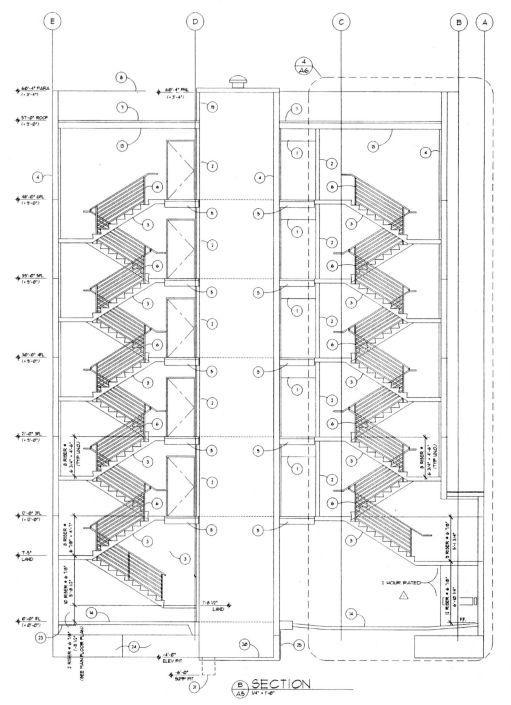

FIGURE 19-26 Placement of dimensions using DIMORDINATE command. All dimensions are placed using only a single extension line to represent each finish floor. All dimensions are referenced from the main floor level. (Courtesy Russ Hanson, HDN Architects, PC)

Dimension text: <10'-0"> [enter]

The leader is a single horizontal line when the Ydatum ordinate is used and a single vertical line for the Xdatum.

Radius Dimensions

The DIMRAD command is useful for many engineering and architectural applications. The way the information is placed will vary depending on the size of the circle or arc to be dimensioned. Figure 19–27 shows each option. The type of dimension also depends on the DIMVAR settings. The command can be selected from the toolbar, pull-down menu, or entered by keyboard by typing **DRA** [enter] **DIMRAD** [enter] or **DIMRADIUS** [enter]. The command sequence is:

Command: **DRA** [enter]
DIMRADIUS
Select arc or circle: *(Select desired arc or circle.)*
Dimension text <10'-6">
Dimension line location (Mtext/Text/Angle): *(Select location.)*
Command:

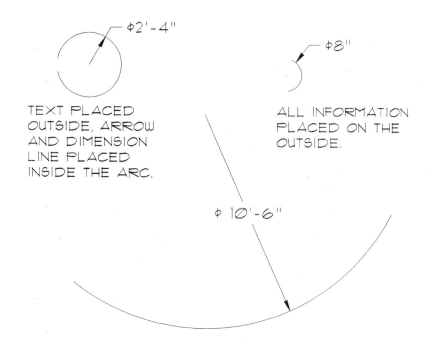

FIGURE 19–27 The DIMRADIUS command allows a radius to be placed using three different techniques.

When you select the circle or arc to be dimensioned, the default dimension is dragged as you move the cursor. Entering **M**, **T** or **A** allows the text or angle value to be altered. Mtext, Text, and Angle function exactly as discussed with other commands. Once the desired Text and Angle prompts have been responded to, the prompt will be returned to locate the dimension line. Figure 19–28 shows an example of the use of the DIMRADIUS command.

Placing Diameters

The DIMDIA command will allow you to specify the diameter of circular objects. The commands can be selected from the Dimension toolbar and pull-down menu or typed at the command prompt. Enter **DDI** [enter] , **DIMDIA** [enter] or **DIMDIAMETER** [enter] by keyboard. Three methods of placing the diameter specifications on a drawing can be seen in Figure 19–29. The default method is to use a leader line to display the dimension. The command sequence is:

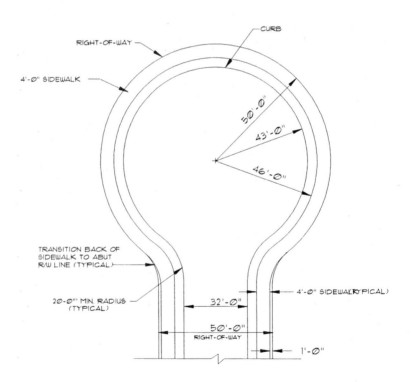

FIGURE 19–28 Placing of dimensions using the DIMRADIUS command. (Courtesy Department of Environmental Services, City of Gresham, OR)

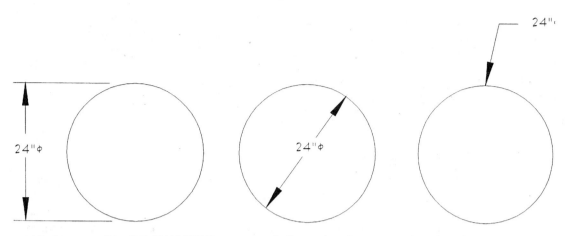

FIGURE 19-29 The DIMDIAMETER command allows the diameter to be placed using three different methods.

Command: **DDI** [enter]
DIMDIAMETER
Select arc or circle:
Dimension text <24.00>
Dimension line location (Mtext/Text/Angle):
Command:

When you select the circle to be dimensioned, the default dimension is dragged as you move the cursor. Depending on the size of the circle, the dimension toggles between being placed inside or outside the circle. Placing the dimension inside or outside the circle depends entirely on the size of the circle and the room available. Chapter 20 will discuss methods for controlling where the text is placed using DIMVARS. Entering **M** [enter], **T** [enter], or **A** [enter] allows the text or angle value to be altered. Text and Angle function exactly as discussed earlier for the DIMLINEAR command. Once the desired prompts have been responded to, the prompt will be returned for locating the dimension line location.

Methods for displaying other circular display options will be discussed later in this chapter as variables are introduced.

Placing a Centerpoint

The DIMCENTER command will provide a centerpoint for a circle or an arc. Two commands can be used to control the centerpoint. The DIMCEN command allows the size of the mark to be altered. The command sequence is:

Command: **DIMCEN** [enter]
New value for DIMCEN <0'-0 1/8">: *(Provide desired value)* **6** [enter]
Command:

As the command prompt is returned, nothing is changed. Selecting Center Mark from the Dimensioning toolbar or pull-down menu will now allow a circle or arc to be selected. The command sequence is:

Command: **DIMCENTER** ⏎
Select arc or circle:
Command:

Labeling Angles

The DIMANG command can be used to label angles formed by selecting two straight nonparallel lines, an arc, a circle, and another point or three points. You can start the command by selecting the Angular Dimension icon from the Dimensioning toolbar, selecting ANGULAR from the tablet menu, or Angular from Dimensioning in Draw of the pull-down menu. To enter the command at the keyboard type **DAN** ⏎, **DIMANG** ⏎, or **DIMANGULAR** ⏎ at the command prompt. This will produce the following prompt:

Command: **DAN** ⏎
DIMANGULAR
Select arc, circle, line, or press ENTER:

Choosing Two Lines.   One of the most common situations for dimensioning angles is describing the angle formed between two intersecting lines. Respond to the prompt by selecting a line. The command sequence continues with:

Command: **DAN** ⏎
DIMANGULAR
Select arc, circle, line, or press ENTER: *(Select first line.)*
Second line: *(Pick second point.)*
Dimension arc line location (Mtext/Text/Angle) *(Select location.)*
Dimension text <44.00>
Command:

As the Dimension line location (Mtext/Text/Angle): prompt is displayed, you will be able to choose the location of the dimension placement relative to the angle. Figure 19–30 shows alternatives for placing the angle dimensions.

Angles Based on an Arc.   When you respond to the initial angle prompt by choosing arc, dimensions can be placed to describe the tangent points of an arc. This could be especially useful when labeling subdivision maps or other large parcels of land. The command sequence is:

Command: **DAN** ⏎
DIMANGULAR
Select arc, circle, line, or press ENTER: *(Select first line.)*

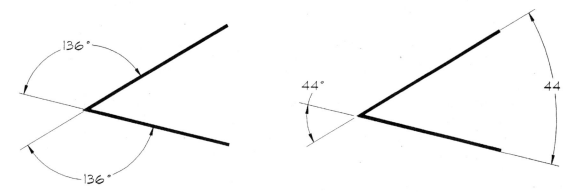

FIGURE 19–30 Alternatives for placement of dimensions using the DIMANG command.

Second line: *(Select second line.)*
Dimension arc line location (Mtext/Text/Angle)
Dimension text <156.00>
Command:

As you select the second line, the angle is displayed. Depending on the location of the cursor, the display can be located in one of four positions. The Text and Angle options are the same as other dimensioning commands.

Figure 19–31 shows an example of how text will be placed.

Angles Based on a Circle. When you respond to the initial angle prompt by choosing Circle, dimensions can be placed to describe angular patterns formed within a circle. This type of dimensioning is often used in specifying steel placement in circular columns and footings. The command sequence is:

Command: **DAN** [enter]
DIMANGULAR
Select arc, circle, line, or press ENTER: *(Select desired arc.)*

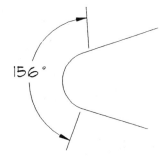

FIGURE 19–31 Dimensioning
of an arc using the DIMANGU-
LAR command.

Dimension arc line location (Mtext/Text/Angle): *(Select desired location.)*
Dimension text <67.00>
Command:

Options for placing the dimension line can be seen in Figure 19–32.

Describing Three Points.   When you respond to the original angular prompt with the [enter] key, you will be allowed to select three different points that can be used to describe an angle. This option can be used to describe the centerpoints of bolts around a centerpoint in a steel connector strap. The command sequence is:

Command: **DAN** [enter]
DIMANGULAR
Select arc, circle, line or press ENTER: *(Pick arc.)*
Angle vertex: *(Pick a point.)*
First angle endpoint: *(Pick a point.)*
Second angle endpoint: *(Pick a point.)*
Dimension arc line location (Mtext/Text/Angle): *(Pick desired location.)*
Dimension text <69>:
Command:

The two options for placing the angle can be seen in Figure 19–33.

Baseline

Occasionally when the spacing of objects is extremely critical, the baseline system of placing dimensions can be used. Baseline dimensions (also called datun dime-

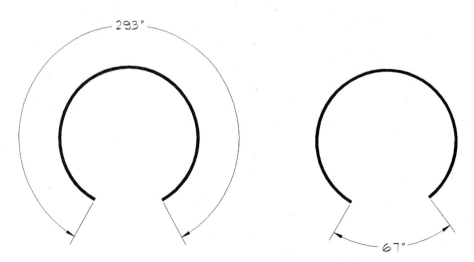

FIGURE 19–32   Placement of dimensions using the circle option.

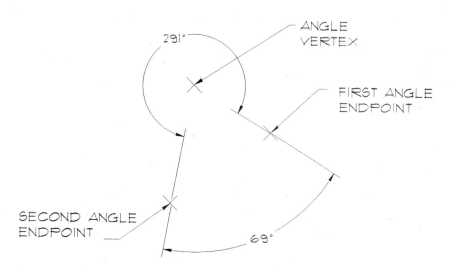

FIGURE 19–33   Placement of dimensions using the three point option.

sions) assume one edge to be perfect and reference all dimensions back to that surface. Figure 19–34 shows an example of baseline dimensions. Although this type of dimensioning is not used in residential design, it is common in some areas of commercial construction where a high degree of accuracy must be achieved. You start the command from the Dimensioning toolbar or from the Dimensioning pull-down menu. The command can also be started by typing **DBA** [enter] , **DIMBASE** [enter] or **DIMBASELINE** [enter] at the command prompt. The DIMBASE must be started by using a linear, angular or ordinate dimension. AutoCAD will use the first extension line from the existing dimension as the base point for other dimensions. The command sequence is:

> Command: **DLI** [enter]
> DIMLINEAR
> First extension line origin or press ENTER to select: *(Select point.)*
> Second extension line origin: *(Select second point.)*
> Dimension line location (Mtext/Text/Angle/Horizontal/Vertical/Rotated): *(Select dimension line location.)*
> Dimension text <30'-0">:
> Command: **DBA** [enter]
> DIMBASELINE
> Specify a second extension line origin or(undo/<Select>): *(Select second point.)*
> Dimension text <50'-0">
> Specify a second extension line origin or (undo/<Select>): *(Select second point.)*
> Dimension text <60'-0">
> Specify a second extension line origin or (undo/<Select>): *(Select second point.)*
> Dimension text <70'-0">

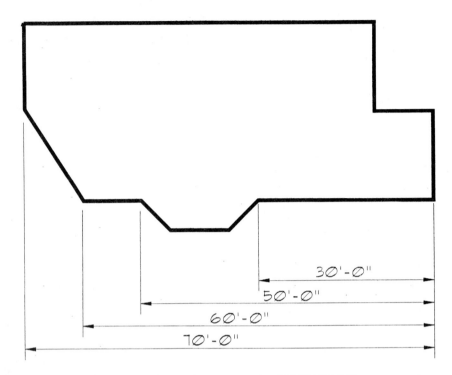

FIGURE 19–34 Placement of dimensions using the DIMBASELINE command.

Specify a second extension line origin or (undo/<Select>): [enter]
Select base dimension: [enter]
Command:

Continuous Dimensions

A series of dimensions is usually needed to dimension a structure. In Figure 19–1, dimension lines extend from the exterior walls, to interior walls, and finally to the exterior wall on the opposite side. These dimensions can be placed with the Continue command. The command sequence starts by using the DIMLINEAR command to place the first dimension. To use DIMCONTINUE, it must be based on a linear, angular or ordinate dimension. You can start the CONTINUE command by picking the Continue Dimension icon from the Dimensioning toolbar or by selecting Continue from Dimensioning of the Draw menu. You can also start the command from the keyboard by typing **DCO** [enter], **DIMCONT** [enter] or **DIMCONTINUE** [enter] at the command prompt. This will produce the prompt:

Command: **DCO** [enter]
DIMCONTINUE
Specify a second extension line origin or (undo/<Select>):

After you select the starting point for the second extension line, AutoCAD uses the second extension line of the DIMLINEAR command sequence as the first extension line for the DIMCONTINUE sequence. Press the ESC key to end the DIMCONTINUE command sequence. Figure 19–35 shows an example of placing dimensions using the Continue command with the default settings. The command sequence to place the dimensions across the bottom is:

Command: **DLI** ⏎
DIMLINEAR
First extension line origin or press ENTER to select: *(Select point.)*
Second extension line origin: *(Select second point.)*
Dimension line location (MText/Text/Angle/Horizontal/Vertical/Rotated):
    (Select dimension line location.)
Dimension text <10'-0">:
Command: **DCO** ⏎

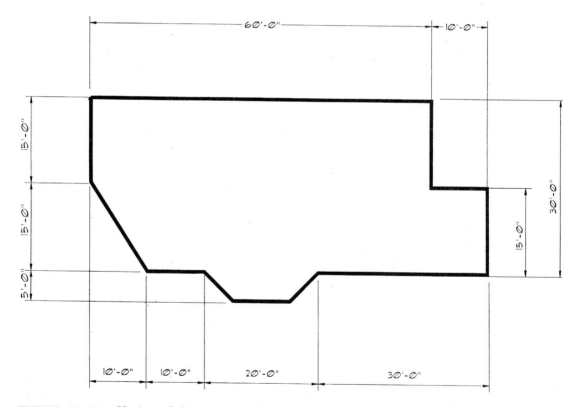

FIGURE 19–35. Placing of dimensions using the DIMLINEAR and DIMCONTINUE commands. The 10'-0" dimension on the bottom left was placed using the DIMLINEAR command. The next 10'-0" dimension was placed using the DIMCONTINUE command. The DIMCONTINUE command uses the second extension line of the preceeding sequence as the first extension line of the DIMCONTINUE sequence.

DIMCONTINUE
Specify a second extension line origin or (undo/<Select>): *(Select second point.)*
Dimension text <10'-0">
Specify a second extension line origin or (undo/<Select>): *(Select second point.)*
Dimension text <20'-0">
Specify a second extension line origin or (undo/<Select>): *(Select second point.)*
Dimension text <30'-0">
Specify a second extension line origin or (undo/<Select>): [enter]
Select continued dimension: [enter]
Command:

LEADER LINES

The LEADER command can be used to place a leader line to connect notes or dimensions to a specific portion of the drawing. Figure 19–36 shows examples of the use of leader lines. Leaders are generally placed at any angle except horizontal or vertical.

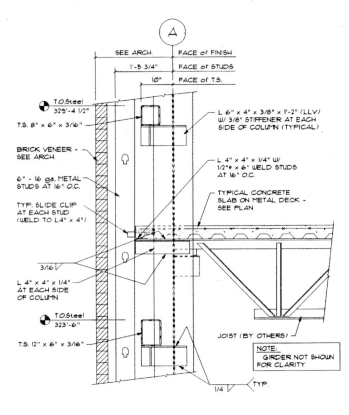

FIGURE 19–36 Leader lines can be used to attach text, dimensions and welding symbols to a specific portion of the drawing. (Courtesy of Van Domelen / Looijenga / McGarrigle / Knauf Consulting Engineers)

Some architectural offices use horizontal and vertical leaders with a custom arrow to duplicate manual practice. A horizontal leg is placed at the end of the leader line to tie the text to the leader. The horizontal leg should extend from the left side of the first line of text or from the right side of the last line of text. Figure 19–37 shows common methods of placing leader lines.

The command can be accessed by selecting LEADER from the Dimensioning toolbar, from the pull-down menus or by typing **LE** [enter], **LEAD** [enter], or **LEADER** [enter] at the command prompt. This will produce the prompt:

Command: **LE** [enter]
From point: *(Select desired point to start leader.)*
To point: (Format/Annotation/Undo) <Annotation>: *(Select desired point to end leader.)*
To point (Format/Annotation/Undo)<Annotation>: *(Press the F8 key to produce a horizontal line and then select the desired ending point.)*

The From point: and To point: prompts are exactly like those of the LINE command. AutoCAD will place a line with the tip of an arrow extending from the From point: to the To point: A prompt for a second To point: is also provided. As shown in Figure 19–38, this second To point: produces a horizontal line. Pressing the F8 key activates ORTHO and produces a horizontal line. Once the final To point: is selected, a To point: prompt is displayed again. This allows for an irregular shaped leader line to be created similar to the leader in Figure 19–36. When the leader is formed, one of the three options can now be selected.

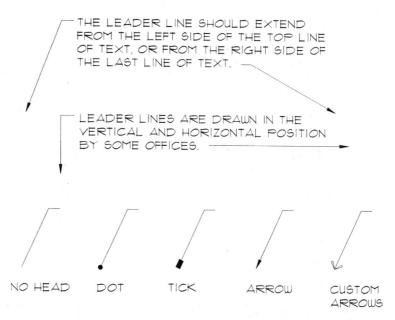

FIGURE 19–37 Common methods of placing and terminating leader lines.

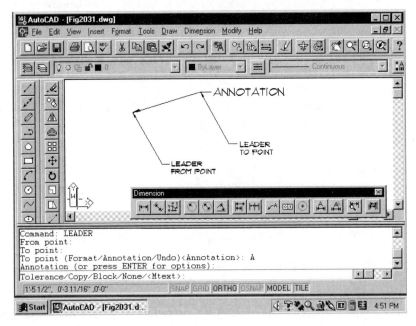

FIGURE 19–38 A leader line consists of one or more angled lines, a horizontal tail and annotation.

Annotation

The Annotation option allows text, numbers or symbols to be placed at the end of the leader line. The prompt is:

> To point (Format/Annotation/Undo)<Annotation>: [enter]
> Annotation (or press ENTER for options): *(Enter desired text, numbers or symbols.)* [enter]
> Mtext: [enter]
> Command:

Pressing the [enter] key produces the Mtext: prompt. This option allows multiple lines of text to be entered.

If [enter] is entered at the prompt in place of annotation, the following prompt will be displayed:

> To point(Format/Annotation/Undo)<Annotation>: [enter]
> Annotation (or press ENTER for options): [enter]
> Tolerance/Copy/Block/None/<Mtext>:

The **Tolerance** option produces the Tolerance Symbol dialog box shown in Figure 19–39. These are tolerance symbols that are used to control the accuracy of a dimen-

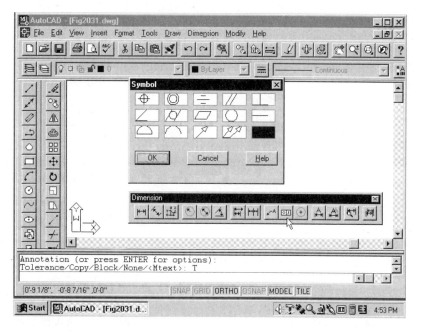

FIGURE 19–39 Selecting the Tolerance option will display symbols which can be used to define the accuracy of a dimension to a specified surface.

sion in relation to a specific surface or condition. These symbols are not usually used with construction drawings.

The **Copy** option allows existing text to be copied to the new leader line. When you enter **C** [enter] at the command prompt, a prompt to Select an object is displayed. When a group of text is selected, it will automatically be copied to the new location.

The **Block** option can be used to insert predefined groups of text by the leader line. Creation of BLOCKS will be discussed in Chapter 21. The option is selected by entering **B** [enter] at the command prompt.

The **None** option exits the command and places the leader line with no annotation. The option is selected by entering **N** [enter] at the prompt.

The **Mtext** default allows multiple line of text to be entered using either the Multi-line Text Editor. The prompt is selected by entering **M** [enter] at the command prompt. Once **M** [enter] is entered at the command prompt, the Multiline Text Editor will be displayed allowing large bodies of text to be easily placed.

Format

The Format option is used to determine the appearance of the leader and the arrow. Selecting the Format option produces the following prompts:

To point(Format/Annotation/Undo)<Annotation>: **F** [enter]
Spline/Straight/Arrow/None/<Exit>:

The **Spline** option changes the lines to curved, which are drawn from the From point: and To point:. It functions similar to the SPLINE command. Select the option by entering **S** [enter] at the prompt.

The **STraight** option allows straight line segments to be drawn. Select the option by entering **ST** [enter] at the prompt.

The **Arrow** option allows an arrow to be drawn at the From point. This is the default. Select the prompt by entering **A** [enter] at the prompt.

The **None** option allows a leader line to be drawn without using an arrow as a terminator. Select the prompt by entering **N** [enter] at the prompt.

The **Exit** option exits the Format menu and returns to the base prompt. Select the prompt by entering **E** [enter] at the prompt.

Undo

Choosing Undo removes the last completed option from the command sequence. Undo functions like the Undo found in other AutoCAD commands.

CHAPTER 19 EXERCISES

E-19-1. Enter a new drawing and set the units to architectural with 1/16 accuracy. Draw a hexagon with a horizontal top and base inscribed in a 4″ diameter circle. Use the DIMLIN command to provide overall dimensions. Save the drawing as E-19-1.

E-19-2. Open drawing E-19-1. Change the hexagon to a layer titled BASE. Change the dimensions to a layer titled DIMEN and turn the layer off. Create a layer titled ANGLE and make it current. Provide dimensions to indicate the angle of the sides. Save the drawing as E-19-2.

E-19-3. Enter drawing E-8-9 and dimension objects a, c, e, and g using two-place decimal dimensions. Place the remaining objects on a layer titled EXTRA, and freeze the layer. Once the dimensions are complete, freeze the existing dimensions and change the UNITS to architectural with $\frac{1}{16}$″ accuracy. Dimension the objects again using alternatives to the methods already used. Save the drawing as E-19-3.

E-19-4. Open drawing E-11-3 and scale the drawing using a scale factor of .5. Create a layer for the dimensions. Completely dimension the drawing so the steel beam can be fabricated. Save the drawing as E-19-4.

E-19-5. Open drawing E-11-5 and create a layer titled DIM1. Rescale the drawing by a factor of .25 and completely dimension the drawing so that the metal hanger can be fabricated. Save the drawing as E-19-5.

E-19-6. Open drawing E-17-4 and create a layer titled DIM1. Change the lines representing the one-story footing to .5 wide polylines. Place the two-story footing on a new layer and set that layer to OFF. Rescale the drawing using a scale factor of .25. Completely dimension the one-story footing and save the drawing as E-19-6.

E-19-7. Open a new drawing and draw the following object. Starting @ 2,2:

B. @1.5,0	K. @ −2,0
C. @ 0,1	L. @ 0,−.75
D. @1.25,0	M. @ −2,0
E. @ .75,.75	N. @ 0,−1.25
F. @ 1.5,0	P. @ −2.5,0
G. @ 0,.75	Q. @ .5,−1.5
H. @ 1.5,0	R. AND THEN CLOSE.
J. @ 0,1.75	

Change the line segments to polylines. Completely dimension the drawing using the Continue option where possible. Save the drawing as E-19-7.

E-19-8. Open drawing E-19-7 and create a new layer for the existing dimensions. Change the dimensions to the new layer that was just created, and set it to OFF. Create another layer for dimensions and completely dimension the object using Baseline dimensioning. Assume the lower left corner as the starting point for Horizontal and Vertical dimensions.

CHAPTER 19 QUIZ

1. What four components comprise dimensioning?

2. How is text meant to be read on architectural drawings?

3. List four options typically used for the intersections of dimension and extension lines.

4. Describe how to place dimensions relative to a concrete wall.

5. List four groups of commands that relate to placing dimensions on a drawing.

_____

_____

_____

_____

6. Describe two methods of locating extension lines for Linear dimensions.

_____

_____

7. List four methods to describe Angular dimensions.

_____

_____

_____

_____

8. Enter one of the drawing exercises that was completed for this chapter, and provide the name of the layer that was added to the drawing base by AutoCAD.

_____

_____

9. List four methods of placing linear dimensions.

_____

_____

_____

_____

10. Explain the difference between aligned and Rotated dimensions.

_____

_____

11. What is the first prompt for DIMLINEAR asking for?

_____

_____

12. Should the default text value always be accepted? Explain your answer.

_____

_____

13. What does the angle option alter in the DIMLINEAR command?

14. What is the effect of entering [enter] at the first DIMLIN prompt rather than selecting a point?

15. Should a dimension be placed inside or outside of a circular object?

16. What is the process for placing a 6" center mark in the center of a circular steel tube?

17. List and explain the options for LEADER annotation.

18. Four lines of text need to be placed by a leader line. What option should be used?

19. Explain the options for ORDINATE dimensioning.

20. List the full name of each of the following commands and give a short description of the effect of each: DAL, DIMANG, DBA, DCE, DCO, DDIA, DLI, DOR, DRA, DVAR.

CHAPTER

20

PLACING DIMENSIONS
ON DRAWINGS

In Chapter 19 you were introduced to dimensioning requirements of different drawings and the methods of describing linear, angular, circular, continuous, and baseline dimensions. Each system was described using the default values. If all of your drawings could fit on a computer screen, everything would be fine. To allow for drawings of various sizes, the spacing of extension and dimension lines and the size of text and arrows must be altered. In older releases of AutoCAD, this wide variety of choices was often very confusing for new users to master. In Release 14, the dimensioning variables can be controlled by dialog boxes. In this chapter you'll learn how to alter dimension variables, establish dimension styles, and edit existing dimensions.

CONTROLLING DIMENSIONS VARIABLES WITH STYLES

As text was explained, you were introduced to the STYLE command. STYLE allows you to group variables such as the font, text height, width, obliquing angle, and orientation so that several different styles of text can easily be created within one drawing. AutoCAD also allows dimension styles to be created within a drawing to meet the needs of various situations. Dimensioning styles are created using the DDIM command. To save time, individual styles as well as each dimensioning variable should be set on the templet drawing on the ANNO-DIMS layer.

Introduction to Dimensioning Styles

A *dimension style* is a set of dimension variables that control various aspects of placing the extension lines, dimension lines, and annotation in relation to these lines.

The type of drawing you are working on will affect how the variables are to be set. Dimensions that are placed on a site plan are often written in engineering units. Dimensions that locate the property lines are placed with no extension or dimension lines. Dimensions placed to locate utilities in an easement are often dimensioned using baseline dimensions expressed in feet and inches. Dimensions written on architectural drawings are written in feet and inches using continuous dimensions. Dimensions on elevations and sections can be placed using ordinate dimensioning. You can create styles that incorporate the requirements for each of these dimensions and save them in a templet drawing. You will be introduced to creating various styles later in this chapter.

Before a style can be created you must have a thorough understanding of dimension variables. *Dimension variables* are the qualities that control how dimensions will appear. You will be introduced to dimensioning variables through the default style of STANDARD. Later you'll be introduced to methods of creating specific styles to meet the needs of each drawing type.

The Dimension Style Family

AutoCAD refers to groupings of dimensioning variables as a *family*. A family of dimension styles consists of a set of dimension styles related to one another by name. A dimension style can have families of different settings so that you can use different settings for different drawing types or different types of dimensions. For example, most construction drawings place the text above the dimension line, but occasionally dimensions are rotated to draw attention to them. This feature could be part of the variables in a family set. Families of styles allow a specific dimension variable, such as the use of arrows or ticks, to be altered depending on whether the dimension is to be placed on a site plan or a floor plan. Although the setting varies from one drawing to another, they are still members of the same family.

All members of a family of styles share the same parent name. For example, a style called ALL can have one member of the ALL family for linear dimensions, and another for radial dimensions. Once these members are established, AutoCAD automatically uses these families whenever a linear or radial dimension is created. Creating styles and families of styles will be discussed in depth later in this chapter.

CONTROLLING VARIABLES USING THE DIMENSION STYLES DIALOG BOX

Dimension variables can be altered by keyboard or by using the Dimension Styles dialog box. Changing one variable can often be done by entering the required command at the keyboard. Changing a whole group of variables, or creating a style can be done quickly using the dialog box. The Dimension Styles dialog box displayed in Figure 20–1 is accessed by selecting Dimension Style from the Dimension toolbar, selecting Dimension Style... from the Format pull-down menu or by typing **D** [enter] or **DDIM** [enter] at the command prompt. The main areas of this box include the Dimen-

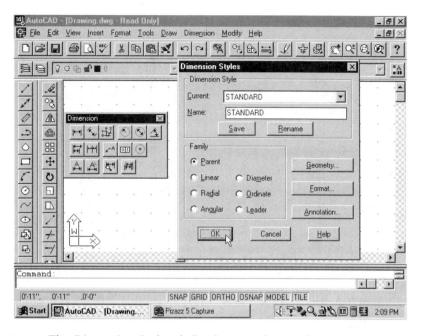

FIGURE 20–1 The Dimension Styles dialog box can be used to create and save dimension attributes.

sion Style box, the Family Box, and the Geometry..., Format..., and Annotation... buttons.

Dimension Styles

Major areas of this box include the Current, Name, Save, and Rename boxes.

Selecting the Current Style. The Current: edit box lists the name of the current style and allows another style to be made current. The current style dictates how dimensions are placed on the drawing. The only style currently available is STANDARD. As changes are made, this style will be altered to become the +STANDARD style. As styles are added to the drawing base, the arrow on the right side of the box can be used to scroll through the listed names.

Naming New Styles. The Name: edit box allows the name of a new style to be entered, or an existing style to be renamed. Once the variables are set to the desired settings, the name of this style can be entered in this box.

Saving Styles. The Save button allows a group of variables to be saved in their present setting as a style family using the name displayed in the Name: edit box.

Changing the Name of Existing Styles. The Rename button assigns the highlighted style family to the name displayed in the Name edit box.

The Family Box

The Family box contains listings for each of the types of dimensions presented in Chapter 19. Selecting one of the radio buttons assigns the current set of values to that family member. Parent is the default setting and contains all of the current variable settings. The other family members contain only the specific dimension variables that are associated with their applications.

The Geometry... Button

Selecting this button displays the Geometry sub-dialog box shown in Figure 20–2. This box controls the placement of dimension and extension lines, the scale, arrowheads and center mark using the sub-boxes with the corresponding name. Each of the options to be discussed can be easily set using the dialog box or can be adjusted from the command prompt. Most users will find the dialog box the quickest, and best method for controlling variables. The command line entry methods are best suited to users of older releases of AutoCAD that predate the dialog boxes.

Arrowheads Box. This box controls the type of dimension line terminator to be used. The current selection for the first and second terminator, the closed filled arrow, is displayed above the menu. By default, AutoCAD uses the terminator selected for the first dimension line used for the second line. The type of terminator can be selected by picking the icon or the arrow on the right side of the 1st: list box.

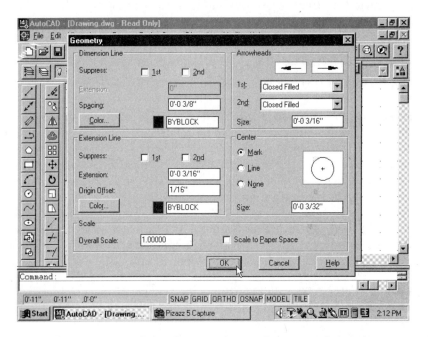

FIGURE 20–2 The Geometry dialog box can be used to control the placement of dimension lines, extension lines, arrows and center marks.

Picking the icon displays the next option in the menu. Selecting the down arrow displays the menu. Making the selection of Architectural Tick, automatically alters the display of each arrow to tick marks. The second terminator can be altered from the first by using the menu for 2nd: menu option. Rarely is this done on professional drawings, and you should not decide to start a new trend! Figure 20–3 shows an example of each terminator.

User Arrow....   Most architectural and engineering offices use the Oblique, Architectural Tick, or User Arrow for a terminator. The user-defined tick can be created

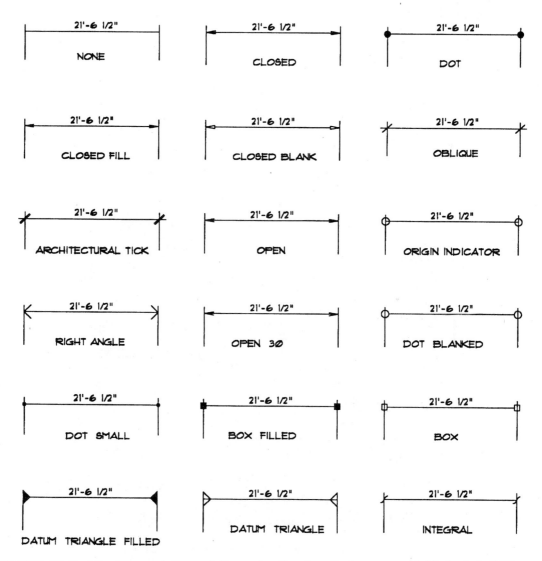

FIGURE 20–3 Nineteen variations of dimension line terminators are available in AutoCAD Release 14.

using the PLINE and BLOCK commands. BLOCK will be discussed in the Chapter 21. Once the tick is drawn, it is named and saved to the drawing base. Selecting User Arrow... displays the dialog box shown in Figure 20–4. The name that was assigned to the block can be assigned in the Arrow Name: list box. The icon tile will now display *USER...* rather than showing an icon.

Controlling the Terminator Height. The Size: box can be used to determine the size of the terminator relative to other dimensioning features. Selecting a tick value that is the same height as the text is a safe guideline to follow. If ⅜" high text is to be used, change the value to .375. The size of the terminator can also be altered by typing **DIMTSZ** ⏎ (DIMension Tick SiZe) at the command prompt. This will produce the prompt:

New value for DIMTSZ <0.1800>: **.125** ⏎
Command:

If arrows are to be used, the variable is DIMASZ (DIMension Arrow SiZe). If the DIMTSZ variable has been set, the DIMASZ value is ignored.

Dimension Line Box. This box contains controls for suppressing the first or second dimension line, setting the distance the dimension line extends past an exten-

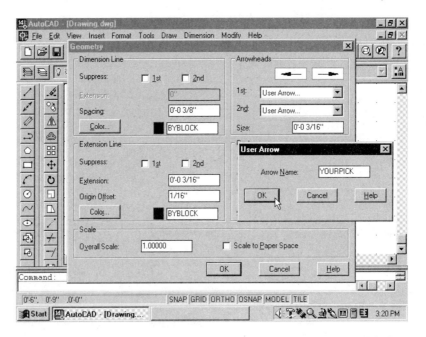

FIGURE 20–4 A user-defined arrow can be created as a BLOCK and added into the drawing. Chapter 21 will introduce BLOCKS.

sion line, setting the distance between rows of dimensions, and the color used to display dimension lines.

Suppressing Dimension Lines.   This option works as a toggle allowing parts of the dimension line to be omitted. The 1st and 2nd boxes allow the first or second portion of the dimension line to be omitted from the drawing. Figure 20–5 shows an example of the first or second dimension line suppressed. The first or second dimension line is relative to the selection of the first and second extension line. This option can also be controlled from the command prompt by typing **DIMSD1** [enter] (Dimension Suppress Dim.1). The default setting is OFF. Selecting the ON option suppresses the first dimension line. Typing **DIMSD2** [enter] allows control of the second dimension line. These options are generally not altered for construction drawings.

Providing Dimension Line Extensions.   By default, this option is only available if Oblique or Architectural Tick has been selected for the terminator. This option controls how far the dimension line will extend past the extension line. Figure 20–6 shows examples of adjusting the extension. Most architectural offices set the extension value to equal the height of the text to be used. This variable can be controlled at the command prompt by entering **DIMDLE** [enter] (DIMension Line Extension). To extend the dimension line, enter a positive value at the prompt:

New value for DIMDLE <0.0>: **.125** [enter]
Command:

Controlling Dimension Line Spacing.   Selecting the Spacing: box allows the distance between lines of dimensions to be set. AutoCAD refers to this as the baseline setting. The default setting is .38. Most professionals use a distance of .5. Figure 20–7 shows an example of the baseline spacing. The setting can also be controlled at the command prompt by typing **DIMDLI** [enter] (DIMension Line Increment). This will produce the prompt:

New value for DIMDLI <0.38>: **.5** [enter]
Command:

FIGURE 20–5 Suppression of dimension lines is generally not required in architectural drafting.

FIGURE 20–6 The default placement of dimension line extension on the left is often altered so that the dimension line extends past the extension line to simulate manual drafting methods.

Setting Dimension Line Color. The final variable to set in the Dimension Line box controls the color of the dimension lines, arrowheads, and dimension leader lines. Selecting the Color... button displays the color palette. Although a different color can be set for each aspect of a dimension, colors are best controlled by the layer that they are to be placed on. The color of dimension lines can be controlled from the command prompt by typing **DIMCLRD** enter. This will produce the prompt:

    New value for DIMCLRD <0>: (Enter a value from 0 to 256.)
    Command:

Basic colors include: 1 = Red, 2 = Yellow, 3 = Green, 4 = Cyan, 5 = Blue, 6 = Magenta, 7 = White.

Extension Line Box. This box has features similar to those in the Dimension Line box except these features control each aspect of the extension line in relation to the object being dimensioned.

Suppressing Extension Lines. This option allows for the suppression of the first and second extension lines as shown in Figure 20–8. By default, extension lines will be displayed at each end of the dimension line. This option works well for establishing overall sizes.

The First extension line can be omitted by picking the 1st box. With this option active, as the location for extension lines is selected, the first location will not receive

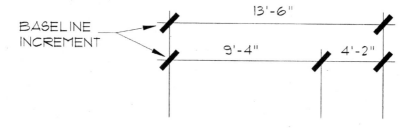

FIGURE 20–7 The baseline increment controls the spacing between stacked lines of dimensions.

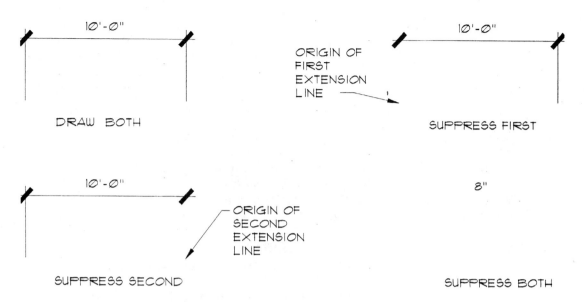

FIGURE 20–8 Suppression of extension lines.

an extension line. This option can also be controlled from the command prompt by typing **DIMSE1** [enter] (DIMension Suppress Extension 1). This will produce the prompt:

New value for DIMSE1 <OFF>:
Command:

Setting the value to ON, suppresses the first extension line. Picking the 2nd box suppresses the second extension line. This option can also be altered at the command prompt by typing **DIMSE2** [enter]. Figure 20–9 shows examples of suppressing extension lines.

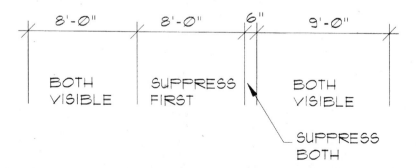

FIGURE 20–9 Practical applications for supressing extension lines.

Picking both boxes suppresses the extension lines at each end of the dimension line. This could also be done by setting DIMSE1 and DIMSE2 to ON at the command prompt. This is useful if a dimension line is to be placed between two existing extension lines. This option also works well when the object is being used as the extension lines. This is typically done when dimensioning between floor and ceiling levels on sections or elevations.

Extension Line Extension. The Extension: option allows the distance that the extension line extends past the dimension line to be controlled. The default value is $\frac{3}{16}$". An extension offset of $\frac{1}{8}$" is typical on most professional drawings. In addition to setting the option in the Extension Line box, the variable can be set at the command prompt by typing **DIMEXE** [enter] (DIMension EXtension Extension). This will produce the prompt:

New value for DIMEXE <0'-0 $\frac{3}{16}$">: **.125** [enter]
Command:

The effect of setting Extension: can be seen in Figure 20–10.

Controlling the Extension Origin Offset. The Origin Offset: option is used to control the gap between the extension lines and the object. The default value is $\frac{1}{16}$" but a space of $\frac{1}{8}$" is common in most offices. This variable can be controlled at the command prompt by typing **DIMEXO** [enter] (Dimension EXtension OFFSET). This will produce the prompt:

New value for DIMEXO <0'-0 $\frac{1}{16}$">: **.125** [enter]
Command:

The effect of setting the Extension line Offset: is shown in Figure 20–11.

Setting Extension Line Color. The final variable to set in the Extension Line box controls the color of the extension lines. The Selecting the Color... button displays the color palette. Although a different color can be set for extension lines, colors are best controlled by the layer that they are to be placed on. The color of

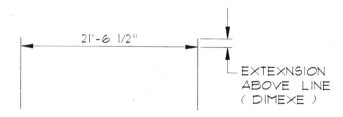

FIGURE 20–10 The extension line is usually set to extend above the dimension line by a distance equal to the text height.

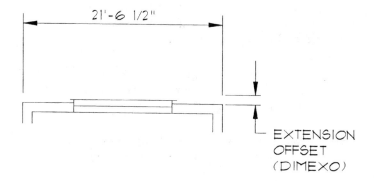

FIGURE 20-11 The extension line should be offset from the object being described.

extension lines can be controlled from the command prompt by typing **DIMCLRE** [enter]. This will produce the prompt:

New value for DIMCLRD <0>: *(Enter a value from 0 to 256.)*
Command:

Center Box. The Center box is used to control how and when the center mark for radial dimensions will be displayed. The center mark is only drawn if the dimensions for DIMRAD or DIMDIA are placed outside the circle or arc being described. Figure 20–12 shows the options for placement. The center mark is controlled by three radio boxes, an edit box, or by typing **DIMCEN** [enter] at the command prompt. Chapter 19 contains information about the center mark.

Displaying the Center Mark. The default setting of Mark displays the center mark using the center Size: value. The value can be controlled at the command prompt by entering a positive value at the DIMCEN prompt.

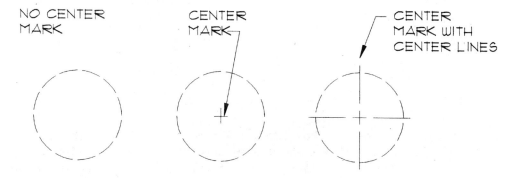

FIGURE 20-12 Practical applications for suppressing extension lines.

Displaying Center Lines. Selecting the Line radio button displays center lines that project beyond the limits of the circle. Centerline placement can be controlled from the command prompt by typing **DIMCEN** [enter] and entering a negative value.

Displaying No Center Marking. Selecting the None radio button will display no markings for the centerpoint of a circle or arc. Placement can be controlled from the command prompt by typing **DIMCEN** [enter] and entering a 0 value.

Controlling the Size of the Center Mark. The Size: edit box can be used to control the size of the center mark. The default setting of ⅟₁₆" is adjusted to ⅛" by many professionals.

Determining the Drawing Scale. The Scale box allows the scale of all dimensioning features to be increased or decreased based on the scale a drawing is to be plotted at. Typically you will be drawing in model space with actual sizes. If you like to think and do lots of extra work, you could go through all of the dimension variables and set them to real sizes. Text that is to be ⅛" high when plotted at a scale of ¼" = 1'-0", will need to be set to 6" in height. You'll have to convert every variable. If you actually have a life, you may want to just set the scale factor to 48. This will control all dimension variables, and leave you a little free time. To find the desired scale factor, take the reciprocal of the drawing scale. A scale of ¼" = 1'-0" is 12/.25 = a scale factor of 48. Other common scale factors include:

ARCHITECTURAL VALUES		ENGINEERING VALUES	
Plotting Scale	Scale Factor	Plotting Scale	Scale Factor
¾" = 1'-0"	16	1" = 1'-0"	12
½" = 1'-0"	24	1" = 10'-0"	120
⅜" = 1'-0"	32	1" = 100'-0"	1200
¼" = 1'-0"	48	1" = 20'-0"	240
³⁄₁₆" = 1'-0"	64	1" = 200'-0"	2400
⅛" = 1'-0"	96	1" = 30'-0"	360
³⁄₃₂" = 1'-0"	128	1" = 40'-0"	480
⅟₁₆" = 1'-0"	192	1" = 50'-0"	600
		1" = 60'-0"	720

The value is stored in the DIMSCALE variable. It is important to remember that only dimensions that are added after the scale is set will be affected by the DIMSCALE variable. The DIMSCALE must be set before the dimensioning process is started.

Overall Scale. The Overall Scale: edit box allows the scale factor to be entered. This will affect the dimensions. The value adjusts sizes and does not affect distances, coordinates, angles, or tolerance. A positive value will increase the size and a negative value will decrease the dimension sizes.

Scale to Paper Space. Selecting the Scale to Paper Space box will apply a scaling factor to dimensions created in paper space (See Chapters 10 and 22). The length scale factor will be adjusted to reflect the zoom scale factor for objects created in model space viewports. If Paper Space is active, the Overall Scale is disabled and the DIMSCALE is set to 0. The value for Paper space length scaling is stored in as the DIMLFAC variable.

OK / Cancel / Help.... It may seem redundant, but if you're going to take the time to adjust the geometry variables, take time to save them. Selecting the OK button will add the variables to the drawing base, and return you to the Dimension Styles box. If you exit the Geometry box using the cancel box, you'll be returned to the Dimension Style box, but none of your changes will be stored.

The Format... Button

Selecting this button displays the Format sub-dialog box shown in Figure 20–13. This box is used to set the variables that control the placement of text, dimension and extension lines, arrowheads and leader lines.

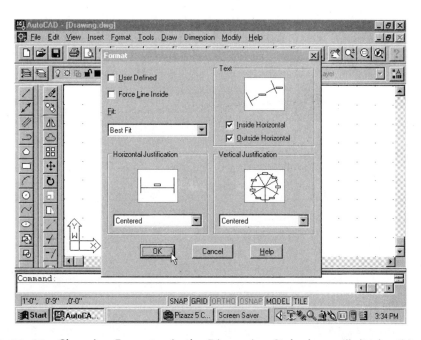

FIGURE 20–13 Choosing Format... in the Dimension Styles box will display this dialog box for the control of the placement of dimensioning features.

User-Defined Text Placement.   Activating the User Defined check box allows the location for dimension text to be specified as you place each dimension regardless of the justification. With this box active, AutoCAD ignores other justification settings and allows you to specify the position where the text will be placed. Adjusting the variable can be useful on radial dimensions as shown in Figure 20–14. The variable can be controlled by keyboard by typing **DIMUPT** [enter] (DIMension User Positioned Text) and entering either **ON** or **OFF**.

Forcing a Line Between Extension Lines.   When a small space is dimensioned, the text will be placed outside of the extension lines. Selecting the Force Line Inside box places a line between the dimension lines when the text is outside. Figure 20–15 shows examples of this option on a linear and radial dimension. The value can also be controlled by typing **DIMTOFL** [enter] (DIMension Text Outside, Force Line Inside) at the command prompt. This produces a prompt requesting an ON/OFF setting with a default setting of OFF.

Adjusting the Fit.   The Fit: menu is used to set the DIMFIT (DIMension Fit Text) variable which controls the placement of text and arrows relative to the extension lines based on the space available. The current setting is displayed in the edit box with a scroll arrow to the right of the edit box. Select the arrow to display the menu shown in Figure 20–16. As a selection is made, the menu is removed, and the selected value is displayed in the Fit: edit box as the current value. Results of each option can be seen in Figure 20–17.

Text and Arrows.   This option displays the text and arrows outside of the extension lines if there is not enough room to place them inside. This can be controlled by typing **DIMFIT** [enter] at the command prompt and entering **0** as the value.

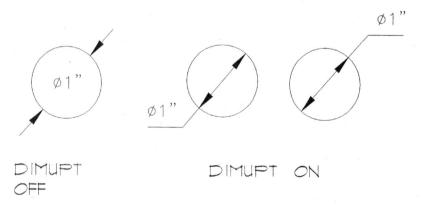

FIGURE 20-14 The DIMUPT variable can be adjusted to control the placement of the text. The default is OFF.

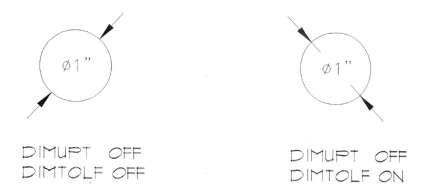

FIGURE 20–15 With DIMUPT off, and DIMTOLF on, a dimension line will be forced inside the circle.

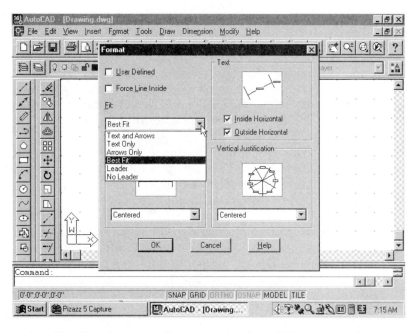

FIGURE 20–16 The Fit sub-menu allows for selection of placement of text, arrows and dimension lines inside of a small space.

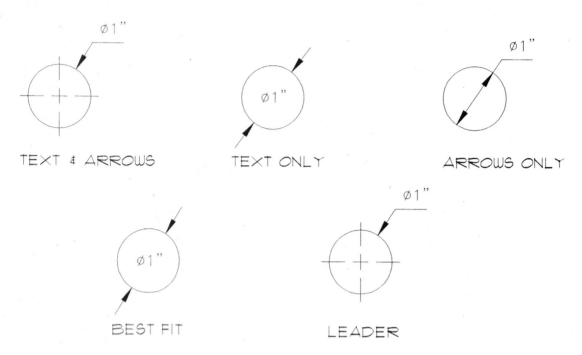

FIGURE 20–17 The Fit sub-menu options. Notice that depending on the size of the space to be dimensioned, some options produce the same results.

Text Only. This option displays the text inside and the arrows outside of the extension lines. If sufficient space is not available between the extension lines, the text will be placed outside the extension lines. This variable is controlled by typing **DIM-FIT** [enter] at the command prompt and entering **1** as the value.

Arrows Only. This option displays the arrows inside and the text outside of the extension lines. If sufficient space is not available between the extension lines, the arrows will be placed outside the extension lines. This variable is controlled by typing **DIMFIT** [enter] at the command prompt and entering **2** as the value.

Best Fit. Selecting the default option will display the arrows and the text inside of the extension lines. If sufficient space is not available between the extension lines, the text is placed between the extension lines and the arrows are placed outside the extension lines. If no space is available for text, only the arrows are placed between the extension lines. This variable is controlled by typing **DIMFIT** [enter] at the command prompt and entering **3** as the value.

Leader. If enough room is available, dimension text is placed between the extension lines. If sufficient space is not available to place text between the extension lines, it is attached to a leader line that attaches to the dimension line. This variable is controlled by typing **DIMFIT** [enter] at the command prompt and entering **4** as the value.

No Leader. When this variable is selected, if enough space is available, text is placed inside, and arrows are placed outside the extension lines. When insufficient space is available for text, the arrowheads are placed inside and the text is placed outside the extension lines. Unlike the leader option, no leader line is provided. If arrows and text must be placed outside, you are allowed to place the text anywhere, independent of the dimension line. This option works well for many architectural features.

Adjusting the Horizontal Justification. The Horizontal Justification box allows the location of text along the dimension line to be set. The box contains an icon and an edit box which can be used to control where the text is placed. Variables can be altered by picking the icon or the down arrow. As the icon is picked, the icon for the next menu listing is displayed and made current. If the arrow is picked, the menu shown in Figure 20–18 is displayed. Picking an option makes it current and removes the menu. Each of the variables for this option can be entered at the command prompt by typing **DIMJUST** [enter] (DIMension JUSTification). Examples of each variable are shown in Figure 20–19.

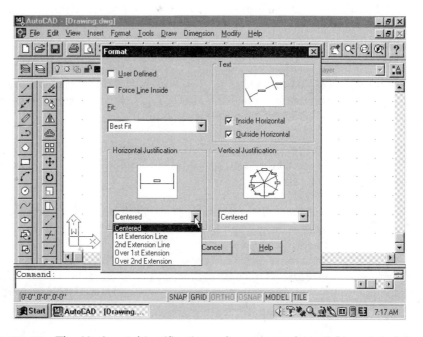

FIGURE 20–18 The Horizontal Justification submenu can be used to control the placement of text along the dimension line. The Centered option is used by most professionals.

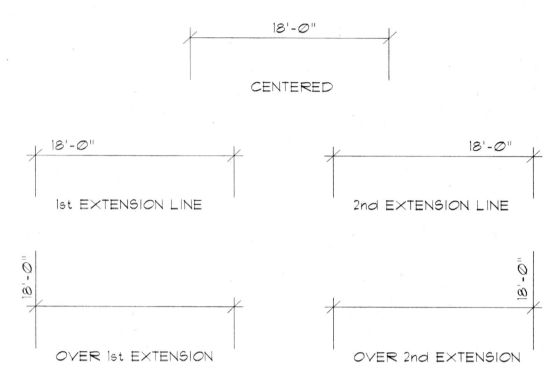

FIGURE 20–19 Placement of the dimension using the Horizontal Justification sub-menu.

Centered. The default value places text centered along the dimension line. This variable is controlled by typing **DIMJUST** [enter] at the command prompt and entering **0** as the value.

1st Extension Line. This option places the text close to the first extension line. This variable is controlled by typing **DIMJUST** [enter] at the command prompt and entering **1** as the value.

2nd Extension Line. This option places the text close to the second extension line. This variable is controlled by typing **DIMJUST** [enter] at the command prompt and entering **2** as the value.

Over 1st Extension. This option places the text parallel and above the first extension line. This variable is controlled by typing **DIMJUST** [enter] at the command prompt and entering **3** as the value.

Over 2nd Extension. This option places the text parallel and above the second extension line. This variable is controlled by typing **DIMJUST** [enter] at the command prompt and entering **4** as the value.

Controlling Text Alignment.   The Text box controls the orientation of text to the dimension and extension lines. The selection can be made by picking the image tile or picking one or both boxes. The default is shown in Figure 20–18. Both boxes are active, which will place inside and outside text horizontal.

Inside Horizontal.   With this box active, the inside text is placed horizontal, regardless of the angle of the dimension line. With the box inactive, inside text is drawn parallel to the dimension line. This variable is controlled by typing **DIMTIH** [enter] (DIMension Text Inside Horizontal) at the command prompt and entering **ON** or **OFF** as the value. The value should be set to OFF for most construction drawings so that the text remains parallel to the dimension line.

Outside Horizontal.   With this box active, text placed outside of the extension lines remains horizontal, regardless of the angle of the dimension line. With the box inactive, text outside the extension lines is drawn parallel to the dimension line. This variable is controlled by typing **DIMTOH** [enter] (DIMension Text Outside Horizontal) at the command prompt and entering **ON** or **OFF** as the value. The value should be set to off for most construction drawings so that the text remains parallel to the dimension line. Figure 20–20 shows examples of the effects of DIMTIH and DIMTOH.

Controlling Vertical Text Placement.   The Vertical Justification portion of the Format box controls the vertical justification of text to be controlled relative to the dimension line. Figure 20–21 shows the menu options. Values that control these

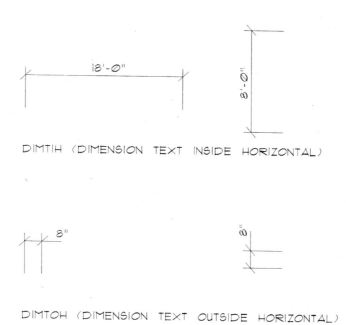

DIMTIH (DIMENSION TEXT INSIDE HORIZONTAL)

DIMTOH (DIMENSION TEXT OUTSIDE HORIZONTAL)

FIGURE 20–20   The results of the DIMTIH and DIMTOH settings.

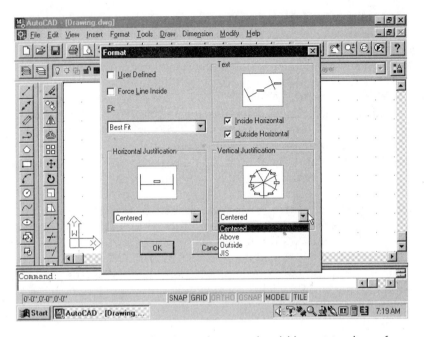

options are stored in the DIMTAD (DIMension Text Above Dimension Line) variable. Figure 20–22 shows each option as it relates to a drawing. Neither the outside or the JIS (Japanese Industrial Standards) are used on construction drawings.

Centered. The default option breaks the dimension lines so that text is centered between the dimension lines. The value can be controlled at the command prompt by entering **0** at the DIMTAD prompt.

Above. This option is often used for architectural and engineering drawings. The option places the dimension above the dimension line. The value can be controlled at the command prompt by entering **1** [enter] at the DIMTAD prompt. The distance from the dimension line to the bottom of the text is the current text gap. This distance is controlled by the Text setting in the Annotation box, which will be discussed later in this chapter.

FIGURE 20–22 The two common setting of the Vertical Justification submenu.

Outside. The Outside option places the dimension text on the side of the dimension farthest away from the defining point.

JIS. This option places dimension text to conform to Japanese Industrial Standards (JIS) representation.

OK/Cancel/Help... As with other dialog boxes, the OK or Cancel button must be used to exit the Format box. Selecting the OK button will add the variables to the drawing base, and return you to the Dimension Styles box. If you exit using the cancel box, you'll be returned to the Dimension Style box, but none of your changes will be stored.

The Annotation... Button

Selecting this button displays the Annotation sub-dialog box shown in Figure 20–23. This box is used to set the variables that control the appearance of text for dimension and leader lines. It includes boxes for Primary units, Alternate Units, Tolerance, and Text.

Primary Units. Most construction-related dimensions consist of primary units with no other information required. Occasionally a tolerance may be added to the primary units. This portion of the Annotation box controls the display of the primary measurement units and any prefixes or suffixes for the dimension text.

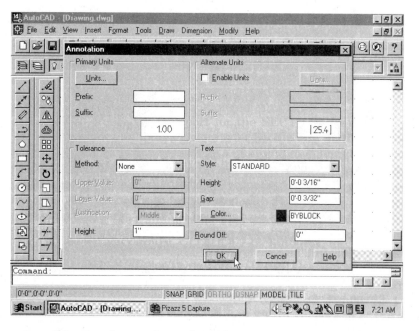

FIGURE 20–23 Selecting Annotation... from the Style dialog produces the Annotation dialog box that is used to control the appearance of text dimensions and leaders.

A prefix or suffix can be added by selecting the appropriate box with the cursor. Once the text is picked, it can be added to the edit box. Common suffixes might include .TYP (typical) , .MM (millimeter), or .MAX (maximum). This can also be done by typing **DIMPOST** [enter] at the command prompt. Because these are so rarely done, prefixes and suffixes will not be addressed in this text. Consult the Help menu for further information.

Controlling the Primary Units. Selecting the Units... button will display the Primary Units dialog box shown in Figure 20–24. This dialog box allows setting of the primary measurement units. The default setting is decimal. Pick the down arrow to display the menu shown in Figure 20–25. This sub-dialog box allows the setting of the format for linear, radial, ordinate, and leader dimension text. The stacked units refer to tolerances that are placed above each other and does not apply to most architectural drawings. The Architectural units should be selected for most drawings, although Engineering units are used on site-related drawings.

In addition to selecting from this menu, the units can be set by typing **DIMUNIT** [enter] at the command prompt. This will produce a prompt requesting a value from 1 to 5. Enter the value at the command prompt to set the units. Enter **1** for Scientific units (1.55E + 01); enter **2** for Decimal units (15.50); enter **3** for Engineering units (1'-3.50"); enter **4** for Architectural units (1'-3½"); and enter **5** for Fractional units (15 ½).

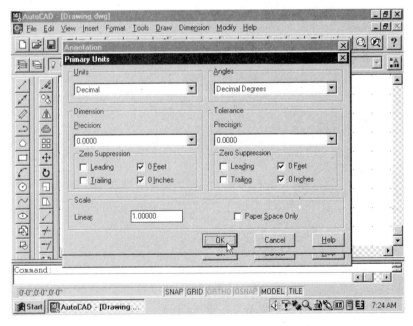

FIGURE 20–24 Selecting Units... in the Annotation dialog box will produce this box for controlling the primary text of the dimension text.

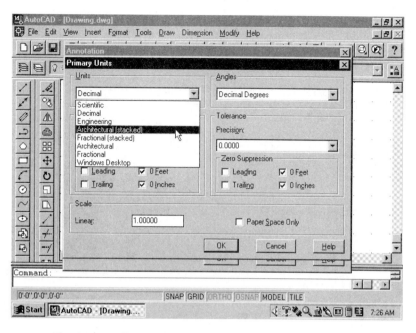

FIGURE 20–25 The Units submenu should be set to Architectural.

Controlling Dimension Precision. Selecting the down arrow of the Precision: box displays the precision of the dimension. Options are shown in Figure 20–26. The value can also be set at the command prompt by entering **DIMDEC** [enter] and a value between 0 and 8. For most dimensions, a setting of 0 (0'-0") will provide needed accuracy. Most engineers and architects determine the fractions based on calculations rather than relying on drawing accuracy. If a fraction needs to be added to the dimension, it can be added as the text is displayed, prior to adding the dimension to the drawing base.

Controlling the Display of Zeros. The Zero Suppression box controls the display of zeros. Four options are available. When an X is in the box, it is activated. These variables can also be changed at the command prompt by typing **DIMZIN** [enter] (DIMension Zero INch).

Leading. With this box active, the zero is suppressed preceding a decimal point in units less than one. Seven tenths will be written .7 rather than 0.7.

Trailing. This box controls the display of zeros after a decimal point. With the box active, no zeros will be displayed after the decimal. Seven tenths will be written .7 rather than .70. The number of zeros written after the decimal point will be written to the accuracy specified in DIMDEC.

0 Feet. A dimension less than one foot may be displayed, as 0'-8" or 8". Professional practice is mixed and the setting for the 0 Feet box will vary from office to office.

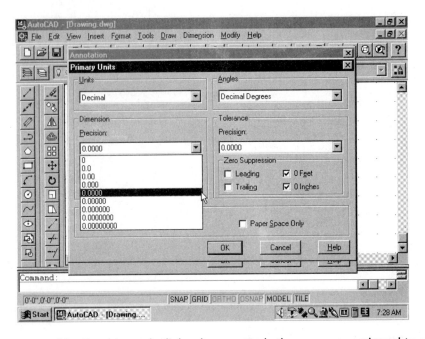

FIGURE 20-26 The Precision sub-dialog box controls the accuracy assigned to a dimension.

Selecting the feet box causes the zero representing feet to be suppressed when the dimension is less than twelve inches.

0 Inches. Most professionals display the zero after the foot symbol. Writing 5'-0" is preferred to writing 5'. With this box active, zeros will be suppressed. This box should be inactive.

Controlling the Display of Angles. The angles box controls the format of angular dimensions. Figure 20–27 shows the menu for controlling angles. The value is saved as the DIMAUNIT (DIMension Angle UNIT) variable.

Controlling Tolerance Accuracy. The Tolerance box controls the precision of the tolerance. An example of a tolerance can be seen in Figure 20–28. Since tolerances are rarely used in construction drawings, this portion will not be discussed in this book. Review the Help... section for further information.

Controlling Scale. The Scale box specifies the scale factor for linear measured distances of a dimension without affecting the component, angles, or tolerance values. Selecting Linear: allows a scale factor to be entered in the edit box. A scale factor of 2, doubles the value of the dimension that is entered on the drawing. With a scale factor of 2, a dimension of 8" will be written when dimensioning a 4" long space.

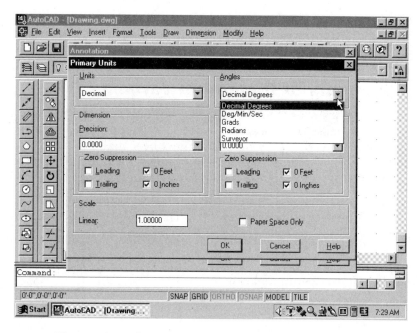

FIGURE 20–27 The Angles submenu can be set to display how angles measurements will be expressed.

Controlling Dimension Text. The Text portion of the Annotation box controls the text style, height and gap.

Selecting a Style. The Style: edit box allows you to select one of the loaded text styles to be used for dimension text. This can be done from the command prompt by entering **DIMTXSTY** [enter] (DIMension TeXt STYle).

Setting the Height of Dimension Text. The Height: box allows you to set the height of the dimension text.

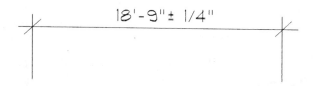

FIGURE 20–28 A tolerance is used to express a range in size. This distance could actually be between 18'-8¾" and 18'-9¼" long. If the dimension falls outside that range, it is unacceptable for that application.

Controlling the Dimension Line Gap. The Gap: box allows you to set the gap between the dimension line and the text. In construction drawings the text is placed above the dimension line so that no gap is required.

A Summary of Dimension Variables

If you're feeling overwhelmed by styles, variables, and values, don't be discouraged. Remember, most of these values will be set on a prototype drawing and left alone. Throughout this chapter comparisons have been made between using the Dimension Style dialog box and its sub-boxes or typing the value at the keyboard. A complete listing of each variable and a summary of its function can be found in the list below.

DIMALT	Alternate units selected	DIMDEC	Decimal places
DIMALTD	Alternate unit decimal places	DIMDLE	Dimension line extension
		DIMDLI	Dimension line spacing
DIMALTF	Alternate unit scale factor	DIMEXE	Extension above dimension line
DIMALTTD	Alternate tolerance decimal places		
		DIMEXO	Extension line origin offset
DIMALTTZ	Alternate tolerance zero suppression	DIMFIT	Fit text
		DIMGAP	Gap from dimension line to text
DIMALTU	Alternate units		
DIMALTZ	Alternate unit zero suppression	DIMJUST	Justification of text on dimension line
DIMAPOST	Prefix and suffix for alternate TEXT	DIMLFAC	Linear unit scale factor
		DIMLIM	Generate dimension limits
DIMASO	Creative associative dimensions	DIMPOST	Prefix and suffix for dimension text
DIMASZ	Arrow size	DIMRND	Rounding value
DIMAUNIT	Angular unit format	DIMSAH	Separate arrow blocks
DIMBLK	Arrow block name	DIMSCALE	Overall scale factor
DIMBLK1	First arrow block name	DIMSD1	Suppress the first dimension line
DIMBLK2	Second arrow block name		
DIMCEN	Center mark size	DIMSD2	Suppress the second dimension line
DIMCLRD	Dimension line and leader color		
		DIMSE1	Suppress the first extension line
DIMCLRE	Extension line color		
DIMCLRT	Dimension text color		

DIMSE2	Suppress the second extension line	DIMTOFL	Force line inside extension lines
DIMSHO	Update dimensions while dragging	DIMTOH	Text outside horizontal
DIMSOXD	Suppress outside dimension lines	DIMTOL	Tolerance dimensioning
DIMSTYLE	Current dimension style	DIMTOLJ	Tolerance vertical justification
DIMTAD	Place text above the dimension line	DIMTP	Plus tolerance
DIMTDEC	Tolerance decimal places	DIMTSZ	Tick size
DIMTFAC	Tolerance text height scaling factor	DIMTVP	Text vertical position
		DIMTXSTY	Text style
DIMTIH	Text inside extensions is horizontal	DIMTXT	Text height
		DIMTZIN	Tolerance zero suppression
DIMTIX	Place text inside extensions	DIMUNIT	Unit format
DIMTM	Minus tolerance	DIMUPT	User positioned text
		DIMZIN	Zero suppression

COMBINING VARIABLES TO CREATE A STYLE

Dimensioning variables can be combined to create a particular style in much the same way that options are combined with Text to create different styles. Styles can be created, saved, and quickly loaded by using the dialog box shown in Figure 20–1.

Creating Dimensioning Styles

Dimensioning styles allow you to save groups of settings for dimension variables. If an office does both architectural and civil engineering projects, two distinct dimensioning styles could be created and saved on prototype drawings to speed the drawing process. Even in a firm that only does architectural drawings, styles can be useful to group variables. For instance, on a multilevel structure, some dimensions will be shown on each floor plan and the framing and foundation plans. These might be placed in a style titled ARCHBASE that uses a certain size, color and style of text with two extension lines. Styles might be created for dimensions that will be used to dimension to the center of interior walls and have the left or right extension line suppressed, with another style created for a style that has both extension lines suppressed. Once created, these styles can be quickly loaded and interchanged. To create a style, start by adjusting the desired variables as required. Variables may be set by using the Geometry..., Format..., and Annotation... boxes of the Dimension Styles dialog box or by using the DIMSTYLE command. The command can be accessed by typing **DST** [enter] , **DIMSTY** [enter] , or **DIMSTYLE** [enter] at the command prompt at the following prompt:

Command: **DST** [enter]
Dimension Style Edit (Save/Restore/STatus/Variables/Apply/? <Restore>:

The Restore Option.   Accepting the default of <Restore> will allow a created style to be made the current style.

Pressing the [enter] key will display the following prompt:

?/Enter dimension style name or return to select dimension:

Pressing the **?** key and [enter] will produce a listing of current styles contained in the drawing base. If you enter the name of an existing style at the prompt, it will be restored as the current style. Press [enter] to select an existing dimension. The style of the selected dimension will become the current style. All dimensions that are placed after this change will reflect the variables of the selected style.

Type the name of an existing style preceded by a tilde at the <Restore>: prompt to display the differences between the current style and the selected style. The command sequence is:

Command: **DST** [enter]
Dimension Style Edit (Save/Restore/STatus/Variables/Apply/? <Restore>: [enter]
?/Enter dimension style name or return to select
    dimension: **~ARCHITECTURAL** [enter]

This will produce a display similar to the following listing.

Differences between ARCHITECTURAL and current setting:

	ARCHITECTURAL	Current Setting
DIMBLK1	_OBLIQUE	
DIMBLK2	_OBLIQUE	
DIMCEN	$-\frac{1}{8}$"	$\frac{1}{8}$"
DIMCLRD	3(green)	1(red)
DIMSAH	On	Off
DIMTAD	1	0
DIMUNIT	6	3

?/Enter dimension style name or return to select dimension: [enter]
Command:

Saving a Style.   Using the Save option in the Dimension Styles box was introduced earlier in this chapter. The Save option of the DIMSTYLE command allows the current setting of variables to be saved as style and makes it the current style. The command sequence is:

Command: **DST** [enter]
Dimension Style Edit (Save/Restore/STatus/Variables/Apply/? <Restore>: **S** [enter]

?/Name for New dimension style: **ARCHSMALL** ⏎
Command:

Determining the Current Status. The STatus option will list all of the dimension variables and their current setting. A possible listing for architectural drawings is listed below:

Variable	Status	Variable	Status
DIMALT	Off	DIMJUST	0
DIMALTD	2	DIMLFAC	1.0000
DIMALTF	25.4000	DIMLIM	Off
DIMALTTD	2	DIMPOST	
DIMALTTZ	0	DIMRND	0"
DIMALTU	2	DIMSAH	On
DIMALTZ	0	DIMSCALE	48.0000
DIMAPOST		DIMSD1	Off
DIMASO	On	DIMSD2	Off
DIMASZ	0	DIMSE1	Off
DIMAUNIT	0	DIMSE2	Off
DIMBLK		DIMSHO	On
DIMBLK1	_Oblique	DIMSOXD	Off
DIMBLK2	-Oblique	DIMSTYLE	ARCHITECTURAL
DIMCEN	⅛"		
DIMCLRD	BYBLOCK	DIMTAD	1
DIMCLRE	BYBLOCK	DIMTDEC	0
DIMCLRT	BYBLOCK	DIMTFAC	1.0000
DIMDEC	0	DIMTIH	Off
DIMDLE	⅛"	DIMTIX	Off
DIMDLI	½"	DIMTM	0"
DIMEXE	⅛"	DIMTOFL	On
DIMEXO	⅛"	DIMTOH	Off
DIMFIT	1	DIMTOL	Off
DIMGAP	0"	DIMTOLJ	1

Variable	Status	Variable	Status
DIMTP	0"	DIMTXT	⅛"
DIMTSZ	⅛"	DIMTZIN	0
DIMTVP	0.0000	DIMUNIT	6
Variable	Status	DIMUPT	Off
DIMTXSTY	Stylus BT	DIMZIN	2

Listing the Variables of Any Style. The Variables option will list all of the dimension variables and their current setting for any of the styles contained in the drawing base. The listing will be similar to the listing shown for the current style.

Applying Options to an Existing Dimension. The Apply option allows you to select a dimension in the drawing base and have the current setting of the dimension variables applied to it. Although the variables will be altered, the listing will be similar to the listing shown above.

EDITING DIMENSION TEXT

Dimensions that have been placed on a drawing can be edited using the DIMEDIT and DIMTEDIT commands or the editing commands found in the Modify and Grip editing modes.

Associative and Normal Dimensions

Dimensions in AutoCAD are placed as either associative or normal using the DIMASO variable. Associative dimensions function as one element, rather than as individual units. When you select a tick mark of an associative dimension, the extension lines, dimension lines, tick and text are all highlighted. Associative dimensions are the default, and are set with a DIMASO value of ON. Associative dimensions can be edited using each of the editing commands found in the Modify menu or using grips. Normal dimensions are placed with the DIMASO variable set to OFF. When you select a tick mark of a normal dimension, only the tick mark will be selected for editing. Associative dimensions can be converted to normal dimensions using the EXPLODE command.

Editing Dimensions Using DIMEDIT

The DIMEDIT command allows dimension text to be edited. The existing text can be edited, rotated, moved, or restored to its original position. The angle of the extension lines can also be altered from perpendicular to oblique. The command is accessed by picking the Dimension Edit icon from the Dimension toolbar, or by typing **DIMED** [enter] or **DIMEDIT** [enter] at the command prompt using the following prompt:

Command: **DIMED** [enter]
Dimension Edit (Home/New/Rotate/Oblique) <Home>: [enter]

Changing the Dimension Text. To alter existing text select the New Option. Selecting New Option will produce the prompt:

Dimension Edit (Home/New/Rotate/Oblique)<Home>:**N**[enter]

This will produce the Multiline Text editor and allow the desired text to be entered. Remove the brackets to replace the existing text. If the new text is placed in front of the brackets, the new text will be placed in front of the existing dimension text. Enter the new text and pick the OK button to remove the editor. The prompt will continue:

Select objects: *(select dimension text to be altered)* 1 found
Select objects: [enter]
Command:

The results of the command can be seen in Figure 20–29.

Rotating Existing Text. Choosing the Rotate option allows existing text to be rotated. Selecting the text in Figure 20–29 to be rotated to a 45 degree angle would require the following command sequence.

Command: **DIMED** [enter]
Dimension Edit (Home/New/Rotate/Oblique) <Home>: **R** [enter]
Enter text angle: **45** [enter]
Select object: *(Select text to be altered.)*
Command:

This would produce text as shown in Figure 20–30.

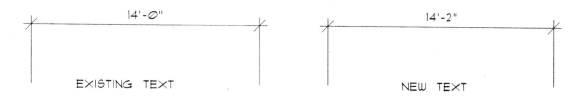

EXISTING TEXT   NEW TEXT

FIGURE 20–29 The New option of the DIMEDIT command can be used to edit exiting dimension text.

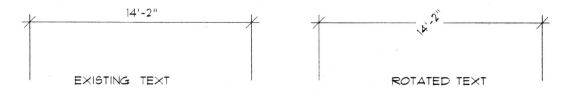

FIGURE 20–30 The Rotate option of the DIMEDIT command can be used to rotate exiting dimension text.

Altering the Extension Line Angle. The angle of existing extension lines can be altered using the Oblique option of DIMEDIT. The command sequence to alter the angle of the extension lines in Figure 20–30 is:

Command: **DIMED** [enter]
Dimension Edit (Home/New/Rotate/Oblique) <Home>: **O** [enter]
Select object: *(Select text to be altered.)*
Enter obliquing angle (RETURN for none): **–75** [enter]
Command:

The results of the sequence can be seen in Figure 20–31.

Returning Home. Using the default option of Home will return the text to its default position.

Editing Dimensions Using DIMTEDIT

The DIMTEDIT command is used to position the text relative to the dimension line. The command can be accessed by selecting the Dimension Text Edit icon from the Dimension toolbar, by selecting Align Text from the Dimension pull-down menu, or by typing **DIMTED** [enter] or **DIMTEDIT** [enter] at the command prompt. Either access method will prompt you to select a dimension to be edited. Once a dimension is selected, the following prompt will be displayed.

Enter text location (Left/Right/Home/Angle):

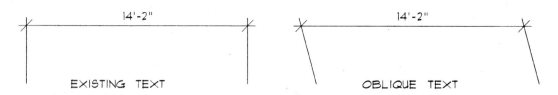

FIGURE 20–31 The Oblique option of the DIMEDIT command can be used to alter the obliquing angle of exiting extension lines.

Choosing the Left option will move the text toward the left end of the dimension line. Choosing Right, will move the text toward the right end of the dimension line. Choosing Angle will allow the text to be rotated, and choosing Home will return the text to its default position. Each option can be seen in Figure 20–32.

Editing Dimension with the Modify Menu

Associative dimensions can be edited using the editing commands contained in the Modify menu as well as the GRIP editing functions. As an object is edited, the dimensions are also edited.

STRETCH.   Dimensions can be stretched in the same way that drawing entities can be stretched. By selecting the dimension with the object, as seen in Figure 20–33, both will be stretched and the dimension text will be revised automatically. (The STRETCH command sequence was introduced in Chapter 12.) The STRETCH command also can be used to stretch the text and dimension line, thereby producing the same results as the TEDIT command.

EXTEND.   The EXTEND command sequence also can be used to extend a dimension line and adjust the text accordingly. Figure 20–34 shows the effects of the EXTEND command on dimensions.

TRIM.   The TRIM command sequence can be used to shorten a dimension line and adjust the text accordingly. Figure 20–35 shows the effects of TRIM on dimensions.

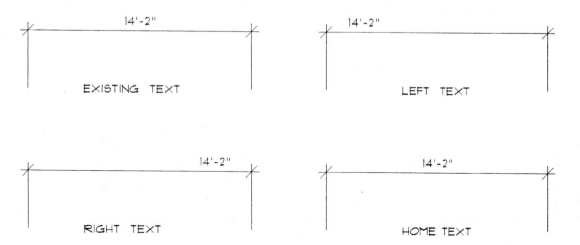

FIGURE 20–32   The four options of DIMTEDIT will alter the location of existing dimension text.

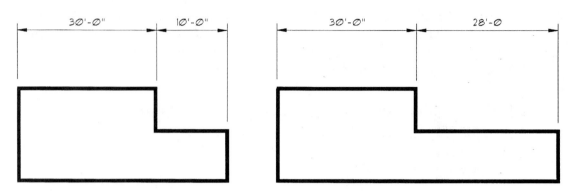

FIGURE 20–33 Using the STRETCH command with dimensions.

OVERRIDING DIMENSION STYLES

The DIMOVERRIDE command can be used to change the dimension variable settings of selected dimensions. The command can be accessed by picking Override from the Dimension pull-down menu or by typing **DOV** ⏎ , **DIMOVER** ⏎ or **DIMOVERRIDE** ⏎ at the command prompt. This will produce the prompt:

Dimension variable to override (or Clear to remove Overrides):

Enter the name of the desired variable to be altered. In the following example, the text style will be altered from archstyl to ROMAND.

Dimension variable to override (or Clear to remove Overrides): **DIMTXSTY** ⏎
Current value <ARCHSTYL> New value: **ROMAND** ⏎
Dimension variable to override (or Clear to remove Overrides): ⏎
Command:

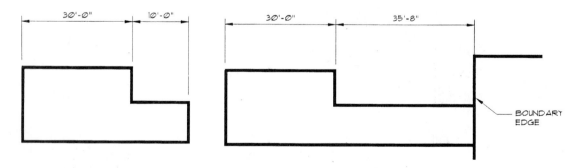

FIGURE 20–34 The effect of the EXTEND command on dimensions.

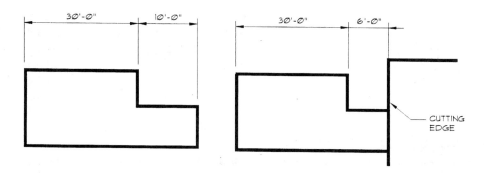

FIGURE 20–35 The TRIM command also affects the dimension line and text.

Instead of typing the name of the dimension variable to override, you can enter **C** [enter] for clear. You will now be allowed to select an object with the pick box.

WHAT IT ALL MEANS

More so than any area of AutoCAD, dimensioning seems to cause new users the most anxiety. You've read through what must seem an unlimited number of commands, options, and variables and may be having a hard time getting all of this information organized. The good news is that once you set the variables using the dialog boxes and store them in templet drawings, most of your work is done. The other thing to keep in mind is that although remembering the names of all of the variables is helpful, it is no longer necessary because of the addition of the dialog boxes.

Alternative Methods

Now that you've heard the good news, you might have guessed that there is some bad news. As good as the software is, it is really intended for mechanical drafters. They're nice people, but they do things so differently. You've probably noticed that all of the extension lines throughout the last two chapters have been created with continuous lines. This is totally unacceptable for many architectural offices. Continuous extension lines have traditionally been used to represent dimensions that extend to the face of a surface such as a stud wall. Center extension lines are traditionally used to represent the center of objects. This shortcoming can be overcome, but at a cost to other dimensioning values.

To take best advantage of the editing qualities, dimensions typically are placed as associative dimensions (DIMASO). Dimension could be set as normal dimension which would allow the extension lines to be edited using the CHANGE command. Another option is to surpress the extension lines for interior objects, and place them in the drawing using the LINE command.

An alternative to giving up the qualities of associative dimensions is to use the EXPLODE command; this command will be covered in detail in Chapter 21. Some

Layer problems will be created using EXPLODE. Basically the command will allow you to select a group of entities that function as one entity, and then allow you to separate that one group. This will allow the majority of dimensions to retain their editing qualities, while allowing the selected dimensions to be edited as individual lines, arrows, and text.

CHAPTER 20 EXERCISES

E-20-1. Open drawing Floor14. Use the appropriate dialog box and a style suitable for dimensions with both extension lines, a style with one dimension line, and a style with no extension lines. Set the text gap for all styles to be ⅟₁₆″, with a baseline increment of ½″. Assign each style a different color. Mark circles with a ⅛″ center mark and provide for centerlines. Use ⅛″ for the feature offset and extension above line settings. Use ³⁄₁₆″ long tick marks and a ⅛″ long tick extension. Set the text height as ⅛″ with the text forced inside, and above the extension line. Set the zero suppression controls so that a zero will be shown following feet, but not preceding inch sizes. Save these styles and then save the drawing as FLOOR14.

E-20-2. Use appropriate dimension techniques to label the floor to plate height of drawing E-18-8. Save the drawing as ELEV.

E-20-3. Use appropriate dimension techniques to label the base footings created for drawing E-16-4. Save the drawing as FOOTINGS.

E-20-4. Use appropriate dimension techniques to label the two-story footing created for drawing E-16-4. Show the concrete as 8″ above grade, with an 8″ stem wall, a 7″ × 15″ footing that extends 18″ below the finish grade. Save the drawing as FOOTINGS.

E-20-5. Use baseline dimension methods with dot line terminators to locate each tree center. Place all dimensions separate from all other information.

E-20-6. Open drawing E-16-9 and provide the needed dimensions to describe the manhole cover created in problem E-14-9. Save the drawing as E-20-6.

E-20-7. Open drawing E-14-7 and provide the needed dimensions to describe the structure. Use the required commands to create the linetypes and text of the original problem. Hatch the 8″ wide wall with ANSI31. Save the drawing as E-20-7.

E-20-8. Open drawing E-20-7 and provide the needed dimensions and notes to describe the foundation plan created for E-11-7. Change the lines representing the beams to 3″ wide lines. Label the main structural members to represent post and beam methods of your region. Save the drawing as E-20-7.

E-20-9. Use the drawing on the next page as a guide, draw and dimension the curb detail. Place dimensions, text, general notes, and hatching symbols on separate layers. Use different text styles for the notes and dimensions. Save the drawing as CURB.

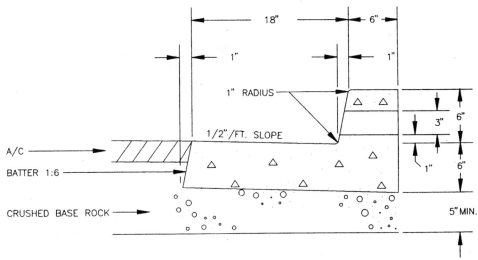

1. PCC SHALL BE 3300 PSI STRENGTH AT 28 DAYS.

2. CONTRACTION JOINTS SHALL BE PLACED AT 15' MAX. SPACING.

3. EXPANSION JOINTS SHALL BE PLACED AT 45' MAX. SPACING AND AT ALL DRIVEWAYS AND CURB RETURNS.

4. CURB EXPOSURE SHALL BE 8" AT ALL CATCH INLETS.

5. DRAIN BLOCKOUT SHALL BE SUFFICIENT FOR 3" DRAIN PIPE AND PLACED WHERE DIRECTED.

6. CRUSHED BASE ROCK SHALL BE COMPACTED TO 95% RELATIVE DENSITY PER AASHTO T−180.

E-20-09

E-20-10. Open a new drawing and use the drawing on the next page as a guide to show the support for two 6¾ × 37½" glu-lam beams. Establish separate layers for the beam, steel, text, and dimensions. Completely label and dimension the drawing. Set the required scale factors for plotting at a scale at 1"=1'-0".

The beams will be supported on a ⁵⁄₁₆" × 6⅞" × 24" base plate with two ⁵⁄₁₆" × 10" × 24" side plates welded to the base plate with a ¼" fillet weld. The base plate will be supported by a 6- × 6- × ¼" steel column and welded with a ⁵⁄₁₆" fillet weld all around. Each beam will be bolted to the side plate with four ⅝" diameter machine bolts. Bolts will be 1½" in from the top and sides and be 3" on center.

A second connector plate will be located with the center of the 3" × 22" × ⁵⁄₁₆ steel plate 8" down from the top of the beam. Two bolts will be used to join the plate to the beam. Bolts will be 2" from the plate end, and 3" on center. Save the drawing as E-20-10

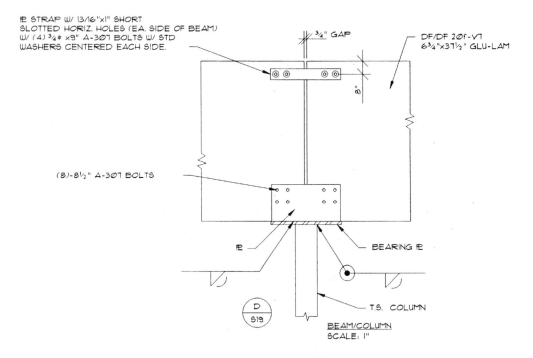

E STRAP W/ 13/16"x1" SHORT
SLOTTED HORIZ. HOLES (EA. SIDE OF BEAM)
W/ (4) ¾φ x9" A-307 BOLTS W/ STD
WASHERS CENTERED EACH SIDE.

¾" GAP

DF/DF 20f-V1
6¾"x37½" GLU-LAM

8"

(8)-8½" A-307 BOLTS

E

BEARING E

E

T.S. COLUMN

D
S19

BEAM/COLUMN
SCALE: 1"

E-20-10

E-20-11. Open a new drawing and use the drawing on the next page as a guide to show the column-to-base-plate connection. Establish separate layers for the steel, concrete, text, hatch, and dimensions. Completely label and dimension the drawing. Set the required scale factors for plotting at a scale of 1½" = 1'-0".

The column will be a 6- × 6- × ⅜" t.s. welded to a 10- × 10- × ⅞" base plate with a ³⁄₁₆" fillet weld all around. Use four 18" long anchor bolts set in an 8½" square grid around the column. Support the base plate on 1" mortar. Save the drawing as E-20-11.

E-20-12. Open a new drawing and use the drawing on the next page as a guide to show the beam-to-beam connection. Establish separate layers for the steel, text, hatch, and dimensions. Completely label and dimension the drawing. Set the required scale factors for plotting at a scale of 1½"=1'-0".

The support beam is a W18 × 76 (18¼" dp.) steel wide flange with a W18 × 46 (18" dp.) on the left and a W18 × 40 (17⅞" dp.) on the right. Use two 4 × 12 × ⁵⁄₁₆" steel connection plates welded to the W18 × 76 with ⁵⁄₁₆ × 12" fillet each side. Use four machine bolts through the plate and beam @ 3" o.c., 1½" from the edge. Save the drawing as E-20-12.

E-20-13. Open drawing FLOOR14 and use the drawing on the next page as a guide to draw the slab-on-grade plan. Establish separate layers for the concrete, text, and dimensions. Completely label and dimension the drawing. Save the drawing as E-20-13.

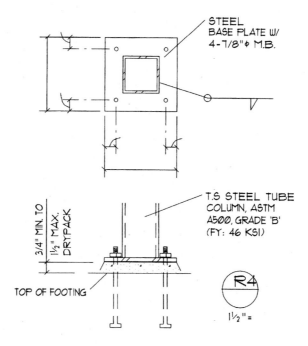

STEEL
BASE PLATE W/
4-7/8"φ M.B.

T.S STEEL TUBE
COLUMN, ASTM
A500, GRADE 'B'
(FY: 46 KSI)

3/4" MIN. TO
1½" MAX. DRYPACK

TOP OF FOOTING

R4

1½" =

E-20-11

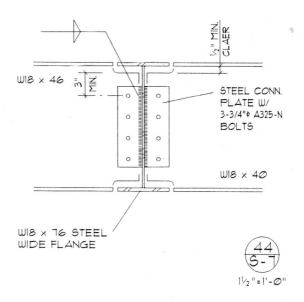

½" MIN.
CLAER

W18 × 46

3" MIN.

STEEL CONN.
PLATE W/
3-3/4"φ A325-N
BOLTS

W18 × 40

W18 × 76 STEEL
WIDE FLANGE

44
S-7

1½" = 1'-0"

E-20-12

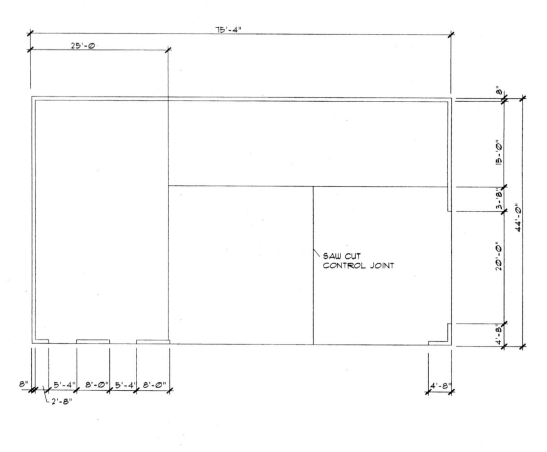

SLAB PLAN
1/8" = 1'-0"

E-20-13

E-20-14. Open a new drawing and use the drawing on the next page as a guide to steel framing elevation. Establish separate layers for the steel, text, and dimensions. Completely label and dimension the drawing. Set the required scale factors for plotting at a scale of ⅜″ = 1′-0″. Establish the eave height as 24 feet. Locate the first Z brace at 7 feet above the floor, with the balance at 72″ o.c. Vertical supports will be set 12″ in from each end of the 100′-0″ structure with interior supports at 25′-0″ o.c. Save the drawing as E-20-14.

E-20-15. Open a new drawing and use the drawing on the next page as a guide to draw the wall and floor detail. Establish separate layers for the wood, concrete, steel, text, and dimensions. Completely label and dimension the drawing. Set the required scale factors for plotting at a scale of ½ = 1′-0″. Establish the upper floor 30″ above the lower floor, and the bottom of the 4 × 10 that supports the lower floor is to be 18″ clear to finish grade. Dimension the height of the rail as 42″. Save the drawing as E-20-15.

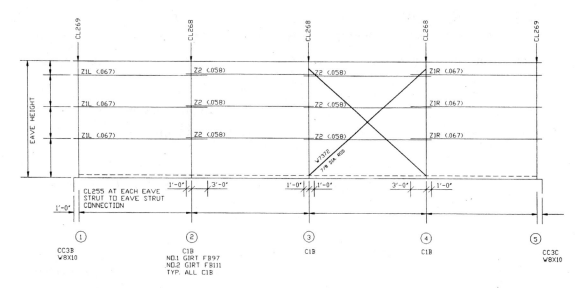

E-20-14

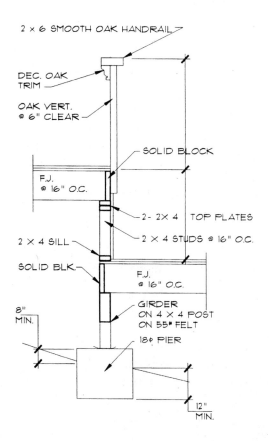

E-20-15

CHAPTER 20 QUIZ

1. A group of dimensioning controls that can be saved for future use is called a

   _____

2. A _____ is used to control individual aspects of dimensioning.

3. What command must be typed to access a listing of dimensioning features?

   _____

4. How can styles be used in an engineering office that does both civil and structural projects?

   _____

   _____

5. What value controls the size of each dimension feature and how should this value be set for a drawing that will be reproduced at a scale of ¼″ = 1′-0″?

   _____

6. What dialog box and option will remove both extension lines from a dimension?

   _____

7. The spacing between two dimension lines is controlled by what menu option?

   _____

8. What does the DIMGAP control?

   _____

   _____

9. List two features that can be displayed to describe round objects.

   _____

   _____

10. List five options for terminating a dimension line.

    _____

    _____

    _____

11. List and describe two features of the extension line as it relates to the dimension line and the object being described and give the variable name for each.

    _____

    _____

12. What dialog box provides options for controlling the height of text and what variable will retain this information?

13. Name the dialog box that can be used to make changes to dimensioning text.

14. What style will be used as a base when creating dimensioning styles?

15. What variable will allow for changing an attribute without changing the entire style?

16. What does the TEDIT command affect?

17. Describe how associative dimensions will affect editing.

18. How can the LEADER command be used to provide two arrows pointing to the same text?

19. What dimensioning command is used if the text is to be placed so that it is not parallel, perpendicular, or aligned with a surface?

20. What command can be used to change existing dimensions to the current style?

21. Without trying to be politically correct, what is a family?

22. Describe the major functions of the Annotation, Format, and Geometry dialog boxes.

23. Sketch examples of right angle and closed terminators.

24. Explain the process for creating a user-defined terminator.

25. What variable controls the dimension line extension?

26. What is DIMCLRE?

27. Explain the options to display a centerline in each circle to be dimensioned.

28. Give the proper scale factor for drawing a site plan to be plotted at a scale of ⅛" = 1'-0" and explain how the value is determined.

29. Explain the process to place text above the dimension line.

30. List and explain the options for controlling zeros in dimensions.

31. Explain the difference between associative and normal dimensions.

32. Explain the options of the DIMEDIT command.

33. Explain the effect of selecting an extension line to be erased.

   _____

   _____

34. Explain the effect of stretching a window that is dimensioned 3'-0" long to a length of 6'-0".

   _____

   _____

35. How can a variable be altered without changing the entire style?

   _____

   _____

SECTION 5

GROUPING INFORMATION

Creating Blocks and Wblocks

Adding Attributes to Enhance Blocks

Combining Drawings Using XREF

···············

CHAPTER

21

CREATING BLOCKS
AND WBLOCKS

····························

You might have been wondering while working through the first 20 chapters why you haven't started drawing some really neat projects. Since Chapter 7 you've had the basic drawing skills to draw most projects you could imagine. Subsequent chapters have helped you to develop skills needed to ease the drawing and design process but still have not equipped you to master a complicated project easily. This chapter will provide the final ingredients for completing complicated projects by introducing you to the BLOCK, INSERT, DDINSERT, MINSERT, BASE, EXPLODE, WBLOCK, and PURGE Commands.

BLOCKS

Drawings are composed of symbols. To show a door on a floor plan, you don't draw a door; instead you use a line and an arc to represent a door. A block is a group of entities that are treated as one entity. The entities used to represent a door symbol can be saved as a block and be inserted throughout the drawing as one entity. The BLOCK command allows symbols that are used repeatedly to be created, saved, and inserted easily in the drawing in which they were created, which provides many benefits for the user.

Benefits of Blocks

Symbols that are used frequently can be saved as a block, which will greatly increase drawing speed and efficiency.

Speed. If you're drawing a typical residential floor plan, doors will need to be represented 20 or 30 times. Rather than draw each door, one door can be drawn and saved as a block. As a block, a door can be inserted quickly and the size altered or the position rotated, as seen in Figure 21–1. With a little more effort, these same qualities can also be accomplished with the COPY, ROTATE, MIRROR, ARRAY, or SCALE commands, but why use several commands when you could use one?

Editing. Throughout the evolution of construction drawings, many changes and revisions will take place. Blocks can greatly aid in this revision process. If a client decides that all of the doors in a building are to be 6 inches wider, several hours might be required to change all of the doors on the various floor plans. If the doors are drawn as a block, the block can be redefined and all references to that block will be updated automatically.

Attributes. Many blocks, such as a door, require text to explain variations. Size, material, quantity, and supplier are examples of information that can be saved as an attribute with a door block. This text is called an attribute and can vary with each application of the block. Attributes can be displayed as text or can be made invisible. Attributes associated with blocks can also be extracted from a drawing and transferred to a data base form, such as a door schedule.

Efficiency. In addition to saving time and providing a quick method for revisions, the BLOCK command combines the storage of groups of entities into one unit. In drawing a door, information must be stored about the start points and endpoints of the original line and arc, the radius of the arc, the scale factor as the symbol is enlarged or reduced for various uses, and the location of each unit as it is displayed throughout the drawing. When the symbol is installed as a block, only a block reference is stored, saving valuable storage space.

Uses for Blocks

Before a possible listing of blocks is considered, it should be remembered that the list is endless. Any symbol that is used repeatedly should be stored in a templet drawing as a block. A sample listing for architectural blocks would include items for the site, floor, and foundation plans; elevations; and sections as shown below.

FIGURE 21–1 Blocks can be inserted into a drawing, altered in size and rotated to meet specific needs.

Site plan: north arrow, trees, shrubs, spas, pools, fountains, curbcuts and driveways, utility symbols, and general notes. Plots for subdivisions might include structure shapes, sidewalks, and decks or patios.

Floor plans: appliances, door and window symbols, heating equipment, heating and cooling symbols, electrical symbols, plumbing symbols.

Framing plans: joist and rafter direction indicators, section tags, general notes.

Foundation plans: piers, post and piers, beam and column symbols, general notes, and foundation details.

Elevations: door styles, window styles, shutters, trim, eave molding, decorative posts, garage doors, gable wall vents, trees and shrubs, lighting fixtures, and people.

Sections: structural shapes such as plates, sills, double sills, beams, columns and piers, foundation components, eave components, insulation, standard dimensions, and common notes.

In addition to drawing simple symbols as a block, several individual blocks can be combined to form a larger block. Figure 21–2 shows an example of a typical truss-to-plate intersection. The block could be inserted into a section drawing, mirrored, and then the truss shape completed by the FILLET and EXTEND commands, as seen in Figure 21–3.

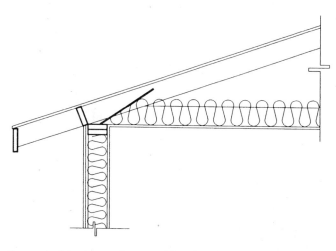

FIGURE 21–2 Several objects can be combined to form a block.

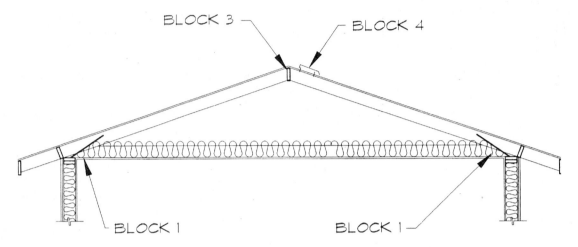

FIGURE 21–3 Using the MIRROR, FILLET, and EXTEND commands, this section was created by inserting and connecting blocks.

Creating Blocks

Because a block is a group of entities that is saved as one entity, there is no special method to drawing the block. The components that will comprise the block can be created using any of the drawing or editing commands. Some consideration should be given to the layer the block is created on, and the size of the block while it is being constructed to avoid some problems when the block is inserted into the drawing.

The Effects of Layer on Blocks. A block will take on the characteristics of the layer on which it is created. A block can also consist of objects that have been drawn on several layers with different linetypes and colors. When the block is inserted into a drawing, the block will retain its original characteristics. If a lavatory is created on a layer titled MPLUMB and inserted into a drawing with the UPPLUMB layer current, the lavatory will be inserted into the drawing on the MPLUMB layer. There are, however, three exceptions to this guideline.

1. Blocks that are created on the 0 layer will be placed on the current layer when placed into a drawing.
2. Blocks drawn with the color created BYBLOCK are drawn with the color of the current layer, but will reflect the color of the layer into which the block is inserted.
3. Blocks drawn with the linetype created BYBLOCK are drawn with the linetype of the current layer, but will reflect the linetype of the layer onto which the block is inserted.

Block Size. Once the block is saved, it can be inserted throughout a drawing. When the INSERT command is introduced later in this chapter, you will have an opportunity to alter the size of the block in the X and Y direction. Altering the size of the block during the insertion process can be easier if the size of the block is considered during creation.

Some blocks, such as toilets, electrical symbols, and appliances, are a standard size. These blocks can be drawn at their actual size and inserted. Other blocks, such as doors, windows, tubs, and showers, vary in size with each application. Drawing these blocks in one-unit squares will allow the size to be adjusted each time it is inserted. Rather than having separate blocks for 4-foot, 5-foot, and 6-foot windows, one window can be created, and the X value controlling its length can be altered.

The size of the unit can vary depending on the drawing the block will be inserted in. On floor plans, you might use 1″ as the unit and scale up, or use 1 foot as the unit and scale down. Figure 21–4 shows a window created using a 12″ unit, and the effects that can be achieved with one block.

Entering the Block Command From the Command Prompt. Keep in mind that although the toilet has been drawn, it is not a block yet. Once the drawing is ready to be reproduced, enter the BLOCK command through the keyboard by entering **–B** [enter] or **BLOCK** [enter] , by choosing the Make Block icon from the Draw toolbar or by selecting Make... from Block of the Draw pull-down menu. Figure 21–5 shows the drawing process to create a toilet, which can then be saved as a block. This can be done by entering the BLOCK command. This will change the prompt to:

> Command: **–B** [enter]
> BLOCK Block name (or ?):

Choosing a Name. A name of up to 31 characters can be used to describe the block. The block name can contain letters, digits, and special characters such as $, - , and _. If a name for an existing block is provided for the new block, the old block will be destroyed and the new block will take its place. The following prompt will be displayed:

> Block <existing name> already exist.
> Redefine it? <N>

Responding [enter] will exit the command and not alter the existing block. Entering **Y** [enter] at the prompt will save the new block and replace the old block.

FIGURE 21–4 BLOCKS should be created using one unit, such as a foot, so the size of the block can be adjusted easily each time the BLOCK is inserted.

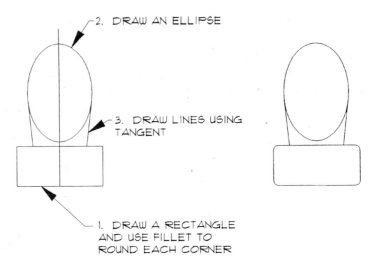

2. DRAW AN ELLIPSE

3. DRAW LINES USING TANGENT

1. DRAW A RECTANGLE AND USE FILLET TO ROUND EACH CORNER

FIGURE 21-5 A toilet can be created by drawing an ellipse and a rectangle and connecting the two with straight lines. The FILLET command was used to round each corner.

If the question symbol is entered, the prompt will change to:

Block(s) to list <*>

By pressing the [enter] key, a list of all existing blocks on the current drawing will be listed. At the bottom of the listing will be the display:

User Blocks	External References	Dependent Blocks	Unnamed Blocks
6	0	0	0

Each of these terms will be explained later in Chapter 23 when XREF drawings are introduced.

Once a block name is entered, the command sequence will continue. The entire command sequence for creating a block is:

Command: **–B** [enter]
Block name (or ?): **TOILET** [enter]
Insertion basepoint: (*Select a point.*)
Select objects: (*Select objects to comprise the block.*)
Select objects: [enter]
Command:

Enter an Insertion Point.   Once a name has been provided for the block to be created, a request for an insertion point will be given. The insertion point of the block will align with a point that will be designated as the block is inserted in a future drawing. Figure 21–6 shows the effect of the insertion point of a block with the insertion point at the destination.

The insertion point should be chosen based on the block's use. Blocks such as a toilet, a lavatory, or a sink will be inserted most easily using a centerpoint. Other objects such as a sill or beam may be best located from an edge. Figure 21–7 shows some typical insertion points for common architectural features. The insertion point also can be used as a rotation point for the block as the block is inserted into its new location. Figure 21–8 shows how a block can be rotated about its insertion point. Other options will be discussed as the INSERT command is explored. Using a centerpoint or the lower left corner of a block will help to keep track of where to insert the block.

Select Objects.   The final step in creating a block is to define the entities that will comprise the block. Entities may be selected by any of the entity selection methods. The selected entities will be removed from the screen once the ⏎ key is pressed at the Select object prompt. When the block is removed you may feel a sense of panic. No need to worry, the block is stored in the drawing base, and the names of each block are recorded and can be viewed by using the question symbol at the original BLOCK prompt.

Creating Blocks Using a Dialog Box.   When the BLOCK command is started using the Make Block option of the Draw pull-down menu or the Draw toolbar, the Block Defination dialog box shown in Figure 21–9 is displayed. Start the process by entering the name of the block to be created by entering the name in the Block name: edit box. Once the name has been provided, press either the Select Point< or Select Objects< button. Each option must be selected, but the order will not effect the outcome of the command. The Select Point< button removes the dialog box from the screen and allows the insertion point to be selected. Selecting the Select Objects< button removes the dialog box from the screen and allows the objects that will make up the block to be selected. Selecting the List Block Names... button will provide a listing of all of the blocks contained in the drawing similar to Figure 21–10. When a

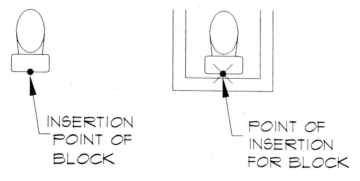

INSERTION POINT OF BLOCK

POINT OF INSERTION FOR BLOCK

FIGURE 21–6   An insertion point defines the location of the block to be used as a handle.

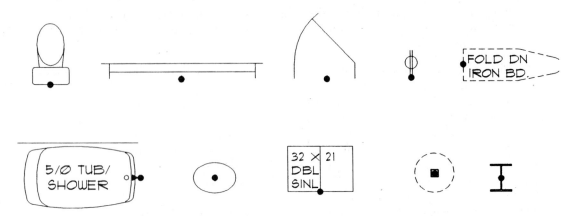

FIGURE 21–7 Typical insertion points for common architectural features.

block is created from the command line, the objects will disappear from the screen. The Block Defination dialog box allows the objects that comprise the block to remain in their current position if the Retain Objects box is activated. Clicking the Retain Objects box so that it is inactive will remove the objects that comprise the block from the screen as the block defination is saved to the drawing. Once all of the components of the block have been set, pressing the OK button will add the block to the drawing.

Inserting Blocks

Once a block has been created and saved, it can be placed in a drawing using the INSERT command. The command can be accessed by typing **–I** [enter] or **INSERT** [enter], by selecting the Insert Block icon from the Draw toolbar, or by selecting the BLOCK... option of the Insert pull-down menu. Using either the toolbar or pull-down menu will produce the Insert dialog box. Using the dialog box will be introduced later in this chapter.

Inserting Blocks Using The Command Prompt.   Blocks can be inserted easily and quickly into a drawing from the command prompt. The command sequence is:

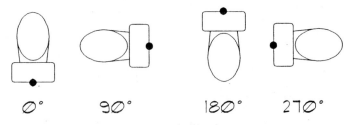

FIGURE 21–8 Rotating a block around its insertion point.

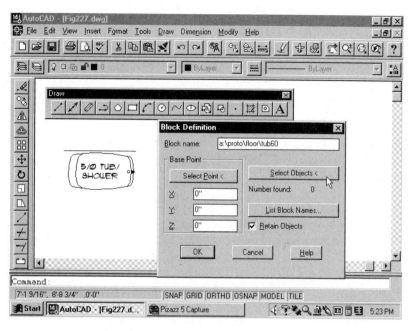

FIGURE 21–9 The Block Definition dialog box can be used to name and create a block. The box is accessed by selecting Make Block from either the Draw toolbar, or from the Draw pull-down menu.

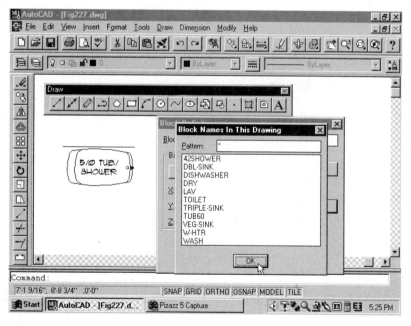

FIGURE 21–10 Selecting the List Block Names... button will produce a listing of all blocks contained in the drawing.

Command: **–I** (enter)
Block name (or ?) **TOILET** (enter)
Insertion pointed: (*Select point.*)
Insertion point: X scale factor <1> / Corner / XYZ: (enter)
Y scale factor (default = X): (enter)
Rotation angle <0>: (enter)
Command:

Block Names. The name of the block to be inserted can be typed at the prompt, or the question symbol can be used to display a listing of the drawing blocks if your memory has failed.

*Names. One of the benefits of using blocks is that they save space because several entities are saved as one. By adding an asterisk to the block title as a block is inserted, the individual entities comprising the block will retain their qualities.

Block Insertion Point. As the block was created, an insertion point was determined. Now you are being asked for a location to place the original insertion point. Figure 21–11 shows the relationship of the block insertion point to the destination insertion point. Once the insertion point is selected, the block will be dragged into the indicated position, and the scale can be altered.

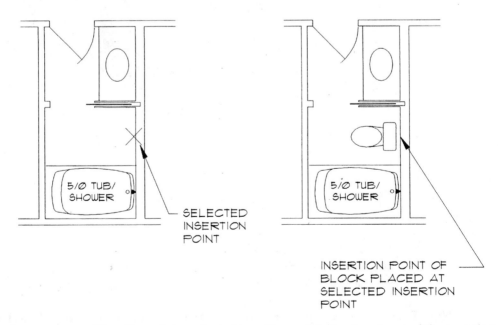

FIGURE 21–11 The effect of the selected insertion point of a drawing and the insertion point of a block.

X/Y Scale Factors. If you have created the block at the size required, pressing the [enter] key twice will bring up the next prompt and allow the block to be rotated. For a block such as a window that needs to be adjusted, enter the desired values. Inserting a window created as a 1-foot block to be a 6-foot-long window in a 6" exterior wall would require an X scale factor of 6, and a Y scale factor of .5. The effects of the scale factor can be seen in Figure 21–12.

Negative Scale Factors. The location of a block can be altered by using negative X and Y values. Using negative values combines the INSERT and MIRROR commands. Figure 21–13 shows an example of negative value alternatives.

Corner Specification of Scale. A final option for inserting a block is to determine the scale by picking one corner of the block and then picking the opposite corner with a window. This option works well for inserting objects if the scale is not critical. When the prompt for X is displayed, use the cursor to indicate a point for the beginning of the box. The second location should be above and to the right of the first point to keep the block in its original orientation.

Rotation Angle. The final prompt to insert a block in a drawing is to specify a rotation angle. Pressing the [enter] key will insert the block at the angle used to create the block. Figure 21–8 shows four options for rotating an object around its insertion point. Any angle can be provided for the rotation angle.

Presetting Block Variables. Values for the X/Y scale and rotation can be preset for blocks before they are dragged into position. By presetting the values, the blocks'

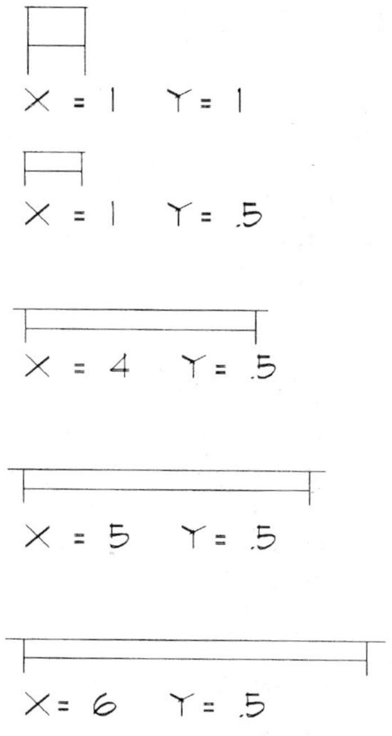

FIGURE 21–12 The effect of X/Y scale factors on a block.

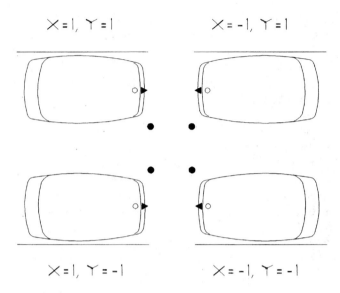

FIGURE 21-13 Altering a block using negative scale factors.

size and rotation are displayed prior to insertion. The following preset values can be entered at the insertion point: prompt.

Scale: Prompts for X, Y, and Z scale factors and the rotation angle can be provided.

Xscale: Similar to scale except that only the X factor is set.

Yscale: Similar to scale except that only the Y factor is set.

Rotate: Prompts for the rotation angle.

PScale: Similar to scale except that the scale factor is only used for display as the block is being dragged into position.

PXscale: Similar to PScale except that only the X factor is affected.

PYscale: Similar to PXscale except that only the Y factor is affected.

PRotate: The same as rotate, except that the angle is used only for display purposes as the block is dragged into position.

After any of these options is entered and the value supplied, the Insertion point: prompt is repeated to allow another option to be entered or to select an insertion point.

Inserting Blocks with a Dialog Box. The Insert dialog box can be used to INSERT blocks into a drawing. The Insert dialog box shown in Figure 21–14 is displayed by selecting the Insert Block icon of the Draw and Insert toolbars, by selecting Block... from INSERT of the Draw pull-down menu, or by typing **I** [enter] or **DDINSERT** [enter] at the command prompt. The dialog box provides the same menu options as the typed menu.

Choosing a Block.   Selecting the Block... box will display the dialog box shown in Figure 21–15. This box will display a listing of all Blocks, similar to the question symbol of the INSERT command.

Selecting Drawing Files for Blocks.   Selecting the File... box of the Insert dialog box will display a dialog box similar to the one in Figure 21–16 showing a listing of existing drawings. These drawings can now be selected to be used as blocks.

Specify Parameters on Screen.   With a block named in the Block... edit box, activating the Specify Parameters on Screen box and picking OK will remove the dialog box from the screen and display the block to be inserted. The insertion point, scale factors, and rotation angle are specified in the same method that was used for the INSERT command. If the box is not activated, the Insertion point, Scale, and Rotation select boxes will be usable, and values can be added to these boxes.

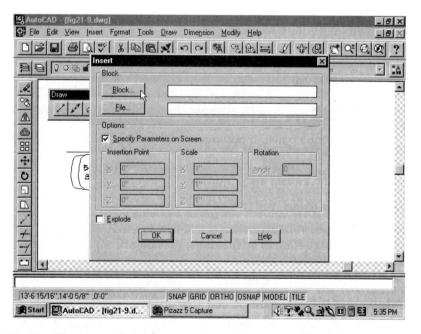

FIGURE 21–14 The Insert dialog box can be used to insert blocks into a drawing. The box can be accessed by picking the Insert Block icon from the Draw toolbar, typing **I** [enter] or **DDINSERT** [enter] at the command prompt, or by picking Block... from the Insert pull-down menu.

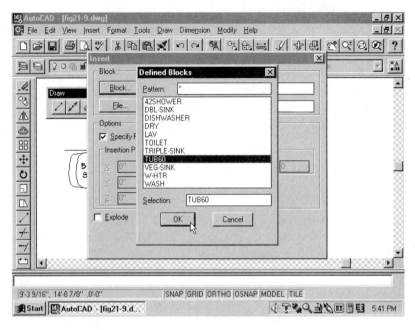

FIGURE 21–15 Selecting the BLOCK... box will display a listing of blocks contained in a drawing.

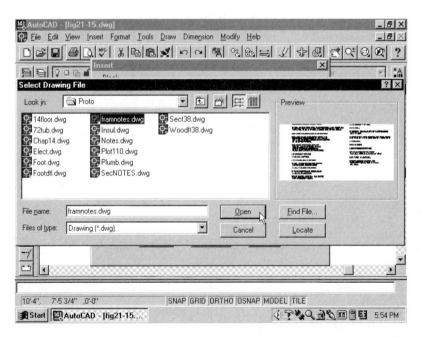

FIGURE 21–16 The Select Drawing File will list drawings that can now be used as a block.

Insertion Point. The Insertion point box allows for defining the X and Y coordinate locations (Z is the vertical reference for 3D) and provides the location for the block insertion point.

Scale. This box allows you to define scale factors for the block to be inserted. Unless a different value is entered, the value entered for X will become the default value for Y. Any dimensions contained in the block will be multiplied by the scale factor as the block is inserted.

Rotation. The final selection box allows you to supply a rotation angle.

Explode. Blocks function as one entity when they are inserted into a drawing. If the Explode box is activated, as the block is inserted into the drawing, each entity of the block will function as separate entities.

MINSERT—Making Multiple Copies

The MINSERT (multiple insert) command combines the features and the command sequences of INSERT and ARRAY. MINSERT will allow blocks to be arrayed in a rectangular pattern typical of what might be found in the interior column supports for multilevel structures, or the supports of a post-and-beam foundation. The MINSERT command can be started by typing **MINSERT** ⏎ at the command prompt. Figure 21–17 shows an example of a multiple insert. The drawing was created using the following command sequence:

> Command: **MINSERT** ⏎
> Block name (or ?): **PIER** ⏎
> Insertion point: **CEN** ⏎
> X scale factor <1> / Corner / XYZ: ⏎
> Y scale factor < default = X>: ⏎
> Rotation angle: **<0>:** ⏎
> Number of rows (—) <1>: **3** ⏎
> Number of columns (|||) <1>: **6** ⏎
> Unit of cell or distance between rows (—): **96** ⏎
> Distance between columns (|||): **48** ⏎
> Command:

The array pattern created with MINSERT has the qualities of a Block except that it cannot be exploded. The EXPLODE command will be introduced later in this chapter.

Making An Entire Drawing A Block

An entire drawing can be used to form a block. When the entire drawing is to be a block, an insertion point other than 0,0 may be desired. The BASE command can be used to define a new base point. The command can be accessed by typing **BASE** ⏎

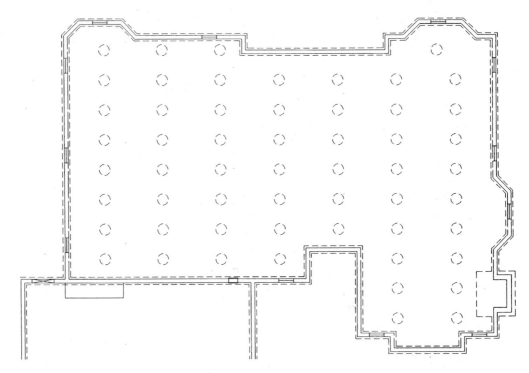

FIGURE 21–17 The MINSERT command allows blocks to be arrayed in a rectangular pattern. (Courtesy Piery & Barclay Designers, Inc., A.I.B.D.)

at the command prompt or by selecting Base from the fly-out menu of Block of the Draw pull-down menu. Each will produce the following prompt:

Command: **BASE** [enter]
Base point: (*Select a point.*)

The base point will now become the insertion point.

A better alternative to this method of using an entire drawing as a block will be introduced in Chapter 23 as externally referenced drawings (XREFS) are introduced.

EDITING BLOCKS

The easiest way to edit a block is to save the block with the Specify Parameters on Screen box of the Insert dialog box. Blocks will also retain their individual characteristics if the * is added preceding the desired title. A block inserted as *TUB would function as individual entities allowing use of each of the edit commands. Blocks may also be edited by using the EXPLODE command or using the Redefine option of the BLOCK command.

EXPLODE

Combining all of the entities of a block as one entity is great if you want to MOVE, ROTATE, or COPY a block. Only one entity has to be picked and the entire block will respond to the command. Using other editing commands such as ERASE to remove one entity also will cause the entire block to be erased. To erase just one entity in the block, you must use the EXPLODE command. The command can be accessed by selecting the Explode icon from the Modify toolbar, by selecting Explode from the Modify pull-down menu, or by typing **X** (enter) or **EXPLODE** (enter) at the command prompt. The command sequence is:

    Command: **X** (enter)
    Select objects: *(Select the desired blocks.)*
    Select object: *(This prompt will continue until all of the desired blocks are selected.)*
    (enter)
    Command:

The block will disappear for an instant and then be redisplayed with no apparent change. Individual components of the block can now be edited.

Redefining a Block

Even if a block has been inserted into a drawing in several locations, it is possible to update all of the blocks without editing them individually. Blocks are edited by redefining a copy of one of the blocks, then saving the edited copy of the block using the existing block name. It's important to work with a copy of the block, since the last step of the BLOCK command will remove the selected block from the screen. It's also important to remember that this process will update ALL blocks of the specified name. There is no "I only wanted a few of them" option. Use the following sequence to redefine a block.

1. Make a copy of the desired block to be edited.
2. EXPLODE the block.
3. Update the block using any of the editing commands.
4. Enter the BLOCK command. The command sequence is:

    Command: **–B** (enter)
    Block name (or ?): **TOILET** (enter)
    Block TOILET already exist.
    Redefine it ? (N): **Y** (enter)
    Insertion basepoint: *(Pick the desired point.)*
    Select objects: **W** (enter)
    First corner: Other corner: 13 found
    Select objects: (enter)
    Block TOILET redefined
    Regenerating drawing.
    Command:

Each toilet used throughout the drawing will now be updated to reflect the current changes. Figure 21–18 shows an example of a redefined block.

CREATING WBLOCKS

Blocks are a great way to make repeated copies of a symbol throughout one drawing. A symbol that is used on a foundation for one structure can be used on the foundation of several other structures. Symbols created with the BLOCK command can be used only inside the drawing where they were created. If a block is saved on a templet drawing, the block can be used each time the templet drawing is used. The WBLOCK command allows blocks to be used in multiple drawings no matter where the drawing was created by saving the WBLOCK as a .DWG file. WBLOCK will be displayed as a .DWG file each time a directory of drawings is made, but will act as a block when inserted into a drawing. Figure 21–19 shows an example of two blocks edited to form a WBLOCK.

The command can be entered by typing **W** (enter) or **WBLOCK** (enter) at the command prompt. The command sequence for WBLOCK is similar to the sequence used for creating a BLOCK except for the naming of files for WBLOCKS. As the WBLOCK command is entered, the Create Drawing File dialog box is displayed. The box displays a listing of the current folder, AutoCAD R14. The box allows the location of the new drawing to be specified. The drive and folder can be selected by moving up and down the directory of your computer. Figure 21–20 shows an example of the Create Drawing File. By selecting the arrow, the wblock to be created will be placed in a subfolder titled SCHOOL in a folder DRAWINGS on the hard drive.

Filename

Because WBLOCKS are treated as a drawing file, a storage location must be given. Once a folder name has been selected, the Create Drawing dialog box will allow for selection of a filename and a location for the file. Responding to the File prompt with a name such as WINDOW will place a .DWG file named WINDOW in the specified Drive and folder. Great care should be given in the naming and storing of wblocks. The name should clearly describe the contents of the drawing. Wblocks should be stored in folders and subfolders that will define the use. For instance, a name of PROTOS could be used for the folder. Subfolder names would include names such as SITE, FLOOR, ELEV, FND, or other common drawings to be used. Each subfolder

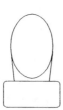
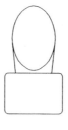

FIGURE 21–18 When a block is redefined, all occurrences of the block in the drawing will be updated.

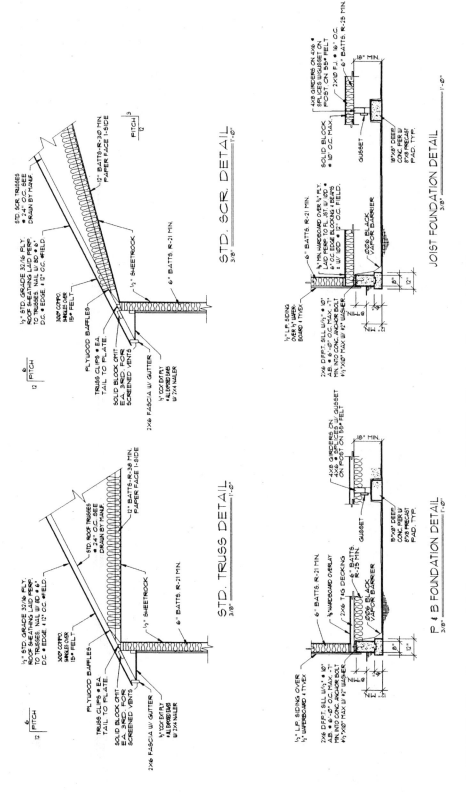

FIGURE 21-19a Common details can be stored as a WBLOCK. (Courtesy Tereasa Jefferis)

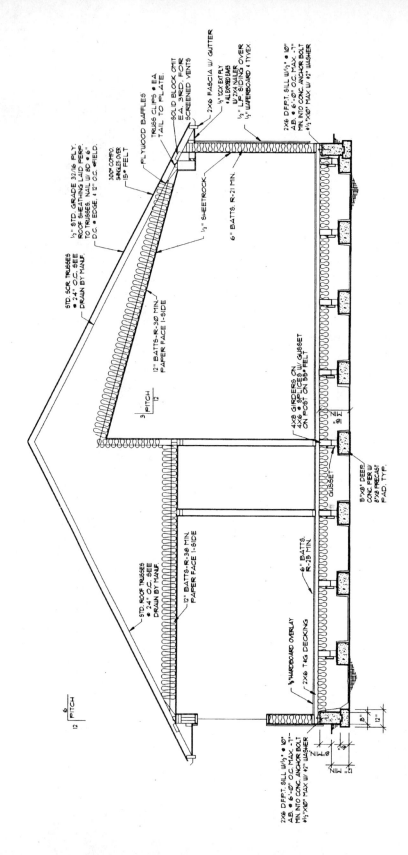

FIGURE 21–19b Wblocks can be combined to make other drawings. (Courtesy Tereasa Jefferis)

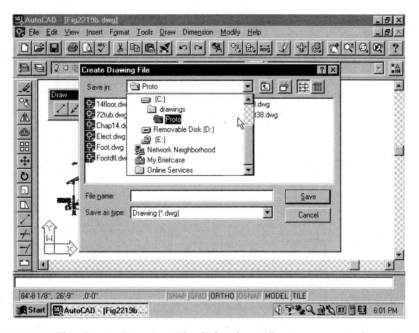

FIGURE 21-20 The Create Drawing File dialog box allows a name to be assigned to a wblock. Selecting the down arrow displays the menu shown in the Explorer.

would contain only wblocks for that specific type of drawing. Figure 21–21 shows a sample listing of wblocks for an architectural firm. Once a filename is entered, the dialog will be removed and prompts will be displayed at the command line. A prompt requesting a Block name: is now displayed at the command prompt.

Block Name

Several options are available for the block name. These include: providing a name, =, *, and the ⏎ key. Responding to the Block name: prompt by pressing ⏎ will allow the command to proceed without a block name to the next prompt. The command will now continue like the BLOCK command and ask for an insertion point.

Provide a Name. A block can be converted to a wblock by typing the block name at the Block name: prompt. To change the TOILET block to a wblock, the sequence is:

```
Command: W ⏎
File name: POTTY ⏎
Block name: TOILET ⏎
Insertion point:
```

The command sequence continues using the same prompts as the BLOCK sequence.

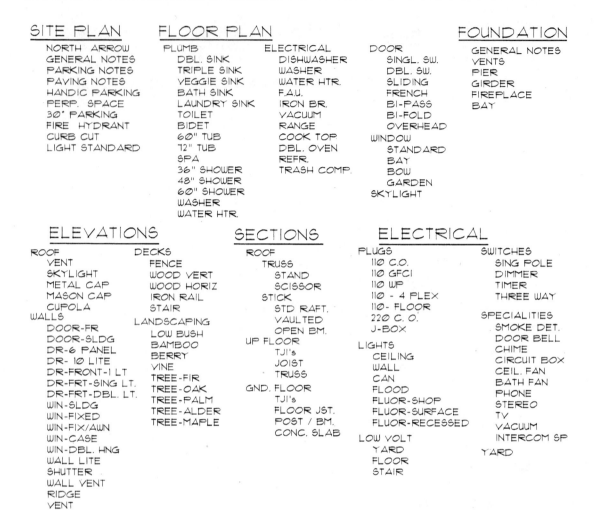

SITE PLAN
NORTH ARROW
GENERAL NOTES
PARKING NOTES
PAVING NOTES
HANDIC PARKING
PERP. SPACE
30° PARKING
FIRE HYDRANT
CURB CUT
LIGHT STANDARD

FLOOR PLAN
PLUMB
  DBL. SINK
  TRIPLE SINK
  VEGGIE SINK
  BATH SINK
  LAUNDRY SINK
  TOILET
  BIDET
  60" TUB
  72" TUB
  SPA
  36" SHOWER
  48" SHOWER
  60" SHOWER
  WASHER
  WATER HTR.

ELECTRICAL
  DISHWASHER
  WASHER
  WATER HTR.
  F.A.U.
  IRON BR.
  VACUUM
  RANGE
  COOK TOP
  DBL. OVEN
  REFR.
  TRASH COMP.

DOOR
  SINGL. SW.
  DBL. SW.
  SLIDING
  FRENCH
  BI-PASS
  BI-FOLD
  OVERHEAD
WINDOW
  STANDARD
  BAY
  BOW
  GARDEN
SKYLIGHT

FOUNDATION
GENERAL NOTES
VENTS
PIER
GIRDER
FIREPLACE
BAY

ELEVATIONS
ROOF
  VENT
  SKYLIGHT
  METAL CAP
  MASON CAP
  CUPOLA
WALLS
  DOOR-FR
  DOOR-SLDG
  DR-6 PANEL
  DR- 10 LITE
  DR-FRONT-1 LT
  DR-FRT-SING LT.
  DR-FRT-DBL. LT.
  WIN-SLDG
  WIN-FIXED
  WIN-FIX/AWN
  WIN-CASE
  WIN-DBL. HNG
  WALL LITE
  SHUTTER
  WALL VENT
  RIDGE
  VENT

DECKS
  FENCE
  WOOD VERT
  WOOD HORIZ
  IRON RAIL
  STAIR
LANDSCAPING
  LOW BUSH
  BAMBOO
  BERRY
  VINE
  TREE-FIR
  TREE-OAK
  TREE-PALM
  TREE-ALDER
  TREE-MAPLE

SECTIONS
ROOF
  TRUSS
    STAND
    SCISSOR
  STICK
    STD RAFT.
    VAULTED
    OPEN BM.
UP FLOOR
  TJI's
  JOIST
  TRUSS
GND. FLOOR
  TJI's
  FLOOR JST.
  POST / BM.
  CONC. SLAB

ELECTRICAL
PLUGS
  110 C.O.
  110 GFCI
  110 WP
  110 - 4 PLEX
  110- FLOOR
  220 C. O.
  J-BOX
LIGHTS
  CEILING
  WALL
  CAN
  FLOOD
  FLUOR-SHOP
  FLUOR-SURFACE
  FLUOR-RECESSED
LOW VOLT
  YARD
  FLOOR
  STAIR

SWITCHES
  SING POLE
  DIMMER
  TIMER
  THREE WAY

SPECIALITIES
  SMOKE DET.
  DOOR BELL
  CHIME
  CIRCUIT BOX
  CEIL. FAN
  BATH FAN
  PHONE
  STEREO
  TV
  VACUUM
  INTERCOM SP
YARD

FIGURE 21–21 Wblocks should be divided into folders and subfolders so that they can be easily accessed. Most offices name folders for the drawings that will use the WBLOCK.

The Equal Symbol. Using the equal symbol as the name will provide a similar result but allows the BLOCK and WBLOCK to have the same name. The command sequence is:

Command: **W** [enter]
File name: **WINDOW** [enter]
Block name: = [enter]
Insertion point:

The Asterisk Symbol.   Using the * symbol at the Block Name: prompt will save the entire drawing giving the same results as if SAVEAS had been used. The benefit of using * is that unreferenced entities, such as blocks, layers, and linetypes that were not used, will be removed from the drawing file. This is an excellent method for reducing the size of a drawing, and is an excellent method of saving a drawing at the end of a work session.

The Command sequence is:

> Command: **W** [enter]
> File Name: **WINDOW** [enter]
> Block name: ***** [enter]
> Command:

Inserting Wblocks.   Once a symbol or drawing has been saved as a WBLOCK it can be used just as any other drawing. It can be retrieved using the drawing Start-up menu as a new drawing session is started, or it can be inserted into an existing drawing using the INSERT or DDINSERT command. The INSERT command will bring an existing drawing into the current drawing from the command prompt. If you know the name of the wblock file to be inserted, the command prompt will provide quick access. The command prompt is:

> Command: **–I** [enter]
> Block name (or ?)<Floor>: [enter]
> X scale factor<1> /Corner / XYZ: [enter]
> Y scale factor (default = X): [enter]
> Rotation angle <0>: [enter]
> Command:

The default name for the wblock to be inserted is the last name assigned as a wblock was created. Each aspect of inserting a WBLOCK is the same as inserting a BLOCK. If you're unsure of the wblock name, or if you've grown accustomed to dialog boxes, use the DDINSERT command instead of INSERT. Typing **I** [enter] or **DDINSERT** [enter] at the command prompt will produce the Insert dialog box shown in Figure 21–14. The dialog box is also displayed if ~ is provided at the prompt for the wblock name. Wblocks are inserted into a drawing using the same methods used to insert a block.

Using Wblocks.   Symbols and construction components that are saved as a wblock can be used to save hours of drawing time on future drawings. Using an existing drawing as a wblock can also save hours on drawings such as elevations or sections that have features that must be located based on their position on the floor plan. By freezing all unnecessary materials and only thawing materials such as the roof outline, walls, windows and exterior doors, a wblock of the floor plan can be created and inserted into the ELEVATION.DWG file. The wblock can then be copied and rotated to be used for projection of each elevation as shown in Figure 21–22. Projecting irregular shaped structures from a wblock of the floor plan allows each surface to be accurately projected. Sizes can be projected to determine both vertical and

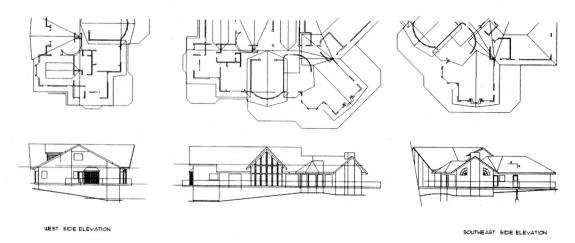

WEST SIDE ELEVATION

SOUTHEAST SIDE ELEVATION

FIGURE 21–22 A block was made of the floor plan; then it was reinserted and rotated to project each elevation. (Courtesy Residential Designs)

horizontal relationships. The roof and wall locations are projected vertically from the plan to the elevation, and heights are projected from one elevation to the next.

Editing Wblocks.   One of the strengths of using an entire drawing as a wblock is that when it is inserted, it will move as one entity. Insert a wblock of a floor/roof plan, and all of the components will move as one entity. To alter the wblock use the EXPLODE command. If a drawing is saved as a wblock using the * option, entities in the wblock will function as individual entities and will not need to be exploded.

PURGE

This command does not create symbols or drawings, but the PURGE command can be a useful tool within a drawing file to eliminate unused items and minimize storage space used. Entities that have been named in a drawing session, such as Blocks, Dimstyles, Layer, Linetypes, shapes, and Text styles, are examples of items that can be purged from a file. PURGE can be used, for instance, if a prototype drawing is used as a base for a new drawing. The base drawing may contain LAYER listings for a three-story structure with prefixes such as UPPER, MAIN, and LOWER. If only a two-level structure is to be drawn, all layers with the MAIN prefix could be purged.

Items to be purged can be eliminated at any time after entering the drawing file. Purge can be started by selecting Purge from Drawing Utilities of the File pull-down menu or by entering **PU** ⏎ or **PURGE** ⏎ at the command prompt. The command sequence is:

Command: **PU** ⏎
Purge unused Blocks/Dimstyles/LAyers/LTypes/SHapes/ STyles/Mlinestyles/All:
    (*Select one option.*) ⏎

One object can be purged by selecting the appropriate option, or all named items can be removed by selecting the ALL option. Once an option has been entered, the name of each unused item will be listed along with verification that the item should be purged.

No matter what option is selected in the PURGE command, unused options will be removed one item at a time, and you will be given the option to purge or keep the listing. The command sequence is:

Command: **PU** ⏎
Purge unused Blocks/Dimstyles/LAyers/LTypes/SHapes/STyles/All: **A** ⏎
Verify each name to be purged? <Y>: ⏎
No unreferenced blocks found.
Purge Layer WOOD? <N>: ⏎
Purge Layer CONC? <N>: **Y** ⏎
Purge Layer STEEL? <N>: ⏎
No unreferenced text styles found.
No unreferenced shape styles found.
No unreferenced dimenstyles found.
No unreferenced Mlinestyles found.
Command:

The command will continue listing each unused item in sequence until each unused block, dimension style, layer, linetype, shape, and style has been listed.

CHAPTER 21 EXERCISES

E-21-1. Open the FOOTINGS drawing that was updated in Chapter 20. Save the drawing as a block titled FND1STORY.

E-21-2. Open drawing FLOOR14 and use the drawing to create a WBLOCK titled BASE14.

E-21-3. Open drawing E-20-8. Save the drawing as a WBLOCK titled FNDPLN. Insert FNDPLN into drawing FND1STORY, rescale the footing and save the entire drawing as FNDDLT.

E-21-4. Open drawing FNDDLT and use it to create a WBLOCK titled FNDPLN1. Insert FNDPLN1 into BASE14.

E-21-5. Use the drawings below as a guide to draw the following symbols and save them as a wblock in a folder titled \PROTO\FLOOR using the designated BLOCK name. Create all blocks on the 0 layer and use 4″ high text.

DESCRIPTION	SYMBOL	BLOCK NAME
BATH		
a. Draw a 60″ × 32″ tub		TUB60
b. Draw a 72″ × 32″ tub		TUB72

DESCRIPTION	SYMBOL	BLOCK NAME
BATH		
c. Draw a 19″ × 16″ lavatory		LAV
d. Draw a toilet		POTTY
e. Draw a 36″ shower		SHOWER36
f. Draw a 42″ shower		SHOWER42
g. Draw a 60″ shower		SHOWER60
h. Draw a 32″ × 21″ double sink with a garbage disposal on either side		DBLSINK
i. Draw a 36″ × 21″ double sink with a garbage disposal on small side		XDBLSINK
j. Draw a 43″ × 21″ triple sink with a garbage disposal in the middle		XXSINK
k. Draw a 16″ × 16″ veggie sink		VEGSINK
l. Draw a dishwasher		DW

DESCRIPTION	SYMBOL	BLOCK NAME
m. Draw a trash compactor	T.C.	TC
n. Draw a lazy susan	L.S	LS
o. Draw a range	RANGE	RANGE
p. Draw a double oven	DBL. OVEN	DBLOVEN
q. Draw a 18″ pantry	PAN.	PAN

UTILITY

r. Draw a washing machine	W	WASH
s. Draw a dryer	D	DRYER
t. Draw a built-in ironing board	FOLD DN IRON BD.	IRON
u. Draw a 21″ × 21″ laundry tray	L.T.	LT
v. Draw a water heater	W.H.	WH
w. Draw a hose bibb	H.B.	HB

E-21-6. Use the drawings below as a guide to draw the following symbols, and save them as a wblock in a folder titled \PROTO\ELECT using the designated BLOCK name. Use a 6″ diameter circle for all electrical symbols with no text. Use an 8″ diameter symbol with 4″ high text for symbols that require text to be placed inside the symbol. Create all blocks on the 0 layer.

DESCRIPTION	SYMBOL	BLOCK NAME
PLUGS		
a. Draw a 110 c.o.		110CO
b. Draw a multi 110 c.o.	4	110CO2
c. Draw a 220 c.o.	220	220CO
d. Draw a 110 half hot c.o.		110HH
e. Draw a 110 waterproof	WP	110WP
f. Draw a 110 ground fault interrupter	GFI	110GFI
g. Draw a floor-mounted 110 c.o.		110F
h. Draw a junction box	J	JBOX
LIGHT FIXTURES		
i. Draw a ceiling-mounted light fixture		CEILLITE

DESCRIPTION	SYMBOL	BLOCK NAME
j. Draw a wall-mounted light fixture		WALLLITE
k. Draw a can light fixture		CANLITE
l. Draw a recessed ceiling light fixture		RECESSLITE
m. Draw wall-mounted spotlights		WALLSPOT
n. Draw a 24″ × 48″ surface-mounted fluorescent fixture		24FLUOR
o. Draw a 48″ × 48″ surface-mounted fluorescent fixture		48FLUOR
p. Draw a 48″ shop fluorescent fixture		48SHOP
q. Draw a 48″ track lite		TRACK

SWITCHES

r. Draw a single-pole switch		SWITCH
s. Draw a three-way switch		SWITCH3

DESCRIPTION	SYMBOL	BLOCK NAME
t. Draw a switch with dimmer		DIMSWITCH

SPECIAL SYMBOLS

DESCRIPTION	SYMBOL	BLOCK NAME
u. Vacuum		VACUUM
v. Intercom		INTERCOM
w. Phone		PHONE
x. Stereo speakers		SPEAKERS
y. Smoke detector		SMOKE
z. Light, heat, fan		LHF

E-21-7. Use the drawings below as a guide to draw the following symbols, and save them as a wblock in a folder titled \PROTO\FLOOR\DOORS using the designated BLOCK name. Create all blocks on the 0 layer using a 12″ block unit.

DESCRIPTION	SYMBOL	BLOCK NAME
a. Draw an exterior door		XDOOR
b. Draw an exterior door with 1 sidelight		XDOOR1LITE
c. Draw an exterior door with two sidelights		XDOOR2LITE

DESCRIPTION	SYMBOL	BLOCK NAME
d. Draw a pair of exterior door with 2 sidelights		PR2DOOR2LITE
e. Draw an interior door		INDOOR
f. Draw an exterior sliding glass door		XSLDOOR
g. Draw an interior by-pass door		BYPASS
h. Draw an interior bi-fold door		BYFOLD
i. Draw an interior folding door		FOLD
j. Draw a window		WINDOW

E-21-8. Open the appropriate templet and use the drawing below to draw the bathroom. Draw the exterior windows 6 feet wide, *except for a 2-foot wide window at the toilet room*, a 4-feet window over the counter, and 3-feet-wide windows in the bedroom. Use a pair of 2-foot-wide doors in the bedroom, a pair of 3-foot doors from the bedroom to the exterior, and a 28″ pocket door into the toilet room. All other doors to be 30″ wide. Create the following layers: WALLS, PLUMB, APPLIANCES, DIMEN, NOTES, DOORWIN and CABS. Create and insert the required blocks to complete the drawing. Provide dimensions and label all fixtures. Save the drawing as a WBLOCK titled BATH.

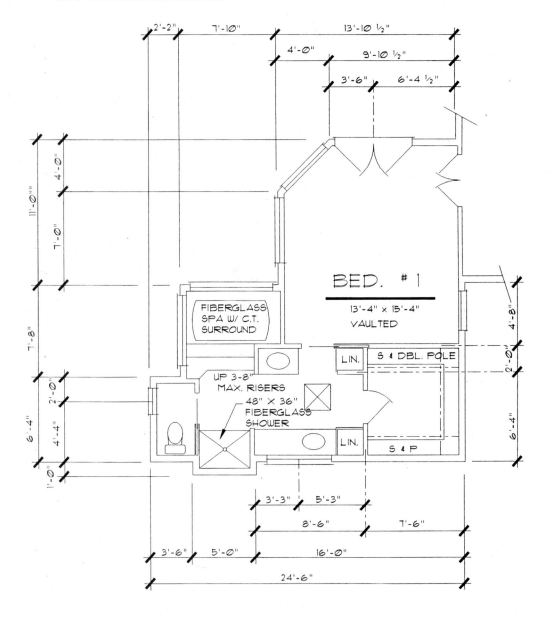

E-21-9. Open the apropriate templet and use the drawing below to draw the kitchen. Design door and windows sizes to fit the space provided. Create layers that will logically separate information for future printing on floor, framing, and electrical plans. Insert the required blocks to complete the drawing. Align the bathroom with the utility room and provide a 42″ fiberglass shower unit. Provide dimensions, determine the square footage of each room, and label the drawing. Turn off all layers containing notes and dimensions and design an electrical plan that assumes all

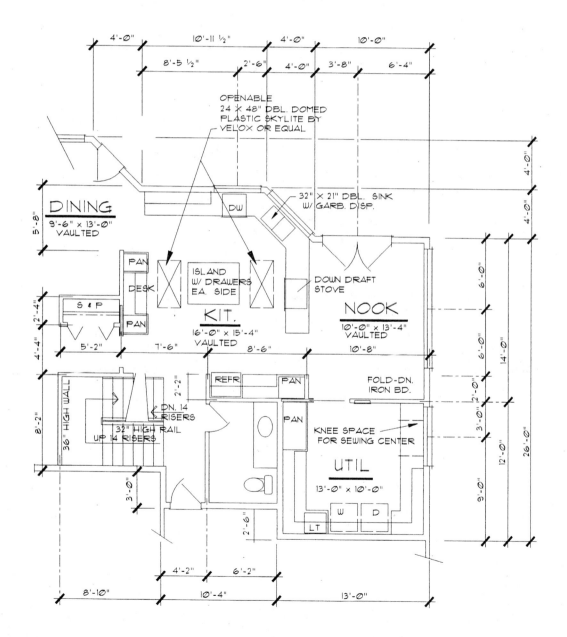

appliances are electrical. Use 2 ceiling-mounted fixtures in the nook, a surface-mounted fixture in the utility room, 8 can lites in the kitchen and 2 in the hallway, a wall-mounted fixture over the bathroom sink, and a ceiling-mounted fixture in the center of the bath. Save the drawing as a WBLOCK titled KITCHEN.

E-21-10. Open the FOOTING drawing and use it to create a WBLOCK drawing titled BASEFTG. Save the file with an appropriate title.

CHAPTER 21 QUIZ

1. How can a wblock increase drawing speed?

2. Why is using blocks more efficient than copying an object?

3. If a block is created on the 0 layer, where will the block be located when it is inserted?

4. When the color of a block is set "BY BLOCK," how will the block be affected?

5. What is the advantage of drawing blocks in 12″ boxes rather than full size?

6. List the prompts displayed to create a block.

7. How does the Insertion Point specified when the block is defined affect the insertion point requested when the block is inserted?

8. Once objects have been selected to form a block, what happens to the selected objects?

9. What effect will placing an asterisk in front of a block name have?

10. What command will display the dialog box for inserting blocks?

11. What command will allow for multiple blocks to be inserted?

12. What command will allow for editing blocks?

13. What is the quickest method for editing the same mistake in several blocks?

14. Explain the difference between a block and a wblock.

15. List the prompts displayed for inserting a Wblock into a drawing.

16. How can a block be mirrored before it is inserted?

17. List the options to preset block variables.

18. What method can be used to insert a block by picking two points?

19. What is the effect of activating the Explode box in the Insert dialog box?

_____

_____

20. What is the effect of FILEDIA on creating a WBLOCK?

_____

_____

_____

22

ADDING ATTRIBUTES
TO ENHANCE BLOCKS

• •

Even with all the qualities that Blocks and Wblocks add to a drawing, they still can be enhanced. Many blocks need text to explain or provide a specification. Text can be added to a block using three methods. In the last chapter, text was added to the drawings using DTEXT, and saved as part of the block. This works well if the text remains constant, such as the specification for a double kitchen sink. By adding the text to the block, the block must be exploded if the text is to be edited. The EXPLODE command allows the text to be edited, but characteristics of the block may be lost. Once a block has been inserted, text can be added to the drawing base to supplement the block by using the TEXT, DTEXT, or MTEXT commands. This option works well only if a few different blocks need written specifications.

This chapter will introduce you to a new, and typically the best method of adding text to a block by using Attributes. An Attribute is a grouping of entities that contains text. An example of a block with Attributes can be seen in Figure 22–1. These text groupings are similar to a block, which is then attached to a drawing block. Attributes have several features that make them superior to text created with TEXT, DTEXT, or Mtext and are well worth the effort to master. Once the benefits have been explored, thought must be given to defining the attribute text, controlling how the attributes will be displayed, changing the values, and extracting attributes from a drawing to compile a written listing of attributes.

THE BENEFITS OF ATTRIBUTES

The two major benefits of adding text through the use of Attributes are control and retrieval of information.

MAGNOLIA-GRAND-FLORA
SOUTHERN-MAGNOLIA
15 GAL. MIN
12' MIN HIGH
20' O.C.
24" MIN.DEEP

FIGURE 22–1 Attributes can be added to a block to explain qualities that could vary with each insertion.

Controlling Text

Using Attributes will allow for assigning prompts to a block to request information each time the block is displayed. For instance, each time the door block is inserted into a drawing, the following prompts could also be provided:

> size, thickness, type, manufacturer, finish, rough opening and hardware requirements.

Default values can be provided, but altered with each insertion. Using attributes also allows for correcting the text that is displayed without altering the block characteristics. These prompts may be hidden during the display or plotting process, even though the block they are attached to is displayed.

Extracting Attribute Information

In the example above, much of the information that will comprise the attribute is displayed in a door schedule. Attributes can be extracted from the drawing, stored in a separate file, and printed using a third party database text editor, such as the Microsoft Notepad. A list of attributes also can be compiled and printed within the drawing.

PLANNING THE DEFINITION

It is easiest to add attributes to a drawing before the drawing has been saved as a block. Once the symbols have been drawn, a decision must be made about what specifications will be associated with the block. For best efficiency this should be done in a planning session prior to entering the ATTDEF (ATTribute DEFined) or DDATTDEF command. For a block of a window, this might include the size, type, frame material, manufacturer, glazing, and rough opening. Other than the actual space within the drawing, there is no limit to the amount of information that can be associated with a block. As the attributes are being defined, an option is even available to keep the attributes invisible.

After the material to be listed with the block has been determined, decide how the prompts for each attribute will be displayed. Attribute prompts usually are in the form of a question, but statements also can be used. For instance, the prompt might read "What is the window size?" or "Provide the window size." Determine the prompts for each attribute that you will be listing.

Entering the Attribute Commands

Several commands will be used as you work with attributes. These include DDATTDEF (ATTribute DEFinition), DDATTE (ATTribute Edit), and DDATTEXT (ATTribute EXTract). Once the block has been drawn and you've decided what attributes are to be assigned, the process of attaching text to the block is started by using the DDATTDEF command. Each command for controlling attributes can also be completed at the command line using the ATTDEF (ATTribute DEFinition), ATTDISP (ATTribute DISPlay), ATTEDIT (ATTribute Edit), and ATTEXT (ATTribute EXTract) commands. Command line entry will be discussed after each dialog box is explored.

CREATING ATTRIBUTES USING A DIALOG BOX

Each of the qualities of an attribute can be controlled using the Attribute Definition dialog box. The box is accessed by selecting Define Attributes... from Block of the Draw pull-down menu, or by entering **AT** [enter] or **DDATTDEF** [enter] at the command prompt. This will display the dialog box shown in Figure 22–2. Major elements of the dialog box include the Mode, Attribute, Insertion Point, and Text Options boxes.

Attribute Display

The Mode box allows the display options for the attribute to be set. The initial prompt for ATTDEF provides four options for how the attributes will be displayed. These include Invisible, Constant, Verify, and Preset. Each of these options serve as a toggle switch with a default setting of Normal. If none of the options is selected the Normal setting will be kept. A Normal setting will prompt you for each attribute and the response will be displayed.

Invisible. Picking the Invisible check box will make the attributes invisible. Attributes will still be attached to the block when it is inserted, but they will not be displayed. If a separate window schedule is to be created, values can be invisible so they are not listed twice. The ATTDISP command, which will be introduced shortly, can be used to override the Invisible mode setting.

Constant. Picking the Constant check box indicates that all uses of the block will have the same attribute value. The value is entered as the Attribute is being defined and set throughout the use of the block. No prompt for new values will be given when

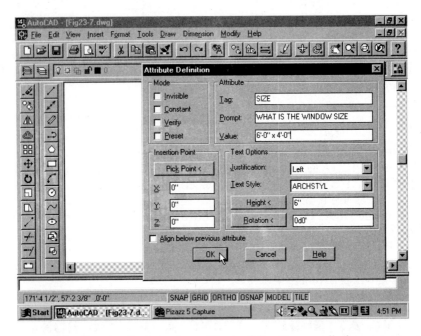

FIGURE 22–2 Selecting Attributes... from Block of the Draw pull–down menu or typing **DDATTDEF** [enter] at the command prompt will produce the Attribute Definition dialog box. The box can be used to design the tag, prompt, value and insertion point of each attribute as well as controlling how each will be displayed.

the block is inserted with this box active. Using the default of Normal will allow variables to be changed at a later time.

Verify.   By defining an attribute with this mode active, you will be given an opportunity to verify that the attribute value is correct during the insertion process.

Preset.   This mode will allow you to create Attributes that are variable but are not requested each time the block is inserted. When a block containing a preset Attribute is inserted, the Attribute value is not requested and is automatically set to its default value.

Assigning Values to Attributes

Once all of the display options are entered, the ATTDEF command will continue with prompts for defining Attribute values. The three prompts are the tag, prompt, and value and can be seen in Figure 22–3.

Attribute Tag.   The tag is used to identify an attribute definition in a similar way that a specific name is used to identify a block. The tag will identify each occurrence of the Attribute throughout the drawing. During the initial planning stage for the

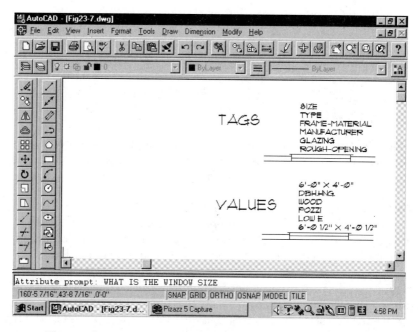

FIGURE 22–3 The attribute tags are displayed as the block is being inserted. The values are displayed once the block has been inserted.

window, it was determined that information regarding the size, type, frame material, manufacturer, glazing, and rough opening would be associated with each window block. Each of these listings can be used as an Attribute Tag, but each tag must be entered individually with the following prompts used to control only this tag. As the Tag is entered, it can be composed of any letters or characters except a blank space. A hyphen can be used between words to create a space. An appropriate response would be to move the cursor to the Tag: edit box and enter a tag such as **SIZE**.

Attribute Prompt.   The Attribute prompt: is the prompt you will see when the block is inserted. As you planned the Attribute, questions or statements were selected. The attribute prompt is where that prompt should be entered. An appropriate response would be to move the cursor to the Prompt: edit box and enter a prompt such as **WHAT IS THE WINDOW SIZE**. If the Prompt: edit box is left blank, the Tag (SIZE) will be used as the attribute prompt. If the Constant mode is active, the Prompt option will be inactive.

Attribute prompt: **WHAT IS THE WINDOW SIZE** ⏎

Default Attribute Value.   Once the tag and the prompt have been specified, the prompt will request a Value. The value that is entered at this prompt will be displayed as the default value as the block is inserted. A specific value such as 6'-0" × 4'-0" can be entered, or the null response can be used. If the Constant mode for the Attribute has been selected, a request for the prompt will not be made, and the prompt "Attribute Value:" will be displayed instead. An appropriate response would be to move the cursor to the Value: edit box and enter a response such as **6'-0" × 4'-0"**.

Text Options

Once the three Attribute prompts have been given, the prompts used to control the text can be set. Options include menus for justification and style, as well as edit boxes for entering the values for height and rotation.

Justification.   Selecting the Justification down arrow will display the menu shown in Figure 22–4. The options are the same as those used with the TEXT command. To replace the Left option, pick the desired option and the box will be closed. The name of the desired option will be placed in the text box.

Style.   Selecting the style option will produce a list of current drawing styles. To alter the style, pick the down arrow to display the style options, and select the name of the desired style. The selected style name will be placed in the text box, the menu will be closed, and the new style will become current.

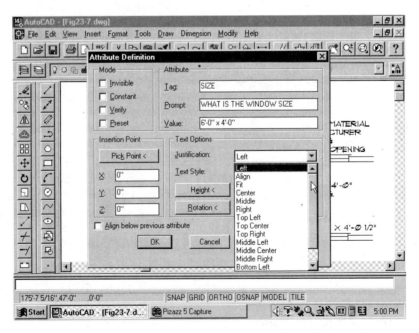

FIGURE 22–4 Options for justification can be displayed by selecting the down-arrow. The options are the same as those of the TEXT command.

Height. The attribute text height can be altered by placing the cursor in the text box and clicking with the mouse. Remove or edit the existing value and then provide the desired value by key board. The height value can also be altered by selecting the Height< button. If the Height< button is selected, the Attribute Definition box is removed and a prompt is displayed to allow points to be selected to indicate the text height. When the points have been provided, the dialog box will be restored and the indicated height will be displayed in the text box.

Rotation. The rotation angle for attributes is set in the same manner as the height. Enter the desired angle in the text box or pick the Rotation< button to pick the angle with the mouse.

Picking the Insertion Point

The Insertion Point box allows the location of the attribute to be selected. Selecting the Pick Point< button removes the dialog box and allows the insertion point to be selected using the mouse. If the exact location is known, coordinates can be entered using the X, Y, and Z text boxes. Once all of the options have been selected, picking the OK button will close the dialog box and display the tag at the selected insertion point similar to the display shown in Figure 22–5. Don't be discouraged at seeing the tags displayed as the command is completed. The tags will not be displayed as the block is inserted. If you are unsatisfied with the tags, they can be altered using the CHANGE or DDEDIT commands.

Aligning Additional Attributes

The Align below previous button will allow additional attributes to be placed below previous attributes using the same justification. The option is active when a check is placed in the box. With the option active, the Text options and Insertion Point areas will be inactive.

Repeating the Process

To add additional Attributes to the block, press the [enter] key and the entire process can be repeated. By repeating the process five more times, the tags for type, frame material, manufacturer, glazing, and rough opening can be added to the block.

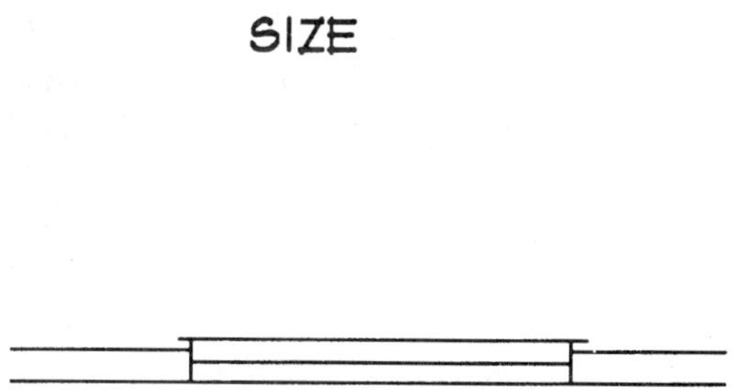

SIZE

FIGURE 22–5 The window block with the SIZE attribute displayed.

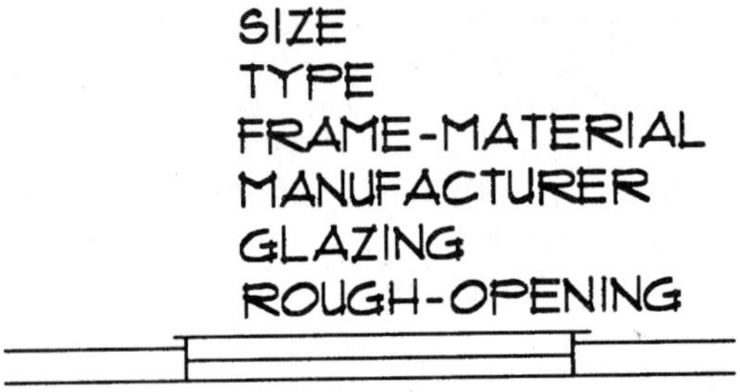

```
SIZE
TYPE
FRAME-MATERIAL
MANUFACTURER
GLAZING
ROUGH-OPENING
```

FIGURE 22–6 Repeating the procedure used to create the SIZE tag allows an unlimited number of tags to be attached to a block.

Repeating the entire sequence will produce a display similar to the one in Figure 22–6.

CREATING ATTRIBUTES AT THE COMMAND PROMPT

An attribute can be created at the command prompt using the ATTDEF command. The command is started by entering **–AT** [enter] or **ATTDEF** [enter] at the command prompt. Prompts will be given to specify the definition. The definition will include a *tag*, a *default* for the attribute value, and information about the placement, size, and style of text used to display the attribute value. The ATTDEF command sequence is:

```
Command: –ATTDEF [enter]
Attribute modes—Invisible: N Constant: N Verify: N Preset: N
Enter (ICVP) to change, RETURN when done:
Attribute Tag:
Attribute prompt:
Default attribute value:
```

Attribute Display

The initial prompt for ATTDEF provides four options for how the attributes will be displayed. These include:

```
Attribute modes—Invisible: N Constant: N Verify: N Preset: N
Enter (ICVP) to change, RETURN when done:
```

Each of the options serves as a toggle switch with the default setting of No. By pressing the RETURN key, the Normal display will be provided showing each attribute, and prompts will be given for each one. If **I, C, V,** or **P** is entered at the command prompt, the corresponding mode is reversed. The attribute mode line is then redisplayed with the N changed to Y and the prompt is repeated. The process can be repeated or a different value can be adjusted. Only one mode can be adjusted at a time. The process is ended by pressing the [enter] key.

Invisible.   Entering **I** [enter] at the command prompt will make the attributes invisible.

Constant.   Entering **C** [enter] at the command prompt indicates that all uses of the block will have the same attribute value. The value is entered as the Attribute is being defined and set throughout the use of the block. No prompt for new values will be given when the block is inserted if you enter **C** for Constant. Using the default of N will allow variables to be changed at a later time.

Verify.   By defining an attribute with this mode set to Y, you will be given an opportunity to verify that the attribute value is correct during the insertion process.

Preset.   This mode will allow Attributes that are variable but are not requested each time the block is inserted. When a block containing a preset Attribute is inserted, the Attribute value is not requested and is automatically set to its default value.

Assigning Values to Attributes

Once all of the display options are entered, the ATTDEF command will continue with prompts for the tag, prompt, and default attribute value.

Attribute Tag.   The tag is used to identify an attribute definition. The command prompt might resemble:

Attribute tag: **SIZE** [enter]

Attribute Prompt.   The Attribute prompt: is the prompt you will see when the block is inserted. The response might resemble:

Attribute prompt: **WHAT IS THE WINDOW SIZE** [enter]

A null prompt can be entered by pressing the [enter] key, which will use the Tag (SIZE) as the Attribute prompt.

Default Attribute Value.   Once the tag and the prompt have been specified, the prompt will request a Default Attribute Value. A specific value can be entered, or the null response can be used. If the Constant mode for the Attribute has been selected, a request for the prompt will not be made, and the prompt "Attribute Value:" will be displayed instead. The prompt line for the block being created could be:

Default attribute value: **6'-0" × 4'-0"** [enter]

Text Prompts. Once the three Attribute prompts have been given, the sequence will continue with three prompts that are used with the TEXT command. These prompts are:

Justify/Style/<Start point>:*(select start point)*
Height <6>:
Rotation angle <0>:

Respond to each of these prompts as you would if you were using the TEXT, DTEXT, or MTEXT commands. The only difference is that after completing these three prompts, you will not be given a prompt for the text to be inserted. This will happen as the block is inserted. As the three prompts are completed, the command prompt will be returned, and the tag will be displayed by the drawing, as seen in Figure 22–7.

Repeating the Process

To add additional Attributes to the block, press the [enter] key and the entire ATTDEF sequence will be repeated. By repeating the process five more times, the tags for type, frame material, manufacturer, glazing, and rough opening can be added to the block. As the process is repeated, pressing the [enter] key at the "Text Start point"

FIGURE 22–7 Attributes can be created at the Command line by using the ATTDEF command.

prompt will automatically place additional Tags directly below existing Tags with the proper spacing. Repeating the entire sequence will produce a display similar to the one in Figure 22–8.

Adding Attributes to Existing Blocks

If a block has already been created, attributes can still be added to it. The original block name must be changed, however, to avoid an error message. If the existing POTTY block is inserted into a drawing, and attributes are added to it, it cannot be saved using the original name of POTTY. Attempting to do so would produce the following message:

Command: **–B** [enter]
Block name (or ?): **POTTY** [enter]
Block POTTY already exist.
Redefine it? <N>: **Y** [enter]
Insertion base point:
Select objects: (*Select desired objects and attributes.*) [enter]
Block POTTY references itself
* Invalid*
Command:

By entering a new name, the existing block can be used with the desired attributes. Many companies will use a name such as POTTY-A using the -A to indicate that the block contains attributes. It's much easier to add the attribute prior to saving the block.

Controlling Attribute Display

Rather than wasting disk space by storing POTTY and POTTY-A to distinguish between blocks with and without attributes, the ATTDISP (ATTribute DISPlay)

FIGURE 22–8 The window on the left shows how the tags will be displayed as the block is created. The window on the right shows how the values will be displayed, once the block is inserted into a drawing

command can be used to hide attributes if you decide that they should not be displayed. This command sequence is:

Command: **ATTDISP** [enter]
Normal/ON/OFF <Normal>:

Normal will display all attributes, ON will display the variables, and OFF will make all attributes invisible.

The ATTREQ (ATTribute REQuest) command will allow attributes to be dislayed, but will suppress the sttribute prompts. Some blocks such as kitchen sink typically retain their default values with very little variation. When the prompt is not required for attributes, the ATTREQ command can be toggled from 1 to 0. The 0 setting will display all attributes in their default setting.

ALTERING ATTRIBUTE VALUES PRIOR TO INSERTION

The Edit Attributes dialog box similar to the box shown in Figure 22–9 can be displayed during the INSERT process by typing **ED** [enter] or **DDEDIT** [enter] at the command prompt by picking the Edit Text icon of the Modify II toolbar or by selecting Text... from Object of the Modify pull-down menu. This dialog box can be used to dis-

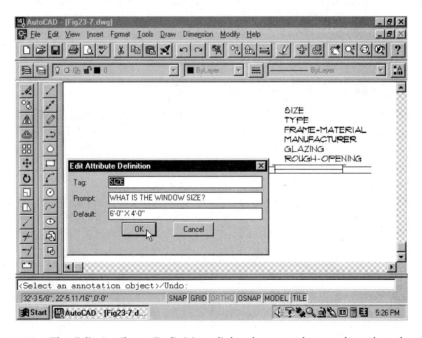

FIGURE 22–9 The Edit Attribute Definition dialog box can be used to alter the tag, prompt, and default value for an attribute before it is saved with a block. The dialog box is accessed by typing **DDEDIT** [enter] at the command prompt.

play and enter Attribute values. The default values are listed to the right side of the current prompts. Displayed values can be altered using methods similar to the methods used for other dialog boxes.

The Modify Attribute Definition dialog box can also be used to edit attributes. The box can be displayed by **MO** [enter] or **DDMODIFY** [enter] at the command prompt, selecting Properties from the Object Properties toolbar, or by selecting Properties... from the Modify pull-down menu. Figure 22–10 shows an example of the Modify Attribute Definition dialog box. It can be used to edit the color, layer, linetype, and each of the options defined as the attribute was created.

Attributes can also be edited prior to saving the block using the CHANGE command. CHANGE allows the insertion point, text style, height, rotation angle, tag, prompt, and default value to be edited. The command is accessed by typing **CH** [enter] or **CHANGE** [enter] at the command prompt. The sequence to edit one or more of the attribute options is:

> Command: **–CH** [enter]
> Select objects: *(Select attribute to be edited.)*
> Select objects: [enter]
> Properties/<Change point>: [enter]

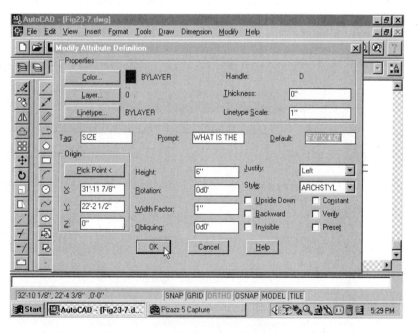

FIGURE 22–10 The Modify Attribute Definition dialog box can be used to edit the color, layer, linetype and each of the options of an attribute. The box is accessed by selecting Properties from the Object Properties toolbar, by selecting Properties... from the Modify pull-down menu or by typing **MO** [enter] or **DDMODIFY** [enter] at the command prompt.

Enter text insertion point: (Pick a new insertion point with the mouse or press
  [enter].)
Text style: STANDARD
New style or press ENTER for no change: (Enter a new style name and press [enter]
  or press [enter] to accept the existing style).
New height: (Enter a new height and press [enter] or press [enter] to accept the
  existing height.)
New rotation angle <0>: (Enter a new angle and press [enter] or press [enter] to
  accept the existing angle.)
New tag <SIZE>: (Enter a new tag and press [enter] or press [enter] to accept the
  existing tag.)
New prompt <WHAT IS THE WINDOW SIZE>: (Enter a new prompt and press
  [enter] or press [enter] to accept the existing prompt.)
New default value <6'–4" × 4'–0">: (Enter a new value and press [enter] or press
  [enter] to accept the existing default value.)

INSERTING A BLOCK WITH ATTRIBUTES WITH A DIALOG BOX

The DDINSERT command can be used to insert Attributes into a drawing. By typing
I [enter] or **DDINSERT** [enter] at the command prompt, a dialog box similar to Figure
22–11 will be displayed. The Insert dialog box can also be displayed by picking the

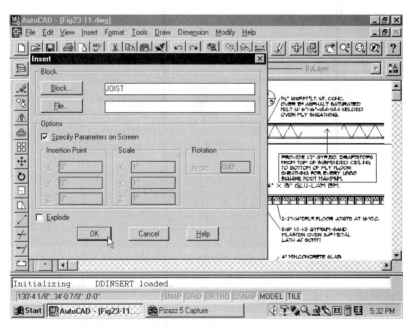

FIGURE 22–11 Entering **I** [enter] or **DDINSERT** [enter] at the keyboard will allow blocks to
be inserted using a dialog box. (Courtesy StructureForm Masters, Inc.)

Insert Block icon from the Draw toolbar or by picking Block... from the Insert pull-down menu. By picking the Block... select box in the Insert dialog box, a listing of blocks contained on the current drawing will be displayed as seen in Figure 22–12. Picking the File... select box will display a listing of drawing files so that Wblocks can be inserted into the current drawing. By entering the name of an existing block or wblock, that name will be displayed in the display box to the right side of the Block... prior to insertion. Notice that the default setting of the Options box is active, indicating "Specify Parameters on Screen" with the Insertion Point, Scale, and Rotation options grayed out. Once the name of the Block to be inserted is displayed, select the OK box. The BLOCK will be dragged into the drawing area, allowing for insertions and specification of the scale, rotation angle, and attribute values. By high-lighting the "Select Parameters on Screen" box and making it inactive, the select boxes for Insertion Point, Scale, and Rotation will switch from gray to black, allowing values for each to be specified using the dialog box.

INSERTING ATTRIBUTES FROM THE COMMAND LINE

Blocks with attributes can be inserted into a drawing from the command line using the INSERT command.

The easiest method of saving attributes is to save them with the block as the block is created. This can be done by drawing all of the entities to be contained in the block

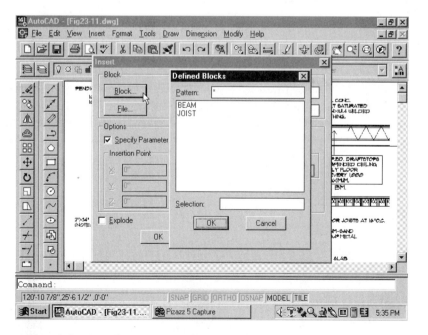

FIGURE 22–12   Selecting the BLOCK... box will display a listing of all blocks contained in the drawing. (Courtesy StructureForm Masters, Inc.)

and then assigning all attribute tags and variables. Once this is done, enter the BLOCK, BMAKE, or WBLOCK commands and select the desired entities and attributes to be included with the BLOCK. With the block saved with its attributes, it can be inserted using a similar insertion sequence for a block without attributes.

As the block is inserted, the initial insertion request of X and Y scale factors, and the rotation angle, will be followed with the prompts that were entered as the Attribute was defined. For the WINDOW block defined earlier in this chapter, the sequence is:

Command: **–I** [enter]
Block name (or ?): **WINDOW** [enter]
Insertion point: *(Select point.)*
X scale factor <1> / Corner /XYZ: [enter]
Y scale factor (Default = X): [enter]
Rotation Angle <0>: [enter]
Enter attribute values
What is the window size <6'-0" × 4'-0">: [enter]
What type window <DBH.HNG.>: [enter]
What is the frame matl. <WOOD>: [enter]
Manufacturer <POZZI>: [enter]
What is the glazing < LOW E>: [enter]
What is the rough opening <6'-0 ½" × 4'-0 ½">: [enter]
Command:

The results of the sequence can be seen in Figure 22–7.

EDITING ATTRIBUTES AT THE COMMAND PROMPT

Once attributes are integrated into a block, they must be edited using the ATTEDIT (ATTribute EDIT) command by typing **ATE** [enter] or **ATTEDIT** [enter] at the command prompt or by selecting Global from the Attribute cascading menu of the Object cascading menu in the Modify pull-down menu. This will produce the prompt:

Command: **–ATE** [enter]
Edit attributes one at a time? <Y>

The Y or N response will determine whether attributes will be edited globally or individually. Either choice will require specifying the set of attributes to be edited. Only attributes that are parallel to the current UCS may be edited. Attempting to edit nonparallel attributes will produce the prompt:

0 attributes selected, N attributes not parallel with UCS
INVALID

Restricting Attribute Editing

After Y or N has been selected for global or individual editing, restrictions can be set for selecting attributes to edit. Restrictions can be set so that only Attributes with a specific tag, value, or name are selected. Restrictions are prompted by:

    Block name specification <*>:
    Attribute tag specification <*>:
    Attribute value specification <*>:

Specific sets of Block names, Attribute tags, and Attribute values, including wild card characters, can be entered to define the selection set for editing. For an attribute to be selected for editing, the Block name, Attribute tag, and Attribute value must all match the name, tag, and value entered for the selection set. Be sure to match the case of type used for the selection sets to the case used when the attributes were entered originally. Matching uppercase titles with lowercase titles will not be accepted as matching sets.

To edit all Attribute tags and values in all defined blocks, press the ⏎ key. Use of a * or ? wild card symbol will allow more than one string to match.

Global Editing

Entering N at the initial prompt sets the Attribute editing to Global. This will allow the same attribute to be edited in all of its applications throughout a drawing. This can be useful if you find an error in a block value that has been inserted several times. Rather than changing each attribute individually, you can change all of them using the global option. The command sequence is:

    Command: –**ATE** ⏎
    Edit attributes one at a time? <Y>: **N** ⏎
    Global edit of attribute values.
    Edit only attributes visible on screen? <Y>:

Entering an N will display the AutoCAD Text window with the following message:

    Drawing must be regenerated afterwards.

Any changes that are made to attributes will not be reflected until the drawing is regenerated. If AUTOREGEN is set to ON, a drawing regeneration will be performed when the editing of this command is complete.

Editing Visible Attributes.   If a Y is accepted at the prompt and only the blocks shown on the screen are to be edited, the prompt will read:

    Select attributes:

Attributes may be selected using Box, Window, Crossing, WPolygon, or CPolygon. Once selected, attributes will appear highlighted. When the selection process is complete, the prompt will change to:

 String to change:
 New string:

The "String to change" prompt initiates a search of all eligible attribute strings for matching strings. Each matching string will be changed to reflect the entry for the "New string" prompt. A block titled FOOTING with a tag of DIA. and a value of 17″ diameter × 8″ could be changed to 18″ by entering:

 String to change: **17** ⌨enter⌨
 New string: **18** ⌨enter⌨

Entering ⌨enter⌨ at the "String to change" prompt will cause the entry at the "New string" prompt to be placed ahead of all eligible Attribute value strings. If a block is specified as SINK with an attribute value of DBL. SINK, it can be changed to 32 × 21 DBL SINK by entering:

 String to change: ⌨enter⌨
 New string: **32 × 21** (*space*) ⌨enter⌨

Be sure to use the space bar to provide a space after the 21 or the new display will read 32 × 21DBL SINK.

Editing All Attributes.   If an N is accepted at the prompt, all blocks in the drawing will be edited.

Individual Editing

Responding to the initial ATTEDIT prompt with a Y will allow individual attributes to be edited. The initial prompt will be followed with a request to

 Select Attributes:

The selection set can be limited by selecting Blocks, Tags, or Values using methods similar to those used for global editing. Each selected attribute will be marked with an X in the lower left corner of the attribute allowing any of its properties to be altered. When the selection set is completed, the prompt will read:

 Value/Position/Height/Angle/Style/Layer/Color/Next<N>:

If the original text was entered using the Aligned or Fit text options, the Angle option will not be displayed. The Height option is omitted if the original text was entered using the Aligned placement method. Any of the applicable options can be selected by entering the first letter of the corresponding option. Accepting the default option of N will move the X to the next Attribute. For each of the other options, you will be prompted for a new value. With the exception of the Value option, the ATTEDIT options are similar.

Value.   This option will allow the value of an attribute to be changed. The prompt is:

Change or Replace? <R> :

By selecting the default option, a second prompt will be displayed asking for:

New Attribute value

allowing for a new value to be substituted for the existing value. Selecting the Change option will allow for minor corrections to be made to the existing value. Entering **C** [enter] at the "Change or Replace?" prompt will produce two additional prompts:

String to change:
New string:

The first prompt allows for specifying the string of characters that will be revised. The second prompt requests how the revised prompt will appear. If the * and the ? symbols are used in the new attribute string, they will be interpreted literally rather than as wild card symbols.

Position/Height/Angle/Style/Layer/Color.   Each of these options can be selected by entering the first letter of the desired option. When the Position option is selected, prompts for a new Starting, Center, or Endpoint will be given, depending on whether the attribute is left justified, centered, or right justified. Both ends of the Attribute line will be requested if Aligned text was used.

Changing the Attribute.   As the Attribute option is provided, the Attribute will be redrawn to reflect the current change, but the option prompt will be repeated in case further changes are desired. The X will remain on the current Attribute until the N option is used to move to a new attribute. Entering the N option after editing the last selected Attribute will exit the ATTEDIT command. ESC also can be used to terminate the command.

EDITING ATTRIBUTES USING A DIALOG BOX

Once the values for an attribute have been assigned to a block they can be altered using the Edit Attributes dialog box. The dialog box is activated by selecting the Edit Attribute icon from the Modify II toolbar, by selecting Single... from Attribute of the Object cascading menu in the Modify pull-down menu, or by typing **DDATTE** [enter]

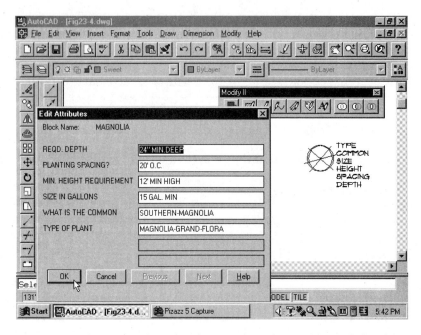

FIGURE 22–13a Attributes can be edited using the Edit Attributes dialog box which is accesed by typing **ATE** (enter) or **DDATTE** (enter) at the command prompt or by picking the Edit Attributes icon from the Attributes toolbar. Notice that the values for the depth, spacing, height and size have been altered.

at the command prompt. As the command is entered, you'll be prompted to select a block to be edited. Once the block is selected the Dialog box shown in Figure 22–13 is displayed to allow each attribute value to be altered. Text is altered by moving the cursor to the desired edit box and inserting the desired text. Selecting the OK button will alter the values displayed on the screen. Selecting the Cancel button will remove the dialog box with no changes made to the drawing base.

FIGURE 22–13b The effects of altering the values in the Edit Attributes dialog box in Figure 22–11a. (Courtesy Tereasa Jefferis.)

EXTRACTING ATTRIBUTES

One of the most useful benefits of using Attributes is the ability to extract information stored as an Attribute and assemble it in a written list. Window, door, and finish schedules typically are associated with architectural drawings. Pier, steel, and beam schedules are but a few of the types of written engineering drawings. Figure 22–14 shows portions of a floor plan and the doors that will be extracted. The doors shown are a fraction of the total number of doors that would be represented on a complete multi–level structure. Each door was inserted into the floor plan as a block with one visible and four invisible attributes. Attributes have been assigned to represent symbol, size, style, manufacturer, price, and quantity. By assigning these values as attributes, an accurate schedule can be maintained and quickly updated. By using the DDATTEXT (Dialog Display ATTribute EXTraction) or ATTEXT (ATTribute EXTraction) command, a complete listing of all of the doors used throughout the project can be written to file. Figure 22–15 shows an example of the Attribute Extraction dialog box. Figure 22–16 shows an example of a sample listing of door values. The heading above each column will not be listed to the extract file by the ATTEXT command. When operated on by a data base program such as Notepad, the values listed under each heading will be extracted and used to sort doors by symbol, size, price, style or manufacturer.

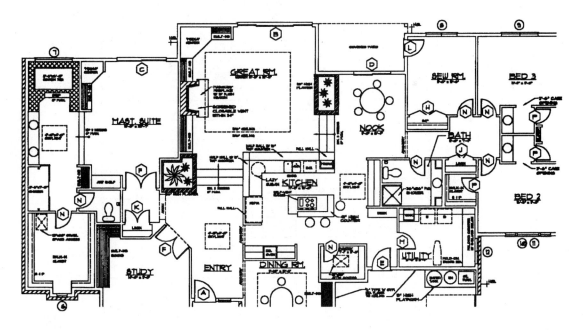

FIGURE 22–14 Extracting attributes can be very useful on drawings containing multiple blocks with attributes such as this floor plan.

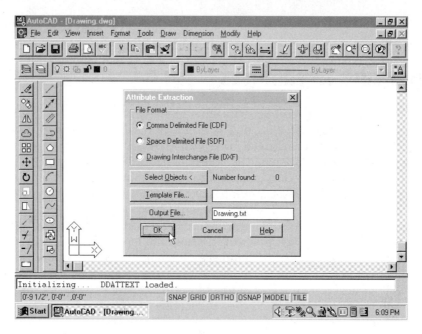

FIGURE 22–15 The Attribute Extraction dialog box can be used to control the steps for extracting attributes from a drawing.

DOOR SCHEDULE			
SYM.	SIZE	TYPE	QUAN.
A	3'-6" x 8'-0"	S.C.,R.P. W/ 24" SIDELIGHT W/ SQ. TRANSEM ABOVE	1
B	12'-0" x 8'-0"	SL. GLASS W/ 12'-0" x 3'-0" SQ. TRANSEM ABOVE	1
C	8'-0" x 6'-8"	SL. FRENCH DRS. W/ 12" SQ. TRANSEM ABOVE	1
D	6'-0" x 6'-8"	SL. GLASS DRS. W/6'-0" x 3'-0" SQ. TRANSEM ABOVE	1
E	3'-0" x 6'-8"	SELF CLOSING, M.I.	1
F	PR 2'-6" X 6'-8"	HOLLOW CORE	2
G	6'-0" x 6'-8"	SL. MIRROR	1
H	6'-0" x 6'-8"	BIFOLD	1
J	4'-0" x 6'-8"	BIFOLD	1
K	PR 2'-0" x 6'-8"	HOLLOW CORE	1
L	2'-8" X 6'-8"	S.C., M.I.	1
M	2'-8" x 6'-8"	HOLLOW CORE	1
N	2'-6" x 6'-8"	HOLLOW CORE	7
P	2'-4" x 6'-8"	HOLLOW CORE	3
Q	16'-0" x 8'-0"	OVERHEAD, GARAGE	1
R	9'-0" x 8'-0"	OVERHEAD, GARAGE	1

FIGURE 22–16 A list of doors values to be extracted using the DDATTEXT command.

Using a Data Base

A data base program can be used to manipulate the data stored in block attributes. A listing of doors or windows for a multi–level structure is a data base that has essential data associated with it—such as size, type, rough opening, etc. Two terms generally associated with each data base are record and field. A list of each of the construction components related to doors would be a record. The door, frame material, hinge type, and finishing method are each examples of record. A door might be swinging, sliding, folding, or overhead. Each listing of a door is a record. All of the doors that are swinging are a field. Swinging doors can be divided by interior, exterior, raised panel, slab, 1-lite, or 10-lite. It is possible to take a listing of doors that are listed by size, and by using a data base, to reorder the listing by type. The ATTEXT command of AutoCAD allows a list to be generated that contains all of the desired objects to be manipulated by the data base. ATTEXT creates an extract file that allows AutoCAD to find and list attributes in the templet file.

Using a Templet

A templet is a file that lists the fields that specifies the tags and determines what blocks will have their attribute data extracted. The templet file can be made using a data base, word processor, or text editing program. The balance of this chapter will rely on Notepad. The templet must be a file on an accessible path with a .TXT extension. If a text editor or word processing program in the ASCII mode is used, you must add this extension when you name the file. A templet is composed of a field name and a character-numeric value.

Field Name.   The Field name must be exactly the same in every respect to the tag in order for that tag to be extracted. If the attribute tag is SLDG, then there must be a field name in the templet titled SLDG. A field name of SLDG will not have the tag of SLIDING extracted.

Character-numeric.   The ATTEXT command uses the templet to classify the data written to a particular field either a numeric or as character type. Characters such as a, b, c, A, #, and " are always character type, but numbers do not have to be numeric type. Characters occupy less memory space than numeric values. Numbers that represent a width or height are better stored as characters, unless they are to be used in mathematical functions. The only significance of numbers as characters is in their order (456...) for sorting purposes by the data base program. Characters have the same significance.

Templet Components.   The extraction templet has two elements for each field. These are the field name and the character-numeric element. Numbers in strategic spaces in the character-numeric element of the templet file specify the number of spaces to allow for the values to be written in the extract file. Others specify many decimal places to carry numeric values. The format is:

fieldname	Nwwwddd	for numeric values with decimal allowance
fieldname	Nwww000	for numeric values without decimal allowance
fieldname	Cwww000	for character values

Each line on the templet is one field. The C and N's represent character or numeric information to be extracted. If a character other than a number is to be extracted, use the C. If a numeric entry such as the price will contain the dollar sign, use a C. The w's and d's represent the necessary digits when the templet is created. The first three numbers after the letter represent the number of spaces that will be reserved for the attribute. The last three numbers specify the number of decimal spaces in the value. A listing of N010000 represents a numeric entry with 10 spaces allotted to the attribute, and no space reserved for decimal points. A entry of N006002 would represent a numeric entry of six spaces with two decimal places. The order of fields listed in the templet do not have to match the order that attributes tags appear in the block. Any group of fields in any order is acceptable. The only requirement for a block to be eligible for extraction by the DDATTEXT or ATTEXT command is that there is a least one tag–field match. A templet file for the example in Figure 22–14 could be written as follows:

SYMBOL	C002000
TYPE	C008000
SIZE	N012000
STYLE	C008000
MANUF	C010000
PRICE	N006002
QUANTITY	C002000

From the list of doors shown in Figure 22–16, a data base program can generate a sorted list with items that correspond to the selected values. To create a door schedule, listings would be sorted based on the symbol. To create a bill of materials to order doors the listing could be sorted by the quantity or the price.

Non–attribute Fields. Additional information is automatically stored with each block as it is inserted. This can be seen by entering the LIST command and selecting a block. Information about the block handle, insertion point for the X, Y and Z coordinate, layer, level, name, number, orientation, scale value for X, Y, and Z, and extrusion direction in X, Y, and Z. These values can be used to manipulate attributes. In addition to listing attributes, the templet file can be designed to extract block properties that include:

Handle–An identifier for a block.

Layer–The name that the layer is inserted on.

Level–Block nesting level.

Name–Block name.

Number–Counts the number of insertions.

Orient–The rotation angle of the block

X, Y, or Z. The appropriate coordinate of the insertion point.

XSCALE, YSCALE, ZSCALE–The appropriate X, Y or Z insertion scale factor.

XEXTRUDE, YEXTRUDE, ZEXTRUDE – the appropriate X, Y, or Z extrusion direction.

In creating a templet, an entry might include BL:ORIENT N010000 to list all block data by orientation, allowing 10 spaces for the width of the field with no decimal places in the value. An entry of BL:PRICE C006002 will sort all blocks by price, allowing six spaces for the value to be written including two decimal places. If the value to be written exceeds the width allowed, AutoCAD truncates the written data, proceeds with the extraction, and displays the following error message:

**Field overflow in record <record number>

Creating the Templet

Now that the tools have been discussed, a templet can be created to manipulate the attributes. Start by opening a drawing with blocks and attributes. From the Start menu, select Programs, Accessories, and then Notepad. The display will now look like Figure 22–17. Enter the required text at the flashing cursor. Use the Tab key to place a space between **each** portion of the listing. The entry will resemble Figure 22–18. When the entry is complete, use the SaveAs option from the File pull-down menu to save the file. Use the Exit option of File menu to exit the Notepad and return to AutoCAD.

Extracting Attributes from the Command Line

Once a templet has been created, attributes can be extracted from blocks using the ATTEXT command. Entering **ATTEXT** ⏎ at the command prompt will produce the following prompt:

Command: **ATTEXT** ⏎
CDF,SDF, or DXF Attribute extract (or Objects)? <C>:

File Format.   The default setting for the file format is the Comma Delimited File (CDF). Additional radio buttons allow for the selection of Space Delimited Files (SDF) or Drawing Interchange Files (DXF). CDF format will generate a file containing no more than one record for each block reference in the drawing. The values written under the fields in the extract files are separated by commas instead of spaces. The character fields are enclosed in single quotes.

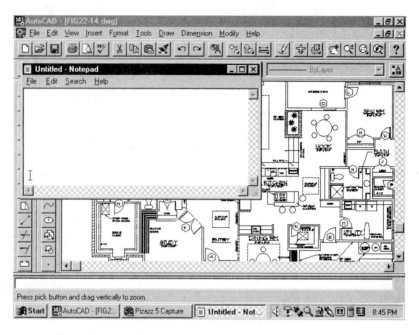

FIGURE 22–17 A word processing or data base program can be used to generate a templet for extracting attributes. Notepad can be used by selecting the Start button, followed by selecting Programs, Accessories, and then Notepad.

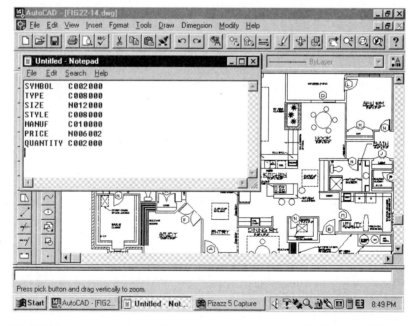

FIGURE 22–18 The templet for attribute extraction must include the field name and a character-numeric entry.

SDF format writes the values lined up in the widths allowed. It is possible for adjacent mixed fields of characters and numeric entries not to have spaces between them. Dummy fields such as BLANK or SPACE can be added to provide space between fields. To select this setting enter **S** ⏎ at the prompt.

DXF format will generate files contain only the block reference, the attribute and end-of-sequence entities. To select this setting enter **D** ⏎ at the prompt.

Because it tends to be the easiest to use, enter **S** ⏎ at the prompt.

>   Command: **ATTEXT** ⏎
>   CDF, SDF, or DXF Attribute extract (or Objects)? <C>: **S** ⏎

Once the format is selected, the Select Templet File dialog box will be displayed allowing the desired .TXT file to be selected. In Figure 22–19, the DOORSYM.TXT file has been selected for extraction. Pressing the ⏎ key will remove the current display and produce the Create Extract File dialog box. Notice in Figure 22–20 that the current drawing name is the default name for saving the file. Alter the name to be used to save the extracted files so that it is different from the templet file. With the name entered, press the ⏎ key to extract the files. The number of records in the Extract file will be displayed, and the command prompt will be restored.

Return to Notepad and open the extract file. Figure 22–21 shows the file that was created throughout this chapter.

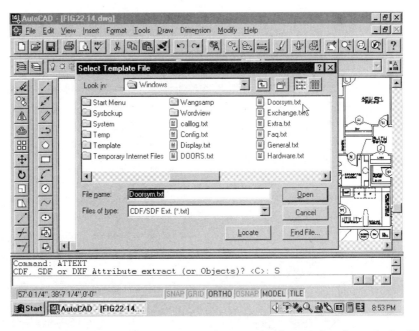

FIGURE 22–19 Using the ATTEXT command, once the format is selected, the Select Templet File dialog box will be displayed.

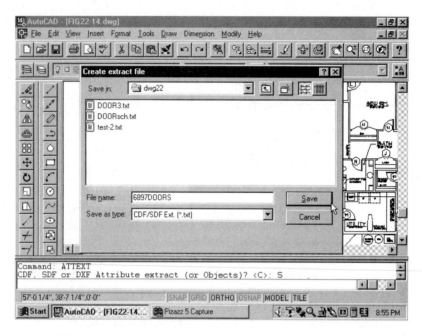

FIGURE 22–20 The Create Extract dialog box allows a new file name or the existing name to be used.

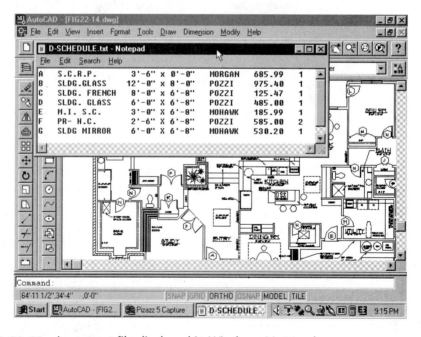

FIGURE 22–21 An extract file displayed in Windows Notepad.

Extracting Attributes with a Dialog Box

Once a templet has been created, attributes can be extracted from blocks using the DDATTEXT command. Entering **DDATTEXT** [enter] at the command prompt will produce the Attribute Extraction Dialog box that was introduced in Figure 22–15. The three major areas of the dialog box allow for selecting the file format, selecting the templet file and the Output file.

Templet Files. Once the format has been selected, the templet file for attribute extraction can be selected using the Selecting the Templet File... button. The button will be deactivated if the DXF format has been selected. Selecting the Templet File... button will display a Templet File dialog box similar to Figure 22–19, allowing the DOOR.TXT templet to be selected. Selecting Open will return the Attribute Extraction dialog box.

Output Files. Selecting the Output... button allows the method of output to be determined. The default file name is shown in the Output file edit box with a .TXT extension. The extension can be altered to .CON for writing to the screen, or .PRN for writing to the printer. A printer must be connected and ready for printing for the .PRN to be used.

Select Objects. With the method of output selected, picking the Select Objects< button will remove the dialog box and return you to the drawing screen. Select the attribute entities, and the dialog box will reappear. The number of blocks you selected is now displayed after the Number Found: entry. If no objects are selected, all blocks in the drawing specified by the templet file will be used. With the process complete, return to notepad to view the extract file that was created.

CHAPTER 22 EXERCISES

E-22-1. Draw an 18″ diameter footing using dashed lines and show a 4 × 4 post in the center of the pier. Use 5″ high text to install the following Attributes.

TAG	PROMPT	VALUE
Dia.	What is the diameter?	18″ dia.
TAG	PROMPT	VALUE
Depth	What is the pier depth?	12″ min. into grade
Strength	What is the conc. strength?	2500 p.s.i.
Rebar	Will the pier be reinforced?	10 × 10-4 × 4 wwm 3″ up
Post	Size of post to be supported?	4 × 4 D.F.L. #1
Anchor	Will an anchor be used?	Simpson CB44 base

Save the drawing as a WBLOCK on a floppy disk as FNDPIER.

E-22-2. Draw an elevation of a 3'-0" × 4'-6" double-hung window and use 5" high text to install the following Attributes. Set the mode to invisible display.

TAG	PROMPT	VALUE
Manuf.	List the manuf.	POZZI
Frame	List the frame size	3'-0" × 4'-6"
Unit	List the unit size	UN=3'-2⅝" × 4'-8⅞"
R.O.	List the rough opening	R0=3'-0½" × 4'-6½"
Frame	Frame material	Kiln-dried fir
TAG	PROMPT	VALUE
Exterior	Exterior finish	Prime ext. face
Sash	Sash material	1¹¹⁄₁₆" thick sash
Glaze	Glazing material	⁹⁄₁₆" air space
Weather	Weather stripping	foam-backed
Hardware	Hardware color	Bronze
Screen	Screen material	Alum.-bronze tint
Assembly	Where assembled	Site assembled
Shipping	Destination	Seattle, WA.

Prior to creating a wblock of this material, change the Frame material to Kiln-dried Western pine. Change the hardware to WHITE. Ship from Bend, OR. Change the Assembly point to Factory. Save the drawing as a WBLOCK on a floppy disk as EL3-46DH.

E-22-3. Open a new drawing and draw a 24" × 24" skylight in plan view. Add the following Attribute tags, prompts, and values.

TAG	PROMPT	VALUE
Size	Unit size	24" × 24"
Manuf.	Who is the manuf.?	Velux
Model	Model number	SF
Frame	Frame material	Alum.-clad frame
Finish	Finish material	Lacquered finish
Glass	Glazing	Double-tempered glazing
Flashing	Flash material	32ga. lacquered alum.

Save the drawing on a floppy disk as a wblock titled SKYLIGHT. Insert the skylight into a drawing three different times. Change the size so that the skylight will be 24 × 24, 24 × 48, and 48 × 48. Change the size value of each skylight to reflect the proper size. Use global editing to change the model value to FS, change the frame to Alum-clad Frame, and change the finish to 22ga. Save the drawing as E-22-3.

E-22-4. Open drawing E-21-5 and insert the POTTY block. EXPLODE the block and add the following Attribute tags, prompts, and values.

TAG	PROMPT	VALUE
MANUF	Who is the manuf?	Am. Std.
Cat-no	What is the Catalog number?	2006.014
Descrip.	What is the description?	Elon. Lexington
Color	What color will be used?	White

Supplier	Who will provide fixture?	General cont.
Installer	Who will install fixture?	Plumber
Hardware	Who will supply hardware?	Owner

Save the drawing as a wblock titled POTTY.

CHAPTER 22 QUIZ

1. How are Attributes saved?

2. What commands can be used to add a Block or Wblock to a drawing?

3. What must be done to the Block to add attributes to an existing block?

4. Describe what an Attribute prompt is.

5. What is an Attribute Tag?

6. List the four modes for Attribute display and give a brief explanation of each.

7. What is the initial step taken to assign Attributes to Blocks?

8. List the commands that produce dialog boxes that can be used with Attributes.

9. List two methods to edit an Attribute.

10. What ATTDIA setting is required to produce dialog box displays?

11. What two commands allow for Attribute editing?

12. How can global editing of Attributes be limited?

13. How do the prompts "String to change" and "New string" affect each other?

14. What options are provided when editing Attributes individually?

15. In addition to AutoCAD, what is needed to extract Attributes?

COMBINING DRAWINGS
USING XREF

• •

Most of the drawings for a structure are worked on by teams of CAD drafters representing several firms. In addition to the architectural and engineering firms responsible for the design of a structure, CAD drafters often work for landscape and interior architects, as well as plumbing, mechanical, electrical and civil firms. Each of these firms must have up to date drawings to ensure an error free construction process. The use of AutoCAD's XREF command is instrumental in assuring quality drawings. This chapter will introduce merging a drawing into the current drawing using the XREF, XBIND and XCLIP commands. The XREF command can also be used to compile a drawing sheet containing details or drawings completed at varied scales.

EXTERNAL REFERENCE DRAWINGS

In Chapter 21 it was suggested that a WBLOCK of a floor plan could be inserted into a drawing and used to form the base of that drawing. For example, the floor plan can form the base drawing for the framing plan. This works well, but it does take up lots of disk space. A more efficient method than inserting a block is to use an externally referenced drawing or XREF. An externally referenced drawing is similar to a WBLOCK in that it is displayed each time the master drawing is accessed. These drawings, however, are not stored as part of the master drawing file. Externally referenced drawings are created through the XREF command. Figure 23–1 shows a drawing that serves as the base drawing. Figure 23–2a shows one of the drawings that makes use of the base drawings.

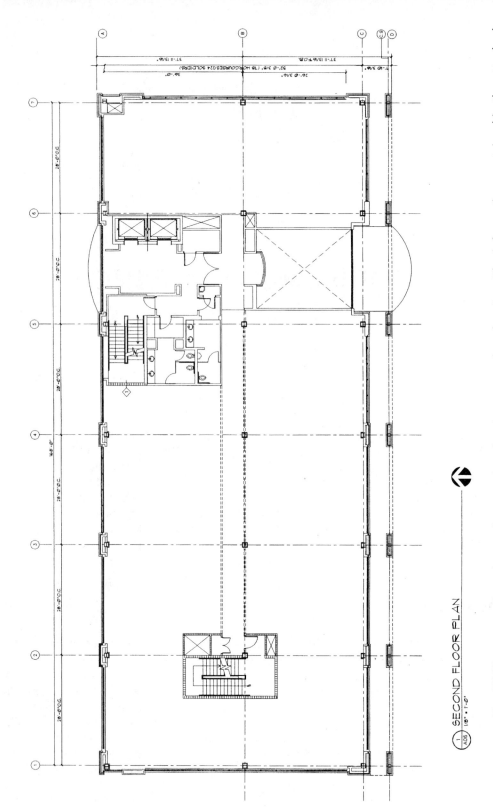

① SECOND FLOOR PLAN
A05 1/8" = 1'-0"

FIGURE 23–1 The XREF command can be used to attach a drawing to another drawing. The information contained in the attached drawing is not stored in the current drawing. Each time the base drawing is updated, all drawings that it is attached to will also be updated. (Courtesy Peck, Smiley, Ettlin Architects)

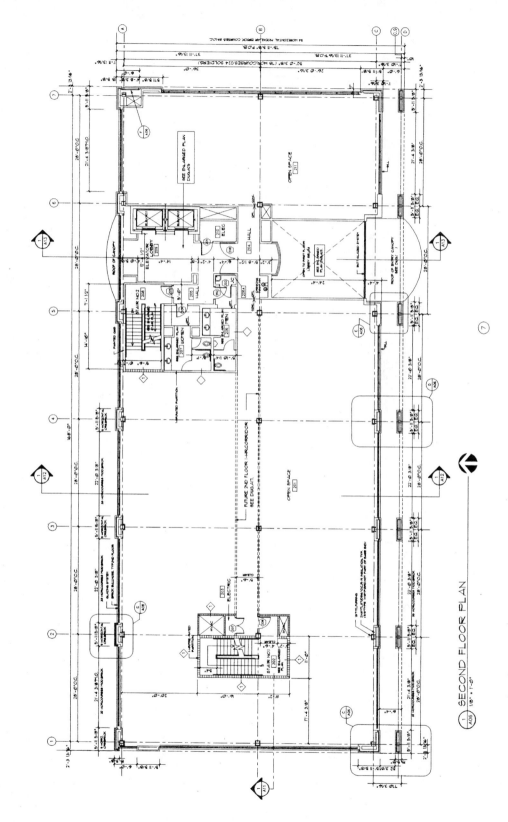

FIGURE 23-2a The drawing shown in Figure 23–1 is used as the base for the floor plan shown here and the drawings prepared by the mechanical, electrical and plumbing subcontractors. (Courtesy Peck, Smiley, Ettlin Architects)

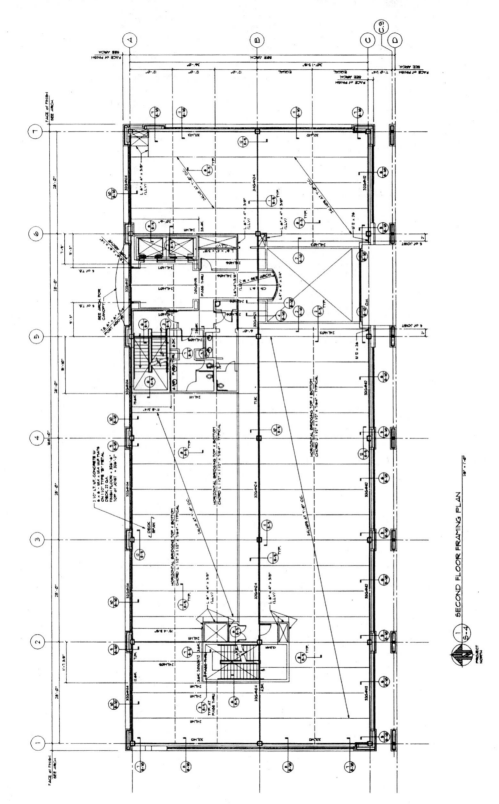

FIGURE 23–2b The framing plan is completed by an engineering firm that uses the base drawing created by the architectural team. The drawing is attached using XREF, and then the needed material is added to the drawing. (Courtesy Van Domelen / Looijeng / McGarrigle / Knauf consulting Engineers)

Saving Space

An xref drawing can be brought into the current drawing for viewing, but it does not become part of the current drawing base and it cannot be altered. An xref is similar to a block, except that no drawing entities are recorded in the drawing base. Only the drawing name and a small amount of information needed to access the drawing is stored in the new drawing file.

Automatic Updating

Xref drawings provide excellent benefits for drawing projects that are being developed by a team of engineers, architects, contractors, and drafters. Because of time constraints, subcontractors often need part of the drawings before they are complete. By using xref drawings, every time the drawing is accessed, the most recently saved version of the external drawing is loaded. Changes made to the xref source file will be updated automatically in every drawing where it is referenced. If one team member is working on the plan views, and another is working on exterior elevations, every time the xref floor plans are loaded for reference for the elevations, they will be updated.

Assembling Multiscaled Details

A common use for xref drawings is to assemble details for a stock detail sheet. In many manual offices, stock sheets of details have been assembled for every possible construction alternative that is likely to be encountered. You can assemble CAD drawings using externally referenced drawings with the knowledge that the most up-to-date version of the details is being provided.

ATTACHING AN XREF TO A NEW DRAWING

Although the XREF command can be completed from the command line, using the External Reference dialog box will greatly aid the use of the command. The dialog box can be accessed by picking the External Reference icon from either the Insert or Reference toolbar, by selecting External Reference... from the Insert pull-down menu, or by typing **XR** enter or **XREF** enter at the command prompt. Each method will produce the External Reference dialog box shown in Figure 23–3. The dialog box allows a drawing to be added to the current file, much like the INSERT command. Figure 23–4a shows a floor plan that will be used as the base drawing in the following example. The base drawing is titled C:\DRAWINGS\ARCH\FLOOR.DWG. Figure 23–4b shows the electrical information that will be attached to the base drawing to make the electrical plan. This drawing is stored as C:\Drawings\SCHOOL\ARCH\ELECT.DWG.

To attach the electrical information to the floor drawing, open the External Reference dialog box and pick the Attach button. This will display the Select Files to Attach dialog box. The box is similar to each dialog box that is used to open a new

FIGURE 23–3 The External Reference dialog box can be used to reference a drawing into a base drawing. The dialog box can be accessed by picking the External Reference icon from either the Insert or Reference toolbar, by selecting External Reference... from the Insert pull-down menu, or by typing **XR** [enter] or **XREF** [enter] at the command prompt.

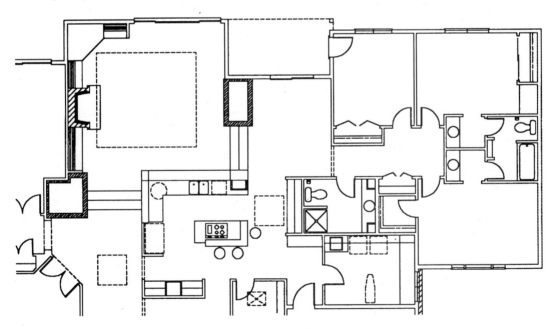

FIGURE 23–4a This portion of a floor plan will serve as a base for the electrical plan shown in Figure 23-4b. The information on the referenced drawing will be displayed as the floor plan is accessed, but it will not be stored in the FLOOR.DWG drawing base. (Courtesy of Tereasa Jefferis)

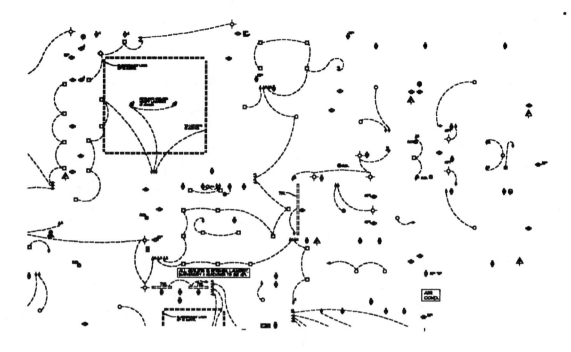

FIGURE 23–4b By using referenced drawings, subcontractors from various firms can each complete their work without having to wait for others to finish. This electrical information was added to the base drawing shown in Figure 23-4a. (Courtesy of Tereasa Jefferis)

drawing. Select the desired folder and file of the drawing to be attached. With the desired file highlighted, picking the Open button will display the Attach Xref dialog box shown in Figure 23–5. This dialog box allows you to control how and where the attached drawing will be located in the base drawing by using the Reference Type and Parameters boxes.

Using the Reference Type Box

This portion of the Attach Xref dialog box provides the options of Attachment and Overlay for deciding how an external drawing will be attached to the base drawing.

Attaching a Drawing. With the default setting of Attachment, picking the OK button will remove the dialog box, return you to the drawing area, and provide a prompt for an insertion point. As the point is selected with the mouse, the drawing will be attached and the command will be closed. Figure 23–6 shows the merged drawings from Figure 23–4a and 23–4b. The insertion point of the lower right corner was used with a running OSNAP of Intersection to assure proper alignment of the two drawings.

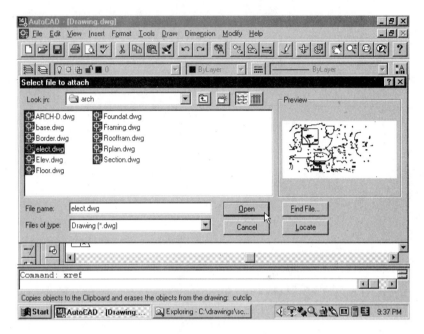

FIGURE 23–5 Picking the Attach… button in the External Reference dialog box will open and display the Attach Xref dialog box. This box can be used to select what drawing will be attached to the base drawing.

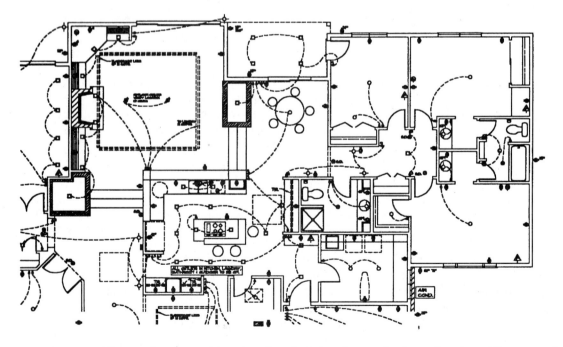

FIGURE 23–6 The results of attaching the electrical drawing to the base drawing. (Courtesy of Tereasa Jefferis)

If a second drawing is to be attached to the base drawing, selecting the Attach... button will no longer produce the Select Files to Attach dialog box. Once one drawing has been attached, when the Attach... button is selected the Attach Xref dialog box is displayed. The Browse... button must now be used to select the drawing to be attached.

Once a drawing is attached, AutoCAD adds dependent symbols in the xref to the base drawing. Dependent symbols are named items such as blocks, dimension styles, layers, linetypes, and text styles. Although these dependent symbols can be viewed, they can't be altered. This is one of the benefits of an xref, you can give a referenced drawing to a subcontractor and not have to worry about the drawing being altered. When a drawing is referenced into a base drawing, you'll notice that the layers names of the xref will be altered. Layer names for attached drawings will be displayed showing the name of the drawing followed by a vertical bar and the layer name.

A name such as 72PLUMB|UPPLUMB would indicate a referenced drawing of 72PLUMB and a layer title of UPPLUMB.

Overlaying an External Drawing. Choosing the Overlay option will produce a drawing that looks like an attached drawing. The overlay option allows you to see how the materials on each drawing align with each other, but you are not allowed to work on the attached drawing. The major difference between an attached and an overlaid drawing occurs when the referenced drawing contains nested reference drawings. A nested drawing is a drawing composed of other drawings. When a nested drawing is overlaid, the nested drawings within the drawing will not be displayed. When a nested drawing is attached, all objects will be displayed. The Overlay option is typically used when you need to view, but not plot the drawing.

Setting Parameters

As the drawing in Figure 23–4b was attached to the drawing in Figure 23–4a, the insertion point was selected prior to attachment. Using the Parameters portion of the Attach Xref dialog box allows the location, scale factors, and rotation angle to be controlled. In the default setting, the Specify On-Screen radio button is active, allowing the insertion point to be selected using the mouse. With the Specify On-Screen radio button deactivated, the At: text box is activated allowing specific X and Y coordinates to be specified for the insertion point. The default values for the X, Y, and Z scale factors are set as 1. Picking the Specify On-Screen radio button for scale factors will deactivate the scale factor values, and allow the values to be adjusted from the command prompt as the drawing is attached.

Including the Path

The final option of the Attach Xref dialog box is the Include Path radio button. This button determines if the full path for the referenced drawing will be saved in the data base of the base drawing. In the default setting, the path will be stored in the base drawing. The inactive option is often used when referenced drawings are shared with sub contractors who will work with the drawings. With this option inactive, as AutoCAD loads the base drawing it will only search for the referenced drawing in the path specified in the Files tab of the Preferences dialog box.

EXTERNAL REFERENCE DIALOG BOX OPTIONS

So far, the External Reference dialog box shown in Figure 23–3 has only been used to attach one or more reference drawings to a base drawing. The display provide in the box will depend on the setting of the button in the upper left corner of the External Reference dialog box. The button is a toggle between List View and Tree View. The display can be toggled by picking the inactive button or by pressing the F3 or F4 key.

List View Options

In the default setting, the dialog box displays in List View, with information related to the reference name, status, size, type, date, and path. Notice in Figure 23–3, that not all of the information is visible. The path scrolls off the edge of the display field. Move the cursor to the line that divides the Reference Name area and the Status area. As the line is touched, the cursor turns to a double arrow, allowing the size of the box to be altered. This can be done to each box to allow a complete display. (See Figure 23–7a.) Specific information related to each area includes:

Reference Name.   This portion of the box list each of the names of referenced drawings that have been attached to the base drawing.

Status.   This box shows the current status of referenced drawings. Options include:

- Loaded—The attached drawing is displayed.
- Not Found—The reference drawing cannot be found.
- Orphaned—The parent of the nested reference drawing has been unloaded.
- Reload—The referenced drawing has been marked to be reloaded. The option re–reads and displays the most recently saved version of the drawing. A referenced drawing is loaded or unloaded after the command is closed.
- Unload—The referenced drawing has been marked to be unloaded.
- Unloaded—The referenced drawing is not displayed.
- Unrefreshed—An unrefreshed drawing is not displayed. The referenced drawing has nested reference drawings that are not found.
- Unresolved— The referenced drawing is missing from the specified path and can't be found.
- Size—The Size column displays the size of the attached drawing.
- Type—Indicates if the referenced drawing option is Attach or Overlay.
- Date—List the last date that the referenced drawing was modified.
- Saved Path—Shows the path of the referenced drawing.

FIGURE 23–7a The space between listings can be reduced to allow longer listings to be completely viewed. By placing the cursor on the lines that divide titles, the display space can be reduced or enlarged.

Tree View Options

With the Tree View active, a hierarchical representation of referenced drawings similar to the listing in the Windows Explorer is displayed. The Tree display method displays nested drawings in their relationship to the attached drawing, as well as the attachment method, and the status. Figure 23–7b shows a listing using the Tree option.

EXTERNALLY REFERENCED DRAWING OPTIONS

Once the drawing is attached, each time the base drawing is opened the attached drawing will be opened and displayed as long as the path to the attached drawing is accessible. In addition to opening the referenced drawing, the other options of the External Reference dialog box will become active. Options include Detach, Reload, Unload, and Bind.

Detach

The Detach option will remove unneeded referenced drawings from the master drawing. Similar to how ERASE will remove a block, DETACH removes copies of the referenced drawing and its definition. A drawing is detached by picking the name of

FIGURE 23–7b Listings in the External Reference dialog box displayed using the Tree option.

the unneeded xref. Picking the name will highlight the listing of the drawings. Picking the Delete button will remove the xref listing from dialog box. If the OK button is selected, the selected xref will be removed from the screen and all references to the drawing will be removed from the base drawing.

Reload

This option allows an attached referenced drawing to be reloaded and updated in the middle of a drawing session. If a coworker revises the referenced drawing while you're working on a drawing that contains the xref, your xref is out of date. The Reload option allows the most current referenced drawing to be used without having to exit the drawing session and reopening the master drawing.

Unload

The Unload option allows a referenced drawing to be temporarily removed from the screen display without removing the xref from the drawing base. The Unload option is similar to using the Freeze option of the LAYER command. The referenced drawing is not displayed or considered in a regeneration increasing the drawing speed. The drawing can be reactivated by using the Reload option.

Bind

The Bind option allows a referenced drawing to become a permanent part of the base drawing. When this option is used, referenced drawings function as a WBLOCK rather than an XREF. When a drawing is referenced to a base drawing, the drawing is displayed, but you can't add or alter the referenced drawing. Using the Bind option allows an external drawing to be edited, but the xref will no longer be updated as the original drawing is edited. The Bind option adds dependent symbols to the drawing base so that they can be used just as any other drawing object. Layer names such as 72PLUMB|UPPLUMB in an attached drawing will be altered to read 72PLUMB0UPPLUMB. This naming system allows you to be able to quickly identify attached and bound layers. The Bind option should only be used when you know for sure that the referenced drawings has completed the review process and is no longer subject to change. Because construction drawings are so subject to change, Bind may not be an acceptable option. The XBIND command will be introduced later in this chapter that will allow specific symbols of a drawing to be permanently attached.

Binding a Drawing

Selecting the Bind... button will produce the Bind Xrefs dialog box shown in Figure 23–8 providing the options of Bind and Insert.

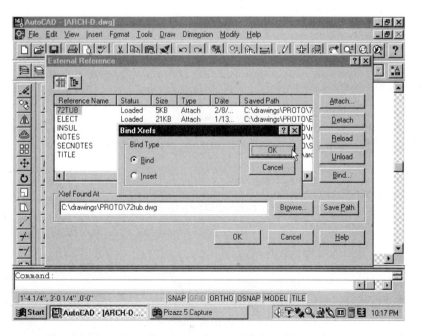

FIGURE 23–8   The Bind option allows a referenced drawing to be permanently attached to a base drawing. Picking the name of an attached drawing from the External Reference dialog box activates each of the XREF options. Selecting the Bind... option displays the Bind Xrefs dialog box.

Bind. The default option of Bind will permanently attach a drawing to the base. A drawing is bound by picking the name of the layer to be bound so that it is highlighted. Picking the OK button will remove the dialog box, bind the attached drawing to the base, and update the names of dependent symbols. Symbols in the bound drawing can now be edited just as any object created in the base drawing.

Insert. If the Insert option of Bind is used, an attached drawing is bound to the base drawing as if a wblock had been inserted into a drawing. Insert will bind the referenced drawing to the current drawing in a method similar to detaching and inserting the reference drawing. A drawing is inserted by picking the name of the layer to be inserted so that it is highlighted. Picking the OK button will remove the dialog box, insert the attached drawing to the base, and update the names of dependent symbols. A layer of 72PLUMB|UPPLUMB would be renamed to UPPLUMB.

CONTROLLING XREFS FROM THE COMMAND PROMPT

As with most commands in AutoCAD that are controlled by a dialog box, the XREF command can also be controlled from the command prompt. The command is started by entering **–XREF** [enter] at the command prompt. The command sequence to attach a drawing is:

Command: **–XREF** [enter]
?/Bind/Detach/Path/Unload/Reload/Overlay/<Attach>: [enter]

In the default setting of FILEDIA (1), the Select file to attach dialog box will be displayed, allowing selection of a stored .DWG file to be attached to the current drawing file. With a FILEDIA setting of 0, the file can be selected by command prompt when the Select file to attach: prompt is displayed.

Once the filename is selected, and the OPEN button is selected, a prompt similar to the WBLOCK prompt will continue. The prompts are:

Attach Xref ARCHBASE: C:\ARCHBASE.DWG
ARCHBASE loaded.
Insertion point: *(Select desired insertion point.)*
X scale factor <1> / Corner / XYZ: [enter]
Y scale factor (default = X): [enter]
Rotation angle <0>: [enter]
Command:

The Question Symbol. The ? symbol will provide a listing of xrefs that already have been attached to the drawing as well as the drawing file associated with each one. A prompt will be displayed:

Command: **–XREF** [enter]
?/Bind/Detach/Path/Unload/Reload/Overlay/<Attach>: [enter]
Xref(s) to list <*>

Pressing the [enter] key will display the AutoCAD Text Window and list all of the xrefs contained on a drawing. Wild card characters also can be used to define the selection set to be listed. A listing of external referenced drawings would resemble the following listing:

```
Command: –XREF [enter]
?/Bind/Detach/Path/Unload/Reload/<Attach>:? [enter]
Xref(s) to list <*>: [enter]
Xref Name          Xref type      Xref path
_____         _____       _____

EAVE-PLATE         Attach         C:\DRAWINGS\Proto\EAVE-PLATE.DWG
FND-1STORY         Attach         C:\DRAWINGS\Proto\FND-1STORY.DWG
FND-2STORY         Attach         C:\DRAWINGS\Proto\FND-2STORY.DWG

Total Xref(s): 3
```

Bind. Selecting the Bind option will allow an external reference to become a permanent part of the drawing. The following prompt will be displayed to start the process:

```
Command: –XREF [enter]
?/Bind/Detach/Path/Unload/Reload/<Attach> B [enter]
Xref(s) to bind: (Provide a name or names.) [enter]
```

Entering an * (asterisk) at the prompt will bind every xref in the drawing.

Detach. The Detach option will remove unneeded external referenced drawings from the master drawing. The following prompt will be displayed:

```
Command: –XREF [enter]
?/Bind/Detach/Path/Unload/Reload/<Attach> D [enter]
Xref(s) to detach: (Provide a name or names.) [enter]
```

Entering an * (asterisk) at the prompt will detach every xref in the drawing.

Viewing and Editing the Path. Because xrefs are an external drawing added to a new drawing, a record of the disk location and path of the source file must be recorded. The Path option will allow changing the present path if the source file is to be moved from its original location or if it has been renamed since being attached to the current drawing. The following prompt will be displayed when this option is selected:

```
Command: –XREF [enter]
?/Bind/Detach/Path/Unload/Reload/<Attach> P [enter]
Edit path for which Xref(s): (Provide a name or names.) [enter]
```

A single xref name, or a list of names separated by commas can be entered at the prompt. The command sequence is:

Command: **–XREF** [enter]
?/Bind/Detach/Path/Unload/Reload/Overlay/<Attach>: **P** [enter]
Edit path for which Xref(s): **ARCHBASE** [enter]
Old path: C:\DRAWINGS\ARCHBASE.DWG
New path: D:\DRAWINGS\ARCHBASE.DWG
Command:

Entering an * (asterisk) at the prompt will list every xref in the drawing.

Unload.   The Unload option will temporarily remove the referenced drawing from the display set. The command sequence is:

Command: **–XREF** [enter]
?/Bind/Detach/Path/Unload/Reload/<Attach> **U** [enter]

Entering an * at the prompt will unload all referenced drawings.

Reloading an Xref.   This option will allow a referenced drawing to be reloaded and updated in the middle of a drawing session if you find that the source drawing has been revised. The following prompt will be displayed when this option is selected:

Command: **–XREF** [enter]
?/Bind/Detach/Path/Unload/Reload/<Attach> **R** [enter]
Xref(s) to reload: (*Provide a name or names.*) [enter]

Entering an * (asterisk) at the prompt will reload every xref in the drawing.

Overlaying an External Drawing

When the Attach option is used, the external drawing is attached to the new drawing. The Overlay option allows an external drawing to be brought into drawing but not remain attached when the master drawing is saved.

THE XBIND COMMAND

The bind option adds dependent symbols to the drawing base, so that they can be used. Automatic updating and other qualities of xref drawings are lost. This can be overcome using the XBIND command. XBIND allows a portion of a referenced drawing such as a block, layer, or linetype to be permanently added to a current drawing, with the balance of the external drawing retaining its qualities. The XBIND command can be accessed by selecting External Reference Bind from the Reference toolbar, by selecting Bind... from External Reference of Object of the Modify pull-down menu, or by typing **XBIND** [enter] at the command prompt. Each option will produce the Xbind dialog box shown in Figure 23–9. The dialog box list

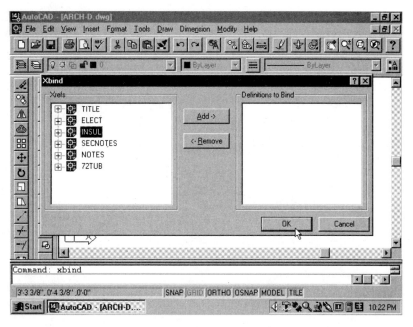

FIGURE 23-9 The Xbind dialog box displays a listing of drawings that are attached to the current drawing. The + symbol preceding AutoCAD R14 icon indicates that the listing can be expanded. Notice the Block, Dimstyle, and Textstyle of each contains additional information.

referenced drawings that are attached to the base drawing. In Figure 23–9, six drawings are attached to the base drawing. Each drawing is preceded by a + and the R14 icon. The + indicates that the listing can be expanded. Picking one of these options and double clicking will display the contents of the drawing. In Figure 23–10a, the INSUL listing was expanded revealing that the drawing contains nested Blocks, Dimstyles, Layers, and Linetypes. Notice in Figure 23–10b that the block INSUL | TANK has been selected to be added using the XBIND command. By picking the desired dependent symbol to bind (INSUL | TANK) and picking the Add> button, the name of the block will be moved from the Xref listing to the Definitions to Bind display. See Figure 23–10c. Picking the OK button will bind the block to the base drawing and close the Xbind dialog box. Picking a listing from the Definitions to Bind display and picking the <Remove button will unbind the definition and restore it to the Xrefs listing.

Dependent symbols can also be added to a base drawing by entering **–XBIND** [enter] at the command prompt. The command sequence is:

Command: **–XBIND** [enter]
Block/Dimstyle/LAyer/LType/Style:

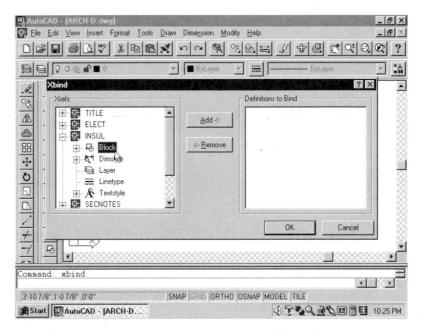

FIGURE 23–10a Picking the INSUL listing reveals nested items in the Block, Dimstyle, and line Textstyle.

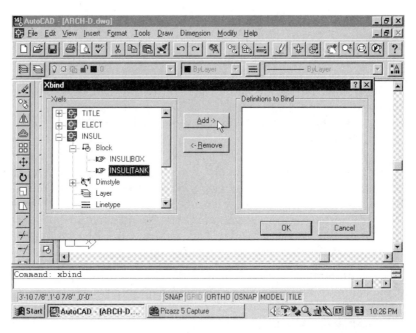

FIGURE 23–10b Picking the INSULITANK block for binding to the referenced drawings.

FIGURE 23-10c Selecting the name and picking the Add > button will permanently bind the selected definition to the base drawing.

Once the type of dependent symbol is selected, you will be prompted for the name of the symbol you want to bind. The command sequence is:

Command: **–XBIND** [enter]
Block/Dimstyle/LAyer/LType/Style: **LA** [enter]
Dependent Layer name(s): **ARCHBASE|SINK** [enter]
Scanning…
1 layer bound
Command:

MOVING WITH XCLIP

The XCLIP command allows portions of an xref drawing that is being attached to be deleted. The command allows a boundary to be created in a referenced drawing so that all material that is outside the boundary is invisible. The command can be accessed by selecting External Reference Clip from the Reference toolbar, by selecting Clip from Object Properties of the Modify pull-down menu, or by entering **XC** [enter] or **XCLIP** [enter] at the command prompt. Each access method will produce a Select object: prompt. Any object selection method can be used to select xrefs to be clipped. The command sequence is:

Command: **XC** [enter]
Select object: (*select xrefs to be clipped*).

1 found
Select objects: [enter]
ON/OFF/Clipdepth/Delete/generate polyline/<New boundary>: [enter]

Accepting the default value allows the clipping boundary to be selected and produces the following prompt:

Select polyline / Polygonal / <Rectangular>

Options for setting the boundary include the default method of forming a rectangular boundary with a selection window, selecting a polyline or by selecting a Polygonal. Selecting polygonal allows a boundary to be set by specifying points for the vertices of a polygon. Accepting the default by pressing the ENTER key produces prompts to select the First and Other corners. As the corners are selected, objects in the xref that are outside of the boundary will become invisible.

Figure 23–11a shows an example of a footing details that have been referenced into a base drawing. Figure 23–11b shows the detail on the right side of Figure 23–11a hidden using the XCLIP command. The complete command sequence to produce Figure 23–11b is:

Command: **XC** [enter]
Select object: *(Select xrefs to be clipped.)*
1 found
Select objects: [enter]
ON/OFF/Clipdepth/Delete/generate polyline/<New boundary>: [enter]
Select polyline/Polygonal/<Rectangular>: [enter]
Select clipping boundary:
First corner: *(Select point.)*

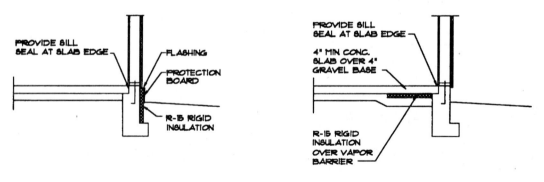

FOOTING INSULATION DETAILS

3/8"=1'-0"

FIGURE 23–11a The XCLIP command allows information that has been permanently attached to a base drawing to be hidden. The command requires that a boundary be selected. Anything outside of the boundary will be made invisible.

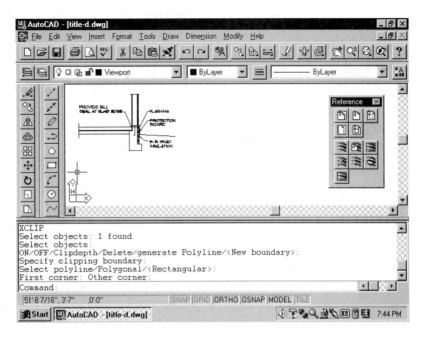

FIGURE 23–11b   The detail on the left side of Figure 23–11a was placed inside of the boundary. Using the XCLIP command, the detail on the right side is still attached to the drawing base, but is removed from the display set.

Other corner: *(Select other corner.)*
Command:

REFERENCING A DRAWING TO A TEMPLET

In Chapter 2 you were introduced to drawing templets that were created in paper space. Throughout the text you've been drawing in model space. The XREF command can be used to attach a drawing such as a floor plan, that is drawn in model space, to a drawing that is drawn in paper space to allow plotting the drawing at a scale of 1/1. Plotting will be covered in the next chapter. The balance of this chapter will help you prepare drawings to be plotted. A step-by-step approach will be used to reference one drawing to the templets. Each step will be explained in detail once you see how painless Release 14 had made the process as multiple viewports are created.

An Overview of Paper Space

To understand the process of attaching a drawing sheet containing the floor plan, Figure 23–6, visualize a sheet of 24″ × 36″ vellum with a title block and border printed on the sheet. Imagine a hole or Viewport cut in the vellum that allows you to look through the paper and see the floor plan. This is a simplified version of what is required to assemble a multiscaled drawing for plotting. Figure 23–12 shows the theory behind attaching a drawing in model space to a templet drawing created in paper space.

Because the drawings you have been creating have been done in Model space, the plan shown in Figure 23–6 is 90'-0" wide. If you were to hold a sheet of D size vellum in front of the plan, the paper would be minute. To make the floor plan fit inside the Viewport on the paper, you're going to have to hold the paper a great distance away from the floor plan until the drawing is small enough to be seen through the hole. This is how the Zoom command affects a drawing. By using the Zoom xp option, the floor plan can be reduced to fit inside the Viewport and maintain a scale typically used in the construction trade.

Compiling a Base Drawing

The initial step for developing a drawing to be plotted is to develop a base sheet. This sheet should be drawn at the size of the paper to be used when plotting using Paper space. Since many drawings for both architectural and engineering projects are

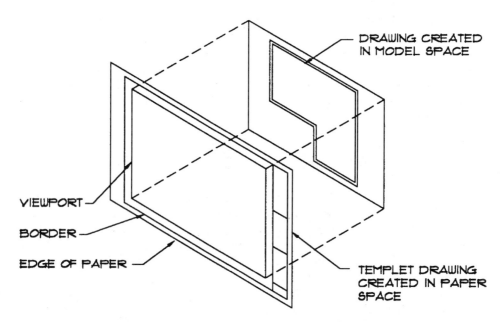

FIGURE 23–12 Attaching a drawing created in model space to a drawing templet displayed in paper space can be a very helpful method of preparing a drawing to be plotted. Think of the templet drawing as having a hole in it. The hole is placed in the templet using the MVIEW command. Holding the drawing containing the floor plan far enough behind the hole in the templet drawing allows the floor plan to be viewed. The XP option of ZOOM determines how far back the floor plan must be placed in order to see the plan at a specific scale.

printed on D size paper, 24″ × 36″ paper will be used in the examples for the balance of this chapter. Typically this can be done by selecting one of the templet drawings contained in AutoCAD Release 14, or by using a templet that you've developed. Figure 23–13 shows an example of the drawing templet that will be used throughout this chapter. To complete the process of attaching or inserting a drawing to a paper space templet, you will need to be familiar with the terms Model Space, Paper Space, Tilemode, and the MVIEW command.

Model Space. All of your drawings are created in model space. When in model space, the MODEL tile in the task bar will be active. Model space can be selected by picking Model Space (Tiled) or Model Space (Floating) from the View pull-down menu, by double-clicking the PAPER tile in the task bar, or enter **MS** [enter] or **MSPACE** [enter] at the command prompt. See Chapter 10 for a review of tiled and floating viewports.

Paper Space. Paper space was first introduced in Chapter 2 as templet drawings were introduced. Paper space allows drawings to be plotted onto paper. Paper space can be selected by double clicking the MODEL tile in the task bar, by selecting Paper

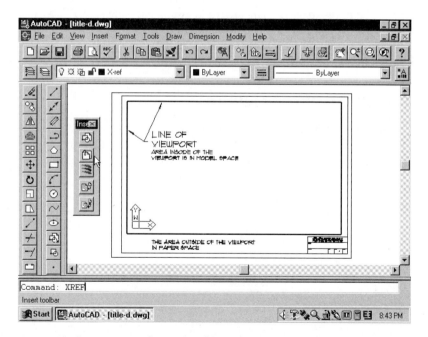

FIGURE 23–13 A viewport can be created in a drawing templet created in paper space by using the MVIEW command. By creating a separate layer for the viewport to be placed on prior to using the MVIEW command, the outline of the viewport can be removed from the plotted drawing. Once the viewport is created, switching to floating model space will place the area inside the viewport in model space.

Space from the View pull-down menu, or by enter **PS** [enter] or **PSPACE** [enter] at the command prompt.

Tilemode. In addition to the methods listed above, adjusting the tilemode is another method to enter model or paper space. TILEMODE is a system variable that toggles between paper and model space.

Mview. The MVIEW command is the computer equivalent to an X-acto® knife. The command is used to "cut" a viewport into the paper space templet so that the drawing behind the templet can be seen. See Figure 23–12. Before the command is used to create a viewport, create a layer to contain the viewport, so that the line that identifies the viewport can be frozen before the drawing is plotted. With this layer now current, a viewport can be created by entering **MV** [enter] or **MVIEW** [enter] at the command prompt. The command sequence is:

Command: **MV** [enter]
ON/OFF/Hideplot/Fit/2/3/4/Restore/<First Point>: *(Select first window corner.)*
Other corner: *(Select other window corner.)*
Command:

Once the desired drawing is inserted or referenced into the viewport, the size of the viewport can be altered if the entire drawing can't be seen. To alter the viewport size, move the cursor to touch the viewport and activate the viewport grips. Select one of the grips to make it hot, and then use the hot grip to drag the window to the desired size.

Attaching the Drawing

Unless you've experimented on your own, drawings throughout this text have been created using Model space. To switch from Model space to Paper space, enter TILEMODE, and change the value to 0. Notice that in Figure 23–14, with a value of

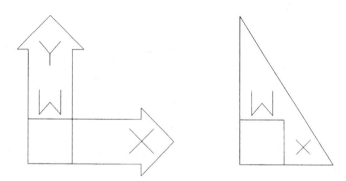

FIGURE 23–14 The UCS icons for model space and paper space.

0, the UCS icon is changed to a triangle to reflect the use of Paper space. The command sequence is:

```
Command: TILEMODE [enter]
New Value for TILEMODE <1>: 0 [enter]
Regenerating papers.
Command:
```

A tile mode of 1 represents model space. A tile mode of 0 represents paper space. Each mode can also be set by picking the desired mode from the View pull-down menu. A drawing created in model space can be inserted or referenced into a templet drawn in paper space using the following procedure:

1. Start a new drawing and use the desired templet.
2. With the desired templet opened, verify that paper space is active.
3. Set the UNITS to match the drawing that will be inserted.
4. Set the LIMITS to match the paper space, in most cases this will be 36, 24.
5. Create a layers to contain the drawing to be attached. In this example FLOOR.
6. Create a layer titled VPORT to contain the viewport, and make this the current layer.
7. Use the MVIEW command and use mouse to indicate the size of the viewport. When only one drawing will be added, the viewport can be set so that it is just inside the border.
8. Make the FLOOR layer current.
9. Switch from paper space to floating model space. The UCS model space icon will now be displayed inside of the viewport.
10. Use the Attach option of XREF to reference the FLOOR drawing to the title block.
11. Pan the drawing as needed.
12. Use the XP option of ZOOM. Since the drawing will be plotted at a scale of 1/4"=1'-0", enter 1/48XP for the zoom factor.
13. Return to paper space.
14. Adjust the size of the viewport as needed.
15. Save the drawing for plotting.

Figure 23–15 shows an example of a floor plan attached to a drawing templet for plotting.

ATTACHING MULTIPLE DRAWINGS AT VARIED SCALES

As seen in Figure 23–16, construction projects typically comprise drawings that are drawn at a variety of scales. These details range in scale from 3/8"=1'-0" to 1 ½" = 1'-0". This sheet of details can not be easily drawn or plotted without the use of multiple viewports and referencing each detail to a base drawing. As with drawings

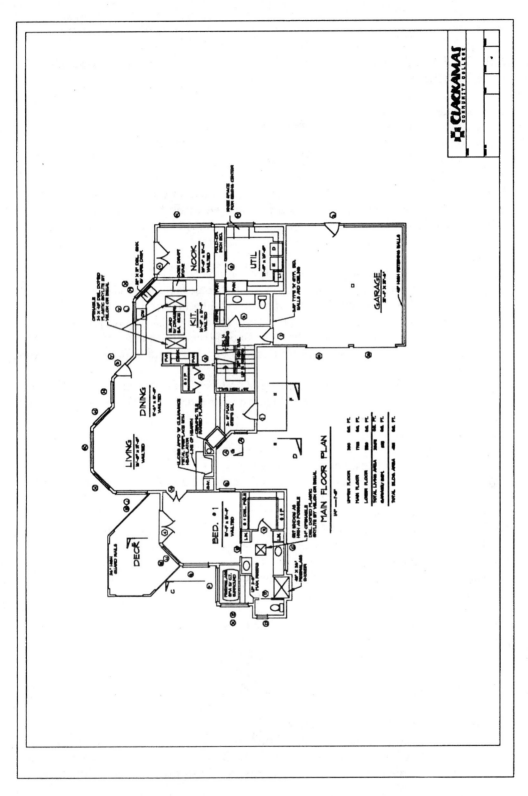

FIGURE 23–15 A floor plan referenced into a drawing templet. As a drawing is attached to the templet, it may not be displayed as the insertion point is selected. Use the Extents option ZOOM to display the attached drawing. Using the ZOOM XP option will display the drawing at the proper size for plotting at a scale of 1/1.

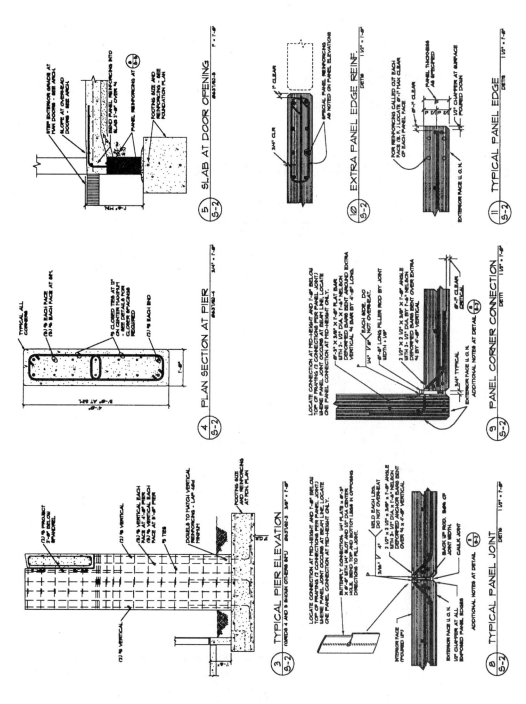

FIGURE 23-16 Details drawn at varied scales can be assembled on one sheet for plotting using the XREF command. (Courtesy Van Domelen / Looijenga / McGarrigle / Knauf Consulting Engineers)

where a single object is attached, the use of scale factors to control text and dimension variables will be critical to the success of the final outcome of the drawing. Figure 23–17 shows a list of common scale factors for architectural and engineering drawings. The process for creating multiple viewports will be similar to the creation of one viewport with the addition of the VPLAYER command. The VPLAYER command is used to control the layer visibility as multiple drawings are attached to the base drawing.

ATTACHING MULTIPLE DRAWINGS FOR PLOTTING

For the example to be used in the balance of this chapter, four drawings will be attached to the base sheet to form a master sheet for plotting. These individual drawings can be seen in Figure 23–18. Enter the base drawing and set the current layer to VIEWPORT. The first drawing to be brought into the base drawing will be the section shown in Figure 23–18. The overall model space size of the section is 60 feet ×

ARCHITECTURAL SCALE FACTORS		ENGINEERING SCALE FACTORS	
SCALE	SCALE FACTOR	SCALE	SCALE FACTOR
1" = 1'-0"	12	1"= 1'-0"	12
3/4"= 1'-0"	16	1"= 10'-0"	120
1/2" = 1'-0"	24	1"= 100'-0"	1200
3/8"=1'-0"	32	1"= 20'-0"	240
1/4" =1'-0"	48	1"= 200'-0"	2400
3/16" =1'-0"	64	1"= 30'-0"	360
1/8" = 1'-0"	96	1"= 40'-0"	480
3/32" = 1'-0"	128	1"= 50'-0"	600
1/16" = 1'-0"	192	1"= 60'-0"	720

FIGURE 23–17 Common scale factors for architectural and engineering drawings. Use the scale factor as the ZOOM XP factor when referencing a drawing to a templet. After the ZOOM Extents option was used to display the floor plan in the viewport, the ZOOM command was used again, and a value of 1/48XP was entered to display the floor plan in the viewport so that it can be plotted. Using this method, the 36" × 24" templet drawing can be plotted at a scale factor of 1/1 and produce a plotted drawing that can be accurately measured using a scale of 1/4" = 1'-0".

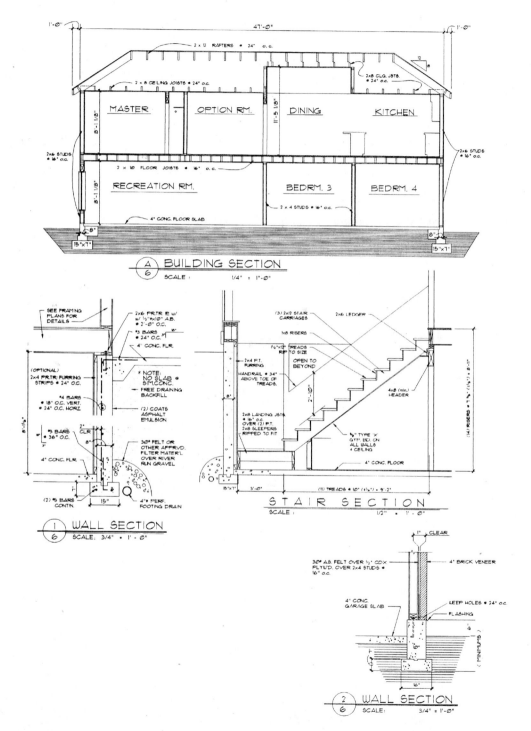

FIGURE 23–18 Drawings to be attached to the basesheet. (Courtesy Piercy & Barclay Designers, Inc., A.I.B.D.)

36 feet. Using a scale of ¼″ = 1′-0″ at plotting will reduce this size to fit in a 15″ × 9″ viewport.

Use the MVIEW command to define the size of the viewport. The sequence is started by typing **MV** [enter] or **MVIEW** [enter] at the command prompt. The command sequence is:

Command: **MV** [enter]
ON/OFF/Hideplot/Fit/2/3/4/Restore/<First Point>: *(Select first corner of viewport.)*
Other corner: **@15,9** [enter]
Regenerating drawing.
Command:

The results of this command can be seen in Figure 23–19 shows the results of this sequence. Each option will be reviewed after a viewport has been created. For now, the coordinates for the viewport can be entered, or specified by indicating the corners of the viewport just as if it were a window used for determining a selection set. Although you should try to size the viewport accurately, it can be stretched to enlarge or reduce its size once the drawing is referenced into it.

Inserting an Xref Drawing

With the viewport drawn, make the xref layer current. This will allow a previously created drawing to be attached to the base drawing. By using the xref command

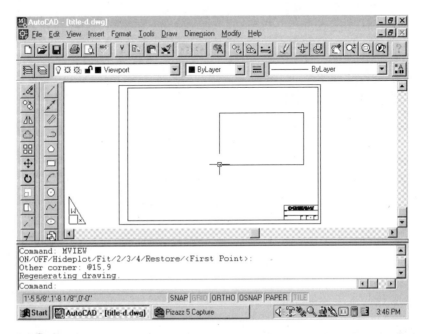

FIGURE 23–19 The MVIEW command is used to create a viewport in the drawing templet.

rather than INSERT, the size of the base drawings will not be increased. With the xref layer current, return to model space by typing **MS** [enter] at the command prompt or double-click the PAPER tile in the task bar. Now the UCS icon for Model space will be displayed in the lower left corner of the viewport. If the cursor is located inside the viewport, the cross-hairs will be displayed. As the cursor is moved out of the viewport, it is changed to an arrow. The command sequence is:

```
Command: MS [enter]
Command: –LAYER [enter]
?/Make/Set/New/ON/OFF/Color/Ltype/Freeze/Thaw/LOck/Unlock : S [enter]
New current layer <VIEWPORT>: [enter]
?/Make/Set/New/ON/OFF/Color/Ltype/Freeze/Thaw/LOck/Unlock: [enter]
Command: –XREF [enter]
?/Bind/Detach/Path/Unload/Reload/Overlay/<Attach>: [enter]
```

By entering Model space, you will maintain all of the characteristics of the drawing being referenced. Once the drawing is attached, layers of each referenced drawing will be displayed in the LAYER dialog box with the drawing name serving as a prefix, followed by a colon and the layer name. Typical layer names might include:

BORDER

SECTION: DIMEN

SECTION: TEXT

SECTION: WALLS

TITLE

XREF

Attaching the Drawing

Use the XREF command to attach the section drawing to the base drawing. Start the sequence by typing **XREF** [enter] at the command prompt. Use the Attach option of the External Reference dialog box as discussed earlier in the chapter. (See Figure 23–3.)

As the Attach option is selected, the Select File to Attach dialog box is displayed. Once the desired drawing name is selected, a prompt will be given for insertion point. Select a point in the center of the viewport.

Viewing the Drawing

Although you have just attached a drawing to the viewport, the display may not be what you expected. Use the ZOOM command and the E (extents) option to ensure that you can see the entire drawing. The command sequence is:

```
Command: ZOOM [enter]
All/Center/Dynamic/Extents/Left/Previous/Vmax/Window/<Scale (X/XP): E [enter]
```

If you omit the ZOOM E option, your drawing may disappear in the next stage of the process. Using ZOOM E will produce a display similar to the one in Figure 23–20.

Setting the Scale

A drawing has now been attached to the base drawing, but the scale still needs to be adjusted so that the drawing will be plotted at a scale of ¼″ = 1′-0″. This can be done by using the XP option of the ZOOM command with the sequence:

Command: **ZOOM** [enter]
All/Center/Dynamic/Extents/Left/Previous/Vmax/Window/<Scale (X/XP):
1/48XP [enter]

As the scale is altered, the drawing may now extend beyond the limits of the viewport. Any part of the drawing that touches or extends beyond the viewport will not be plotted. The STRETCH command can be used to enlarge the viewport. Changes to the location or the size of the viewport must be made while you are in Paper space.

All of the drawing commands may be used inside the viewport, but try to limit editing to changes that will only affect this plot. Changes to the drawing, which will be needed on future drawings, should be made to the original drawing. Remember,

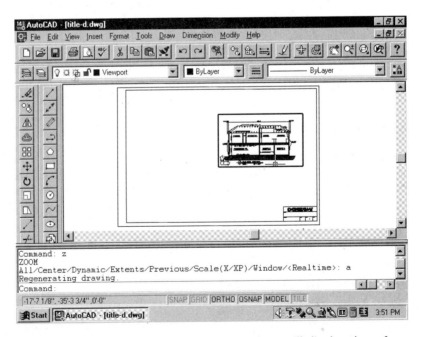

FIGURE 23–20 Using the ZOOM All or Extents options will display the referenced drawing in the viewport. Once the entire drawing is displayed, use the PAN command to center the drawing in the viewport.

because the original is attached to this drawing by XREF, the changes will be reflected in all uses of the original. Enter Model space prior to making changes to the drawing in the current viewport. This will allow you to add detail numbers and drawing titles that change with each use of the drawing.

Controlling the Display of New Viewports

In its current state, if a new viewport is created, the display from the existing viewport will be displayed in the new viewport as well. To avoid this problem, the layers of the drawing in the first viewport must be frozen. This can be done by using the Vpvisdflt (ViewPort VISibilty DeFauLT) option of the VPLAYER (Viewport Layer) command. All of the layers of the SECTION drawing that is attached to the base drawing can be frozen. When you enter **V** [enter] you can handle all layers of a referenced drawing as one layer in subsequent viewports. The command sequence is:

```
Command: VPLAYER [enter]
?/Freeze/Thaw/Reset/Newfrz/Vpvisdflt: V [enter]
Layer names(s) to change default viewport visibility <>: SECTION* [enter]
Change default viewport visibility to <Frozen/Thawed>: F
?/Freeze/Thaw/Reset/Newfrz/Vpvisdflt: [enter]
```

Entering the name as SECTION* will freeze all of the layers of the SECTION drawing in other viewports. Forgetting to enter the * after the name will cause SECTION to be displayed in each viewport. The setting of the VISRETAIN variable can also affect the layers that are displayed in a viewport. The Variable should be set to ON so that layers are displayed in only one viewport.

Repeating the Process for Additional Viewports

Once the layers of the previous viewport have been displayed, additional viewports can be created. The process to create the viewport for the 8FTWALL drawing is the same process used to create the viewport for the SECTION drawing. If you have planned the sheet contents well, the viewports for BRICK1ST and STAIR can be created also. Placing all of the viewports at once will eliminate having to keep switching from PSPACE to MSPACE. The commands needed to repeat the process are:

```
Command: PS [enter]
Command: -LAYER [enter]
?/Make/Set/New/ON/OFF/Color/Ltype/Freeze/Thaw/LOck/Unlock: S [enter]
New current layer <BORDER>: VIEWPORT [enter]
?/Make/Set/New/ON/OFF/Color/Ltype/Freeze/Thaw/LOck/Unlock: [enter]
Command: MV [enter]
ON/OFF/Hideplot/Fit/2/3/4/Restore/ <first point>: (Enter desired point.)
Regenerating drawing.
Command: MS [enter]
Command: -LAYER [enter]
```

?/Make/Set/New/ON/OFF/Color/Ltype/Freeze/Thaw/LOck/Unlock: **S** `[enter]`
New current layer <VIEWPORT>: **8FTWALL** `[enter]`
?/Make/Set/New/ON/OFF/Color/Ltype/Freeze/Thaw/LOck/Unlock: `[enter]`

Command: **XREF** `[enter]`
?/Bind/Detach/Path/Reload/<Attach>: `[enter]`
Xref to Attach: (*Select desired drawing to be attached, in this case 8FTWALL.*)
Insertion point: (*Enter the coordinates or pick a point.*)

Command: **ZOOM** `[enter]`
All/Center/Dynamic/Extents/Left/Previous/Vmax/Window/<Scale (X/XP): **E** `[enter]`

Command: **ZOOM** `[enter]`
All/Center/Dynamic/Extents/Left/Previous/Vmax/Window/<Scale (X/XP):
    1/16XP `[enter]`

Command: **VPLAYER** `[enter]`
?/Freeze/Thaw/Reset/Newfrz/Vpvisdflt: **V** `[enter]`
Layer names(s) to change default viewport visibility: **SECTION*** `[enter]`
Change default viewport visibility to <Frozen/Thawed>: **F** `[enter]`
?/Freeze/Thaw/Reset/Newfrz/Vpvisdflt: `[enter]`
Command:

If the process is repeated, the STAIR and BRICK1ST details also can be attached to the drawing. Once all of the desired details have been added to the drawing, the viewport outlines can be Frozen. Enter the LAYER command and select VIEWPORT as the layer to be edited, then select the F option. Freezing the viewports will eliminate the line of the viewport being produced as the drawing is reproduced. A box still surrounds the viewport when the drawing is in Model space. To verify that all viewports have been frozen, switch to Paper space. Any viewports that still exist were created on a layer other than VIEWPORT. Use the CHANGE command to place the viewport on the correct layer.

Plotting

Once all of the viewports are frozen and all additions have been made to drawing, the PLOT sequence described in Chapter 24 can be used. The scale factor is the only variation that will be affected. Since the scale of each viewport was entered using the ZOOM XP option, the scale factor required for the plot sequence is now 1 = 1. The resulting plot will produce a drawing similar to Figure 23–21.

EXPLORING THE REMAINING MVIEW COMMAND OPTIONS

The MVIEW command was introduced during the process of creating viewports for multiscaled plots. Reviewing the menu in Figure 23–19 shows that several other options are available through the MVIEW command. These options include:

ON/OFF/Hideplot/Fit/2/3/4/Restore/ <first point>:

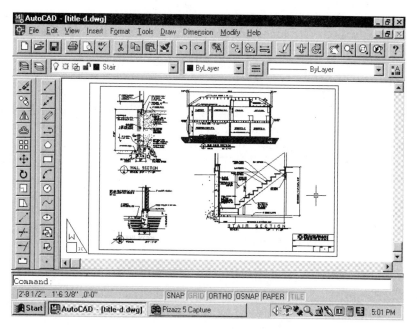

FIGURE 23-21 A templet drawing with four drawings at varied scales. The XREF command allows this drawing to be plotted at a scale factor of 1/1. As multiple viewports are created, the object in the first viewports will be displayed in other viewports. The first drawing, and each successive drawing added to other viewports, can be limited to display in only one viewport by using the Vpvisdflt option of the VPLAYER command.

ON/OFF

ON and OFF serve as toggle switches to display or hide the contents of the individual viewports. Selecting either option will provide a prompt to "Select objects" so the desired viewport may be controlled. This option can be very useful when plotting stock sheets of details, if one or two Attached Details do not apply to the current use. Setting these viewports to OFF will eliminate the viewport from the plot. This option is very useful for controlling REGEN time also. Inactive viewports can be set to OFF while the current viewport is edited. If all viewports are set to OFF, you will not be allowed to work in Model space until a new viewport is created.

Hideplot

This option will allow hidden lines to be removed from selected viewports when plotting in Paper space. This option generally is not required in architectural and engineering drawings.

Fit

This option will create a viewport to fit the current screen display. The size of viewport will be determined by the dimensions of the Paper space view. In the example used throughout this chapter, the viewports were created first in base drawings and then drawings were inserted into a viewport. The FIT option will allow a viewport to be created around an existing drawing or a portion of a drawing.

2/3/4

Selecting this option will allow for creating two, three, or four viewports without having to reissue the command. The required number of viewports is entered, followed by specifications for how to arrange the viewport. Each option is similar to the arrangements used for VPORTS in Chapter 10. The results of each option can be seen in Figure 10–19.

> 2: Specifying two viewports will produce the prompt:
>
> Horizontal <Vertical>:
>
> allowing you to place the division between the two viewports to be created.
>
> 3: Specifying three viewports will produce the prompt:
>
> Horizontal/Vertical/Above/Below/Left/<Right>:
>
> Choosing the Horizontal or Vertical option will allow you to divide the viewport into thirds. Other options will create two small viewports Above, Below, or to the Left or Right of one large viewport.
>
> 4: Choosing this option will create four viewports equal in size by dividing the specified viewport horizontally and vertically.

Restore

This option can be used to change the viewport created in VPORTS into individual viewports in Paper space. The prompt will be:

> ?/Name of window configuration to insert <default name>:

The name of a saved view can be entered. Typing the question symbol (?) will provide a list of View names.

EXPLORING THE REMAINING VPLAYER COMMAND OPTIONS

The VPLAYER (ViewPortLayer) command allows you to control the viewing of a viewport individually. This command will allow for displaying a layer in one viewport but not in others. This feature would be useful in viewing two different options side by side, at the same time. To use the VPLAYER command the TILEMODE must be set to 0. The initial prompt includes options for:

?/Freeze/Thaw/Reset/Newfrz/Vpvisdflt:

?. When you use the question option, the prompt will be given to name a viewport; then you will be provided a listing of frozen layers in that viewport.

Freeze

Selecting this option will allow for one or more layers to be frozen in a selected viewport. A single layer name may be entered, or multiple names can be entered with each name separated by a comma.

Thaw

This option will allow previously frozen layers to be thawed. Layers to be thawed can be selected using the same methods that were used for Freeze.

Reset

This option will allow for controlling the visibility of layers in specified viewports. The option can be used to reverse the effects of Vpvisdflt.

Newfrz

This option can be used to create new layers that are frozen in all viewports. Selected layers created with Newfrz can be thawed in a desired viewport to allow editing.

Vpvisdflt

No, this isn't a foreign language. The Vpvisdflt option will allow for setting the View-Port VISability DeFauLT. If a layer is frozen in a viewport but visible in another, you can use this option to determine what layers will be seen if a new viewport is opened.

CHAPTER 23 EXERCISES

E-23-1. Open the FOOTING drawing and use it to create an xref drawing titled BASEFTG. Open drawing FLOOR14 and attach BASEFTG to the drawing. Save the drawings as XREFTEST.

E-23-2. Open a drawing templet containing the border and title block. Attach any four drawings in your library. Each drawing is to be shown at a different scale when plotted. Provide a title and the scale used by the original drawing prior to plotting.

E-23-3. Open any three drawings and attach them to a drawing base. Save the drawing as XREF.

CHAPTER 23 QUIZ

1. Describe how an xref can be used in an office to complete a project.

2. List and briefly define the six XREF options.

3. How does the TILEMODE setting affect multiscaled plots?

4. It was suggested that two specific layers be added to the drawing base when creating multiscaled drawings. What were these layers and how will having them current affect the drawing?

5. What command and process is used to create a viewport? Which layer should be current?

6. A drawing titled UPFLOOR and containing the layers DIMEN, WALLS, DOORWIN, TEXT, and CABS is to be attached to a base drawing containing the layers suggested in this chapter for multiscaled plots. List the layers that would be displayed in the Layer Control dialog box.

_____

_____

_____

7. List the command, the option required, and the prompts to attach an XREF drawing to a base drawing.

_____

_____

_____

_____

8. List the effect of ZOOM E and ZOOM XP on the process of attaching a drawing.

_____

_____

_____

_____

9. What is a dependent symbol and how does it affect a drawing?

_____

_____

_____

_____

10. Explain the effects of overlaying a drawing.

_____

_____

_____

_____

11. The bind option of XREF has been used. How will the option affect the external drawing?

_____

_____

_____

12. Explain the effects of XBIND on a referenced drawing.

_____

_____

_____

13. How does the XCLIP command affect an attached drawing?

_____

_____

_____

14. What are the benefits of using paper space?

_____

_____

_____

15. How can the display in the second viewport be altered so that it does not display the same drawing as another viewport?

_____

_____

_____

SECTION 6

EXPANDING AUTOCAD APPLICATIONS

Plotting Hardcopies

Oblique and Isometric Drawing

AutoCAD and the Internet

CHAPTER

24

PLOTTING HARDCOPIES

As computers become more common in contractors' and subcontractors' offices, more construction projects will be transmitted between offices by modem or floppy disk. This allows the designers for the subcontractors to have access to areas of the project that require contributions from them. In addition to providing access to the drawings, electronic communication can save many hours by eliminating the need to produce paper copies. Unfortunately, some offices have not made the transition to CAD and still require paper copies. This chapter will introduce methods of controlling the plotting commands from the dialog boxes and command prompt.

SETTING PLOT SPECIFICATIONS USING DIALOG BOXES

The dialog box that controls plotting can be accessed by picking the Print icon from the Standard toolbar by selecting Print… from File of the screen menu, or by typing **PLOT** [enter] at the command prompt. Each menu can be seen in Figure 24–1. Each access methods will produce the Print/Plot Configuration dialog box that can be used to control either a printer or a plotter. Printing was introduced in Chapter 7. The major areas of the dialog box include the Device and Default Information box, the Pen Parameters assignment selection box, the Additional Parameters selection box, the Paper Size and Orientation selection box, the Scale, Rotation, and Origin selection box, and the Plot Preview selection box.

Device and Default Selection Box

This selection box displays the plotting devices that are currently being used. Notice that in the top half of the box shown in Figure 24–2a, the current plotter is listed as System Printer. This would be a printer or plotter that is configured in the Printer

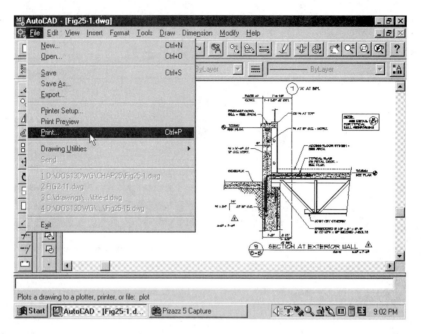

FIGURE 24–1 Plot commands can be accessed by picking the Print icon from the Standard toolbar, by selecting Print... from File of the pull-down menu, or by keyboard. (Courtesy Van Domelen / Looijenga / McGarrigle / Knauf Consulting Engineers)

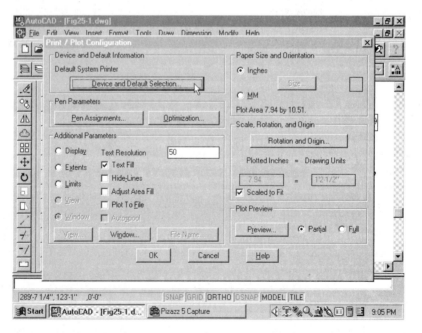

FIGURE 24–2a The Plot Configuration dialog box is used to control printing and plotting.

folder of My Computer or added by selecting Printer from the Settings menu. Selecting the Add Printer option from the Printer Menu of Windows 95 or Windows NT will produce a prompt for Add Printer Wizard and allow a printer to be added and accessed by other programs. The Device and Default Information box of the Print/Plot Configuration dialog box allows for choosing between printers that have been previously configured to be selected. Notice in Figure 24–2b, that the current plotter is the DraftPro DXL. Methods of configuring a printer or plotter and adding it to this list will be introduced later in this chapter. The lower half of the box is a selection box that will show a listing of all plotting devices that are currently configured. When you pick the "Device and Default Selection . . ." box, a display similar to Figure 24–3 will be displayed. The box will allow for viewing or changing the plotting device and the plotting configuration.

Select a Device Configuration. As your computer is configured initially, the selection box will be empty. Plotting devices are listed in the selection box by picking Printer Setup... in the File pull-down menu or by selecting Preferences... from the Tools pull-down menu. Each method will produce the Preferences dialog box. Selecting the Printer tab in the Preference dialog box shown in Figure 24–4 allows printers or plotters to be added, modified or removed from the list of plotting devices.

AutoCAD has the ability to have the configuration for more than one plotting device stored for use. The information for each plotting device is stored in a Plot Configuration Parameters (.PCP) file and can be accessed through the Select Device Box. In Figure 24–3, two plotting devices have been configured, with the current device highlighted. A different device can be selected by moving the arrow to the desired listing and pressing the pick button. Selecting another plotting device will change the settings of other parameters in the plotting dialog box.

Configuration File. This portion of the Device and Default Selection dialog box saves or retrieves complete or partial configuration files. Each type of file allows plotting parameters to be saved or recalled for future use. A partial configuration is given a .PCP file extension. PCP files store plotting parameters such as pen assignments, plot area, scale, paper rotation, and size. Each of the values were introduced in Chapter 7 as printing was considered. Each of these values will be introduced throughout the balance of this chapter as they relate to plotting. As a drawing is pre-

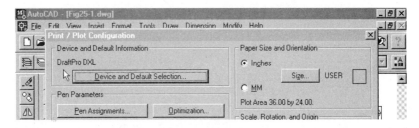

FIGURE 24–2b The Device and Default Selection portion of the Plot Configuration dialog box can be used to switch between two or more configured plotting devices.

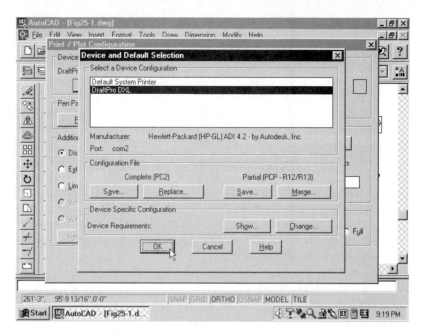

FIGURE 24–3 Selecting the Device and Default Selection... button from the Plot Configuration dialog box will display this box. The dialog box can be used to configure and save plot parameters from the current drawing for future plotting sessions.

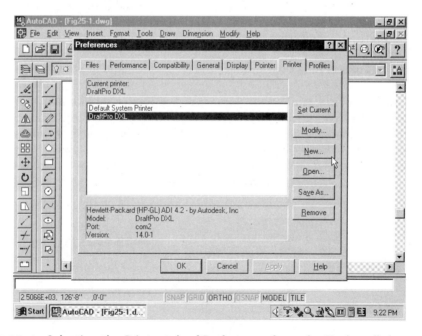

FIGURE 24–4 Selecting the Printer tab of Preferences from the Tools pull-down menu allows a configured plotter or printer to be made active. Plotting devices are added to the list by using the Add Printer option of the Printer folder in My Computer or by selecting Add Printer from Printer of the Settings menu.

pared for plotting, each of the values can be saved in a .PCP file for future plotting sessions. If the drawing needs to be replotted, the .PCP file can be retrieved by picking the Merge... button. The plotting parameters stored in the .PCP file will be merged with the current plot configuration settings.

A complete plot configuration file is given a .PC2 file extension by AutoCAD. PC2 files contain each of the plot parameters contained in a .PCP file as well as device-specific information to be used throughout the plot. PC2 files allow multiple prints to be made on the currently configured plotting device using the BATCHPLT command. Once the plotting parameters have been saved, they can be retrieved for future plotting of the file using the Replace option. Selecting the Replace... button allows the PC2 file to be retrieved and substituted for the current plot settings.

Device Specific Configuration. This area is divided into the Show Device Requirements... and the Change Device Requirements... buttons. These options will allow specific plotting device requirements to be adjusted. These boxes may not be active, depending on which plotting device is current. If the boxes are gray, specific requirements for the plotter are not required. Figure 24–5a and 24–5b show the Show Device Requirements dialog box for each plotting device. The display will vary greatly, depending on what plotting device is configured. For the current examples, a Hewlett Packard DraftPro DXL plotter is being used. The dialog box can be exited by picking the OK button. Either of these defaults can be changed by selecting the

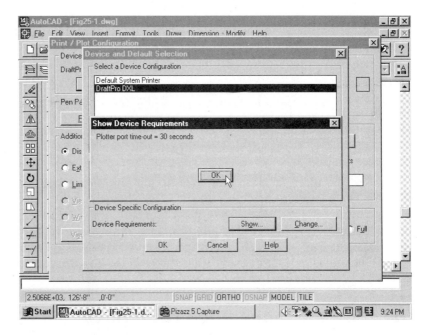

FIGURE 24–5a Selecting the Show... button will display the plotting requirements for the active plotting device.

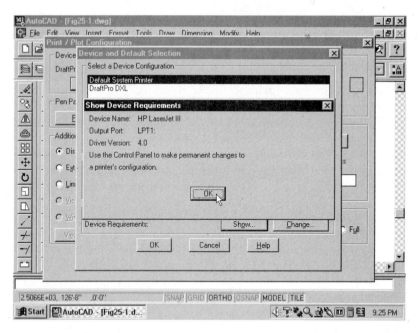

FIGURE 24–5b Switching the active plotting device alters the plotting requirements.

Change… button. This button should only be used to make long term changes to the plotting configuration. Changes that are required by specific plotting projects can be more easily controlled from other areas of the Plot Configuration dialog box.

This dialog box can be used to select which printer or plotter will be used, what size paper will be used, the paper source, and how the image will be arranged on the paper. Using a printer configured through the Windows Control Panel will produce a display similar to Figure 24–6. When DraftPro DXL set as the current choice, the display will resemble Figure 24–7. The DXL was selected by picking the down arrow beside the Printer Name: edit box. With the plotter as the current choice, setup options can be altered. With the current plotter, paper size and source can be selected. The paper and device options can also be established by picking the Properties… button. Figure 24–8 shows the Paper tab of the Properties dialog box. Figure 24–9 shows the Device Options tab. This dialog box can be used to set and control specific functions of the plotter that will tend to remain constant. Each property will be introduced throughout the balance of the chapter for controlling individual plots.

The Pen Parameter… Selection Box

The Pen Parameter box contains the Pen Assignment… and Optimization… buttons. Each can be used to control and enhance the plotting process.

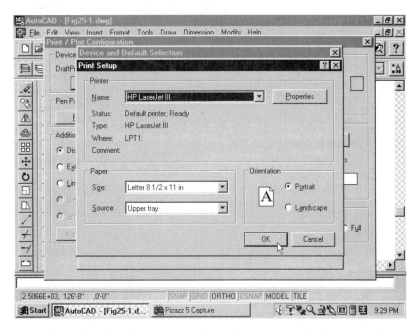

FIGURE 24–6 Selecting the Change... button displays the setup options for the configured plotting device. This display is for the system printer represented in Figure 24–2a.

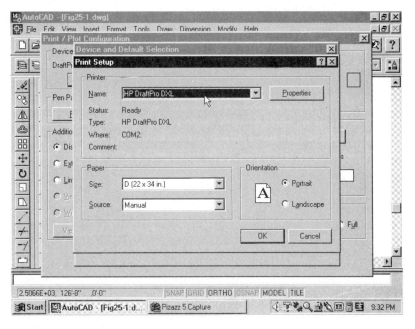

FIGURE 24–7 The setup menu for the DXL plotter.

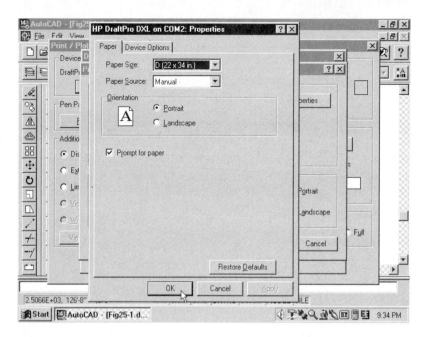

FIGURE 24–8 Selecting the Properties button in the Print Setup dialog box allows the paper options of the plotting process to be configured. This dialog box should be used when a .PCP or .PC2 file is being created. The Paper Size and Orientation portion of the Plot configuration dialog box should be used when configuring the plotter for a single plot.

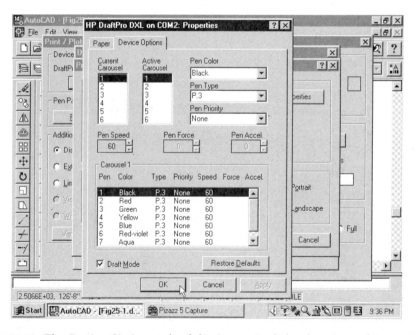

FIGURE 24–9 The Device Options tab of the Property dialog box is used to control the plotter properties when a .PCP or .PC2 file is being created. The Pen Size Assignments portion of the Plot configuration dialog box should be used when configuring the plotter for a single plot.

The Pen Assignments... Button

Selecting the Pen Assignments... button from the Plot Configuration Dialog box will produce a display similar to Figure 24–10. This dialog box can be used to set plotting values for individual plotting sessions. Plot values assigned using the Pen Assignment box will over ride values set in the Properties box for specific plotters. The dialog box is divided into three major parts, including a display of the current pen settings; the Feature Legend... , which contains a display of linetypes; and a box for modifying existing pen defaults.

In Chapter 15 drawing entities were assigned to various layers and colors. Each drawing color can now be assigned a different plotting color, linetype, pen speed, and line weight. An electrostatic plotter can plot lines of varied widths. Even though it has no physical pens, pen width assignments can be used to adjust line widths or line weights. Using a pen plotter, you can plot lines of varied color and width. If your current plotting device does not support multiple pens, the list box will be disabled and the message:

Not available for this device.

will be displayed. Pen plotting variations include color, pen number, linetype, pen speed, and pen width. Any parameters of the Pen Assignments box not supported by

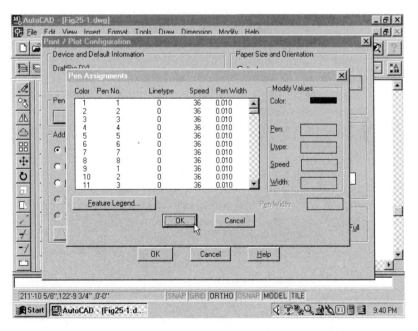

FIGURE 24–10 Selecting the Pen Assignment... button from the Pen Parameters area of the Plot Configuration dialog box will produce this display for controlling the plotting device. The dialog box is used for controlling the pen to be used, its speed, and its width.

the current plotter will be grayed out and unavailable for selection. For instance, if the current plotter is a laser, the Pen Speed is gray and cannot be adjusted. The display seen in Figure 24–10 is for a Hewlett Packard Draftpro DXL plotter. Your display may be slightly different.

Color.   The color category refers to the color for each layer of the drawing to be plotted. For instance, in Figure 24–10, all drawing entities drawn with color 1 (red) will be plotted with pen number 1 (the pen in the number 1 position of the plotter pen carousel) with a 0 linetype at speed 36. As a quick review, the basic colors numbers are:

1-red   5-blue

2-yellow   6-magenta

3-green   7-white

4-cyan

By default, colors 1-7 are assigned to pens 1-7 respectively. Typically, colors will be assigned by pen widths rather than by pen color. For instance, all lines that are to be bold can be assigned the color red, and all red lines can then be assigned to pen 2. All other colors can be assigned to pen 1 so that they can be thin lines.

Pen No.   This option designates the use of a pen in a specified position of the plotter pen carousel to be used for plotting. Pens for plotting are specified by a number that represents the pen width and by a color name. If a color plot is being produced, matching the color of the pen to the specific entity color will help as you plan the plot. For a single-color plot, each layer color needs to be plotted with the pen in holder number 1. Multicolor plots can be achieved with a single pen plotter by changing the pen after each color is plotted. Once a color is plotted, the plotting sequence will pause to allow you to install a new pen. A prompt will be displayed, such as:

Install pen number 2, color 4 (cyan)
Press RETURN to continue:

When plotting drawings on Mylar® or vellum to be used as an original, many offices will plot using two pens. For drawings such as a floor or foundation plan, a thin pen could be used for pen number 2, and a thick pen could be used for pen number 1. Thus pen number 1 could be used with all red lines, and pen number 2 could be used for all other colors.

Linetype.   The linetype display shows the current linetype that will be used to plot each layer color. The current linetypes in Figure 24–10 are listed as type 0, which is a continuous line. Typically, linetypes will be assigned within each drawing and will require no adjustment. A centerline from the drawing will be drawn as a centerline, even though the display shows that all lines are type 0 lines. These linetypes do not have anything in common with the linetypes you've specified in a drawing file.

Many plotting devices have the ability to generate internal plotter linetypes. The numbers that are displayed reflect the plotter settings. For the plotter to dictate the linetype, all drawing lines should be continuous. With the possibilities that are available in AutoCAD, best results will be obtained by leaving the plotter setting at 0, and adjusting the linetype within the drawing.

Pen Speed.   Pen speed should be based on the type of pen being used and the type of material that the drawing will be plotted on. Using too fast a speed for the pen will limit the amount of ink that can be plotted and will affect line density. Plotting with speeds that are too slow will extend the plotting time needlessly. The default pen speed is 36. General guidelines for plotting include:

- **Fiber-tip Pens (P).** Typically used for doing plots on plotter paper for a check print or for printing on glossy paper with a speed of 40cm/s. Effective speeds can range from 5 to 80 cm/s. These pens are not suitable for vellum, polyester film, or transparency film.
- **Transparency Pens (T).** These pens are recommended for transparency films and are excellent for making overlays or overhead projections using a speed of 10 cm/s. Speeds can range from 5 to 15 cm/s. These pens are not suitable for plotter paper, vellum, or polyester film.
- **Disposable Drafting Pens (V).** These pens are recommended for making final plots on plotter paper and vellum using a pen speed of 20 cm/s. Effective speeds can range from 10 to 30 cf/s. These pens are not suitable for polyester film, glossy paper, or transparency film.
- **Disposable Drafting Pens (F).** These pens are recommended for polyester film using a speed of 20cm/s. Effective speeds can range from 10 to 30 cf/s. These pens are not suitable for plotter paper, transparency film, glossy paper, or vellum.
- **Refillable Drafting Pens.** These pens are recommended for making final plots on polyester film and vellum using a pen speed of 15 cm/s. Effective speeds can range from 5 to 15 cf/s. These pens are not suitable for transparency film, glossy paper, or plotter paper. Values also will vary based on the quality of the pens.

Pen Width.   The pen width setting allows a plotter to work efficiently as the pen fills Solids, wide Polylines, or Traces by eliminating unnecessary raising and lowering of the pen. By telling the plotter the size of the pen being used, the number of pen passes can be limited and the plotting time decreased. Plotter pen widths are often described in millimeters, with common sizes of 0.20, 0.35, 0.50, and 0.70. If the plotter scale is to be entered in inches, the pen width should be adjusted, and the new value entered in inches. Many professionals prefer to use the default setting of 0.01 to gain accuracy in the intersection of thick lines while sacrificing some speed.

Feature Legend ... Selection Button.   If plotting linetype values are supported by the current plotter, the available linetypes can be viewed by picking the Feature Legend... selection button. This will produce a display similar to the one in Figure 24–11. If the plotter linetypes are to be used, the number of the desired linetypes can be entered.

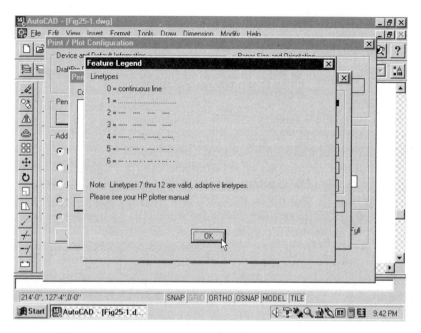

FIGURE 24–11 AutoCAD allows for altering the linetypes, although best results will be achieved if the linetypes are controlled within the drawing file.

Modify Values Selection Box. The Modify Values area of the Pen Assignments dialog box will allow the Pen, Linetype, Speed, and Pen width values to be adjusted easily. Pick the entry to be altered by moving the arrow to the desired line of values. In Figure 24–12, color number 7 is being adjusted. As the line is selected with the pointing arrow, the entire line will be highlighted, and the values will be displayed in the Modify Values box. The desired feature can now be edited by picking the appropriate feature box with the arrow cursor. The flashing line cursor will be displayed in the selection box, allowing a new value to be entered. If the wrong line of values is selected, it can be removed from the editing set by repicking the line before trying to alter any of its features. Multiple lines of values can be altered as one selection set. Although all of the selection sets will appear highlighted, only the values for the first line selected will be displayed. When all desired values are set, pick the OK button to return to the Plot Configuration dialog box.

The Optimizing Pen Motion Selection Box. Selecting the Optimizing Pen Motion selection box will display the dialog box seen in Figure 24–13. The default settings that are displayed will depend on what plotting device is configured. These optimization choices are intended to minimize wasted pen motion when the pen is in the up position so that plot time can be reduced. The first option is no optimization by the software, although many plotters have their own internal optimization programs. The last two options are features that usually are required when plotting 3D objects.

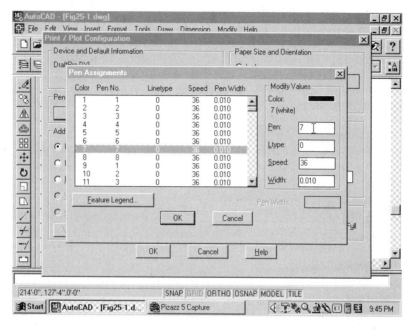

FIGURE 24–12 Selecting a pen number will display the parameters for that pen in the Modify Values box. The desired features can be altered by entering the desired value in the appropriate box.

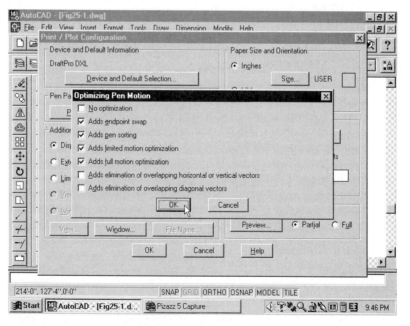

FIGURE 24–13 Optimizing pen motion cab reduce plotting time.

The Additional Parameters Selection Box

This selection box provides options for describing what portion of the current drawing will be plotted, options for describing how specific entities will be handled during the plotting process, and a box for choosing how the file will be stored. Because the entire drawing often does not need to be plotted, five options are given to decide what portion of the drawing will be plotted.

Display.   The default setting is to plot only the material that is currently displayed on the screen. If you have zoomed into a drawing, only that portion of the drawing will be plotted.

Extents.   Choosing this option will plot the current drawing with the extent of the drawing entities as the maximum limits that will be displayed in the plot. The option is similar to the Extents option of the ZOOM command. Using this option may eliminate entities at the perimeter of the drawing since the plotter cannot reach the edge of the plotting surface.

Limits.   Using this option will plot all of the drawing entities that lie within the current drawing limits that were defined during the drawing setup.

View.   If specific Views have been saved using the VIEW command, the View option for plotting will be activated. This option will allow specific views to be plotted as they were saved. If TILEMODE is off, any saved View from paper or model space may be plotted. In the initial setting, the view radio button is inactive. When a drawing contains named views, the View... button will be active. By picking the View... box, a listing similar to Figure 24–14 showing each saved view name will be displayed in the View Name dialog box. With the desired View to be plotted highlighted, selecting the OK button will activate the View radio button and allow the Plot menu to continue. Because the current drawing shown in Figure 24–1 had no named Views, the View radio button is not available.

Window.   This option will allow a specific area of a drawing to be defined for plotting. In the way that a window is selected to zoom or select objects for editing, a window can be specified here to define the selection plot. Initially the Window radio box is inactive. By selecting the Window... button, the Window Selection dialog box shown in Figure 24–15 will be displayed. The window can be specified by entering the drawing coordinates that will define the corners of the window. This method works well if the area to be plotted is not currently on the screen. Typically the Pick option will be used. Selecting the Pick < button will remove the dialog box from the screen and the command line will show the prompt:

    First corner:
    Other corner:

Once the window is defined, the Window Selection box will be redisplayed, allowing the coordinates to be confirmed or edited. Once approved, the Plot Selection dialog

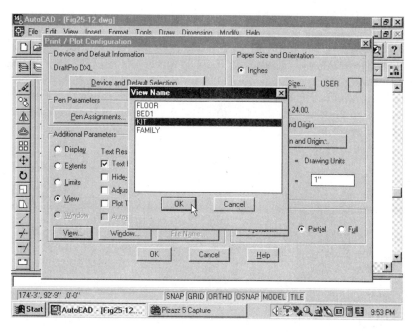

FIGURE 24–14 A named view may be printed be selecting the View... button.

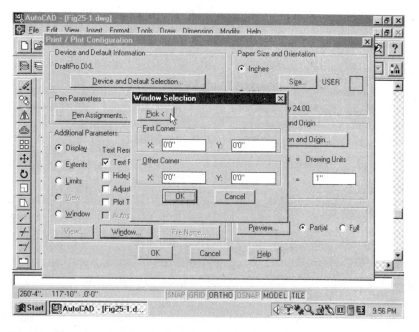

FIGURE 24–15 The size of the window for plotting can be selected by entering drawing coordinates.

box will be returned to the screen. The Window radio button will be activated, allowing other plotter options to be set.

Text Resolution.   The value in the text resolution edit box controls the resolution in dots per inch of TrueType fonts in the plotting process. The default value is 50. Lowering value decreases the resolution, but increases the plotting speed. Increasing the value will improve the quality of the text, and increase plotting time.

Text Fill.   With this option active, TrueType fonts will be filled. With the option deactivated, TrueType letters will be displayed as an outline rather than a solid character.

Hide Lines.   This option can be used to remove hidden lines from the finished plot of objects drawn in paper space. In paper space, hidden lines for each viewport will be processed based on the Viewport's Hideplot setting. Removing the hidden lines from a complex drawing can reduce plotting time.

Adjust Area Fill.   This option can be used if a high degree of accuracy is required. Since construction drawings usually are not scaled, this option does not usually have to be adjusted. If the radio box is activated, the plotting pen will be moved inward a distance of one half of the pen width when drawing Polylines, Trace, and Solids.

Plot to File.   When you complete all of the plotter dialog boxes, the information will be sent to the plotting device and the information will be processed. Depending on the size of the drawing file, the type of plotter, and the speed of the pens, the plot may take a few minutes or an hour to be completed. If the Plot to File radio button is selected, the File Name... button will be activated. Selecting File Name... button will display the Create Plot File dialog box similar to the box shown in Figure 24–16. The dialog box allows the storage location of the plot file to be specified using methods similar to that used by other dialog boxes for saving files. The file will be stored to disk with the indicated name followed by the .PLT extension. The disk then can be inserted into a spooler and supplied to a plotter. A spooler is not supplied with the plotter and must be purchased separately.

If your plotter is not connected to a spooler, leave the button inactive.

The Paper Size and Orientation Selection Box

This selection box allows the plotting units and orientation to be specified. Most construction projects will use the default setting of Inches for the plotting standard. The two other features of this box are the Icon box, and the Size...box.

The box on the right side of the selection box is the orientation icon and not a selection box. The current orientation of the box is based on the configured plotter and indicates the drawing will be plotted in the normal orientation for construction drawing. This setting is referred to as "Landscape." When the plot is rotated so that it is read from the titleblock end, the plot is referred to as a Portrait plot. A portrait

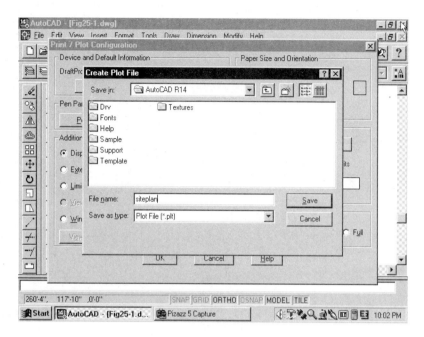

FIGURE 24-16 A drawing file can be saved as a plot file by making the Plot to File plotting option active.

plot is more likely to be used with laser printers. The orientation of the drawing to the paper can be adjusted by picking the appropriate radio button in the Rotation and origin box.

Size... Selection.   Picking this button will produce a dialog box similar to the display seen in Figure 24–17. This display indicates the standard plotting sizes that can be accommodated by the current plotting device. The MAX size is the maximum size the current plotter can process. The right side of the box will allow user-supplied paper sizes to be installed and recalled for future use. User 1 indicates a paper size of 24" × 36" has been selected. When the Width and Height values are entered, they will be stored as part of the standard size menu and can be selected for use. Once the desired value is entered, the paper size in the Paper Size and Orientation box will be updated.

Scale, Rotation, and Origin Selection Box

The Plot Configuration dialog box controls settings for the Scale, Rotation, and Origin of plot. The current plotting scale is indicated beneath the Rotation selection box.

Rotation and Origin . . .   Picking this selection button will display a dialog box similar to the display seen in Figure 24–18. The Plot Rotation radio buttons will allow for rotating the plot in a clockwise direction on the paper.

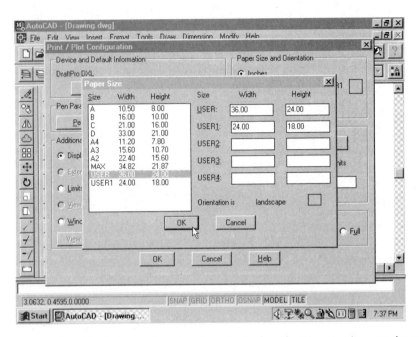

FIGURE 24–17 Standard paper sizes can be selected or the User option can be used to add additional paper sizes.

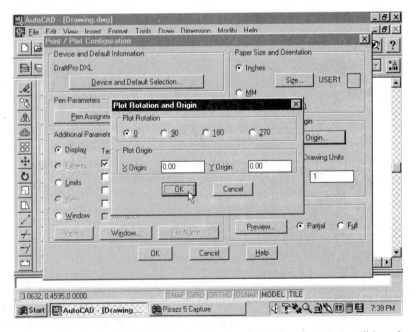

FIGURE 24–18 The Plot Rotation box determines how the drawing will be placed on paper. Plots can be rotated in 90° increments in a clockwise direction.

Plot Origin. The default setting for plotting origin is X = 0 – Y = 0. For most plotters this will be the lower left-hand corner of the paper. By selecting the X or Y origin box, new values can be entered. Entering an X value of 12 and a Y value as 10 would place the new origin 12 units to the right and 10 units above the original plotting origin.

Selecting a Scale. As drawings have been created throughout this text, they've been drawn at real size rather than at a reduced scale, as would be done in manual drawing. In Chapter 23 you were introduced to methods of referencing a drawing to a paper space drawing template. If you're using methods introduced with xrefing, you can set the scale factor at 1" + 1". If you skipped over externally referenced drawings and want to make a plot of a drawing created in model space, you'll need to provide a scale factor for plotting. Dimensions and text were both multiplied by a scale factor in anticipation of the scale at which the drawing would be plotted. By entering values in the Plotted Inches and Drawing Units boxes, the scale at which the drawing will be plotted will be determined. The plotting values will be displayed reflecting the units that were entered for paper size. Figure 24–19 shows a list of common scale factors that should be used during the drawing setup to produce a plot at a given size. Refer to Chapters 18 and 20 for a review of how scale factors will affect a drawing in model space.

An alternative to picking a precise scale is to activate the Scaled to Fit box. This option will allow the desired plot to be sized to fit within the parameters of the paper. The resulting scale will be displayed above the Fit box. Drawings that are Scaled to Fit on a printer are useful as check prints to determine completeness, but are not suitable for most projects within the construction industry.

The Plot Preview Selection Box

The final dialog box to be displayed for plotting is accessed from the Plot Preview button. Selecting Preview... allows the outcome of the current plot settings to be viewed prior to the actual plot. Two options are available for previewing the plot— Partial or Full Preview.

Partial Preview. The default setting of a preview is partial. Selecting the Preview... button with Partial active will produce a display similar to the one seen in Figure 24–20. The paper size is represented by a red line and the limits of the effective plotting area are specified on the screen by a blue line. If either of the two sizes matches, a dashed red and blue line will be used to represent the plot. A warning also will be given of any problems that may be encountered during the plot. Common warnings displayed in the warning box include:

    Effective area too small to display
    Origin forced effective area off display
    Plotting area exceeds plotter maximum

A small triangular Rotation Icon is displayed in the lower left-hand corner to represent a 0 rotation. As the rotation is set to 90 degrees, the icon is moved to the upper

PLOTTING SCALE FACTORS

ARCHITECTURAL SCALE FACTORS

SCALE	LIMITS		TEXT SCALE	LTSCALE
	18 × 24	24 × 36		
1" = 1'-0"	18 × 24	24 × 36	12	6
3/4"= 1'-0"	24 × 32	32 × 48	16	8
1/2" = 1'-0"	36 × 48	48 × 72	24	12
3/8"=1'-0"	48 × 64	64 × 96	32	16
1/4" =1'-0"	72 × 96	96 × 144	48	24
3/16" =1'-0"	96 × 128	128 × 192	64	32
1/8" = 1'-0"	144 × 192	192 × 288	96	48
3/32" = 1'-0"	192 × 256	256 × 384	128	64
1/16" = 1'-0"	288 × 384	384 × 576	192	96

ENGINEERING SCALE FACTORS

SCALE	LIMITS		TEXT SCALE	LTSCALE
	18 × 24	24 × 36		
1"= 1'-0"	18 × 24	24 × 36	12	6
1"= 10'-0"	180 × 240	240 × 360	120	60
1"= 100'-0"	1800 × 2400	2400 × 3600	1200	600
1"= 20'-0"	360 × 480	480 × 720	240	120
1"= 200'-0"	3600 × 4800	4800 × 7200	2400	1200
1"= 30'-0"	540 × 864	864 × 1080	360	180
1"= 40'-0"	720 × 960	960 × 1440	480	240
1"= 50'-0"	900 × 1200	1200 × 1800	600	300
1"= 60'-0"	1080 × 1440	1440 × 2160	720	360

NOTE: ALL LIMIT SIZES ARE GIVEN IN FEET.

FIGURE 24–19 Common scale factors to be used during drawing setup.

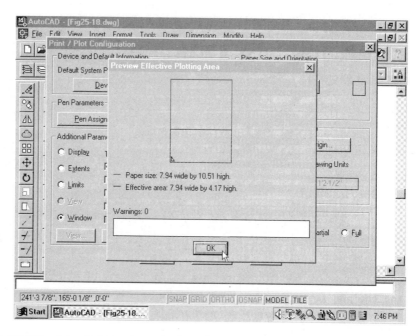

FIGURE 24-20 A partial preview displays the paper size in red and the effective plotting area in blue.

left-hand corner, to the upper right-hand corner for 180 degrees, and to the lower right-hand corner for 270-degree rotation.

Full Preview. Choosing this option prior to picking the Preview... button will produce a graphic display of the drawing as it will appear when plotted. Figure 24–21 shows an example of a full preview. A few seconds will elapse between picking the Preview... box and the display of the proposed plot. An outline of the paper size will be drawn on the screen, which displays the portion of the drawing to be plotted. Once the drawing is displayed, the Real time Zoom icon is displayed. The Pan and Zoom option can be quite useful in viewing detailed areas of the drawing a final time prior to plotting. Once the indicated Pan or Zoom is performed, press [enter] to restore the Plot Configuration dialog box.

Producing the Plot

Seven areas of the Plot Configuration dialog box have been examined. If you change your mind on the need for a plot, select the Cancel button; the dialog box will be terminated, and the drawing will be redisplayed. If you are satisfied with the plotting parameters that have been established, select the OK box. The dialog box will be removed from the screen and the command prompt will display the effective plotting area. You'll also be given a prompt to remind you to position paper in the plotter. Don't laugh! After laboring through the Plot Configuration dialog box the first few

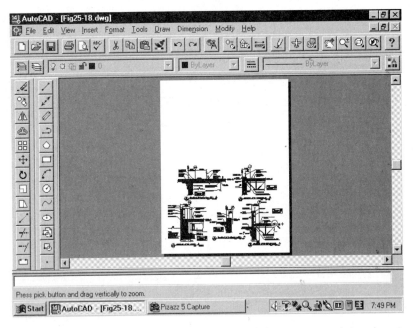

FIGURE 24–21 A full preview will show the limits of the paper and the drawing to be plotted. (Courtesy StructureForm Masters, Inc.)

times, simple things like turning the plotter on, loading the pens, and installing the paper may tend to be forgotten. The final prompt is:

Press RETURN to continue or S to Stop for hardware setup.

Selecting the S option will give you a prompt:

Do hardware setup now

Pressing S is really an unnecessary step. If you need to set up the plotter, you can do it before you press RETURN. Your plotter may have additional features, such as pen pressure and speed, to be adjusted. The completed plot can be seen in Figure 24–22.

PLOTTING FROM THE COMMAND LINE

For experienced AutoCAD users, the command line prompts are still available, but once you've been exposed to the dialog box, you may be quite happy to forsake the menus. The choice between the dialog box or the command line prompts is controlled by the CMDDIA system variable. With the CMDDIA variable default setting of 1, the dialog box will be displayed. If the system variable is set to 0, the plotting

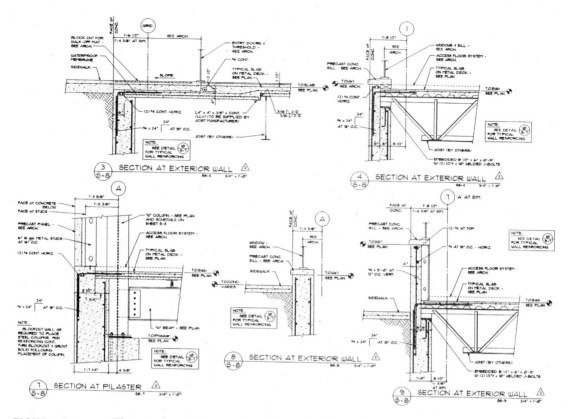

FIGURE 24-22a The results of the preview shown in Figure 24–21. (Courtesy Van Domelen / Looijenga / McGarrigle / Knauf Consulting Engineers)

prompts will be displayed. Typing **PLOT** ⏎ at the command prompt will produce the following prompts at the command line:

What to plot—Display, Extents, Limits, View or Window <D>:

As the selection option is completed, several lines of text will scroll through the command line and the Text window will show the following display.

Plotter Port time-out = 30 seconds

Plot device is Hewlett-Packard (HL-GL) ADI 4.2-by Autodesk, Inc.
Description: DraftPro DXL
Plot optimization level = 4
Plot will NOT be written to a selected file
Sizes are in Inches and the style is landscape
Plot origin is at (0.00, 0.00)
Plotting area is 36.00 wide by 24.00 high (USER size)
Plot is NOT rotated

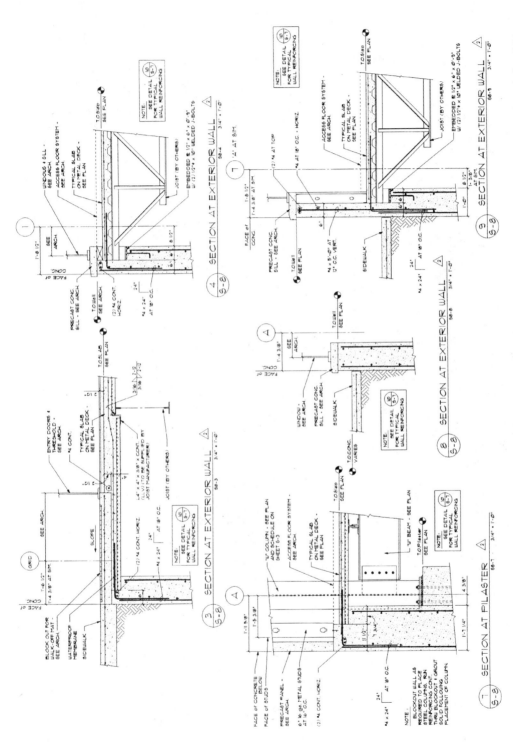

FIGURE 24-22b The results of the preview shown in Figure 24-21. The End Preview button was selected and the 90° rotation was picked prior to selecting the OK button to produce the print. (Courtesy Van Domelen / Looijenga / McGar-rigle / Knauf Consulting Engineers)

Hidden lines will NOT be removed
Scale is 1" = 1"

0. No changes, proceed to plot
1. Merge partial configuration from .PCP file
2. Replace configuration from .PC2 file
3. Save partial configuration as .PCP file
4. Save configuration as .PC2 file
5. Detailed plot configuration
Enter choice, 0-5 <0>:
Effective plotting area: 35 wide by 23 high
Plot complete
Command:

Each of the six options completes the equivalent of its dialog box counterpart.

MAKING MULTIPLE PLOTS WITH BATCH PLOT

New to Release 14 is the BATCHPLT plotting utility that can be used to construct a list of AutoCAD drawings to be plotted. BACTHPLT could be especially helpful when all of the drawings for a construction project are complete, and plots of each drawing must be assembled for printing. Rather than opening one drawing at a time and making a plot, each drawing to be plotted can be assigned to a plotting list, and AutoCAD will run the plots while you do something more productive. Once the list is compiled, the utility controls AutoCAD and plots each of the listed drawings.

Before the utility is used, each file must be prepared for plotting. The drawing should be saved so that the required view to be plotted is displayed as the drawing is saved. You also need to verify that all required reference drawings, linetypes, text fonts, and layer properties are available so that they can be loaded as needed. Also, you need to set and save the needed plot configuration as a .PC2 files. The utility is accessed by selecting Batch Plot Utility from the AutoCAD program group from Programs of the Start menu. As the program is entered, the Batch Plot Utility dialog box shown in Figure 24–23 will be displayed. Pick the Add Drawing... button to add a drawing to the plotting list. Picking the button will produce the Add Drawing File dialog box shown in Figure 24–24. This box is used to select the file, and is removed from the screen as the OK button is selected. Additional drawing files can be added to the list by repeating the steps used to add the first drawing.

With the list complete, highlight the first drawing file title and then pick the Associate PCP/PC2 button. This will produce the Associate PCP/PC2 File dialog box shown in Figure 24–25. The dialog box can be used to select the plot configuration file that was stored for this drawing. The process must be repeated for each drawing in the list. When each file has been added and the plotting devices activated, pick the Plot button to begin the plotting process. Because construction drawings often have to be revised and replotted, saving the batch plot list can be extremely useful.

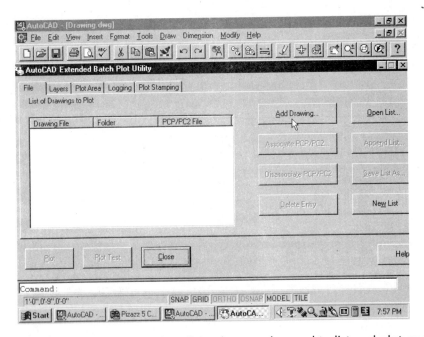

FIGURE 24–23 The Batch Plot Utility dialog box can be used to list and plot multiple drawings.

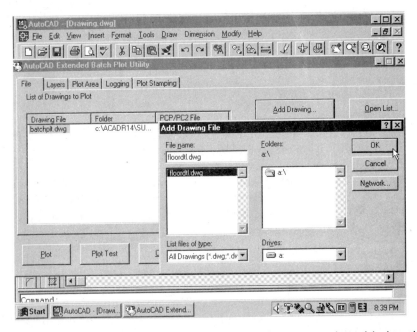

FIGURE 24–24 The Add Drawing dialog box allows drawings to be added to the plot-ting list.

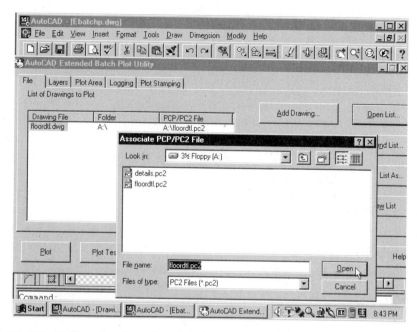

FIGURE 24–25 Picking the Associate PCP/PC2 button will produce this box and allow the plotting list to be altered.

The list can be saved by picking the Save List As... button. The list is saved using methods similar to saving a drawing. A .BP2 extension will be automatically assigned to the plot file as it is saved.

CHAPTER 24 EXERCISES

E-24-1. Open drawing E-11-7. Assume that an HP LaserJet III with 2 MB of additional memory will be used. Use the appropriate dialog box to set the following plot parameters.

> The appropriate plotter to be added to the Config. file.
> Plot the extents.
> Scale = fit.
> Full preview.
> Five copies on letter-size paper.
> No pen optimization.
> Paper size = A with Portrait orientation.
> Save the information to a plot file named E-24-1.

E-24-2. Use a laser or plotter and set the required parameters to plot E-20-8 at a scale of ¼″ = 1'-0″.

E-24-3. Open drawing E-14-4. Plot the drawing at a scale of ⅜" = 1'-0" using at least two different pen sizes or colors.

E-24-4. Open drawing E-16-10. Use the copy command and create a second copy of the elevation. Design a second elevation using a different architectural style and materials. Save the drawing file as E-24-4. Plot your prototype drawing containing your border and title block to a C size sheet of vellum. Plot E-25-4 at a scale of ¼" = 1'-0" neatly centered between the borders.

E-24-5. Plot your drawing templet containing the border and title block onto a D size sheet of paper. Create a new base drawing that will fit within the limits of the borders. Attach any four drawings in your library. Each drawing is to be shown at a different scale when plotted. Provide a title and the scale used by the original drawing prior to plotting.

CHAPTER 24 QUIZ

1. How many colors can be configured for plotting?

2. List the three major components of the Pen Assignment Box.

3. Is it possible to have more than one plotting device connected to a single computer? Explain your answer.

4. What options are available to preview the information to be plotted prior to plotting, and how are they accessed?

5. List the major areas of the Plot Configuration dialog box.

6. How can pen widths be used when using a printer?

_____

_____

7. What major factor influences the setting for pen speed?

_____

_____

8. What are pen optimization controls designed to do?

_____

_____

9. What line number is a continuous line?

_____

_____

10. Describe the meaning of the lines and colors used in a partial plot display.

_____

11. What pens are best for plotting on vellum?

_____

_____

12. How can a plot be produced by command line entry methods?

_____

_____

13. What dialog box contains a display of the current plotting scale?

_____

_____

14. What dialog box can be used to control how much of the current drawing will be displayed?

_____

_____

15. What direction is a drawing turned as it is rotated for plotting?

_____

_____

16. What system variable and setting is required to display dialog boxes?

_____

_____

17. List five options for controlling what portion of the drawing will be plotted.

_____

_____

18. Describe the typical origin location for a plotter and a printer.

_____

_____

19. What is the advantage of saving the plot information to a "Plot to File"?

_____

_____

20. How does the TILEMODE setting affect multiscaled plots?

_____

_____

_____

_____

25

OBLIQUE AND ISOMETRIC DRAWING

The drawings that have been examined throughout this text have been orthographic drawings. These are drawings that show an object by looking directly at it. Objects are assumed to be located in a glass box, with the sides of the object parallel to the sides of the glass box. Orthographic drawings comprise the majority of construction drawings. Oblique and isometric drawings are used in some disciplines of engineering and architectural drawings to show specific details. This chapter will introduce the creation of oblique and isometric drawings, the skills necessary to produce these drawings, and specific commands that can be used to aid in drawing construction.

OBLIQUE AND ISOMETRIC DRAWINGS

Before examining what oblique and isometric drawings are, it is important to understand what they are not. These drawings appear to be three-dimensional, but are really two-dimensional drawings using lines that are created using X and Y coordinates.

3D Drawings

One of the major features of AutoCAD Release 14 is its ability to create true 3D drawings. Three-dimensional drawings require creation of a 3D model. For instance, in drawing a floor plan, heights are provided in addition to the normal information that would be drawn. Once the model is created, AutoCAD can rotate the model in space, allowing the viewing point to be altered. You've most likely seen commercials on television with structures growing out of a floor plan and progressing through the entire construction process. These drawings could be created starting with 3D models. Individual points, lines, and planes comprising the drawings are stored and become part of the drawing base. Figure 25–1 shows examples of 3D drawings created using AutoCAD and the add-on program, BIG D Rendering Software. Although these drawings are both beautiful and very useful, how they are created will not be discussed in this text.

Oblique Drawings

Oblique drawings show height, width, and depth of an object. Three common methods of oblique drawings include cavalier, cabinet, and general drawings. Examples of each can be seen in Figure 25–2. Each system places one or more surfaces parallel to the viewing plane, and shows the receding surface at an angle. Each drawing method may be drawn at any scale.

Cavalier Drawings. Cavalier drawings use the same scale for the front and receding surfaces. The front side is drawn parallel to the viewing plane. The receding surface is drawn at a 45° angle to show depth. See Figure 25–3.

Cabinet Drawings. Cabinet drawings are similar to cavalier drawings but show the receding surfaces using a scale that is half of the scale used on the front surfaces. The receding surfaces are also shown at a 45° angle. See Figure 25–4.

General Drawings. General drawings represent the receding surfaces using a scale that is three-quarters of the scale used to draw the front surfaces. The angle used to create the receding edge can be any angle except 45 degrees. Figure 25–5 shows an example of a general oblique drawing.

Isometric Drawings

Isometric drawings are the most common form for viewing three surfaces of an object. Rather than having one surface resting squarely on a base, the object is rotated at 35° and 16' from a horizontal plane. This results in a view of the object from the front where each edge is drawn at 30° to a horizontal base. All vertical lines remain vertical, and an equal scale is used to represent all surfaces. Figure 25–6 shows the basic construction methods of an isometric drawing. Lines that are parallel to one of the three axes will remain parallel. Inclined surfaces that are not parallel to an axis will need to be projected from a point that lies on a line that is parallel to an axis. Figure 25–7 shows the steps for blocking out an irregularly shaped detail.

FIGURE 25–1 3D drawings rendered on AutoCAD often resemble a photograph. (Courtesy Big D Rendering Software by Graphics Software Inc.)

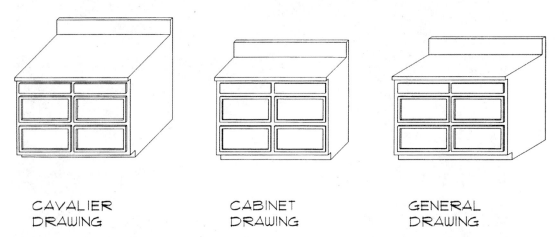

CAVALIER
DRAWING

CABINET
DRAWING

GENERAL
DRAWING

FIGURE 25-2 Although lacking the details of 3D drawings, three types of oblique drawings are often used to show height, width, and depth of an object.

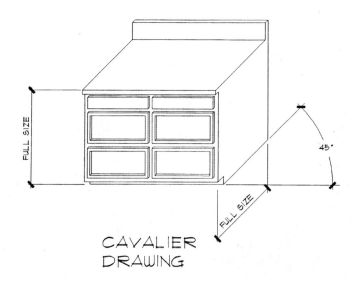

CAVALIER
DRAWING

FIGURE 25-3 Cavalier drawings show all surfaces in full scale, with the receding side drawn at a 45° angle.

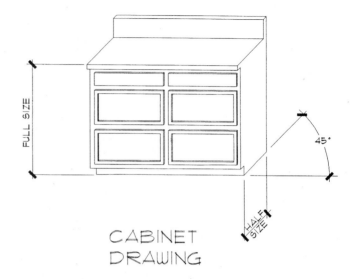

CABINET
DRAWING

FIGURE 25–4 Cabinet drawings show the front view at full scale, with the receding side shown at half scale and a 45° angle.

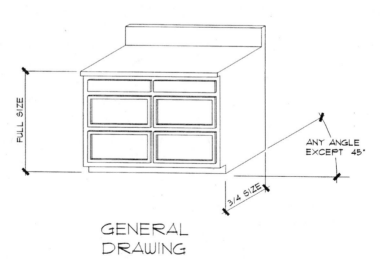

GENERAL
DRAWING

FIGURE 25–5 General drawings show the front view at full scale, with the receding side shown at three-quarter scale and any angle except 45°.

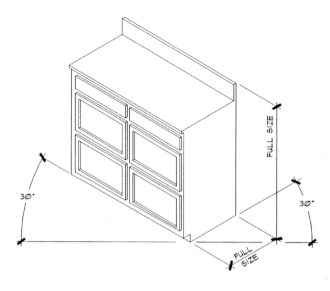

ISOMETRIC DRAWING

FIGURE 25–6 Isometric drawings show all sides at full scale, and each side at a 30° angle.

DRAWING TOOLS FOR PICTORIAL DRAWINGS

The drawing tools used to create two-dimensional drawings also can be used to create pictorial drawings. Commands such as SNAP, GRID, ISOPLANE, and ELLIPSE can be used to aid in drawing setup.

SNAP

Pictorial drawings can be created easily by entering the STYLE option of the SNAP command and picking Isometric. The command sequence is:

Command: **SN** [enter]
Snap spacing or ON/OFF /Aspect/Rotate/Style<0'-1 ">: **S** [enter]
Standard/Isometric <S>: **I** [enter]
Vertical spacing <0'-1 ">: **.25** [enter]
Command:

Returning to the drawing display will display an isometric grid and cursor, as seen in Figure 25–8. The snap setting can also be set using the Drawing Aids dialog box by selecting Drawing Aids... from the Tools pull-down menu or by typing **RM** [enter] or **DDRMODES** [enter] at the command prompt. Each method will produce the display shown in Figure 25–9. The isometric snap grid is activated by picking the On check box in the Isometric Snap/Grid box. Once On is activated, the X spacing for snap and grid will be deactivated.

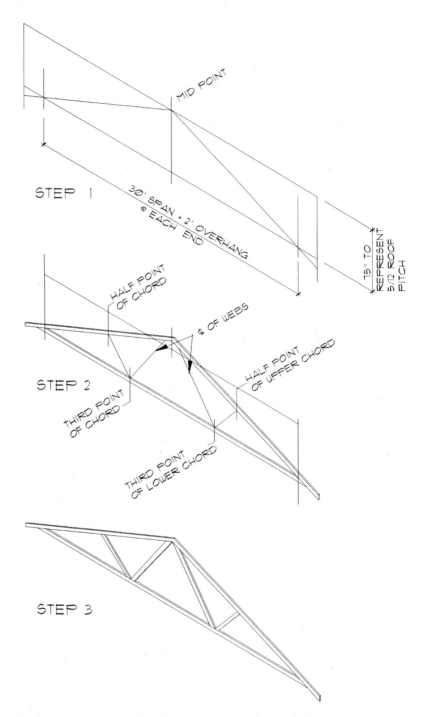

FIGURE 25–7 Blocking irregular-shaped objects in rectangles will aid in drawing layout.

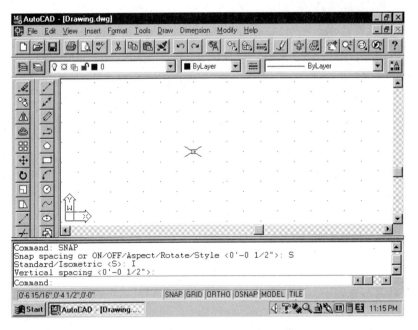

FIGURE 25–8 An isometric grid can be established by using the ISOMETRIC setting of the STYLE option of the SNAP command.

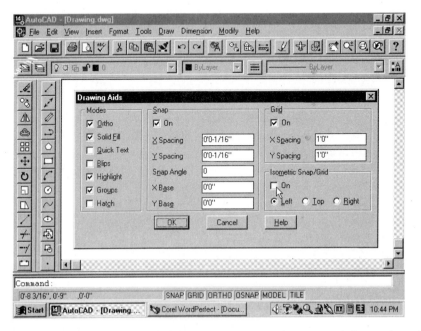

FIGURE 25–9 The Drawing Aids dialog box can be used to control the options for isometric drawings. Access the box by typing **RM** [enter] or **DDRMODES** [enter] at the command prompt or by selecting Drawing Aids... from the tools pull-down menu.

Cross-Hair Orientation

Three surfaces are seen in pictorial drawings and are referred to by AutoCAD as Isoplane Left, Isoplane Top, and Isoplane Right. These three orientations are also available for the cross-hairs to ease drawing construction. Figure 25–10 shows the three isometric planes and the options for cross-hair positioning. The cross-hair orientation can be changed using one of several methods including a Function key, the Isoplane command, the Drawing Aids dialog box, or by using Control keys.

Function Key

The easiest toggle between the three isoplane options is to use the F5 function key. Each time the F5 key is pressed, the cross-hairs will toggle to the next isoplane and the current setting will be displayed at the command line.

The ISOPLANE Command. The cross-hair position also can be controlled using the ISOPLANE command. The command sequence is:

Command: **ISOPLANE** [enter]
Left/Top/Right/<toggle>: **T** [enter] (*Enter either option.*)
Current Isometric plane is: Top
Command:

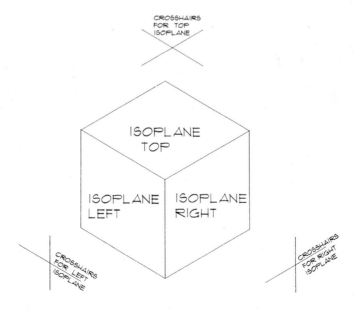

FIGURE 25–10 AutoCAD refers to the three isometric surfaces as Isoplane left, Isoplane top, and Isoplane right.

The letter representing an option can be entered, followed by the [enter] key. Pressing the [enter] key at the toggle prompt will change the cross-hairs automatically to the next option and restore the command prompt.

The Drawing Aids Dialog Box.   The isoplane can be set using the Drawing Aids dialog box by selecting one of the three buttons in the Isometric Snap/Grid box shown in Figure 25–9.

Control Key Control.   The cross-hairs also can be controlled by using the CTRL + E keys. Pressing CTRL + E with the cross-hairs in the current setting will display a prompt of:

    Command: <Isoplane Top>:

Repeated pressing of the CTRL + E keys will toggle transparently the cross-hair between the Left, Top, and Right settings.

Drawing Isometric Circles

Circles in pictorial drawings are represented by ellipses.With the SNAP option set to ISOMETRIC, enter the ELLIPSE command and select the ISOCIRCLE option; an ellipse will be drawn automatically in the current ISOPLANE setting, as seen in Figure 25–11. The command sequence is:

    Command: **EL** [enter]
    Arc/Center/Isocircle/<Axis endpoint 1> **I** [enter]
    Center of circle: (*Specify desired point.*)
    <Circle radius>/Diameter:
    Command:

A centerpoint may be selected and Drag may be used to place the circle, or a Radius or Diameter may be specified.

One of the drawbacks to using pictorial drawings to show construction details is that circular objects often appear distorted. Round objects will appear as an ellipse in an isometric drawing. Figure 25–12 shows the proper orientation of circles for each plane of an isometric drawing. Notice that the center line of the circle is parallel to the edge of one of the three surfaces.

Isometric Text

No special commands are required to place text on pictorial drawings. On small details, text is best placed in its normal position and connected to the detail with leader lines. On larger drawings, text can be placed parallel to the Isoplane on which it will appear by adjusting the Obliquing Angle in the TEXT or DTEXT command. Figure 25–13 shows an example of text placed at 30°, –30°, 90°, and –90°. This can best be done by setting up text styles for each angle.

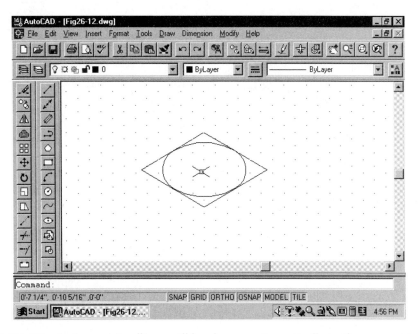

FIGURE 25–11 An isometric ellipse will be drawn automatically in the current isoplane setting using the ISOCIRCLE option of the ELLIPSE command.

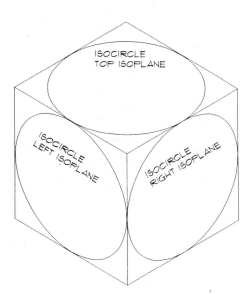

FIGURE 25–12 Proper orientation of isometric circles.

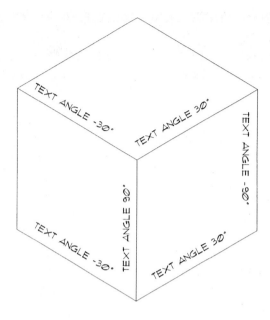

FIGURE 25–13. Text can be placed in each plane of an isometric drawing using the TEXT or DTEXT command and rotating the angle.

Isometric Dimensions

Although no specific command exists to create isometric dimensions, dimensions created using the Aligned can be adjusted quickly to conform to an isometric drawing. Figure 25–14 shows an example of dimensions added to an isometric drawing using the Aligned Dim option. Once created, enter Oblique from the DimEdit: prompt. You will be asked to select a dimension to be edited, and then prompted to specify an obliquing angle. Use either 30° or –30° for the angle. The command sequence is:

> Command: **DIMALI** [enter]
> First extension line origin or RETURN to select: *(Select first point.)*
> Second extension line origin: *(Select second point.)*
> Dimension line location (Mtext/Text/Angle) : *(Select point.)*
> Dimension text <2 3/4">: [enter]
> Command:
> Command: **DIMEDIT** [enter]
> Dimension Edit (Home/New/Rotate/Oblique) <Home>: **O** [enter]
> Select objects: *(Select object.)*
> Select object: [enter]
> Enter obliquing angle (RETURN for none): **30** [enter]
> Command:

The results of this sequence can be seen in Figure 25–15.

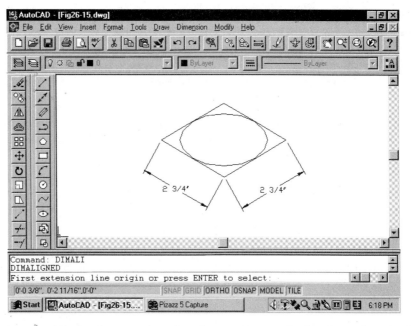

FIGURE 25–14. Dimensions can be placed on an isometric drawing using the DIMALI command.

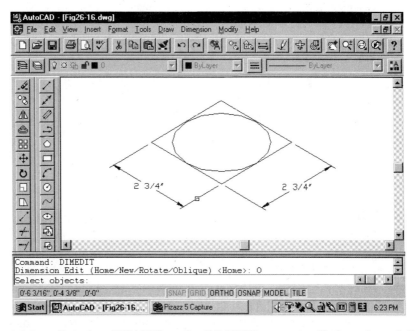

FIGURE 25–15. Entering OBLIQUE at the DIMEDIT prompt will alter aligned dimensions to isometric dimensions.

CHAPTER 25 EXERCISES

E-25-1. Open a new drawing and draw a cube that is 8′ wide, 6′ high, and 4′-6″ deep using each of the oblique drawing methods. Place each drawing on a separate layer. Save the drawing as E-25-1.

E-25-2. Open a new drawing and draw a cube that is 8′ in each direction using isometric techniques. Place an isometric circle in each plane so that the circle is tangent to the edges of each plane. Save the drawing as E-25-2.

E-25-3. Open a new drawing and draw an isometric cube that is 6′ in each direction. Show the bottom, left, and right planes. Place an isometric circle in each plane so that the circle is tangent to the edges of each plane. Save the drawing as E-25-3.

E-25-4. Use the drawing below and draw an isometric drawing showing the framing of a 5′-0″ × 3′-6″ window. Use 2 × 6 studs. Label the top plates, a 4 × 8 header, 2 × 6 sill, and 2 × 6 studs, kingstuds, trimmers, and cripples. Save the drawing as E-25-4.

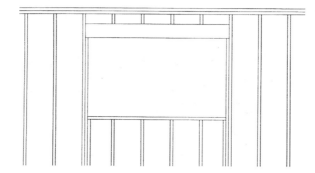

E-25-5. A window supplier would like to update its catalogue. Provide an isometric drawing of a 4′-0″ wide × 3′-6″-high vinyl window with a half-round window above. Show the window with a grid in each slider, and a radial pattern in the upper window. Draw the frame as 1″ aluminum, and the grid as ½″ wide material. Place the grid on a separate layer from the window. Save the drawing as E-25-5.

E-25-6. Use the following floor plan and draw an isometric drawing showing the left wall with the refrigerator and double oven. Design and show drawers and cabinet drawers. Use glass in some of the doors and show the shelving. Consult vendor catalogues for specific sizes of appliances and cabinets. Omit the refrigerator. 8′-0″ high ceilings. Show the drawing as E-25-6.

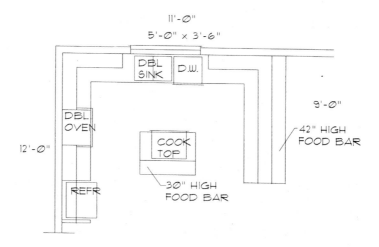

E-25-7. Use the floor plan for drawing E-25-6 and the drawing started in exercise E-24–6 to draw an isometric of the complete kitchen. Draw the cabinets over the food bar as 30″ high. Save the drawing as E-25-7.

E-25-8. Use the drawing drawing on the right to draw an isometric drawing showing the metal beam seat. Show the bracket, but omit the beam. Make the side plates from ½″ steel and the base plate from ⅞″ × 7″ wide steel plates. Place bolts at 2″ in from each edge and at 3″ o.c. Dimension the size of the steel plates. Save the drawing as E-25-8.

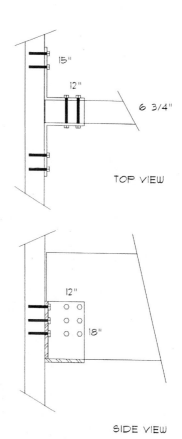

E-25-9. Use the following drawing as a guide to draw an isometric view of the truss. Assume that the top and bottom chords are made from two 2 × 3s, and the webs are 1″ in diameter. Show the distance from the top of the top chord to the bottom of the metal bracket as 3½″. Show the total depth as 36 inches, and lay out the webs at 45 degrees from the top chord. Save the drawing as E-25-9.

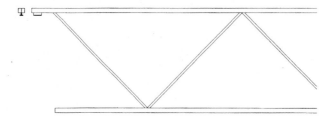

E-25-10. Use the drawing below as a guide to draw an isometric view of the 6¾ × 13½″ glu-
lam beam supported on two 6 × 6 posts at 10 feet o.c. on 18″ diameter concrete piers.
Show piers extending 8″ minimum above the finish grade. Show the wall framed
with 2 × 6 studs at 16″ o.c. Frame the floor with 2 × 10 floor joist at 12″ o.c. 12′ above
the grade covered with ¾″ plywood floor sheathing and an 18″ cantilever. Use a

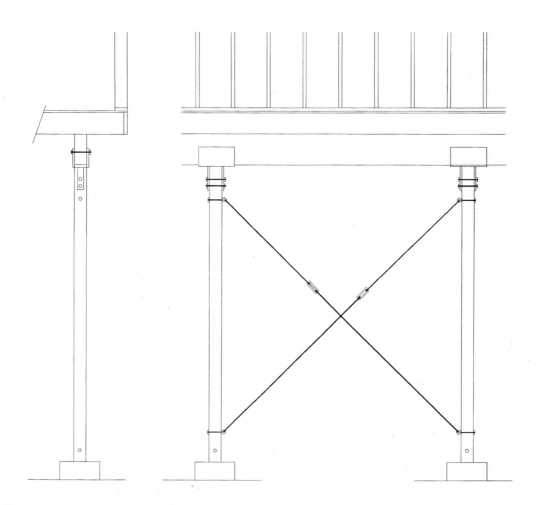

13″ × 8″ metal hanger with 2″ × 10″ legs to post. Show two bolts through the hanger into each beam, and two bolts through the hanger legs and post. Place bolts 1½″ from steel edges and 3″ o.c. Between each post, forming an "X," use ½″ steel cables with turnbuckles. Attach the cable to each post with eyebolts 6″ up and from the bottom. Place the top eyebolt 3″ below the bottom hanger's leg bolt. Label major materials and dimension the cantilever and post spacing. Since your drawing will be used to assemble this project and not to manufacture each item, great accuracy is not required on items such as the turnbuckle and eyebolts. Save the drawing as E-25-10.

CHAPTER 25 QUIZ

1. What are the differences between cabinet and general oblique drawings?

2. Compare the differences between cavalier and cabinet drawings.

3. Describe the relationship of horizontal and vertical lines in an isometric drawing.

4. What command and option adjust the grid for isometric drawings?

5. List the three options for selecting the isometric SNAP options.

6. Explain the easiest way to construct isometric circles.

7. List common angles for placing text on isometric drawings.

8. Sketch and label the three positions of the cross-hairs in isometric drawings.

9. What is the major difference between oblique and 3D drawings?

10. Explain the process for placing dimensions on an isometric drawing.

CHAPTER

26

AUTOCAD AND THE INTERNET

The Internet is an important way to exchange information. AutoCAD allows you to interact with the Internet by launching a Web browser from within AutoCAD and by creating drawing Web format (DWF) files for viewing drawings on Web pages. This chapter will introduce you to methods for launching a Web browser, help you use the Uniform Resource Locator (URL), and introduce the drawing Web file format (DWF).

INTERNET COMMANDS

You are probably already familiar with the best-known uses for the Internet: email (electronic mail) and the Web (short for "World Wide Web"). Email lets users exchange messages and data at very low cost. The Web brings together text, graphics, audio, and movies in an easy-to-use format. Other uses of the Internet include FTP (file transfer protocol) for effortless binary file transfer, Gopher (presents data in a structured, subdirectory–like format), and USENET, a collection of more than 10,000 news groups.

AutoCAD allows you to interact with the Internet in several ways. Release 14 is able to launch a Web browser from within AutoCAD. Release 14 can create DWF (short for "drawing Web format") files for viewing drawings in 2D format on Web pages. AutoCAD can open, insert, and save drawings to and from the Internet.

LAUNCHING A WEB BROWSER

The BROWSER command lets you start a Web browser from within AutoCAD. By default, the Browser command uses whatever brand of Web browser program is reg-

* This chapter reproduced from *Using AutoCAD R14,* by Ralph Grabowski © 1998, Autodesk press.

istered in your computer's Windows operating system. AutoCAD lists the name of the browser before prompting you for the uniform resource locator (URL). The URL is the Web site address, such as **http://www.autodesk.com.** URLs will be introduced in the next section. The BROWSER command is used to launch a Web browser from within AutoCAD and requires that Netscape Communicator or Microsoft Internet Explorer be installed on your computer. The command can be accessed by picking the Launch Browser icon from the Standard toolbar or by typing **BROWSER** [enter] at the command prompt. The command can also be used in scripts, toolbar or menu macros, and AutoLISP routines to automatically access the Internet.

Command: **BROWSER** [enter]
Location <http://www.autodesk.com/acaduser>:

Selecting [enter] will accept the default URL, Autodesk's own Web site. It is listed as <http://www.autodesk.com/acaduser>:. The name and path of a different browser can be entered at the prompt in place of picking the [enter] key. After you type the URL and press [enter], AutoCAD launches the Web browser and contacts the Web site, such as Netscape Communicator and the Autodesk Web site shown in Figure 26–1. If the Browser command (or any other command mentioned in this chapter) does not work,

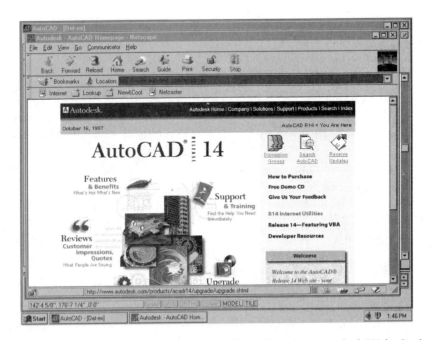

FIGURE 26–1 The Netscape Communicator displaying the Autodesk Web site is accessed by picking the Launch Browser icon from the standard toolbar or by typing **BROWSER** [enter] at the command prompt.

you must load it into AutoCAD. The Browser command is located in the program file called *Browser.Arx*. Use the AppLoad command to locate and load the program, as shown in Figure 26–2.

Four programs contain Internet-related commands for AutoCAD Release 14:

ObjectARx Program Name	Command Names
Browser.Arx	BROWSER
Dwfin.Arx	ATTACHURL
	DETACHURL
	LISTURL
	SELECTURL
Dwfout.Arx	DWFOUT
	DWFOUTD
Internet.Arx	INETCFG
	INETHELP
	INSERTURL
	OPENURL
	SAVEURL

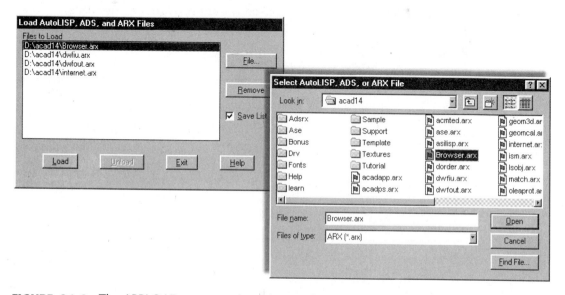

FIGURE 26–2 The APPLOAD command can be used to locate and load the Browser.Arx.

THE UNIFORM RESOURCE LOCATOR (URL)

The uniform resource locator, known as the *URL*, is the file naming system of the Internet. The URL system allows you to find any resource (a file) on the Internet. Example resources include a text file, a Web page, a program file, an audio or movie clip—in short, anything you might also find on your own computer. The primary difference is that these resources are located on somebody else's computer. A typical URL looks like the following examples:

Example URL	Meaning
http://www.autodesk.com	Autodesk Primary Web Site.
http://data.autodesk.com	Autodesk Data Publishing Web Site.
news://adesknews.autodesk.com	Autodesk News Server.
ftp://ftp.autodesk.com	Autodesk FTP Server.
http://www.autodeskpress.com	Autodesk Press Web Site.

Note that the prefix **http://** is not required. Most of today's Web browsers automatically add in the routing prefix, which saves you a few keystrokes. URLs can access several different kinds of resources—such as Web sites, email, news groups—but always take on the same general format:

scheme://netloc

The *scheme* accesses the specific resource on the Internet, including these:

Scheme	Meaning
file://	File is located on your computer's hard drive or local network.
ftp://	File Transfer Protocol (used for downloading files).
http://	Hyper Text Transfer Protocol (the basis of Web sites).
mailto://	Electronic mail (email).
news://	Usenet news (news groups).
telnet://	Telnet protocol.
gopher://	Gopher protocol.

The characters :// indicate a *network address*. Autodesk recommends these formats for specifying URL-style filenames with the Browser command:

Web Site	http://servername/pathname/filename
FTP Site	ftp://servername/pathname/filename
Local File	file:///drive:/pathname/filename
or	file:///drive I /pathname/filename
or	file://\\localPC\pathname\filename
or	file:////localPC/pathname/filename
Network File	file://localhost/drive:/pathname/filename
or	file://localhost/drive I /pathname/filename

The terminology can be confusing. The following definitions will help to clarify these terms.

Term	Meaning
servername	**www.autodesk.com**
pathname	the same as a subdirectory or folder name.
drive	the driver letter, such as **C:** or **D:**.
local file	a file located on your computer.
local host	the name of the network host computer.

If you are not sure of the name of the network host computer, use Windows Explorer to check the Network Neighborhood for the network names of computers.

How URLs are Used in AutoCAD

URLs are used indirectly by the Web browser in several ways. The first method places URLs in the drawing for use by the browser when the drawing is exported in DWF format. To help make the process clearer, here are the steps that you need to go through to use URLs:

Step 1: Open a drawing in AutoCAD.

Step 2: Place URLs in the drawing with the AttachURL command.

Step 3: Export the drawing with the DwfOut command.

Step 4: Copy the DWF file to your Web site.

Step 5: Start your Web browser with the Browser command.

Step 6: View the DWF file and click on a hyperlink spot.

The second method uses URLs to access drawings over the Internet. For example, the InsertURL command uses the URL you type to locate and insert a drawing as a block. If the Web site contains both the DWF files (that you are viewing) and the original DWG file, then you can drag the DWG file into the current AutoCAD drawing.

A third purpose of URLs is to let you create links between files. By simply clicking on a link, you automatically access additional information. For example, clicking on the parts list in the drawing might bring up the original Excel file used to create the part list. Clicking on a standard detail might bring up the local building code. Clicking on a side view might bring up the 3D perspective view.

There is one significant drawback to using URLs in AutoCAD. You cannot use a URL directly within a drawing to create hyperlinks inside AutoCAD. The previous examples only work when you are viewing the drawing in DWF format with a Web browser.

When you are working with URLs in AutoCAD, you may come across these other limitations:

- AutoCAD does not check that the URL you type is valid.
- If you attach a URL to a block, be aware that the URL data is lost when you scale the block unevenly, stretch the block, or explode it.
- You cannot attach a URL to rays and xlines since the URL would be infinitely long, something the DWF format cannot handle.
- Wide polylines have a one-pixel wide URL and not the full width of the polyline.

As you're getting acquainted with using AutoCAD and the Internet you may feel overwhelmed with all of the commands and acronyms. When you need on-line help with AutoCAD's Internet commands, use the Inethelp command or click the Help icon on the Internet Utilities toolbar. The Internet toolbar is activated by selecting inet.mns from the Menu Customization dialog box. The dialog box is accessed by selecting Customize Menus... from the Tools pull-down menu. Once you've accessed the Menu Customization display, selecting the Browse... button will produce the Select Menu File display. The inet.mns option can now be accepted by picking the Open and Load buttons. The command can also be accessed from the command line. The command sequence is:

Command: **INETHELP** `enter`

AutoCAD displays the AutoCAD Internet Utilities help window shown in Figure 26–3. Click one of the gray buttons for help on one of the topics.

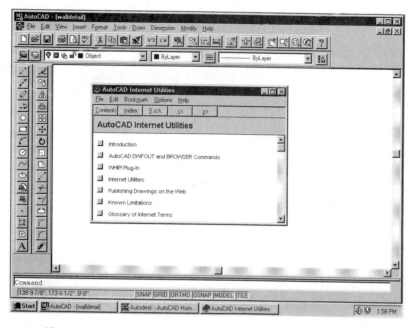

FIGURE 26-3 The AutoCAD Internet Utilities help window is accessed by picking the Help button from the Internet Utilities toolbar or by entering **INETHELP** [enter] at the command prompt.

Attaching a URL to the Drawing

The ATTACHURL command allows you to attach one or more URLs to objects and rectangular areas in a drawing. Note that you cannot use URLs in a drawing. Instead, after one or more URLs are inserted in the drawing, you must export the drawing in DWF format using the Dwfout command. Then, when the DWF file is displayed by a Web browser, the URL locations take on special meaning. The location of the URL is sometimes called a *hyperlink*. When you click on the hyperlink, the Web browser automatically accesses the related URL file location.

The ATTACHURL command lets you attach URLs to objects or rectangular areas. The *object* URL is best for attaching hyperlinks to one or more objects. The *area* URL is best for placing a hyperlink around a group of objects or in an area where there are no objects.

Command: **ATTACHURL** [enter]
URL by (Area/<Objects>): **O**
Select objects: (*Pick an object.*)
1 found Select object: ([enter].)
Enter URL: (*Type a URL and press* [enter].)

AutoCAD gives no indication that an object has a URL attached. Note that AutoCAD's ATTACHURL command lets you pick more than one object for the URL.

Attaching a URL to an object is sometimes called a *1D link*. It is a one-dimensional link because you pick a single object to activate it within the Web browser. Technically, the URL is stored in that object's extended entity data or *xdata* for short.

Area Object.   Since AutoCAD gives no visible indication which objects contain a URL, you might prefer to place the URL in a rectangular area. This will show up in the drawing as a *red rectangle*. To create a rectangular URL that covers an area, use the A option of the Attachurl command:

> Command: **ATTACHURL** [enter]
> URL by (Area/<Objects>): **A** [enter]
> First corner: (*Pick.*)
> Other corner: (*Pick.*)
> Enter URL: (*Type a URL and press* [enter].)

Attaching the URL to an area is sometimes called a *2D link*. It is a two-dimensional link because you can pick anywhere in that area to activate the hyperlink. When the URL is an area, AutoCAD creates a layer named *URLLAYER* (with the color red), places the rectangle, and stores the URL as xdata of that rectangle.

Be careful that you do not overlap URLs (either objects or areas) since you could hyperlink to the wrong URL!

If you find the URL rectangles distracting, turn off layer URLLAYER. Do not erase or freeze that layer, since it would not be exported by the Dwfout command if it were erased or frozen.

Highlighting Objects with URLs

Although you can see the rectangle of area URLs, object URLs and the URLs themselves are invisible. For this reason, AutoCAD has the SELECTURL command, which highlights all objects and areas that have URLs attached. The command is accessed by selecting the Select URLs icon or at the command prompt by entering:

> Command: **SELECTURL** [enter]

AutoCAD highlights all objects that have a URL, including URL area rectangles. Depending on your computer's display system, the highlighting shows up as dashed lines or another color.

Listing URLs in Objects

Now that you know *where* the objects with URLs are located, you can use the LISTURL command to find out *what* the URLs are. The command is accessed by picking the List URLs icon or by typing **LISTURL** [enter] at the command prompt.

> Command: **LISTURL** [enter]
> Select objects: (*Pick.*)

1 found Select Objects: (*Press* enter.)
URL for selected object is: **http://www.autodesk.com**

You cannot use the LIST command since it does not display xdata. If you really want to see the full details, use the XDLIST command found in the Bonus | Tools | List Entity Xdata menu. This program lists all extended entity data found in the selected objects. The command is started by typing **XDLIST** at the command prompt.

Command: **XDLIST** enter
Select object: (*Pick object.*)
Application name <*>: (*Press* enter.)
* Registered Application Name: **PE_URL**
* Code 1000, ASCII string: **http://www.autodesk.com**
Object has 16355 bytes of Xdata space available.

Removing URLs from Objects

To remove a URL from an object, use the DETACHURL command. The process is started by picking the Detach URL icon, or by typing **DETACHURL** enter at the command prompt.

Command: **DETACHURL** enter
Select objects: (*Pick.*)
1 found Select Objects: (*Press* enter.)

When you select the rectangle of an area URL, AutoCAD erases the rectangle and reports, "DetachURL, deleting the area." If there are no more area URLs remaining, AutoCAD also purges the URLLAYER layer name.

THE DRAWING WEB FORMAT

To display AutoCAD drawings on the Internet, Autodesk invented a new file format called *drawing Web format* (DWF). The DWF file has several benefits and some drawbacks over DWG files. The DWF file is compressed as much as eight times smaller than the original DWG drawing file so that it takes less time to transmit over the Internet, particularly with the relatively slow telephone modem connections. The DWF format is more secure, since the original drawing is not being displayed; another user cannot tamper with the original DWG file.

However, the DWF format has some drawbacks:

- You must go through the extra step of translating from DWG to DWF.
- DWF files cannot display rendered or shaded drawings.
- DWF is a flat, 2D file format; therefore, it does not preserve 3D data, although you can export a 3D view.
- The earlier versions of DWF (version 2.x and earlier) did not handle paperspace objects.

To view a DWF file on the Internet, your Web browser needs *a plug-in*—a software extension that lets a Web browser handle a variety of file formats. Autodesk makes the DWF plug-in freely available from its Web site at **http://www.autodesk.com.** It's a good idea to regularly check that URL for updates to the DWF plug-in, which is updated about twice a year. (Because Autodesk changes its Web site around every year, the exact location of the DWF plug-in is not given. Use the Web site's Index to search for the plug-in.)

To view AutoCAD DWG and DXF files on the Internet, your Web browser needs a DWG-DXF plug-in from a third-party developer since Autodesk does not make one available. At the time of publication, the plug-ins were available free for non-commercial use from the following vendors:

SoftSource: **http://www.softsource.com/**

California Software Labs: **http://www.cswl.com**

Creating a DWF File

To create a DWF file, select File | Export... | DWF from the Standard toolbar or type **DWFOUT** [enter] at the command prompt. Selecting the command from the toolbar will produce the Export Data dialog box shown in Figure 26–4. Picking the Save as type: button will display the saving options. Picking the DWF option will activate

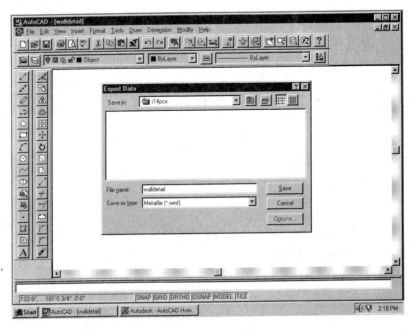

FIGURE 26–4 A DWF file can be created using The Export Data dialog box. The box is displayed by selecting DWF from Export... of the File pull-down menu.

the Options... button and display the DWF Export Options dialog box shown in Figure 26–5. Entering the command from the keyboard will produce the Create DWF File dialog box shown in Figure 26–6. Selecting the Options... button displays the DWF Export Options dialog box shown in Figure 26–5. The Options... button lets you choose the following:

Precision. Unlike AutoCAD DWG files, which are based on real numbers, DWF files are saved using integer numbers. The Low precision setting saves the drawing using 16-bit integers, which is adequate for all but the most complex drawings. The file is about 40 percent smaller than High precision, which means a 40 percent faster transmission time over the Internet. Medium precision saves the DWF file using 20-bit integers. High precision saves using 32-bit integers.

Use File Compression. Compression further reduces the size of the DWF file. You should always use compression, unless you know that another application cannot decompress the DWF file.

As an alternative, you can use the DWFOUTD command, which does not display the file dialog box. DWFOUTD is meant for use in scripts, toolbar or menu macros, and AutoLISP routines. It does not give you the opportunity to select any output options.

AutoCAD itself cannot display DWF files, nor can DWF files be converted back to DWG format without using file translation software from a third-party vendor.

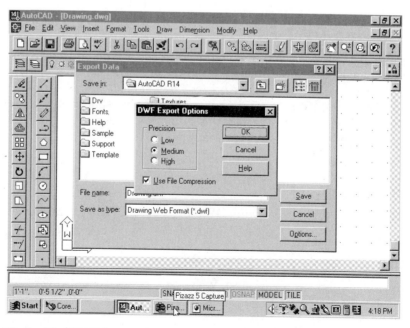

FIGURE 26–5 The DWF Export Options dialog box allows the precision of the DWF file to be controlled.

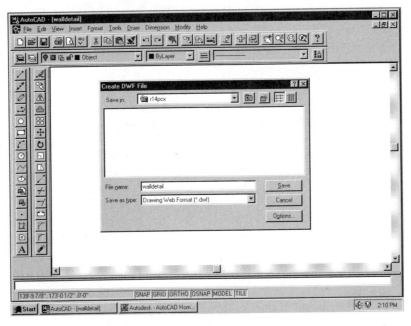

FIGURE 26–6   Entering **DWFOUT** [enter] at the command prompt will produce the Create DWF File dialog box.

Viewing DWF Files

In order to view a DWF file, you need to use a Web browser with a special plug-in that allows the browser to correctly interpret the file. (Remember: you cannot view a DWF file with AutoCAD.) Autodesk has named their DWF plug-in "Whip!," short for "Windows HIgh Performance." The plug-in should not be confused with the AutoCAD display driver, even though both are named "Whip!" The following table summarizes the difference between the two products:

Feature	Whip Plug-in	Whip Display Driver
Meant for:	Viewing DWF files.	Displaying AutoCAD drawings.
Works in:	Web browsers.	AutoCAD Release 13 and 14.
DWF file:	Read, view, and print.	Export only.
Available from:	Autodesk Web site.	Included with AutoCAD.

To help reduce confusion, we'll call the Whip plug-in the "DWF plug-in."

Autodesk updates the DWF plug-in approximately twice a year. Each update includes some new features. In summary, all versions of the DWF plug-in perform the following functions:

- Views DWF files created by AutoCAD within a browser.
- Right-clicking the DWF image displays a cursor menu with commands.
- Real-time pan and zoom lets you change the view of the DWF file as quickly as a drawing file in AutoCAD R14.
- Embedded hyperlinks let you display other documents and files.
- File compression means that a DWF file appears in your Web browser faster than the equivalent DWG drawing file would.
- Print the DWF file alone or along with the entire Web page.
- Works with Netscape Communicator 4.03 or Microsoft Internet Explorer 3.02. A separate plug-in is required, depending upon which of the two browsers you use.

At the time of writing this book, DWF plug-in Release 2 was current and adds these features:

- Views DWF files created by AutoCAD R14 and R13.
- Allows you to "drag and drop" a DWG file from a Web site into AutoCAD R14 as a new drawing or as a block.
- Views percentage- or pixel-specified DWF files; views DWF file in a specified browser frame; views a named view stored in the DWF file; and can specify a view using x,y- coordinates.
- Turns the user interface on and off.
- Sends DWF file information to CGI scripts.
- Supports Netscape Communicator v4.0 (previously named Navigator) and Microsoft Internet Explorer v4.0 Web browsers. A separate plug-in is required for either browser.
- Raster images are supported in DWF files; images are rendered as a purple outline during pans and zooms to help speed up the display.
- TrueType fonts are displayed as fonts, instead of as lines and arcs.

To use the Whip plug-in with a Web browser, your computer should have a fast CPU, such as an 80486, Pentium, or Pentium Pro. Your computer must be running Windows 95, Windows NT v3.51 (with service pack 5), or Windows NT v4.0. The computer should display at least 256 colors. The Web browser must be at least Netscape v3.0x or Explorer v3.0x.

If you don't know whether the DWF plug-in is installed in your Web browser, select Help | About Plug-ins from the browser's menu bar. You may need to scroll through the list of plug-ins to find something like this:

```
WHIP!®
File name: C:\INTERNET\NETSCAPE\PROGRAM\plugins\npdwf.dll
Autodesk Drawing Web Format File
Mime Type        Description                    Suffixes    Enabled
Drawing/x-dwf    Drawing Web Format file        dwf         Yes
```

If the plug-in is not installed, or is an older version, then you need to download it from Autodesk's Web site at:

http://www.autodesk.com/products/autocad/whip/whip.htm

For Netscape users, the file is 3.5MB and takes about a half-hour to download using a typical 28.8Kbaud modem. The large size is due to Autodesk including all of the Windows support files, even if your system already has them. The file you download from the Autodesk Web site is a self-extracting installation file with a name such as Whip2.Exe. After the download is complete, start the program and follow the instructions on the screen. If your computer has an older version of the DWF plug-in for Netscape, you must uninstall it before installing the newer version.

If the Netscape Web browser is running, close it before installing the DWF plug-in.

For Internet Explorer users, the DWF plug-in is an ActiveX control. Explorer's auto-download feature automatically installs the control the first time your browser accesses a Web page that includes a DWF file. The download time is about ten minutes with a 28.8Kbps modem.

DWF Plug-in Commands

To display the DWF plug-in's commands, move the cursor over the DWF image and press the mouse's right button. This displays the cursor menu with commands, such as Pan, Zoom, and Named Views shown in Figure 26–7. To select a command, move the cursor over the command name and press the left mouse button.

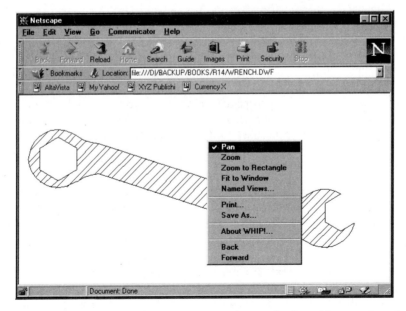

FIGURE 26–7   The Whip DWF Plug-in commands are displayed by moving the cursor over the DWF image and pressing the right mouse button.

Pan is the default command. Press the left mouse button and move the mouse. The cursor changes to an open hand, signaling that you can pan the view around the drawing. This is exactly the same as real-time panning in AutoCAD. Naturally, panning only works when you are zoomed in; it does not work in Full View mode.

Zoom is like the Zoom Realtime command in AutoCAD. The cursor changes to a magnifying glass. Hold down the left mouse button and move the cursor up (to zoom in) and down (to zoom out).

Zoom to Rectangle is the same as Zoom Window in AutoCAD. The cursor changes to a plus sign. Click the left mouse button at one corner, then drag the rectangle to specify the size of the new view.

Fit to Window is the same as AutoCAD's Zoom Extents command. You see the entire drawing.

Full View causes the Web browser to display the DWF image as large as possible all by itself. This is useful for increasing the physical size of a small DWF image. When you have finished viewing the large image, right-click and select the Back command or click the browser's Back button to return to the previous screen.

Named Views works only when the original DWG drawing file contained named views created with the View command. Selecting Named Views displays a *non-modal* dialog box that allows you to select a named view. (A non-modal dialog box remains on the screen; unlike AutoCAD's modal dialog boxes, you do not need to dismiss a non-modal dialog box to continue working.) Double-click a named view to see it; click OK to dismiss the dialog box.

Highlight URLs displays a highlight box around all objects and areas with URLs in the image. This helps you see where the hyperlinks are. To read the associated URL, pass the cursor over a highlight area and look at the URL on the browser's status line. (A shortcut is to hold down the [shift] key, which causes the URLs to highlight until you release the [shift] key).

Print prints the DWF image alone. To print the entire Web page (including the DWF image), use the browser's Print button.

SaveAs saves the DWF file in three formats to your computer's hard drive:

- DWF
- BMP (Windows bitmap)
- DWG (AutoCAD drawing file) works only when a copy of the DWG file is available at the same subdirectory as the DWF file.

You cannot use the Saveas command until the *entire* DWF file has been transmitted to your computer.

About WHIP displays information about the DWF file, including DWF file revision number, description, author, creator, source filename, creation time, modification time, source creation time, source modification time, current view left, current view right, current view bottom, and current view top. Figure 26–8 shows an example of the About WHIP! dialog box.

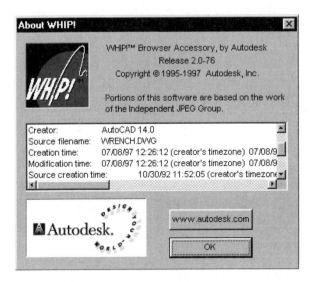

FIGURE 26–8 The About Whip dialog box displays information about the DWF file.

Back is almost the same as clicking the browser's Back button. It works differently when the DWF image is displayed in a frame.

Forward is almost the same as clicking the browser's Forward button. When the DWF image is in a frame, only that frame goes forward.

Drag and Drop from Browser to Drawing

The DWF plug-in allows you to perform several "drag and drop" functions. Drag and drop is when you use the mouse to drag an object from one application to another.

Hold down the ⌊cntrl⌋ key to drag a DWF file from the browser into AutoCAD. Recall that AutoCAD cannot translate a DWF file into DWG drawing format. For this reason, this form of drag and drop only works when the originating DWG file exists in the same subdirectory as the DWF file. (This may change in a future release of the DWF plug-in when DWG directory and DWF directory options are implemented.) Note that this drag and drop function works only for AutoCAD Release 14; it does not work with AutoCAD Release 13 or earlier.

Another drag and drop function is to drag a DWF file from the Windows Explorer (or File Manager) into the Web browser. This causes the Web browser to load the DWF plug-in, then display the DWF file. Once displayed, you can execute all of the commands listed in the previous section.

Finally, you can also drag and drop a DWF file from Windows Explorer (in Windows 95) or File Manager (in Windows NT) into AutoCAD. This causes AutoCAD to launch another program that is able to view the DWF file. This does not work if you do not have other software on your computer system capable of viewing DWF files.

Embedding a DWF File

To let others view your DWF file over the Internet, you need to embed the DWF file in a Web page. There are several approaches to embedding a DWF file in a Web page.

Step 1. This hyper text markup language (HTML) code is the most basic method of placing a DWF file in your Web page:

> `<embed src="filename.dwf">`

The **\<embed\>** tag embeds an object in a Web page. Next, **src** is short for "source." Replace *filename.dwf* with the URL of the DWF file. *Remember to keep the quotation marks in place.*

Step 2. HTML normally displays an image as large as possible. To control the size of the DWF file, add the Width and Height options:

> `<embed width=800 height=600 src="filename.dwf">`

The **Width** and **Height** values are measured in pixels. Replace 800 and 600 with any appropriate numbers, such as 100 and 75 for a "thumbnail" image, or 300 and 200 for a small image.

Step 3. To speed up a Web page's display speed, some users turn off the display of images. For this reason, it is useful to include a description, which is displayed in place of the image:

> `<embed width=800 height=600 name=description src="filename.dwf">`

The **Name** option displays a textual description of the image when the browser does not load images. You might replace *description* with the DWF filename.

Step 4. When the original drawing contains named views created by the View command, these are transferred to the DWF file. Specify the initial view for the DWF file:

> `<embed width=800 height=600 name=description namedview="viewname" src="filename.dwf">`

The **namedview** option specifies the name of the view to display upon loading. Replace *viewname* with the name of a valid view name. When the drawing contains named views, the user can right-click on the DWF image to get a list of all named views.

As an alternative, you can specify the 2D coordinates of the initial view:

> `<embed width=800 height=600 name=description view="0,0 9,12" src="filename.dwf">`

The **View** option specifies the x,y-coordinates of the lower-left and upper-right corners of the initial view. Replace *0,0 9,12* with other coordinates. Since DWF is 2D only, you cannot specify a 3D viewpoint. You can use View or Namedview but not both.

Step 5. Before a DWF image can be displayed, the Web browser must have the DWF plug-in called "Whip!". For users of Netscape Communicator, you must include a description of where to get the Whip plug-in when the Web browser is lacking it.

```
<embed pluginspage=http://www.autodesk.com/products/autocad/whip/
whip.htm width=800 height=600 name=description view="0,0 9,12"
src="filename.dwf">
```

The **pluginspage** option describes the page on the Autodesk Web site where the Whip-DWF plug-in can be downloaded.

Step 6. The code listed above works for Netscape Navigator. To provide for users of Internet Explorer, the following HTML code must be added:

```
<object classid ="clsid:B2BE75F3-9197-11CF-ABF4-08000996E931"
codebase = "ftp://ftp.autodesk.com/pub/autocad/plugin/
whip.cab#version=2,0,0,0" width=800 height=600>
<param name="Filename" value="filename.dwf">
<param name="View" value="0,0 9,12">
<param name="Namedview" value="viewname">
<embed pluginspage=http://www.autodesk.com/products/autocad/whip/
whip.htm width=800 height=600 name=description view="0,0 9,12"
src="filename.dwf">
</object>
```

The two **<object>** and three **<param>** tags are ignored by Netscape Navigator; they are required for compatibility with Internet Explorer. The **classid** and **codebase** options tell Explorer where to find the plug-in. Remember that you can use View or Namedview but not both.

Step 7: Save the HTML file.

Opening a Drawing from the Internet

When a drawing is stored on the Internet, you access it from within AutoCAD using the OPENURL command. Instead of specifying the file's location with the usual drive-Folder-file name format, such as C:\ACAD14\FILENAME.DWG, use the URL format. Recall that the URL is the universal file naming system used by the Internet to access any file located on any computer hooked up to the Internet. The command is accessed by selecting the Open from URL icon from the Internet Utilities toolbar or by entering **OPENURL** [enter] at the command prompt. This will produce the Open DWG from URL dialog box shown in Figure 26–9.

Command: **OPENURL** [enter]

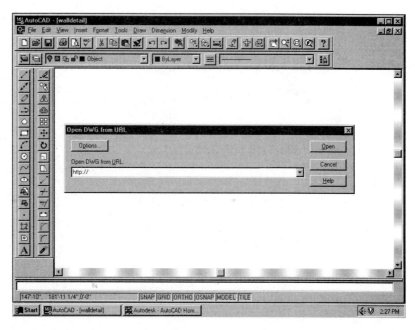

FIGURE 26–9 A drawing stored on the Internet can be accessed using the Open DWG from URL dialog box. The box is accessed by selecting the Open URL icon or by entering **OPENURL** [enter] at the command prompt.

The dialog box prompts you to type the URL. For your convenience, AutoCAD fills in the preliminary *http://*, which is used for accessing Web pages. If you plan to access a different resource on the Internet, erase the *http://* and replace it. For example, you might want to replace it with *ftp://* to access a drawing at an FTP site.

The following table gives templates for typing the URL to open a drawing file:

Drawing Location	Template URL
Web or HTTP Site	http://servername/pathname/filename.dwg
FTP Site	ftp://servername/pathname/filename.dwg
Local File	file:///drive:/pathname/filename.dwg
Network File	file://localhost/drive:/pathname/filename.dwg

Options. Click the Options button only when you want AutoCAD to deal with "secure" sites that ask you for a username and password. (This Internet Configuration dialog box shown in Figure 26–10 is the same one that appears when you type the INETCFG command.) Very often, you do not need them. For example, most FTP sites allow "anonymous" logins where you need no password and the username is

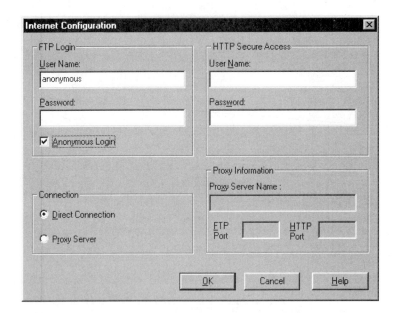

FIGURE 26–10 The Internet Configuration dialog box is used to alter the name, password, and security of an Internet connection.

"anonymous." Most Web sites require no login at all. For security reasons, the passwords are not retained once you exit AutoCAD.

To access a secure FTP site, turn off **Anonymous Login** and type your username and password.

To access a secure Web site (also called an HTTP site), type your username and password. If you prefer, you can leave this blank and AutoCAD will display a User Authentication dialog box later, where you can enter the username and password.

Select **Direct Connection** when your computer connects with the Internet via an Internet service provider. When your computer connects to the Internet through a proxy server, select Proxy Server. (Most typically, larger firms have a "proxy server" that acts as a gateway between the firm's internal network and the external Internet.) Ask your system administrator for the details of how to configure the proxy server, including the proxy server name, default FTP port, and default HTTP port.

Click **OK** to dismiss the Internet Configuration dialog box.

Open. To open the drawing, click Open. If necessary, AutoCAD prompts you to save the current drawing before opening the drawing from the Internet. During the file transfer, AutoCAD displays the Remote Transfer in Progress dialog box to report the progress. (See Figure 26–11). If your computer uses a 28.8Kbps modem, you should allow about ten minutes per megabyte of drawing file size. If your computer

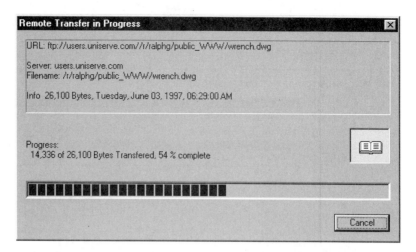

FIGURE 26–11   The Remote Transfer in Progress dialog box displays the progress of a drawing transfer using the Internet. (Courtesy Using AutoCAD R14, Autodesk Press.)

has access to a faster T1 connection to the Internet, you should expect a transfer speed of about one minute per megabyte.

It may be helpful to understand that OPENURL command does not copy the file from the Internet location directly into AutoCAD. Instead, it copies the file from the Internet to your computer's designated Temporary subdirectory, such as C:\WIN95\TEMP, and then loads the drawing from the hard drive into AutoCAD. This is known as *caching*. It helps to speed up the processing of the drawing, since the drawing file is now located on your computer's fast hard drive, instead of the relatively slow Internet.

Inserting a Block from the Internet

When a block (symbol) is stored on the Internet, you can access it from within AutoCAD using the INSERTURL command. The INSERTURL command works just like the OPENURL command, except that the external block is inserted into the current drawing. The command is accessed by selecting the Insert from URL icon, or by typing **INSERTURL** ⏎ at the command prompt. Each method will produce the Insert DWG from URL dialog box similar to Figure 26–12.

The dialog box prompts you to type the URL. After you click the Insert button, AutoCAD retrieves the file and continues with the Insert command's familiar prompts.

> _.insert Block name (or ?): **c:\win95\temp\filename.dwg** ⏎
> Insertion point: (*Pick insertion point.*)
> X scale factor <1> / Corner / XYZ: (*Press* ⏎.)
> Y scale factor (default=X): (*Press* ⏎.)
> Rotation angle <0>: (*Press* ⏎.)

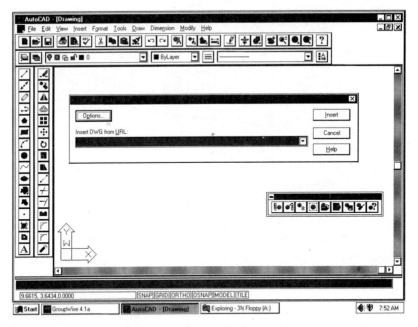

FIGURE 26–12 The Insert DWG from URL dialog box can be used to access a block stored on the Internet. The display is accessed by selecting the Insert URL icon or by entering **INSERTURL** ⏎ at the command prompt.

Saving a Drawing to the Internet

When you are finished editing a drawing in AutoCAD, you can save it to a file server on the Internet with the SAVEURL command. If you inserted the drawing from the Internet (using INSERTURL) into the default Drawing.Dwg drawing, AutoCAD insists you first save the drawing to your computer's hard drive and displays the warning shown in Figure 26–13. The command is accessed by picking the Save to URL icon or by entering SAVEURL ⏎ at the command prompt. Each method will produce the dialog box shown in Figure 26–14.

The dialog box prompts you to type the URL. For your convenience, AutoCAD fills in the preliminary *ftp://*, which is used for writing a file to an FTP site. Because you have to write a file to Web site using FTP (file transfer protocol), *http://* is not displayed. Nor will AutoCAD allow you to use *file://* or other URL prefixes.

When a drawing of the same name already exists at that URL, AutoCAD warns you, just as it does when you use the Saveas command. (See Figure 26–15). Recall from the OPENURL command that AutoCAD uses your computer system's Temporary subdirectory, hence the reference to it in the dialog box.

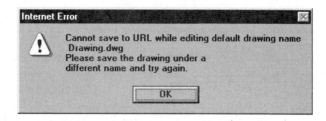

FIGURE 26–13 The Internet Error dialog box will be displayed if the URL cannot be saved.

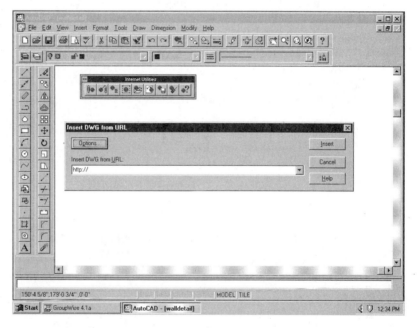

FIGURE 26–14 The Save DWG to URL dialog box.

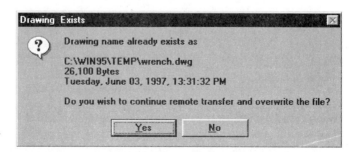

FIGURE 26–15 If a drawing of the same name already exist the Drawing Exists dialog box will be displayed.

CHAPTER 26 EXERCISE

In this exercise, you use the AutoCAD URL tools to place several URLs in a drawing. The URLs are listed, one is deleted, then the drawing is exported in DWF format. If your computer system has a Web browser installed with the DWF plug-in, then in this project you drag a DWF file into the browser. Finally, if your computer can access the Internet, this project has you open a drawing from the Autodesk Press Web site.

1. Start AutoCAD and open any drawing file.
2. Place an area URL anywhere in the drawing.

   Command: **ATTACHURL** [enter]
   URL by (Area/<Objects>): **A** [enter]
   First corner: (*Pick.*)
   Other corner: (*Pick.*)
   Enter URL: **http://www.autodesk.com** [enter]

AutoCAD places a red rectangle in the drawing.

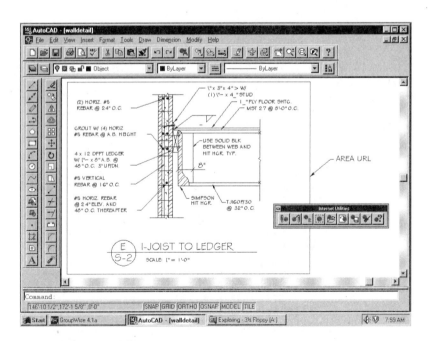

FIGURE 26–16 The Area URL.

3. Now attach a URL to an object in the drawing.

Command: **ATTACHURL** [enter]
URL by (Area/<Objects>): **O** [enter]
Select objects: (*Pick any object.*)
1 found Select object: (*Press* [enter].)
Enter URL: **http://www.autodeskpress.com**

4. You cannot see which objects have a URL attached. Use the SELECTURL command to make them visible.

Command: **SELECTURL** [enter]

AutoCAD highlights the area URL and the object URL, as shown in Figure 26–17.

5. If you don't remember the names of the URLs, use the Listurl command.

Command: **LISTURL** [enter]
URL for selected object is: **http://www.autodesk.com**
URL for selected object is: **http://www.autodeskpress.com**

Since the URL objects were already selected (highlighted), the LISTURL command did not prompt you to select objects.

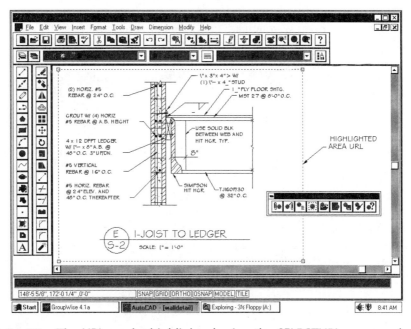

FIGURE 26–17 The URL can be highlighted using the SELECTURL command.

6. Press [esc] twice to remove the highlighting.
7. If you don't need a URL, remove it.

   Command: **DETACHURL** [enter]
   Select objects: (*Pick an object.*)
   1 found Select Objects: (*Press* [enter]*.*)

8. Now export the drawing as a DWF file.

   Command: **DWFOUT** [enter]

9. When AutoCAD displays the Create DWF File dialog box, click the Options button.
10. In the DWF Export Options dialog box, select Low precision.
11. Click OK twice. AutoCAD saves the drawing as a DWF file.
12. If your Web browser has the DWF plug-in installed, start the Web browser now.
13. Switch to Windows Explorer (in Windows 95) or File Manager (in Windows NT) and look for the DWF file you created in step 2. If necessary, use the Search function to find *.DWF.
14. Drag the DWF file from Windows Explorer File Manager into the browser. The browser will take several seconds to first load the plug-in, then load and display the DWF file.
15. Place the cursor in the image and right-click the mouse button to bring up the cursor menu.
16. Practice selecting several commands, such as Zoom, Pan, and Print.
17. To remove the image, click the browser's Back button.
18. Start a new drawing with the New command. When the Start New Drawing dialog box is displayed, click Start From Scratch and OK.
19. If your computer is not logged on to the Internet, do so now. For example, if you access the Internet through an Internet Service Provider, use the Dial-up Networking feature of Windows now. Click the Connect button.
20. Switch back to AutoCAD and open a drawing using a URL.

    Command: **OPENURL** [enter]

21. When you invoke the command, the Open DWG from URL dialog box appears. If your computer has access to the Internet, type the following URL:

    http://www.practicewrench.autodeskpress.com [enter]
22. Click Open. AutoCAD displays the Remote Transfer in Progress dialog box. The file is 25KB in size and should take a few seconds to download.
23. Once the Wrench drawing appears, it looks just like any drawing you open from your computer's hard drive.

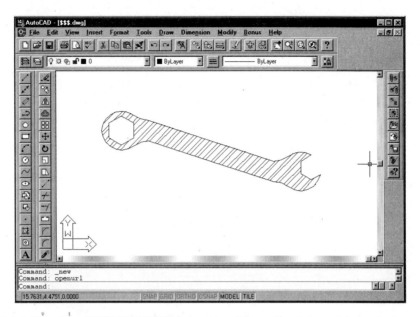

FIGURE 26–18 The WRENCH.DWG is accessed from the Internet by entering **www.practicewrench.autodeskpress.com** [enter] in the Open DWG from URL dialog box. (Courtesy Using AutoCAD R14, Autodesk Press.)

CHAPTER QUIZ

1. Can you launch a Web browser from within AutoCAD?

2. What does DWF mean?

3. The purpose of DWF files is to view?

4. What is URL short for?

5. Which of the following URLs are valid:

   www.autodesk.com

   http://www.autodesk.com

   Both of the above.

   Neither of the above.

6. What is the purpose of a URL?

7. FTP is short for:

   _____

8. What is a "local host"?

   _____

9. Can URLs be used in an AutoCAD drawing?

   _____

10. The purpose of URLs is to let you create _____ between files.

11. When you attach a URL to a block, the URL data is _____ when you scale the block unevenly, stretch the block, or explode it.

12. Can you attach a URL to rays and xlines?

    _____

13. The ATTACHURL command allows you to attach a URL to _____ and
    _____.

14. To see the location of URLs in a drawing, use the _____ command.

15. Rectangular URLs are stored on layer_____ .

16. The LISTURL command tells you _____ .

17. The DETACHURL command removes a _____ from an object.

18. Compression in the DWF file causes it to take (less, more, the same) time to transmit over the Internet.

19. A DWF is created from a _____ file using the _____ or
    _____ command.

20. What does a "plug-in" let a Web browser do?

    _____

21. Can a Web browser view DWG drawing files over the Internet?

    _____

22. _____ is an HTML tag for embedding graphics in a Web page.

23. Are the Whip plug-in and the Whip display driver the same?

_____

24. A file being transmitted over the Internet via a 28.8Kpbs modem takes about _____ minutes per megabyte.

25. To open a drawing located on the Internet, use the _____ command.

26. An "anonymous" FTP login requires _____ for the password and _____ for the username.

27. Are passwords retained when you exit AutoCAD?

_____

28. The SAVEURL command saves the _____ to the Internet via _____.

2D COMMAND ALIASES

• •

The following is a listing of the command aliases provided in the standard *acad.pgp* file. Command aliases are shortcuts for commands that you enter at the keyboard.

Command	Alias	Command	Alias
ALIGN	AL	BLOCK	-B
APPLOAD	AP	BMAKE	B
ARC	A	BOUNDARY	BO
AREA	AA	BOUNDARY	-BO
ARRAY	AR	BREAK	BR
ASEADMIN	AAD	CHAMFER	CHA
ASEEXPORT	AEX	CHANGE	-CH
ASELINKS	ALI	CIRCLE	C
ASEROWS	ARO	COPY	CO or CP
ASESELECT	ASE	DDATTDEF	AT
ASESQLED	ASQ	DDATTE	ATE
ATTDEF	-AT	DDCHPROP	CH
ATTEDIT	-ATE	DDCOLOR	COL
BHATCH	BH OR H	DDEDIT	ED

Command	Alias
DDGRIPS	GR
DDIM	D
DDINSERT	I
DDMODIFY	MO
DDRENAME	REN
DDRMODES	RM
DDSELECT	SE
DDUCS	UC
DDUCSP	UCP
DDUNITS	UN
DDVIEW	V
DIMALIGNED	DAL, DIMALI
DIMANGULAR	DAN, DIMANG
DIMBASELINE	DBA, DIMBASE
DIMCENTER	DCE
DIMCONTINUE	DCO, DIMCONT
DIMIDIAMETER	DDI, DIMDIA
DIMEDIT	DED
DIMEDIT	DIMED
DIMILINEAR	DLI, DIMLIN
DIMORDINATE	DOR, DIMORD
DIMOVERRIDE	DOV, DIMOVER
DIMRADIUS	DRA, DIMRAD
DIMSTYLE	DST, DIMSTY
DIMTEDIT	DIMTED
DIST	DI
DIVIDE	DIV
DONUT	DO
DRAWORDER	DR
DSVIEWER	AV
DTEXT	DT
DVIEW	DV

Command	Alias
ELLIPSE	EL
ERASE	E
EXPLODE	X
EXPORT	EXP
EXTEND	EX
FILLET	F
FILTER	FI
GROUP	G
-GROUP	-G
HATCH	-H
HATCHEDIT	he
IMPORT	IMP
INSERT	-I
INTERSECT	IN
LAYER	LA
-LAYER	-LA
LEADER	LE, LEAD
LENGTHEN	LEN
LINE	L
LINETYPE	LT
-LINETYPE	-LT
LIST	LI, IS
LTSCALE	LTS
MATCHPROP	MA
MEASURE	ME
MIRROR	MI
MLINE	ML
MOVE	M
MSPACE	MS
MTEXT	T, MT
-MTEXT	-T
MVIEW	MV
OFFSET	O
OSNAP	OS

Command	Alias
-OSNAP	-OS
PAN	P
-PAN	-P
PASTESPEC	PA
PEDIT	PE
PLINE	PI
PLOT	PRINT
POINT	PO
POLYGON	POL
PREFERENCES	PR
PREVIEW	PRE
PSPACE	PS
PURGE	PU
QUIT	EXIT
RECTANG	REC
REDRAW	R
REDRAWALL	RA
REGEN	RE
REGENALL	REA
REGION	REG
RENAME	-REN
REVOLVE	REV
ROTATE	RO
SCALE	SC
SCRIPT	SCR
SETVAR	SET
SLICE	SI

Command	Alias
SNAP	SN
SOLID	SO
SPELL	SP
SPLINE	SPL
SPLINEDIT	SPE
STRETCH	S
STYLE	ST
SUBTRACT	SU
TABLET	TA
THICKNESS	TH
TILEMODE	TI
TOLERANCE	TOL
TOOLBAR	TO
TRIM	TR
UNION	UNI
UNITS	-UN
VIEW	-V
VPOINT	-VP
WBLOCK	W
XATTACH	XA
XBIND	XB
-XBIND	-XB
XCLIP	XC
XLINE	XI
XREF	XR
-XREF	-XR
ZOOM	Z

B

AUTOCAD COMMAND REFERENCE

· · · · · · · · · · · · · · · · · · · ·

The following is an alphabetical list of the AutoCAD 2D commands, with brief descriptions and a summary of options. The commands flagged with a ' prefix in this list can be used transparently. For a complete list of commands see the Help menu.

Command	Description
'ABOUT	Displays a dialogue box with the AutoCAD version and serial numbers, a scrolling window with the text of the *acad.msg* file, and other information.
ACISIN	Imports an ACIS file
ACISOUT	Exports AutoCAD solid objects to an ACIS file
ALIGN	Moves and rotates objects to align with other objects
'APERTURE	Controls the size of the object snap target box
'APPLOAD	Loads AutoLISP, ADS, and ARX applications
ARC	Draws an arc of any size. The default method is to specify two endpoints and a point along the arc, but several other methods are available
AREA	Computes the area of a polygon, polyline, or circle
ARRAY	Makes multiple copies of selected objects in a rectangular or circular pattern
ASEADMIN	Performs administrative functions for external database commands
ASEEXPORT	Exports link information for selected objects
ASELINKS	Manages links between objects and an external database
ASEROWS	Displays and edits table data and creates links and selection sets
ASESELECT	Creates a selection set from rows linked to textual selection sets and graphic selection sets

Command	Description
ASESQLED	Executes Structured Query Language (SQL) statements
ATTDEF	Creates an Attribute Definition entity for textual information to be associated with a Block Definition. See also DDATTDEF.
'ATTDISP	Controls the visibility of Attribute entities on a global basis
ATTEDIT	Permits editing of Attributes
ATTEXT	Extracts Attribute data from a drawing.
ATTEDEF	Redefines a block and updates associated attributes
AUDIT	Invokes drawing integrity audit
'BASE	Specifies origin for subsequent insertion into another drawing
BHATCH	Fills an automatically defined boundary with a hatch pattern through the use of dialogue boxes. Also allows previewing and repeated adjustments without starting over each time
'BLIPMODE	Controls display of marker blips for point selection
BLOCK	Forms a compound object from a group of entities
BMAKE	Defines a block using a dialog box
BMPOUT	Saves selected objects to a file in device-independent bitmap format
BOUNDARY	Creates a Polyline of a closed boundary
BREAK	Erases part of an object, or splits it into two objects
BROWSER	Launches the default Web browser defined in your system's registry
CAL	Evaluates mathematical and geometric expressions
CHAMFER	Creates a chamfer at the intersection of two lines
CHANGE	Alters the location, size, orientation, or other properties of selected objects. Especially useful for Text entities
CHPROP	Modifies properties of selection objects. See also DDCHROP.
CIRCLE	Draws a circle of any size. The default method is by centerpoint and radius, but other methods are available
'COLOR or 'COLOUR	Establishes the color for subsequently drawn objects
COMPILE	Compiles shape and font files
CONFIG	Displays options in the text window to reconfigure the video display, digitizer, plotter, and operating parameters
CONVERT	Converts 2D polylines and associative hatches to the optimized Release 14 format
COPY	Draws a copy of selected objects
COPYCLIP	Copies objects to the Clipboard
COPYHIST	Copies the text in the command line history to the Clipboard
COPYLINK	Copies the current view to the Clipboard for linking to other OLE applications
CUTCLIP	Copies objects to the Clipboard and erases the objects from the drawing

Command	Description
DBLIST	Lists database information for every entity in the drawing
DDATTDEF	Displays a dialogue box that creates an Attribute Definition entity for textual information to be associated with a Block Definition. See also ATTDEF
DDATTE	Allows Attribute editing via a dialogue box
DDATTEXT	Displays a dialogue box that extracts data from a drawing. Available formats are DXF, CDF, SDF, or selected entities. See also ATTEXT
DDCHPROP	Displays a dialogue box that modifies the color, layer, linetype, and thickness of selected objects. See also CHPROP
DDCOLOR	Sets color for new objects
DDEDIT	Allows Text and Attribute Definition editing via a dialogue box
'DDEMODES	Sets entity properties (current layer, linetype, elevation, thickness, and text style) via a dialogue box
'DDGRIPS	Allows you to enable grips and set their colors and size via a dialogue box
DDINSERT	Displays a dialogue box that inserts a copy of a previously drawn part or a drawing file into a drawing, and lets you set an insertion point, scale, rotate, or explode the part. See also INSERT
'DDLMODES	Sets layer properties (New, Current, Rename, On/Off, Thaw/Freeze, Unlock/Lock, Current Viewport, New Viewport, Color, Linetype, Filters) via a dialogue box. See also LAYER
DDLTYPE	Loads the Layer & Linetype Properties dialog box.
DDMODIFY	Controls object properties
'DDOSNAP	Displays a dialogue box for setting running osnaps for the Endpoint, Midpoint, Center, Node, Quadrant, Intersection, Insertion, Tanget, Nearest and Quick object snap modes. Also lets you set the size of the cross-hair target box aperture. See also OSNAP
'DDPTYPE	Specifies the display mode and size of point objects
DDRENAME	Displays a dialogue box that renames text styles, layers, linetypes, Blocks, views, user coordinate systems, viewport configurations, and dimension styles. See also RENAME
'DDRMODES	Sets drawings aids via a dialogue box
'DDSELECT	Displays a dialogue box that sets entity selection modes, the size of the pickbox, and the entity sort method
DDUCS	Displays a dialogue box for control of the current user coordinate system in the current space. See also UCS
DDUCSP	Selects a preset user coordinate system
'DDUNITS	Displays a dialogue box that sets coordinate and angle display formats and precision. See also UNITS
DDVIEW	Creates and stores views
'DELAY	Delays execution of the next command for a specified time. Used with command scripts
DIM	Invokes Dimensioning mode, permitting many dimension notations to be added to a drawing. See also 'DDIM

Command	Description
DIMALIGNED	Generates a linear dimension with the dimension line parallel to the specified extension line origin points. This lets you align the dimensioning notation with the object
DIMANGULAR	Generates an arc to show the angle between two nonparallel lines or three specified points
DIMBASELINE	Continues a linear dimension from the baseline (first extension line) of the previous or selected dimension
DIMCENTER	Draws a Circle/Arc center mark or center lines
DIMCONTINUE	Continues a linear dimension from the the second extension line of the previous or selected dimension. In effect, this breaks one long dimension into shorter segments that add up to the total measurement
DIMDIAMETER	Dimensions the diameter of a circle or arc
DIMEDIT	Edits dimensions
DIMLINEAR	Creates linear dimensions
DIMORDINATE	Creates ordinate point associative dimensions
DIMOVERRIDE	Overrides a subset of the dimension variable settings associated with a selected dimension entity
DIMRADIUS	Dimensions the radius of a circle or arc, with an optional center mark or center lines
DIMSTYLE	Switches to a new text style
DIMTEDIT	Allows repositioning and rotation of text items
'DIST	Finds distance between two points
DIVIDE	Places markers along a selected object, dividing it into a specified number of equal parts
DOUGHNUT or DONUT	Draws rings with specified inside and outside diameters
'DRAGMODE	Allows control of the dynamic specification (dragging) feature for all appropriate commands
DRAWORDER	Changes the display order of images and other objects
DSVIEWER	Opens the Aerial View window
DTEXT	Draws text items dynamically
DXBIN	Inserts specially coded binary files into a drawing
DXFIN	Loads a drawing interchange file
DXFOUT	Creates a drawing interchange file of the current drawing
ELLIPSE	Draws ellipses using any of several specifications
ERASE (E)	Erases entities from the drawing
EXPLODE	Shatters a Block or Polyline into its constituent parts
EXPORT	Saves objects to other file formats
EXTEND	Lengthens a Line, Arc, or Polyline to meet another object

Command	Description
'FILL	Controls whether Solids, Traces, and wide Polylines are automatically filled on the screen and the plot output
FILLET	Constructs an arc of specified radius between two lines, arcs, or circles
FILTER	Creates list to select objects based on properties
'GRAPHSCR	Flips to the graphics display on single-screen systems. Used in command scripts and menus
'GRID	Displays a grid of dots, at desired spacing, on the screen
GROUP	Creates a named selection set of objects
HATCH	Performs cross-hatching and pattern-filling
HATCHEDIT	Modifies an existing hatch object
'HELP or '?	Displays a list of valid commands and data entry options or obtains help for a specific command or prompt
HIDE	Regenerates a 3D visualization with hidden lines removed
'ID	Displays the coordinates of a specified point
INTERSECT	Creates composite solids or regions from the intersection of two or more solids or regions
IMPORT	Imports various file formats into AutoCAD
INSERT	Places a named block or drawing into the current drawing
INSERTOBJ	Inserts a linked or embedded object
'ISOPLANE	Selects the plane of an isometric grid to be the current plane for an orthogonal drawing
'LAYER (LA)	Creates named drawing layers and assigns color and linetype properties to those layers
LENGTHEN	Lengthens an object
'LIMITS	Changes the drawing boundaries and controls checking of those boundaries for the current space
LINE (L)	Draws straight lines of any length
'LINETYPE	Defines linetypes (sequences of alternating line segments and spaces), loads them from libraries, and sets the linetype for subsequently drawn objects
LIST	Lists database information for selected objects
LOAD	Loads a file of user-defined Shapes to be used with the Shape command
LOGFILEOFF	Closes the log file opened by LOGFILEON
LOGFILEON	Writes the text window contents to a file
'LTSCALE	Sets scale factor to be applied to all linetypes within the drawing
MASSPROP	Calculates and displays the mass properties of regions or solids
'MATCHPROP	Copies the properties from one object to one or more objects
MEASURE	Places markers at specified intervals along a selected object
MENU	Loads a file of commands into the menu areas (screen, pull-down, tablet, and button)

Command	Description
MENULOAD	Loads partial menu files
MENUUNLOAD	Unloads partial menu files
MINSERT	Inserts multiple copies of a Block in a rectangular pattern
MIRROR	Reflects designated entities about a user-specified axis
MLEDIT	Edits multiple parallel lines
MLSTYLE	Defines a style for multiple parallel lines
MOVE (M)	Moves designated entities to another location
MSLIDE	Makes a slide file from the current display
MSPACE (MS)	Switches to model space
MTEXT	Creates paragraphs of text
MTPROP	Changes paragraph text properties
MULTIPLE	Causes the next command to repeat until canceled
MVIEW	Creates and controls Viewports
MVSETUP	Sets up the specifications of a drawing
NEW	Creates a new drawing
OFFSET	Allows the creation of offset curves and parallel lines
OLELINKS	Updates, changes, and cancels existing OLE links
OOPS	Restores erased entities
OPEN	Opens an existing drawing
'ORTHO	Constrains drawing so that only lines aligned with the grid can be entered
'OSNAP	Enables points to be precisely located on reference points of existing objects. See also DDOSNAP
'PAN (P)	Moves the display window
PASTECLIP	Inserts data from the Clipboard
PASTESPEC	Inserts data from the Clipboard and controls the format of the data
PEDIT (2D)	Permits editing of 2D polylines
PLINE (PL)	Draws two-dimensional polylines (connected line arc segments, with optional width and taper)
PLOT	Plots a drawing to a plotting device or a file. The CMDDIA system variable controls whether you use the dialogue boxes or the command line to plot
POINT	Draws single points
POLYGON	Draws regular polygons with the specified number of sides
PREFERENCES	Customizes the AutoCAD settings
PREVIEW	Shows how the drawing will look when it is printed or plotted
PSDRAG	Controls the appearance of an imported PostScript image that is being dragged (that is, positioned and scaled) into place by the PSIN command

Command	Description
PSFILL	Fills 2D Polyline outlines with PostScript fill patterns defined in the AutoCAD PostScript support file (*acad.psf*)
PSIN	Imports Encapsulated PostScript (EPS) files
PSOUT	Exports the current view of your drawing to an Encapsulated PostScript (EPS) file
PSPACE (PS)	Switches to paper space
PURGE	Removes unused Blocks, text styles, layers, linetypes, and dimension styles from the drawing
QSAVE	Saves the drawing without requesting a filename
QTEXT	Enables Text entities to be identified without drawing the text detail
QUIT	Exits AutoCAD
RAY	Creates a semi-infinite line
RECOVER	Attempts to recover damaged or corrupted drawings
RECTANG	Draws a rectangular polyline
REDEFINE	Restores a built-in command deleted by UNDEFINE
REDO	Reverses the previous command if it was U or UNDO
'REDRAW (R)	Refreshes or cleans up the current viewport
'REDRAWALL	Redraws all viewports
REGEN	Regenerates the current viewport
REGENALL	Regenerates all viewports
'REGENAUTO	Controls automatic regeneration performed by other commands
REGION	Creates a region object from a selection set of existing objects
REINIT	Allows the I/O ports, digitizer, display, plotter, and PGP file to be reinitialized
RENAME	Changes the names associated with text styles, layers, linetypes, Blocks, views, User Coordinate Systems, viewport configurations, and dimension styles. See also DDRENAME
'RESUME	Resumes an interrupted command script
ROTATE	Rotates existing objects
RSCRIPT	Restarts a command script from the beginning
RULESURF	Creates a 3D polygon mesh approximating a rules surface between two curves
SAVE	Requests a filename and saves the drawing
SAVEAS	Same as SAVE, but also renames the current drawing
SCALE	Alters the size of existing objects
'SCRIPT	Executes a command script
SELECT	Groups objects into a selection set for use in subsequent commands
'SETVAR	Allows you to display or change the value of system variables
SH	Allows access to internal operating-system commands

Command	Description
SHAPE	Draws predefined shapes
SHELL	Allows access to other programs while running AutoCAD
SHOWMAT	Lists the material type and attachment method for a selected object
SKETCH	Permits freehand sketching
'SNAP	Specifies a round-off interval for digitizer point entry so that entities can be placed at precise locations easily
SOLID	Draws filled-in polygons
SPELL	Checks spelling in a drawing
SPLINE	Creates a quadratic or cube spline (NURBS) CURVE
SPLINEDIT	Edits a spline object
'STATUS	Displays drawing statistics and modes
STRETCH	Allows you to move a portion of a drawing while retaining connections to other parts of the drawing
'STYLE	Creates named text styles, with user-selected combinations of font, mirroring, obliquing, and horizontal scaling
SUBTRACT	Creates a composite region or solid by subtraction
SYSWINDOWS	Arranges windows
TABLET	Aligns the digitizing table with coordinates of a paper drawing to accurately copy it with AutoCAD
TEXT	Draws text characters of any size, with selected styles
'TEXTSCR	Flips to the text display on single-screen systems. Used in command scripts and menus
'TIME	Displays drawing creation and update times, and permits control of an elapsed timer
TOOLBAR	Displays, hides, and customizes toolbars
TRACE	Draws solid lines of specified width
TRIM	Erases the portions of selected entities that cross a specified boundary
U	Reverses the effect of the previous command
UCS	Defines or modifies the current user coordinate system. See also DDUCS
UCSICON	Controls visibility and placement of the user coordinate system icon, which indicates the origin and orientation of the current UCS. The options normally affect only the current viewport
UNDEFINE	Deletes the definition of a built-in AutoCAD command
UNDO	Reverses the effect of multiple commands, and provides control over the undo facility
'UNITS	Selects coordinate and angle display formats and precision. See also DDUNITS.
'VIEW	Saves the current graphics display and space as a Named View, or restores a saved view and space to the display

Command	Description
VIEWRES	Allows you to control the precision and speed of Circle and Arc drawing on the monitor by specifying the number of sides in a Circle
VPLAYER	Sets viewport visibility for new and existing layers
VPOINT	Selects the viewpoint for a 3D visualization
VPORTS	Divides the AutoCAD graphics display into multiple viewports, each of which can contain a different view of the current drawing
VSLIDE	Displays a previously created slide file
WBLOCK	Writes selected entities to a disk file
WMFIN	Imports a Windows metafile
WMFOPTS	Sets options for WMFIN
WMFOUT	Saves objects to a Windows metafile
XATTACH	Attaches an external reference to the current drawing
XBIND	Permanently adds a selected subset of an Xref's dependent symbols to your drawing
XLINE	Creates an infinite line
XREF	Allows you to work with other AutoCAD drawings without adding them permanently to your drawing and without altering their contents
XREFCLIP	Inserts and clips an XREF
'ZOOM (Z)	Enlarges or reduces the display of the drawing

APPENDIX

C

PULL-DOWN MENUS

· ·

The following pages illustrate the AutoCAD pull-down menus.

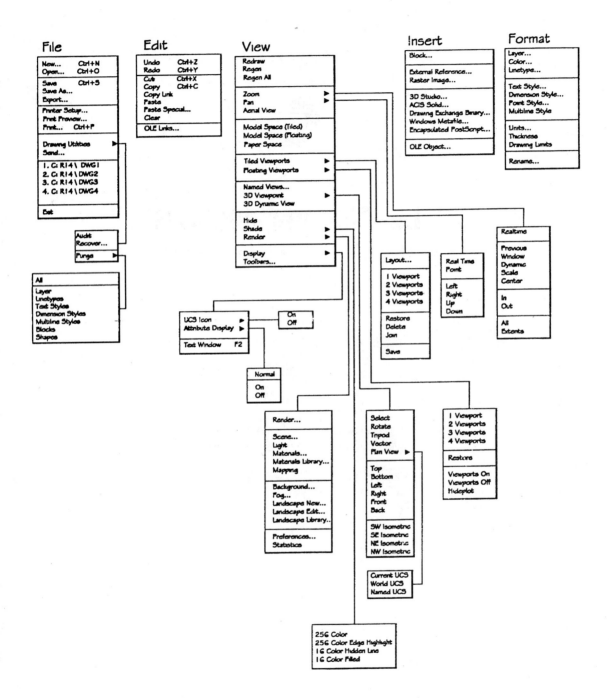

File

New...	Ctrl+N
Open...	Ctrl+O
Save	Ctrl+S
Save As...	
Export...	
Printer Setup...	
Print Preview...	
Print...	Ctrl+P
Drawing Utilities	▶
Send...	
1. C: R14\DWG1	
2. C: R14\DWG2	
3. C: R14\DWG3	
4. C: R14\DWG4	
Exit	

Audit
Recover...
Purge ▶

All

Layer
Linetypes
Text Styles
Dimension Styles
Multiline Styles
Blocks
Shapes

Edit

Undo	Ctrl+Z
Redo	Ctrl+Y
Cut	Ctrl+X
Copy	Ctrl+C
Copy Link	
Paste	
Paste Special...	
Clear	
OLE Links...	

View

Redraw
Regen
Regen All

Zoom ▶
Pan ▶
Aerial View

Model Space (Tiled)
Model Space (Floating)
Paper Space

Tiled Viewports ▶
Floating Viewports ▶

Named Views...
3D Viewpoint ▶
3D Dynamic View

Hide
Shade ▶
Render ▶

Display ▶
Toolbars...

UCS Icon ▶
Attribute Display ▶

Text Window F2

On
Off

Normal
On
Off

Render...

Scene...
Light
Materials...
Materials Library...
Mapping

Background...
Fog...
Landscape New...
Landscape Edit...
Landscape Library..

Preferences...
Statistics

Select
Rotate
Tripod
Vector
Plan View ▶

Top
Bottom
Left
Right
Front
Back

SW Isometric
SE Isometric
NE Isometric
NW Isometric

Current UCS
World UCS
Named UCS

256 Color
256 Color Edge Highlight
16 Color Hidden Line
16 Color Filled

Insert

Block...

External Reference...
Raster Image...

3D Studio...
ACIS Solid...
Drawing Exchange Binary...
Windows Metafile...
Encapsulated PostScript...

OLE Object...

Layout...

1 Viewport
2 Viewports
3 Viewports
4 Viewports

Restore
Delete
Join

Save

Real Time
Point

Left
Right
Up
Down

1 Viewport
2 Viewports
3 Viewports
4 Viewports

Restore

Viewports On
Viewports Off
Hideplot

Format

Layer...
Color...
Linetype...

Text Style...
Dimension Style...
Point Style...
Multiline Style

Units...
Thickness
Drawing Limits

Rename...

Realtime

Previous
Window
Dynamic
Scale
Center

In
Out

All
Extents

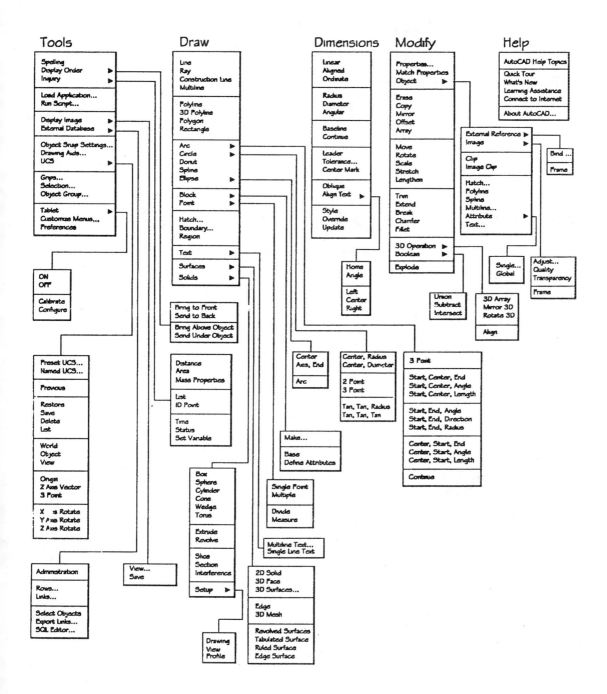

Tools

Spelling
Display Order ▶
Inquiry ▶

Load Application...
Run Script...

Display Image ▶
External Database ▶

Object Snap Settings...
Drawing Aids...
UCS ▶

Grips...
Selection...
Object Group...

Tablet
Customize Menus...
Preferences

ON
Off

Calibrate
Configure

Preset UCS...
Named UCS...

Previous

Restore
Save
Delete
List

World
Object
View

Origin
Z Axis Vector
3 Point

X Axis Rotate
Y Axis Rotate
Z Axis Rotate

Administration

Rows...
Links...

Select Objects
Export Links...
SQL Editor...

View...
Save

Draw

Line
Ray
Construction Line
Multiline

Polyline
3D Polyline
Polygon
Rectangle

Arc ▶
Circle ▶
Donut
Spline
Ellipse ▶

Block ▶
Point ▶

Hatch...
Boundary...
Region

Text ▶
Surfaces ▶
Solids ▶

Bring to Front
Send to Back
Bring Above Object
Send Under Object

Distance
Area
Mass Properties

List
ID Point

Time
Status
Set Variable

Box
Sphere
Cylinder
Cone
Wedge
Torus

Extrude
Revolve

Slice
Section
Interference

Setup ▶

Drawing
View
Profile

Make...

Base
Define Attributes

Single Point
Multiple

Divide
Measure

Multiline Text...
Single Line Text

2D Solid
3D Face
3D Surfaces...

Edge
3D Mesh

Revolved Surfaces
Tabulated Surface
Ruled Surface
Edge Surface

Dimensions

Linear
Aligned
Ordinate

Radius
Diameter
Angular

Baseline
Continue

Leader
Tolerance...
Center Mark

Oblique
Align Text ▶

Style
Override
Update

Home
Angle

Left
Center
Right

Center
Axis, End

Arc

Center, Radius
Center, Diameter

2 Point
3 Point

Tan, Tan, Radius
Tan, Tan, Tan

Modify

Properties...
Match Properties
Object ▶

Erase
Copy
Mirror
Offset
Array

Move
Rotate
Scale
Stretch
Lengthen

Trim
Extend
Break
Chamfer
Fillet

3D Operation ▶
Boolean ▶

Explode

Union
Subtract
Intersect

3 Point

Start, Center, End
Start, Center, Angle
Start, Center, Length

Start, End, Angle
Start, End, Direction
Start, End, Radius

Center, Start, End
Center, Start, Angle
Center, Start, Length

Continue

3D Array
Mirror 3D
Rotate 3D

Align

Help

AutoCAD Help Topics

Quick Tour
What's New
Learning Assistance
Connect to Internet

About AutoCAD...

External Reference ▶
Image ▶

Clip
Image Clip

Match
Polyline
Spline
Multiline
Attribute
Text... ▶

Bind ...

Frame

Single...
Global

Adjust...
Quality
Transparency

Frame

963

D

STANDARD HATCH PATTERNS

· ·

The following pages illustrate the standard hatch patterns supplied in file acad.pat.

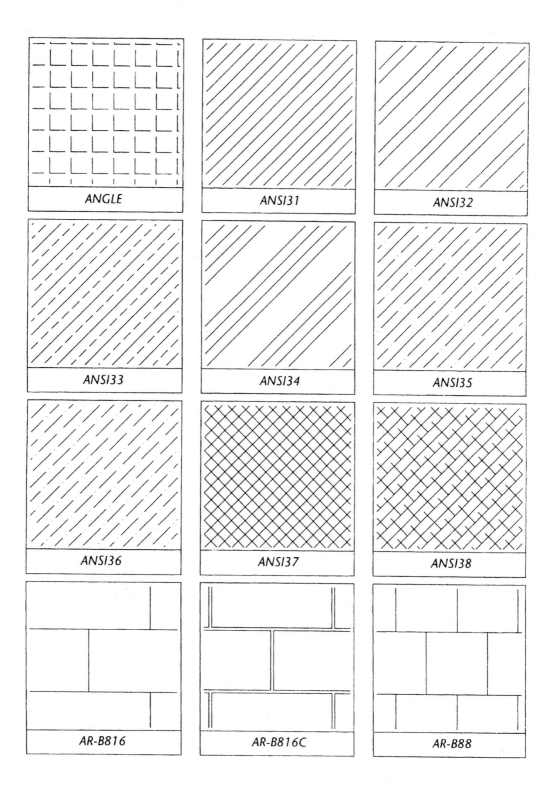

ANGLE

ANSI31

ANSI32

ANSI33

ANSI34

ANSI35

ANSI36

ANSI37

ANSI38

AR-B816

AR-B816C

AR-B88

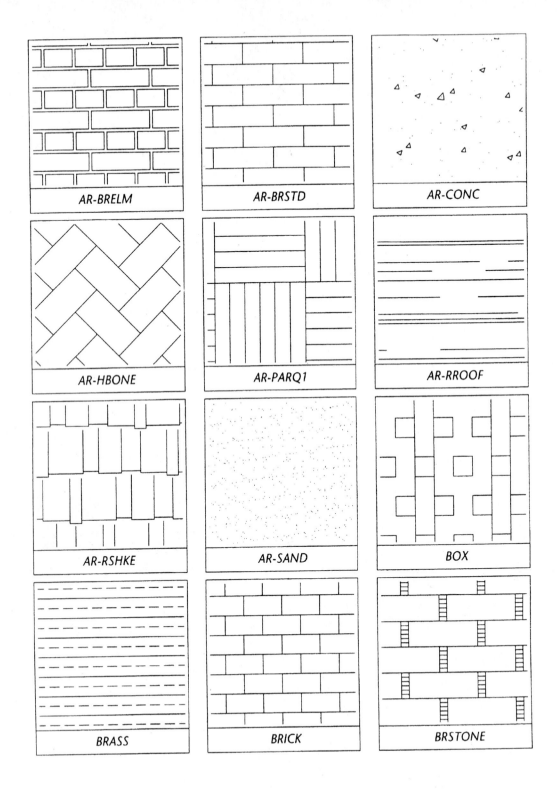

AR-BRELM

AR-BRSTD

AR-CONC

AR-HBONE

AR-PARQ1

AR-RROOF

AR-RSHKE

AR-SAND

BOX

BRASS

BRICK

BRSTONE

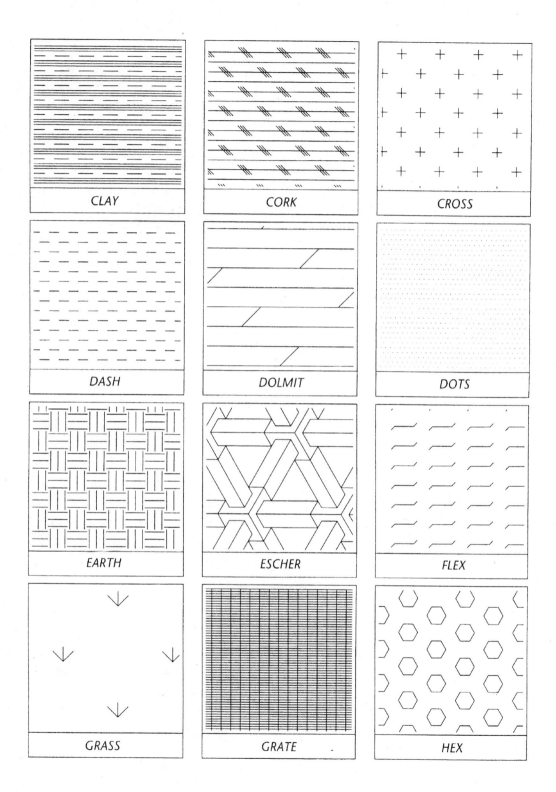

CLAY

CORK

CROSS

DASH

DOLMIT

DOTS

EARTH

ESCHER

FLEX

GRASS

GRATE

HEX

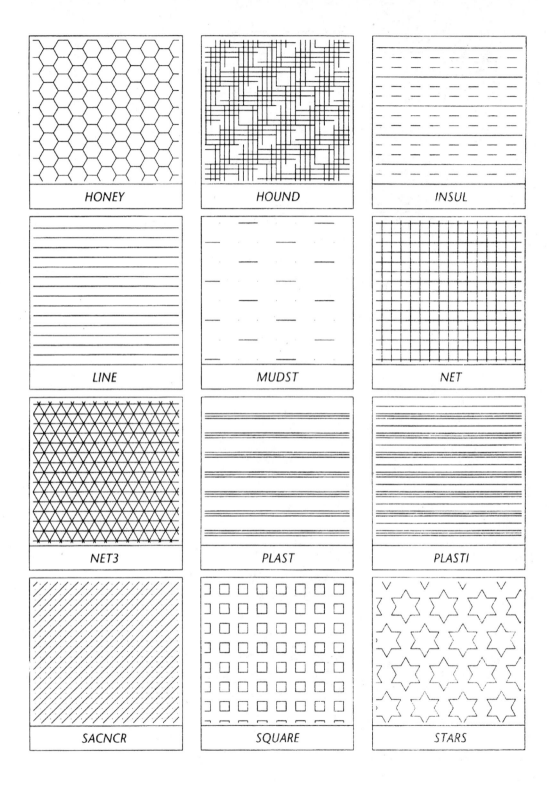

HONEY	HOUND	INSUL
LINE	MUDST	NET
NET3	PLAST	PLASTI
SACNCR	SQUARE	STARS

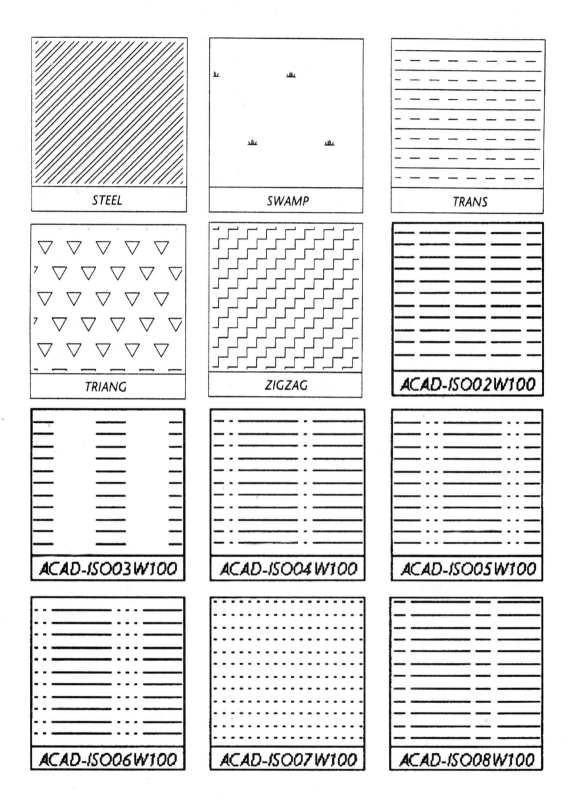

STEEL

SWAMP

TRANS

TRIANG

ZIGZAG

ACAD-ISO02W100

ACAD-ISO03W100

ACAD-ISO04W100

ACAD-ISO05W100

ACAD-ISO06W100

ACAD-ISO07W100

ACAD-ISO08W100

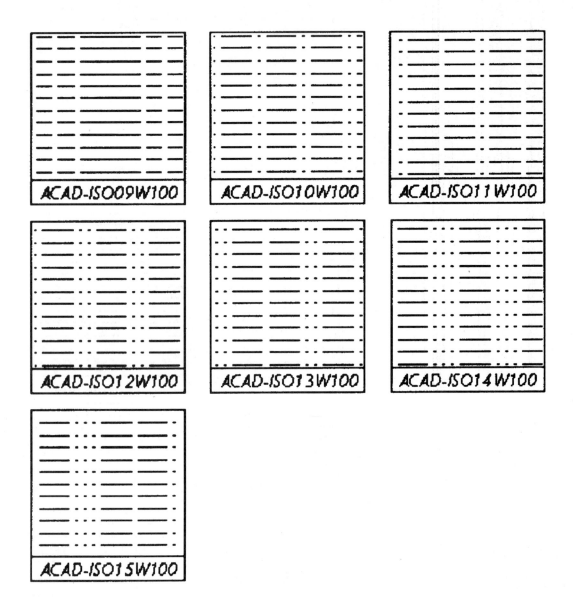

ACAD-ISO09W100

ACAD-ISO10W100

ACAD-ISO11W100

ACAD-ISO12W100

ACAD-ISO13W100

ACAD-ISO14W100

ACAD-ISO15W100

E

STANDARD LINETYPES

Border	— — — — · —
Border2	— — - - — · - — · —
BorderX2	— — — · —
Center	— — - — - — - —
Center2	— - — - — - —
CenterX2	— — — — —
Dashdot	- · — · — · — · —
Dashdot2	· - · - · - · - · - · -
DashdotX2	— · — · — · —
Dashed	- — — — — —
Dashed2	- - - - - - - -
DashedX2	— — — —
Divide	— · · — · · —
Divide2	— · · — · · — · · —
DivideX2	— · · — · · —
Dot	· · · · · · · · · ·
Dot2	·················
DotX2	· · · · · ·
Hidden	- - - - - - - - - - -
Hidden2	- - - - - - - - - - - - - -
HiddenX2	— — — — — — — —
Phantom	—— — — —— — —
Phantom2	— - - — - - —
PhantomX2	—— — — —
ACAD_ISO02W1000	— — — — —
ACAD_ISO03W1000	— — — — —
ACAD_ISO04W1000	— · — · — ·
ACAD_ISO05W1000	— · · — · · —
ACAD_ISO06W1000	— · · · — · · ·
ACAD_ISO07W1000	· · · · · · · · · · · ·
ACAD_ISO08W1000	— — — — —
ACAD_ISO09W1000	— — — — —
ACAD_ISO10W1000	— · — · — · — ·
ACAD_ISO11W1000	— — — —
ACAD_ISO12W1000	— · · — · · —
ACAD_ISO13W1000	— — — —
ACAD_ISO14W1000	— · · — · · —
ACAD_ISO15W1000	— — · · — · · —

INDEX